Restoration and Management of Tropical Eutrophic Lakes

Restoration and Management of Tropical Eutrophic Lakes

Editor
Mallapureddi Vikram Reddy

Science Publishers, Inc.
Enfield (NH), USA   Plymouth, UK

SCIENCE PUBLISHERS, INC.
Post Office Box 699
Enfield, New Hampshire 03784
United States of America

Internet site: *http://www.scipub.net*

sales@scipub.net (marketing department)
editor@scipub.net (editorial department)
info@scipub.net (for all other enquiries)

Library of Congress Cataloging-in-Publication Data

International Workshop on Restoration and Management of Eutrophic Lakes (2001 : Kunming Shi, China).

Restoration and management of tropical eutrophic lakes/editor, M. vikram Redy. p. cm.

Based on revised papers from the International Workshop on Restoration and Management of Eutrophic Lakes, held in Kunming, China, 2001.

Includes bibliographical references and index.

ISBN 1-57808-370-2

1. Lake ecology—Tropics-Congresses. 2. Lake renewal-Tropics-Congresses. 3. Lakes—Tropics-Management Congresses. 4. Eutrophication-Tropics-Congresses. I. Reddy, M. V. II. Title.

QH84.5.I585 2001
577.63'158-dc22

2004059426

ISBN 1-57808-370-2

Published by Science Publishers Inc., Enfield, NH, USA
Printed in India

Foreword–I

Travel, it is said, is good for the mind. This phrase might be challenged if the travel is only in the cocooned environment of airports, international hotels, government and company offices. But if travel means seeing the wonderful detail of different and locally characteristic land- water- and townscapes and the rich diversity of unfamiliar organisms and cultures, then there is no better addition to one's personal education.

For Auguste Thienemann in the 1930s, it was an expedition to the Dutch East Indies, now Indonesia that undermined his confidence in his and Einaar Naumann's recently proposed concept of eutrophic and oligotrophic lakes. Thienemann and Naumann were among the first and most distinguished European limnologists, the counterparts of E.A. Birge and Chancey Juday in the United States. They worked on benthos in the deep alpine lakes (Thienemann) and the phytoplankton of the shallower lakes of the Baltic lowlands (Naumann). In their brilliant initial synthesis, oligotrophic lakes were supposed to have low production and oxygenated hypolimnia; eutrophic lakes were characterized by high production and deoxygenated hypolimnia. But then the urbane Thienemann visited Southeast Asia. Some of the unproductive Indonesian lakes had strongly deoxygenated hypolimnia—it was the end of simple classification and the start of understanding in terms of continua. Also, it was part of the process by which understanding increases. Lakes do not come in simple categories; each has a unique combination of millions of features. Classification is helpful so long as this is realized and we can abandon ideas that are too simple and once they start to hold back further understading.

Another classification that we might examine closely is that of ecosystems into polar, temperate and tropical categories. We could also look at what we mean by the teem ecosystem rather more closely too! In the temperate region lakes that are deep with mountainous catchments, lakes that are larger than many small countries, lakes that are saline, lakes that are dominated by macrophytes, lakes that are fishless, lakes that are of recent origin and lakes that are hugely modified by the works of man may all be found in continua extending to their ultimate antitheses. It is so also in the Tropics. Think only of the Nal Saruvar in India, a shallow slightly salty basin with diverse wading birds and migratory herders with their water buffalo; the small, jade gem of Lake Sonachi in Kenya, a natural monospecific culture system for *Spirulina*, closely surrounded by the walls of a volcanic crater and the teeming jacarZ and capybara in the meadows of floating plants in the lakes of the Pantanal in Brazil. The variety is endless. The lessons that can be learned are infinite in number.

So, why does the title of this book contain two terms, tropical and eutrophic, that might be seen as too general on the one hand and outdated on the other? It is because they remain useful as abbreviations. One reminds us that lakes outside the limnologically well worn laager of Europe and north America are much less studied yet contain the seeds of a much more comprehensive understanding of limnology. The other is simply alarming. It has come (I would argue from a sloppy use of language) to replace an earlier term 'eutrophicated' because lakes all over the world are so polluted with nutrients that a term once used by Thienemann and Naumann to mean naturally relatively fertile has been appropriated to mean those we have damaged.

Lakes are only rarely very fertile in their pristine states. The endorheic lakes of some arid areas where a lack of throughflow allows nutrients to be accumulated, are the exceptions. Available phosphorus and combined nitrogen are naturally scarce elements for land vegetation as well as aquatic. Undisturbed systems on the land have many devices to retain as much as feasible of their nutrients for growth in the next season. Rivers and lakes used to get only the crumbs from the table. Now whole banquets are swilled down. The remarkable thing is that although there is palaeolimnological evidence for eutrophication by human activities in the ancient past, it became a general phenomenon as opposed to a local curiosity only in the latter half of the twentieth century. It is yet another consequence of the product of increasing population, increased resource use and waste production, and technology of increased sophistication.

The consequences of eutrophication in costs to water supply, public health and fisheries—let alone amenity and conservation—are by now well documented. The main nutrient flows urban wastewater, stock husbandry and cultivation are also understood. In this book, you will find the familiar problems of Europe and North America manifested in new places and in lakes that until even the last thirty years were seen as relatively invulnerable. You will also read of the approaches taken to solve these problems in the western world—control of phosphorus and biomanipulation. But you will also see that although the problems in warm lakes may be similar to those in cooler ones, the solutions may not.

Nitrogen is more likely to be the nutrient that needs to be controlled in many tropical systems; the food webs of tropical lakes are richer in omnivorous fish, whose reproduction several times a year may preclude the straightforward biomanipulation used widely in European lakes. The high temperatures may naturally favour a greater proportion of cyanophytes (blue-green algae) in tropical lakes and make definition of targets for restoration more difficult. The principles of limnology are universal but the situations to which they must be applied are everywhere different. Tropical limnologists must, therefore, not be deceived into believing that though the errors of caucasian environmental management have led to parallel problems, the solutions will always be the same.

Travel is not only good for the mind; it must occur in the mental processes too.

Brian Moss

School of Biological Sciences,
Derby Building, University of Liverpool, Liverpool L 69 3 GS,
U.K.

Foreword–II

If humankind is to leave a habitable planet for its descendants, then at some point, there must be, at a minimum, a balance between ecological destruction and repair. One of the emerging goals of human society is the sustainable use of the planet, which requires that all practices be continued indefinitely, failing which they will become unsustainable. Since the beginning of the agricultural revolution some ten thousand years ago and the more recent industrial revolution, ecological degradation has far exceeded the amount of ecological repair. In the last two centuries, human society has refashioned the planet's biospheric life support system of which *Homo sapiens* is a part. Although *Homo sapiens* has always altered the planet's environment, the rate of alteration in the twentieth century occurred on an unprecedented scale and speed. Humankind has—while pursuing such goals as economic development—undertaken a gigantic uncontrolled experiment with the entire planet. Ecological restoration is, arguably, the most significant manifestation of a new mutualistic relationship between human society and natural systems. Ultimately, we must ensure the preservation and restoration of natural capital and also the delivery of the valuable ecosystem services it provides. Both ecological restoration and management must focus on these goals in order to achieve sustainable use of the planet.

Although freshwater lakes comprise only a tiny percentage of the water in the hydrologic cycle and are comparatively small in relation to the terrestrial ecosystems that surround them, they are key ecological components whose importance far outweighs their size. Furthermore, it is well established that the tropics contain most of the world's species, all of which depend in some way upon the hydrologic cycle of which lakes are an important part. Species are the essential components of the interdependent web of life, which constitutes the biospheric life support system. It is not clear how many of these species can be lost without resulting in a major ecological disequilibrium, which would have adverse effects upon both natural capital and ecosystem services.

In order to retain as many lake species as possible, it is important that there be a range of lake quality from oligotrophic to eutrophic and the intermediate stages between these two extremes. Restoring tropical eutrophic lakes will provide a more suitable range of habitat for maintaining a diverse array of species rather than if all the lakes are eutrophic in nature. If this diverse array is to persist, there must exist management practices that will ensure both successful ecological restoration and maintenance of a variety of lake water quality conditions.

Both tropical lake restoration and management will require limnological monitoring to ensure that both of these objectives have been met. Taken together, this requires a long-term relationship between human society and tropical lakes.

Since tropical lakes are markedly affected by ecological events on the terrestrial systems that surround them, restoring and managing these lakes will require a substantial knowledge of the activities on adjacent terrestrial systems and a holistic management of both terrestrial and aquatic ecosystems in order to achieve the stated goals and objectives.

Books such as this one are an essential knowledge base for both ecological restoration and management. Although tropical lakes are not identical and, therefore, require individually developed and restoration and management practices; there are general principles in both restoration and management that can be derived from the case histories in this book and the limnological literature in general. Leaving a habitable planet for our descendants will be a task of unprecedented complexity, but the alternative—living unsustainably—will produce effects not pleasant to contemplate. I applaud the nature of this book as an essential component of the overall quest for sustainable use of the planet, which will require compassion not only for those of our own species presently alive but also future generations of our own species and present and future generations of the planet's other life forms. This seminal volume on restoration and management of tropical eutrophic lakes is an important first step on the path to sustainable use of the planet, and one for which generations to come will be grateful.

John Cairns, Jr

University Distinguished Professor of Environmental Biology
Emeritus and Director Emeritus,
University Center for Environmental and Hazardous Materials Studies
Virginia Polytechnic Institute and State University
Blacksburg, Virginia 24061 USA

Preface

Water, one of the most vital resources, is intrinsic to all kinds of lives as also to the ecosystems on this planet. According to the "Dublin Principles" (1992 International Conference on Water and Environment—ICWE held in Dublin), freshwater is a finite and valuable resource essential to sustain life, development and environment. It is not only used for drinking and domestic purposes (its domestic use being about six percent) but also used for agriculture (i.e. irrigation accounting for about 70% of all water withdrawals worldwide) and industrial processes (about 22% of the total water use in the world). Besides, freshwater also serves other purposes such as recreation, transportation, aesthetics, etc. Though agriculture accounts for the highest consumption of freshwater, drinking water has been the principal critical factor for survival of all kinds of life on the earth. The freshwater ecosystems range from small to large and shallow to deep, many of them being lotic while the others are lentic in nature, and contribute actively to the economy upto about $14,780 per hectare.

The freshwater available on earth is 36 million cu. km and constitutes only 2.57% of the global water (the volume of water available on the earth being 1.4 billion km^3). Of the total quantity of water, only 0.57% (eight million km^3) is available as freshwater found in rivers, lakes, reservoirs and groundwater. The freshwater lakes cover 1,26,000 km in volume, and comprise 0.009% of the total water in the biosphere. Lakes are one of the important sources of drinking water for human population; for example, the largest lake in Japan, Lake Biwa serves as a freshwater source for 14 million people or about 10% of the population of Japan. Today, the world's lakes are in crisis because of increasing pressure caused by the burgeoning human population and anthropogenic disasters such as pollution, accelerated allochthonous nutrient input, toxic contamination and climate change. All these factors seriously undermine the sustainability of many lakes in the developing as well as developed countries. Today's water shortage is directly related to the disappearance of the lakes, as many of the urban lakes are reclaimed for dumping municipal solid wastes, particularly in the developing countries.

According to the UN Population Fund, the world will begin to run out of freshwater by 2050 if pollution keeps pace with population growth, and will lose 18,000 km sq of freshwater, and its biodiversity. Hence, there is an imperative need for the conservation of these aquatic bodies, the available freshwater and biodiversity.

Various kinds of anthropogenic activities impair the natural water quality, and often change the water so completely that it becomes of minimal use or even unusable. Water is perturbed not because it is used for some specific purpose but because it serves as a sink for byproducts of various activities of the human society, such as waste and wastewater and other types of contaminants, the point and non-point sources of present-day pollution. The discharge of untreated and insufficiently-treated domestic and industrial wastewater is one of the most important causes of polluting the water bodies.

The use of aquatic ecosystems as sinks of sewage and effluents is by no means new. It finds mention even in Greek and Roman mythology. Contamination of freshwater with waste and wastewater is recorded in historic documents, for example, in the Bible. Aquatic pollution became widespread because of urbanization and concurrent industrialization. Urban conglomerations were established in the areas where there was sufficient supply of freshwater. As such, industries had developed rapidly along the rivers and other large water bodies such as lakes and reservoirs all over the globe and their effluents were dumped indiscriminately into these water bodies. Presently, in densely populated countries like India, on account of disposal of waste and wastewater in sewers, blockage and over-flowing has become a common feature, which has caused widespead pollution. Polluted water worldwide has been estimated to be 12,000 km^3. Municipal and agricultural runoff causes "Cultural Eutrophication" that begins with the increased concentration of nutrients particularly nitrogen and phosphorus in aquatic bodies, subsequently followed by algal blooms and growth of other macrophytes and episodes of oxygen depletion, and fish-kills.

Such perturbed waters became responsible for myriads of mortalities caused by the outbreak of diseases such as cholera and typhoid fever. The municipal raw sewage that includes untreated human waste is one of the most dangerous pollutants, according to the UNEP (United Nations Environmental Program) report on safeguarding the world's reserves of potable water. Four out of every five common diseases are caused by polluted water or lack of sanitation in the developing countries. The UNEP report further states that waterborne diseases cause an average of about 25,000 deaths a day in the third world countries, and 6,000 people mostly children under the age of five die from diarrhoeal diseases everyday. The overall water quality of many lakes and reservoirs have drastically deteriorated. Thus, there is an urgent need for careful management of these aquatic ecosystems, particularly those adjacent to human settlements, in order to prevent them from pollution, which needs to be dealt with a holistic approach with reference to the entire watershed. The processes that contribute to the quality of water can best be understood if these are examined in the context of the land use pattern in the watershed. Watersheds consist of the aquatic bodies—lakes or rivers together with their associated drainage basin and are integrated complex of both abiotic and biotic (including man) components with their intricate interactions.

Information related to the tropical eutrophic lakes and their restoration and management is, however, meagre. Therefore, attempts were made to assemble and integrate such scattered information in the present edition, which I hope will help in understanding these perturbed ecosystems holistically and that, in turn, may be useful for their effective management. This edition, with "Forewords" written by two eminent scholars in the subject, and a renowned researcher introducing the theme, comprises 15 chapters contributed by experts from different countries. These chapters are spread over three parts, with the first part outlining the different aspects of the abiotic environment and water quality of the tropical eutrophic lakes, while the second part addresses the spatio-temporal variations in different parts of the biotic component starting from planktons to fishes projecting them as bioindicators of eutrophic water quality. The third part deals with the topics of restoration and management of these lake ecosystems.

I am indeed extremely thankful to the contributors of the various chapters of this edition for honouring my invitation and contributing their research findings on tropical and sub-tropical lakes and reservoirs, and for revizing their typescripts with patience. I am very much beholden to Dr J. R. B. Alfred (Director, Zoological Survey of India, Kolkata) and Professor R. G. Michael for giving me an impetus to think over the subject of lake ecosystem ecology over the years. I am also thankful to the organizer, Prof. Song Xueliang of the International Workshop on Restoration and Management of Eutrophic Lakes (RMEL–2001) held at Kunming (P.R. China) and to Prof. Michio Kumagai (Lake Biwa Research Institute, Japan) for inviting me as a Committee member of the Workshop that helped in consolidating this volume. Thanks are due to the various publishers for allowing reproductions and use of their published information in various chapters of this book. My special thanks are due to my beloved wife for her encouragement while I was working for this edition.

Professor Dr. M. Vikram Reddy
Head and Chairman
Salim Ali School of Ecology and Environmental Sciences
Pondicherry Central University, Pondicherry, India

Contents

PART 2—BIOTIC ENVIRONMENT

PART 3—RESTORATION AND MANAGEMENT

List of Contributors

de Anda, Jose
Centro de Nvestigacion y
Assistencia en Tecnologia y
Diseno del Estado de Jalisco,
A.C., Normalistas 800 S.H.,
Guadalajara, Jalisco,
Mexico, C.P. 44270

del Anguilla, Laura Rull
Programa de Posgrado em Ecologia y Oceanografa,
Agencia Espanhola de Cooperacion Internacional,
AECI, Alicante, Espana (Spain)
E-mail: *laura.rull@ua.es*

Bezerra – Neto, Jose Fernandes
Programa de pos-graduacao em Ecologia,
Conserracao e Manejo da Vida Silvestre,
PG-ECMVS, UFMG, Belo Horizonte,
Minas Gerais, Brazil
E-mail: *joseneto@icb.ufmg.br*

Branco, Christina C.
Universidade do Rio de Janeiro,
Depto. Ciencias Naturais,
Av. Pasteur 458,
22290-240 Rio de Janeiro,
Brazil

Cairns, John Jr
University Center for Environmental and Hazardous
Material Studies,
Virginia Polytechnic Institute and State University

Blacksburg, Virginia 24061,
USA
E-mail: *jcairns@vt.edu*

Denny, Patrick,

Department of Environmental Engineering,
IHE Delft, Westvest 7, Po Box 3015,
2601 DA Delft, The Netherlands

Everard, Mark

2 Hollow Street, Great Somerford
Wettshire SN15 5JD, United Kingdom
E-mail: *mark.everard@environment-agency.gov.uk*

Hamilton, David P.

Department of Biological Sciences,
The University of Waikato,
Private Bag 3106,
Hamilton, New Zealand
E-mail: *davidh@waikato.ac.nz*

Harper, David

Ecology Unit, Department of Biology
Leicester University,
Leicester LE1 7RH, United Kingdom

Hecky, Robert E.

Biology Department
University of Waterloo
200 University Avenue West
Waterloo, Ontario N2L3G1, Canada

Huszar, Vera

Universidade Federal do Rio de Janeiro
Museu Nacional, Departmento de Botanica,
Laboratorio de Ficologia, Quinta da Boa Vista,
Sao Cristovao, Rio de Janeiro
Brazil, 20940-040

Jeppesen, Esik

National Environmental Research Institute,
Dept of Freshwater Ecology,
Vejlsovej 25, DK – 8600 Silkebory, Demark
Dept of Botanical Ecology,

University of Aarhus, Nordlandsvej 68,
DK 8230 Risskov, Denmark
E-mail: *ej@DMU.dk*

Kaushik, N.K.
Department of Environmental Biology,
University of Guelph,
Ontario, Canada N1G2W1
E-mail: *nkaushik@uoguelph.ca*

Kumar, Rajesh
Water and Stream Chemistry Lab.,
BARC Facilities,
Kalpakkam-603 102,
Tamil Nadu, India

Lazzaro, Xavier
Institut de Recherche Pour le Development (IRD),
UR 098/FLAG, Bel-Air,
BP 1386, Dakar, Senegal,
E-mail: *lazzaro@ird.sn*

Marshall, B.E.
Department of Biological Sciences,
University of Zimbabwe,
P.O. Box MP 167, Mount Pleasant,
Harare,
Zimbabwe
E-mail: *bmarshall@science.uz.ac.zw*

Mazzeo, Nestor
Seccion Limnologia,
Departmento de Ecologia,
Facultad de Ciencias
Igua 4225, Montevideo
Uruguay

Van dar Meer, H.J.
Institute of Biology Leiden,
University of Leiden,
P.O. Box 9511, 2300 RA Leiden,
The Netherlands

Meerhoff, Mariana
Seccion Limnologia
Departamento de Ecologia,
Facultad de Ciencias, Igua 4225, Montevideo
Uruguay

Moss, Brian,
School of Biological Sciences
Derby Building, University of Liverpool
Liverpool L69 3GS, United Kingdom
E-mail: *brmoss@liverpool.ac.uk*

Nair, K.V.K.
National Institute of Ocean Technology
Narayanapuram
Chennai, Tamil Nadu, India
E-mail: *nairkvk@niot.res.in*

Peimin, PU
Nanjing Institute of Geography & Limnology,
Academic Sinica,
Taihu Laboratory for Lake Ecosystem Res.,
73 East Beijing Road
Nanjing 210008, China
E-mail: *pupm@public1.ptt.js.cn*
pupm1@yahoo.com

Pinto – Coelho, Ricardo Motta
Departamento de Biologia Geral,
Instituto de Ciencias Biologicas,
Universidade federal de Minas Gerais
Belo Horizonte, Brazil
E-mail: *rmpc@mono.icb.ufmg.br*

Rajamohan, R.
Water and Stream Chemistry Lab.,
BARC facilities
Kalpakkam – 603 102
Tamil Nadu, India

Reddy, M. Vikram (Editor)
Salim Ali School of Ecology and
Environmental Sciences,
Pondicherry Central University,
Kalapet, Pondicherry-605 014, India.
E-mail: *mallapureddymv@yahoo.com*

Rutjes, H.A.
Institute of Biology Leiden,
University of Leiden,
P.O. Box 9516, 2300 RA Leiden,
The Netherlands

Scasso, Flavio
Seccion Limnologia,
Department de Ecologia,
Facultad de Ciencias,
Igua 4225, Montevideo,
Uruguay

Sekar, R.
Max Plank Institute for Marine Microbiology,
Celsiusstrasse 1, 28359-Bremen,
Germany

Shear, Harvey
Environment Canada – Ontario Region,
4905 Dufferin Street,
Toronto, Canada M3H 5T4
Email: *harvey.shear@ec.gc.ca*

Silva, E.I.L.
Institute of Fundamental Studies,
Hantana Road, Kandy
Sri Lanka
E-mail: *sil@ifs.ac.lk*

Sondergaard, Martin
NNERI, Dept. of Freshwater Ecology,
Vejlsovej 25, DK – 8600 Silkeborg,
Denmark

Eduardo von Sperling
Department of Sanitary and
Environmental Engineering
Federal University of Minas Gerais,
Av. Contorno 842/701 30.110-060,
Belo Horizonte Brazil
E-mail: *eduardo@desa.ufmg.br*

Starling, Fernando L.R.M.
Universidade Catolica de Brasilia (UCB)/
Companlia de Saneamento do Distrito Federal (CAESB),
SAIN-RIETA Brasilia, CEP 70620-002,
Brasilia – DF, Brazil
E-mail: *starling@pos.ucb.br*

Van den Thillart, G.E.E. J.M.
Institute of Biology Leiden,
University of Leiden,
P.O. Box 9516, 2300 RA Leiden
The Netherlands

Venugopalan, V.P.
Water and Stream Chemistry Laboratory,
BARC Facilities,
Kalpakkam-603 102
Tamil Nadu, India.
Email: *vpv@igcar.ernet.in*

Vijaykumar, A.
Kendriya Vidyalaya
No. 2, Golconda
Hyderabad - 500028, A.P., India

Wanink, J.H.
Institute of Biology Leiden,
University of Leiden,
P.O. Box 9516, 2300 RA Leiden,
The Netherlands

Witte, F.
Institute of Biology Leiden,
University of Leiden,
P.O. Box 9516, 2300 RA Leiden,
The Netherlands
E-mail: *witte@rulsfb.leidenuniv.nl*

Williams, Adrian, E.
APEM Ltd., Enterprise House,
Manchester Science Park,
Lloyd Street North, Manchester,
M15 6SE
United Kingdom
Email: *a.williams@apemltd.co.in*

Introduction

David P. Hamilton
Environment Bay of Plenty Chair in Lakes Management and Restoration
Department of Biological Sciences, The University of Waikato, Private Bag 3105,
Hamilton, New Zealand

TROPICAL EUTROPHIC LAKES: THEIR RESTORATION AND MANAGEMENT

Tropical biomes represent some of the most biodiverse and ecosystem-rich habitats in the world. More than 70% of the world's biota occurs in tropical biomes, including around 50% of the world's human population. Indonesia alone contains some 10% of the world's plant and animal species. Large and rapidly expanding human populations in many tropical biomes have the potential to severely threaten the biodiversity and sustainability of these systems through large-scale plant and animal extinctions along with habitat fragmentation and degradation. For example, of the 16 Asian countries examined by Jalal and Rogers (2002) 13 are expected to undergo rapid environmental degradation relative to economic growth. Arguments have been put forward that extinction of species, deforestation and pollution of air and water are inevitable but temporary consequences of progressive improvements in the living standard of developing countries, most of which are in tropical regions (Lomborg, 2001). This argument is predicated upon assurances that there exists knowledge, ability and willingness to reconstruct habitats and restore biodiversity in these countries (Dudgeon and Lam, 1994; Hamilton and Oldham, 2001). Currently, however, there are few good models of ecosystem restoration in tropical regions; there is increased exploitation by developed countries of the still abundant resources of developing countries; and there are ongoing questions about the successfulness of restoration technology transfer from developed countries and its relevance to tropical regions.

Global ecosystem assessments point to inland waters being the most valuable non-marine ecosystems (e.g. Constanza et al., 1997) and the most threatened and sensitive of all ecosystems to human pressures (Anon., 2002). In addition, tropical biomes are undergoing more rapid rates of environmental deterioration than any other major biome on earth. Evidence of both the conservation value and precarious current standing of tropical inland waters is given in reviews of Asian inland waters by Dudgeon (2000, 2002) as well as in a series of edited reviews for developing countries (Wetzel and Gopal, 1995, 1999, 2001). While it is still debated as to whether tropical inland waters have increased biodiversity compared with temperate regions (Lewis, 2000), comparisons of ecosystem changes across latitudes and biomes suggest that there is a pressing need for management of water quality and biodiversity of tropical lakes.

THREATS TO TROPICAL LAKES

Kira (1997) lists the six major threats to world lakes, namely flow regulation, excess siltation, eutrophication, contamination by toxic chemicals, acidification and general collapse of lake ecosystems. Flow regulation and over-abstraction from inflows, outflows or standing waters of lakes are already a major issue for many developing countries in arid tropical and subtropical regions and will increase as the water demand further exceeds its supply (Stikker, 1998). Acidification, together with salinization, are mostly direct consequences of flow regulation and over-abstraction, and again, are particularly evident in arid tropical and subtropical regions, notably Pakistan, where >6 million hectares of land is considered to be salt-affected (Rafiq, 1990).

Excess siltation in developing countries is closely linked with land use changes, particularly deforestation, establishment of agriculture, major flood control works and floodplain modification. In many cases, the causes of siltation are closely interrelated. For example, deforestation and establishment of agriculture or plantations often precede expensive flood control works. In this context, the services provided by natural ecosystems in developing countries need to be assessed, particularly in the case of highly threatened inland waters.

Cultural eutrophication is one of the major causative factors leading to what has been described by Kira (1997) as a "complete collapse of lake ecosystems". This collapse or catastrophic shift (Scheffer et al., 2001) has important implications for lake management because of the marked hysteresis between systems that are in a transitional state of deterioration in biodiversity and water quality, and those where restoration is attempted *after* a catastrophic shift or collapse (Harris, 1999). Restoration of the former transitional state is far more readily achieved rather than once the lake ecosystem has collapsed (Scheffer et al. 1993). Lakes in developing countries urgently need to be assessed from this perspective in order to develop realistic restoration goals and implement management strategies to reverse and improve systems that are in transitional states of deterioration.

A further major threat to lakes in developing countries is the introduction of invasive exotic species. Perhaps the most conspicuous of these is the introduction of the predatory Nile perch to Lake Victoria (Kaufman 1992), the largest tropical lake and the second largest lake in the world by area. Nile perch have had a severe impact on the food web in Lake Victoria, leading to extinction of at least half of its 300 cichlid species, and to devastation of local fisheries. Cultural eutrophication has combined with ecosystem effects due to Nile perch, to have major negative impacts on the water quality of the lake (Hecky 1993). While Lewis (2000) indicates that there is little evidence of a gradient in biodiversity from temperate to tropical inland waters—as is evident in terrestrial systems—he notes specifically that large, ancient lakes of the tropics, notably African rift valley lakes, have highly endemic fauna. Indeed, these lakes are centres of biodiversity for both aquatic invertebrates and vertebrates (Serrya and Pollingher, 1983) and introduction of invasive species is likely to severely threaten their unique character. The problem of invasive species, however, is confined neither to fish nor to African lakes. Amongst the introduced weeds, for example, *Salvinia molesta* and *Eichhornia crassipes* have had major ecosystem and water usage impacts in countries such as Papua New Guinea (Osbourne, 1995), Sri Lanka (Costa and De Silva, 1995), Indonesia (Lehmusluoto et al., 1999) and Ethiopia (Tudorancea et al. 1995).

LIMNOLOGY OF TROPICAL LAKES

Lewis (2000) presents a comprehensive review of the limnology of tropical lakes as a basis for their management. Here I reiterate some aspects of his review, while presenting additional evidence of the resultant implications for lake management. Lewis (2000) suggests that the fundamental lake-mixing regime in tropical lakes is warm monomictic, with shallow lakes tending to be polymictic, similar to temperate regions. However, there can be relatively wide variations in the dynamics of the surface mixed layer (Lewis, 1987), with intermittent erosion of the metalimnion when there is net heat loss from the lake, as well as frequent development of secondary thermoclines as a result of a reversal in the heat flux. MacIntyre and Melack (1982) suggest that stratification regimes in tropical lakes tend to be more transient, complex and exhibit greater diurnal variation than in lakes of temperate zones. When making generalities about the thermodynamics of tropical lakes, it is also relevant to note that river lakes are relatively abundant as compared with their temperate counterparts (Lewis, 1996) and, thus, river inflows may have a disproportionately large influence on lake hydrology and thermodynamics. Hydrology of subtropical and tropical regions tends to be dominated more by monsoonal or irregular events, leading to far greater variability of river discharges (Gupta, 1988). The net result is that natural tropical lakes may be subject to wide variations in water residence times and water depths, which also influence the stratification regime (MacIntyre and Melack, 1998), as well as variations in stability attributable directly to river inflows (Tundisi, 1994).

Microbial processes occur faster at elevated temperatures in tropical regions, and together with reduced partial pressures of gases and increased duration of thermal stratification in monomictic tropical lakes, result in dissolved oxygen being an even more important determinant and constraint on the geographic distribution of aquatic fauna. Water temperatures in the hypolimnion of temperate monomictic lakes are often at least 10-15°C cooler than in tropical lakes. Thus, rates of microbial metabolism, assuming a Q_{10} of 2.0 (i.e. that metabolic rates will double with each 10°C increase in temperature), are likely to be two to three times higher in the hypolimnion of a tropical lake. Concurrently, however, saturation concentrations of dissolved oxygen will decline, e.g. from 11.5 mg L^{-1} at 10°C to 8.4 mg L^{-1} at 25°C, assuming there is no influence of salinity or water depth. The following assumptions have been made in developing an idealized case to compare the dissolved oxygen depletion rates in the hypolimnion of monomictic lakes of tropical and temperate regions:

- a typical duration of stratification is 10.5 months in the tropical lake and 7.5 months in the temperate lake (Lewis, 2000);
- dissolved oxygen concentrations are at saturation just prior to development of seasonal stratification;
- hypolimnion temperatures are 25°C and 10°C in the tropical and temperate lake, respectively;
- metabolic rates of microbial populations are 2.84 times higher in hypolimnion of the tropical lake (equivalent to Q_{10} = 2.0).

A volumetric oxygen demand of 0.0186 g m^{-3} day^{-1}, referenced at a temperature of 20°C, is sufficient to produce onset of anoxia (i.e. zero dissolved oxygen) at the end of the stratified period in the hypolimnion of a tropical lake. In the temperate lake, the corresponding dissolved oxygen concentration at the end of the stratified period would be 9.4 mg L^{-1}. This hypothetical example clearly demonstrates why tropical lakes are far more likely to experience seasonal anoxia of bottom waters, with greater potential for internal nutrient loading and high sensitivity to eutrophication or organic loading.

Rectifying oxygen deficits in bottom waters and preventing the development of phytoplankton blooms has been attempted through artificial destratification in economically important tropical impoundments (e.g. Hawkins and Griffiths, 1993). However, the high and relatively uniform heat fluxes throughout the year (Straškraba and Gnauck, 1985) at low latitudes mean that considerable investment of energy is required to artificially maintain destratification or at least to induce sufficient periodic mixing so as to prevent anoxia deeper in the water column. Even though temperature gradients may be only around 3°C (e.g. 25-28°C) through the water column of a tropical lake, the equivalent density gradient gives a surface water temperature of >16°C in the hypothetical temperate lake (see above) with a hypolimnion temperature of 10°C.

High prevalence of anoxia and warm temperatures in the hypolimnion and bottom sediments of tropical lakes provide highly suitable conditions for loss of inorganic nitrogen via denitrification (Lewis, 2000). There is no comparable process for the loss of phosphorus and, in fact, liberation of phosphorus from bedrock material is likely to be enhanced by warmer temperatures at low latitudes (Compton et al., 2000), as will atmospheric losses of nitrogen from terrestrial environments. Further, greater extent and duration of anoxia can be expected to promote increased release of Fe- and Mn-bound phosphorus from lake bottom sediments into the water column (Mortimer, 1971). These factors, together with the relatively high prevalence of cyanobacteria with nitrogen-fixing capabilities in enriched tropical waters (Lewis, 2000), suggest that there may be high potential for limitation by nitrogen over phosphorus compared with lakes at temperate latitudes (Talling and Lemoalle, 1998). Limitation by other macronutrients (e.g. inorganic carbon) or micronutrients may occur in exceptional cases, but the circumstances leading to such cases may not necessarily be attributable to latitudinal gradients.

Economic development in tropical regions will eventually lead to reductions in point sources of nutrients entering rivers and lakes, through improved wastewater treatment and management techniques. Diffuse sources of nutrients, however, may prove more difficult to control, particularly with transformations of remnant forest cover or low intensity agriculture into intensive agriculture or plantation forestry in which higher nutrient demand is met through fertilizer additions. The use of phosphatic fertilizers will need to be carefully assessed and balanced against terrestrial plant demands for several reasons:

(1) Potential for erosive losses of sediment-bound phosphorus associated with heavy rains in tropical regions;
(2) The relatively high proportion of river-lakes at low latitudes;
(3) High prevalence of nitrogen-fixing cyanobacteria in tropical waters, leading to a strong likelihood that phytoplankton biomass may ultimately be controlled primarily by phosphorus, as atmospheric fixation by cyanobacteria will meet their demands for nitrogen;
(4) High atmospheric losses of inorganic nitrogen as a result of high rates of denitrification that are supported by high temperatures of terrestrial and aquatic systems at low latitudes, and low concentrations of dissolved oxygen at sediment-water boundaries. This is likely to favour the proliferation of nitrogen-fixing cyanobacteria, many of which are bloom-forming, and will be selectively advantaged in phosphorus-replete environments.

GENERAL ASPECTS OF TROPICAL LAKE MANAGEMENT

The classical approach to lake management has generally been to attempt to deal with identified problems in water quality and biodiversity on an individual basis. It is clear from the above information, however, that many of the causative factors of deterioration are closely interrelated and interdisciplinary in nature (Somlyódy, 1995). For example, changes in river basin hydrology, land use and climate have downstream impacts on lakes which may be difficult to separate from internally mediated changes such as invasive species that may predispose lakes to eutrophication (e.g. introduction of Nile perch or collapse of invasive weed beds). Interdisciplinary studies of river basins are, therefore, a fundamental prerequisite of sound lake management, particularly in view of the high number of river lakes in tropical regions. The lake restoration process requires a sound integration of science and engineering, together with a commitment and willingness of management agencies to engage communities in the lake restoration process.

Tropical lake management and restoration represents a major challenge on three main fronts. First, while many of the fundamental theories and applications of lake management that have arisen out of

Europe and North America (e.g. Cooke et al., 1993) represent universally sound lake management practices, the great diversity of tropical lake environments suggests a far greater variety of responses to restoration procedures. These occurrences suggest that a broader and more flexible approach may need to be developed for the management and restoration of tropical lakes, particularly in light of economic limitations.

The second challenge to the management of tropical lakes is communication between scientists, engineers, governmental jurisdictions, and stakeholders and citizens. Some of the difficulties of scientific communication in developing countries have been alluded to previously by Wetzel and Gopal (1999) and Williams (1994), and also include limitations in infrastructure to support limnology. At a larger scale, however, institutional communication and collaborative arrangements present a major obstacle to developing comprehensive river basin management directives and adopting scientifically supported best management practices (Borre et al., 2001). In this context, financial resources are often limited and political desire weak, relative to the complexity of interdisciplinary lake restoration practices. However, a recent example of a successful restoration project is provided for a Thai lake by Chufamanee and Lønholdt (2001). The procedure involved the establishment of a comprehensive database, application of an integrated surface water model and a participatory environment for stakeholders, the community and government.

The third challenge to tropical lake restoration is related to the fact that the high rate of deterioration of water quality and biodiversity of many tropical lakes means that remediation efforts will often need to be implemented alongside development of scientific data and knowledge of systems. Researchers in developing countries are unlikely to have the benefits often available in developed countries where there are well-established lake monitoring programmes. While this may be viewed as a constraint, it may also present opportunities to rapidly adopt the available knowledge into management plans, conceptual and theoretical models of lake ecosystem function, and to optimize the available resources for monitoring, management and restoration of tropical lakes.

REFERENCES

Anon., 2002. World Resources 2000-2001—People and ecosystems: The fraying web of life. United Nations Development Programme, United Nations Environment Programme, World Bank and World Resources Institute.

Borre, L., D. R. Barker and L. E. Duker, 2001. Institutional arrangements for managing the great lakes of the world: Results of a workshop on implementing the watershed approach. Lakes and Reservoirs: Research and Management 6(3): 199-210.

Chufamanee, P. and J. Lønholdt, 2001. Application of integrated environmental management through the preparation of an environmental action programme: Case study from the Songkhla Lake Basin in southern Thailand. Lakes & Reservoirs: Research and Management 6(4): 323-334.

Compton, J., D. Mallinson, C. R. Glenn, G. Filippelli, K. Follmi, G. Shields and Y. Zanin, 2000. Variations in the global phosphorus cycle. In: Marine Authigenesis: From Global to Microbial, SEPM Special Publication No. 66, ISBN 1-56576-064-6, pp. 21-33.

Costa, H. H. and P. K. De Silva, 1995. Limnological research and training in Sri Lanka: State of the art and future needs. In: Limnology in Developing Countries, Vol. 1., R. G. Wetzel and B. Gopal (eds). International Association for Limnology, pp. 63-104.

Cooke, G. D., E. B. Welch, S. A. Peterson and P. R. Newroth, 1993. Restoration and Management of Lakes and Reservoirs, 2nd Edition. Lewis Publ., Boca Raton.

Constanza, R., R. D'Arge, R. De Groot, S. Farber, M. Grasso, B. Hannon, K. Limburg, S. Naeem, R. V. O'Neill, J. Paruelo, R. G. Raskin, P. Sutton and M. Van den Belt, 1997. The value of the world's ecosystem services and natural capital. Nature 387: 253-260.

Dudgeon, D., 2000. Conservation of freshwater biodiversity in Oriental Asia: Constraints, conflicts and challenges to science and sustainability. Limnology 1: 237-243.

Dudgeon, D. 2002. The most endangered ecosystems in the world? Conservation of riverine biodiversity in Asia. Verh. Internat. Verein. Limnol. 28: 59-68.

Dudgeon, D. and Lam, P. K. S., 1994. Inland waters of tropical Asia and Australia: Conservation and management. Mitt. Internat. Verein. Limnol. 24: 1-3.

Gupta, A., 1988. Large floods as geomorphic events in the humid tropics. In: Flood Geomorphology, V. R. Baker, R. C. Kochel and P. C. Patton (eds). John Wiley and Sons, pp. 301-315.

Hamilton, D. P. and C. E. Oldham, 2001. Integration of ecology and engineering in the aquatic environment. Ecological Engineering 16(3): 319-322.

Harris, G. P., 1999. Comparison of the biogeochemistry of lakes and estuaries: Ecosystem processes, functional groups, hysteresis effects and interactions between macro- and microbiology. Mar. Freshwat. Res. 50: 791-811.

Hawkins, P. R. and D. J. Griffiths, 1993. Artificial destratification of a small tropical reservoir: effects upon the phytoplankton. Hydrobiologia 254: 169-181.

Hecky, R. E., 1993. The eutrophication of Lake Victoria. Verh. Internat. Verein. Limnol. 25: 39-48.

Jalal, K. F. and P. P. Rogers, 2002. Measuring environmental performance in Asia. Ecological Indicators 2: 39-59.

Kaufman, L. S., 1992. Catastrophic change in species-rich freshwater ecosystems, the lessons of Lake Victoria. BioScience 42: 846-852.

Kira, T., 1997. Survey of the state of the world lakes. In: Guidelines Of Lake Management, Vol. 8: The world's lakes in crisis. S. E. Jørgensen and R. A Vollenweider (eds). United Nations Environment Program/International Lakes Environment Committee, pp. 147-155.

Lehmusluoto, P., B. Machbub, N. Terangna, F. Achmad, L. Boer, S. S. Brahmana, B. Setiadji, B. Priadie, K. H. Timotius and F. Goeltenboth, 1999. Limnology in Indonesia: From the legacy of the past to the prospects for the future. In: Limnology in Developing Countries, Vol. 2., R. G. Wetzel and B. Gopal (eds). International Association for Limnology, pp. 119-234.

Lewis Jr, W. M., 1987. Tropical limnology. Ann. Rev. Ecol. Syst. 18: 159-184.

Lewis Jr, W. M., 1996. Tropical lakes: How latitude makes a difference. In: Perspectives in Tropical Limnology, F. Schiemer and K. T. Boland (eds). SBP Academic Publishers, Amsterdam, pp. 43-64.

Lewis Jr, W. M., 2000. Basis for the protection and management of tropical lakes. Lakes and Reservoirs: Research and Management 5: 35-38.

Lomborg, B., 2001 The Skeptical Environmentalist: Measuring the Real State of the World. Cambridge University Press.

MacIntyre, S. and J. M. Melack. 1982. Meromixis in an equatorial African soda lake. Limnol. Oceanogr. 27: 595-609.

MacIntyre, S. and J. M. Melack. 1988. Frequency and depth of vertical mixing in an Amazonian floodplain lake (L. Calado, Brazil). Verh. Internat. Verein. Limnol. 23: 80-85.

Mortimer, C. H., 1971. Chemical exchanges between sediments and water in the Great Lakes - speculations on probable regulatory mechanisms. Limnol. Oceanogr. 16: 387-404.

Osbourne, P. L., 1995. Limnology in the wet tropics: Papua New Guinea. In: Limnology in Developing Countries, Vol. 1., R. G. Wetzel and B. Gopal (eds). International Association for Limnology, pp. 121-162.

Rafiq, M., 1990. Soil resources and soil related problems in Pakistan. In: Soil Physics—Application Under Stress Environments, M. E. Akhtar and M. I. Nizami (eds). Pakistan Agricultural Research Council, Islamabad, pp. 16-23.

Scheffer, M., S. H. Hosper, M. L. Meijer, B. Moss and E. Jeppesen, 1993. Alternative equilibria in shallow lakes. Trends in Ecology and Evolution 8: 275-279.

Scheffer, M. and S. Carpenter, J. Foley, C. Folkes and B. Walker, 2001. Catastrophic shifts in ecosystems. Nature 413: 591-596.

Serruya, C. and U. Pollingher, 1983. Lakes of the Warm Belt. Cambridge University Press, England: 569pp.

Somlyódy, L., 1995. Water quality management: can we improve integration to face future problems? Water Science Technology 31(8): 249-259.

Stikker, A., 1998. Water today and tomorrow. Futures 30(1): 43-62.

Straškraba, M. and A. Gnauck, 1985. Aquatic Ecosystems. Modelling and Simulation. Elsevier, Amsterdam, 309pp.

Talling, J. F. and J. Lemoalle, 1998. Ecological Dynamics of Tropical Inland Waters. University Press, Cambridge.

Tudorancea, C., Z. G. Mariam and E. Dadebo, 1999. Limnology in Ethiopia. In: Limnology in Developing Countries, Vol. 2., R. G. Wetzel, and B. Gopal (eds). International Association for Limnology, pp. 119-234.

Tundisi, J. G. (1994) Tropical South America: Present and perspectives. In: Limnology Now: A Paradigm of Planetary Problems, R. Margalef (ed.). Elsevier, Amsterdam.

Wetzel, R. G. and B. Gopal (eds), 1995. Limnology in Developing Countries, Vol. 1. International Association for Limnology.

Wetzel, R. G. and B. Gopal (eds), 1999. Limnology in Developing Countries, Vol. 2. International Association for Limnology.

Wetzel, R. G. and B. Gopal (eds), 2001. Limnology in Developing Countries, Vol. 3. International Association for Limnology.

PART 1

Abiotic Environment

CHAPTER

1

A Habitat-based Approach to Catchment Assessment and Sustainable Development in East African Freshwater Systems

Mark Everard[1], David Harper[2] and Patrick Denny[3]

[1]2 Hollow Street, Great Somerford, Wiltshire SN15 5JD, UK

[2]Senior Lecturer in Ecology, Ecology Unit, Department of Biology, Leicester University, Leicester LE1 7RH, UK

[3]Professor of Aquatic Biology, Department of Environmental Engineering, IHE Delft, Westvest 7, PO Box 3015, 2601 DA Delft, The Netherlands

ABSTRACT

The contribution of tropical wetlands to sustainability may be highly significant for developing countries, although their extent and functional value is generally poorly understood or recorded. Low-intensity monitoring techniques are required to quantify the conservation value of both extant and historic wetlands. Habitat features may best meet these requirements. Uganda, a tropical East African country rich in wetlands, has pressing development needs which, undirected, threaten to compromise the country's catchment integrity. Appropriate and sustainable development remains a pressing priority. This has been aided by a computer-based expert system which advises on sustainable uses of individual wetlands, based upon their physical characteristics and current utilization. Efforts to achieve 'Wise Use' of Lake Naivasha, Kenya's second

Ramsar Site, have focussed on activities close to the lake shore since its declaration in 1995. Evaluation of the environmental impact of lakeshore horticulture is not possible without any quantification of the influence from the wider catchment. Habitat assessment of the river systems reveals a diversity of habitats that may influence lake quality, including the adverse effects stemming from the intensification of land use, primarily subsistence farming. Protection of certain river corridor habitat features may be critical to the continuing health of both the lake and its catchment. A number of principles are identified to guide the application of habitat-based assessment. These include the need to record and interpret habitat features consistently and appropriately in order to ensure that both site- and habitat-scale processes can be determined. Appropriate and sustainable development of wetlands then rests upon understanding their characteristics and processes, and the needs and culture of the local people. Where the former extent and nature of degraded wetlands can be determined or reconstructed from evidence, it may be possible to estimate their contribution to catchment functioning, 'carrying capacity' and sustainability. Methods developed for the SWAMP expert system may be potentially applied to these degraded wetland areas so as to restore at least some of the beneficial ecosystem functions from which broader-scale social, economic and environmental benefits may flow. Developed countries, and the multinational companies based in them who are the commercial customers of agriculture in tropical countries, have an interest in promoting sustainable forms of development in industrializing nations. Implementation of habitat-based assessment schemes may provide a robust and cost-effective means to achieve this, backed up by education to increase local capacity to determine sustainable uses of habitat within catchments.

INTRODUCTION

The contribution of tropical wetlands to the interconnected ecological, societal and economic attributes of sustainability may be highly significant for developing countries. However, their extent, and the value of their various 'ecosystem functions' to catchment integrity and human well being, are often poorly understood or recorded. Robust monitoring techniques are, therefore, necessary to begin to quantify the conservation value of both extant and historic wetlands, supporting prioritization and benefit-cost justification for their protection, appropriate use and/or restoration. Resource constraints, both human and economic, determine that such methods need to be low in intensity. Relatively stable habitat features, as compared to more variable water quality, hydrological or ecological measures, may best meet these requirements. In this chapter, we evaluate this proposition in an East African context on the basis of former studies by the authors in Kenya and Uganda.

Rivers and their associated wetlands are fully interdependent elements of catchments (river basins), which are dynamic land-water systems comprising interconnected water courses, water bodies, groundwaters and land up to the watershed (Newson 1994, Calder 1999). Although historically subject to parochial management decision-making, many of the ecosystem functions occurring in rivers and wetlands operate at the scale of entire catchments and depend upon the ecological integrity of the complete system (Calder 1999, Everard and Powell 2002). Wetland functions—hydrological, geomorphological, physico-chemical, ecological, etc.—provide goods and services supporting human well being, economic activities and 'quality of life' (Dugan 1990). Catchment-scale ecosystem processes, including the activities of the human population, contribute to shaping the characteristic features of river and wetland habitats (Boon et al., 2000; Everard et al., 2001).

Since river and wetland habitats integrate the influences of changing land use within catchments (Newson 1994), river monitoring can prove to be a cost-effective method for indications of whether sustainable catchment management is being achieved (Ormerod and Jüttner 1998). These same habitat features are indicative of the supportive capacities of ecosystems and can, therefore, also be interpreted as the basis for providing guidance on sustainable uses that do not simultaneously systematically degrade the catchment processes (Everard et al. 1995).

Since the human population is a fully interdependent component of catchments (Everard 1997a and 1997b, Boon et al., 2000), wetland conservation can only be realistically achieved through sustainable use rather than by preventing human use. For many non-industrialized countries, political, population and economic pressures make development a necessity that will proceed, whether guided or not, all too commonly resulting in the over-exploitation of natural resources (Denny 1985 and 1989; Dugan 1990). Wetlands are particularly vulnerable because of their potential land value, high productivity, yet frequent perception as 'waste places' (Purseglove 1988). Tropical wetlands have all too frequently been developed for short-term economic gains that degrade or destroy their ecological values and environmental services in the long term (Dugan 1990, Denny 1995). Many examples exist of their wide-scale clearance, drainage, intensive cultivation or damming in sub-Saharan Africa (as reviewed by Davies and Day 1998 and Pacini and Harper 2000), where ill-advised, prestigious aid programmes have sometimes aggravated the situation (IIED 1994). Political, population and economic pressures also frequently leave insufficient capital available for adequate investment in environmental training and education, resulting in a serious shortfall in the availability of information, knowledge and funding to inform long-term sustainable decision making (Everard et al., 1995, Denny 2001). Thus, assessment methodologies have to be cost-effective as well as robust, as their effective implementation will be, of necessity, less routine than in the developed world.

The arid nature of much of East Africa means that water quality, one of the primary indicators used to assess the quality of aquatic systems throughout much of Europe and North America, is less relevant as the unpredictability and episodic nature of rainfall are amongst the overriding physical processes determining the character and ecology of tropical freshwaters (Talling and Lemoalle 1998). As relatively static features reflecting the cumulative effects of the longer-term processes that have shaped them, interpretation of habitat features may enable low-frequency monitoring and consequent management support (Raven et al., 1998). Habitat assessment and interpretation, therefore, has a key role to play in measuring and guiding sustainable development in the region. On the basis of two case studies in East Africa, one each in Uganda and Kenya, we seek to learn lessons about the value of habitat-based assessment of aquatic resources to determine system integrity, and to guide sustainable forms of development.

SWAMP AND THE WETLANDS OF UGANDA

The political, demographic and economic pressures upon Uganda are intense, further compounded by the legacy of 18 prior years of civil war and the AIDS epidemic (Everard et al. 1995). Ugandan wetland ecosystems are both productive and fragile and, in common with those in many developing nations, have often been the victims of over-exploitation (Denny 1991), including their destructive drainage for intensive agriculture yielding only short-term benefits (Denny 1989). Nevertheless, since the mid-1980s, Uganda has had the foresight to make a commitment to the principles of sustainable development, which has included the regulation of wetland development through a national Wetlands Conservation and Management Programme (the 'Wetlands Programme') reflecting the conflict between their vulnerability and their value for the production of food and other goods and services. Amongst the objectives of the Wetlands Programme is the implementation of combinations of sustainable uses of wetlands to secure their Wise Use whilst securing a sound economic future, consistent with the aims of the Ramsar Convention (Ramsar 1988). Appropriate and sustainable development retains the integrity of wetlands, together with their ecological and hydrological functions, providing long-term economic stability and supporting the material needs of local communities.

The development of a computer-based expert system known as SWAMP (Sustainable Wetland Areas Management Project), supporting the work of the Ugandan Wetlands Programme, provided a vehicle to systematize the expertise of a range of experts supporting decisions about what would constitute sustainable use of East African (specifically Ugandan) wetlands (Everard et al. 1995). This expertise was not merely technical but also sociological, as recommended strategies have a better chance of success if local communities, with their wealth of traditional management techniques and specific needs, are involved from the very first stages of development rather than being imposed upon by developed world opinions (IIED 1994). Social implications arising from the ethnic and cultural acceptability of proposals are vital considerations and, where possible, it is important to offer alternative strategies, enabling local people to select the pattern of development appropriate to their local needs. Since development was imperative and would occur anyway, whether guided or not, the Wetlands Programme recognized that some wetland areas are of relatively low conservation value (perhaps already degraded by prior unwise use) and may not be capable of supporting a diversity of human uses or wildlife. Therefore, where they are appropriate for the purpose, it is sensible to concentrate uses which are not intrinsically sustainable at these locations, thereby avoiding damage to wetland areas of higher ecological value and preventing interference with other potential sustainable uses. Based largely upon the recommendations by Denny and Turyatunga (1992), the nineteen potentially sustainable uses encoded into SWAMP (Table 1.1) represent a continuum from the least intrusive (for example, Conservation/Tourism or Fishing) through to inherently sustainable uses tuned to particular habitat types (for example, Agro-fish Polyculture or Flood-retreating Cultivation) and also unsustainable uses to be targeted on the least valuable wetlands (for example, Commercial Rice or Brick-making).

SWAMP asks questions of local users, via textual or graphics interfaces, about six major groups of wetland features: physical characteristics of the wetland; water chemistry; soil properties; current wetland usage; other human exploitation within the immediate catchment area; and general considerations such as altitude or major vegetation type. Reasoning functions within the expert system minimize the number and select the sequence of questions, utilizing internal reasoning to suggest combinations of recommended uses together with appropriate warnings (i.e. about water-borne diseases) on the basis of wetland characteristics. The final acceptability of the courses of

Table 1.1 The nineteen potentially sustainable uses for Ugandan wetlands encoded into SWAMP (from Everard et al., 1995)

CONSERVATION/ WILDLIFE	Conservation of the biodiversity of a habitat and ecosystem, including the native flora and fauna. Managed tourism is a possible non-intrusive source of income in this situation. The term 'conservation' is being used in something of a limited sense here, meaning the identification of an area of high wetland biodiversity value.
FISHING	Low-intensity capture of fish from the native stock to provide food for the local population using local methods such as trapping, netting, hook and line fishing, digging of finger ponds, etc.
GRAZING	Medium- to high-intensity grazing of a native standing crop of emergent and littoral vegetation.
SEASONAL PASTURES	Grazing of emergent vegetation and flooded areas exposed by a seasonally retreating water level.
LOW FREQUENCY IRREGULAR CROPPING	Occasional and opportunist collection of plant material from the wetland for village use such as plant biomass for matting, thatching, etc., and plant products for food and medicines.
FLOOD RETREATING CULTIVATION	The cultivation of the fringes of the wetlands for catch crops during the dry season when water levels are low. There can be small localized management of the hydrology in the area, though this should not interfere significantly with the overall hydrology of the wetland. (Catch cropping is the opportunist seasonal cultivation and harvesting of a crop.)
MARGINAL RIDGE AND FURROW AGRICULTURE	The opportunist seasonal or non-seasonal cultivation of the wetland margin by creating raised ridges or mounds above the water table without affecting the natural drainage. This is akin to shifting cultivation, and relies on fallow recovery periods of the land before reuse.
FISH PONDS	Managed rearing of fish in artificial ponds as a source of food to fulfil commercial or local needs.
AGRO-FISH POLYCULTURE	Integrated fish and agricultural production utilizing the land/ water interface of wetlands.
PLANT BIOMASS PRODUCTION	Sustainable harvesting of a high-yield standing crop of native emergent vegetation for the production of briquettes, matting and thatching, biogas, etc. For all harvesting techniques, the safety of the people is paramount and it is assumed that harvesting will only take place under appropriate conditions.
RICE (VILLAGE)	Small-scale farming of rice in a wetland margin, fulfilling the needs of small local communities.
RICE (COMMERCIAL)	Intensive production of rice for commercial purposes, involving substantial degradation of the wetland habitat. This use is not generally considered to be sustainable and is only rarely recommended in those circumstances where the wetland does not appear to be performing vital functions within catchments, and where no other sustainable use can be identified.

(Table 1.1 Contd.)

(Table 1.1 Contd.)

EXTRACTION OF WATER BY LOCAL PEOPLE	Local people extract a small volume of water for use in homes.
EXTRACTION OF WATER BY PIPED SYSTEM	Water is extracted for a piped water system to factory, town or village. The water must be treated using appropriate purification technology, such as passing through a settler tank and/or by chlorination.
INDUSTRIAL WATER SUPPLY	High volume supply of water for industrial purposes such as mining (for example washing ores), steel processes, wood pulping, cane sugar processing, etc.
DAMMING	Retention of water behind dams for the purpose of small-scale water storage.
WASTE (DOMESTIC)	Discharge of household or suburban waste into a wetland.
WASTE (INDUSTRIAL)	Discharge of waste of industrial origin into a wetland.
BRICKS	Production of building bricks from the clay sediment of the wetland. Once again, this use is not generally viewed as being sustainable, and is only rarely recommended.

development is left to the decisions of the local people. Habitat is a key determinant of inherent conservation value and processes occurring within wetlands, and is therefore a key factor utilized in inferences within SWAMP. This leads to recommendations for potentially sustainable patterns of human use and development.

THE USE OF RHS IN THE NAIVASHA CATCHMENT, KENYA

The ecological value of Lake Naivasha (0°45′S, 36°26′E; altitude 1,890 metres), Kenya, has been acknowledged by its designation as a Ramsar site (Kenya's second Ramsar site after Lake Nakuru) in 1995. Uniquely for a Ramsar site, the Kenyan government delegated responsibility for the management of the site to the non-statutory Lake Naivasha Riparian Association (LNRA), which is developing *The Lake Naivasha Management Plan* through a process of consensus-building (Anon, 1999). This plan reflects the conflict between protection of the lake's ecology, tourism as a source of foreign revenue, industrial development such as geothermal power, and the growing demand for irrigated land that has hastened the degradation of the riparian ecotone (Enniskillen, 2002). Inevitably, there has been a demographic and ecological change in addition to economic benefits (Henderson and Harper, 1992). The area of land used for intensive horticulture around the lake shore is now approximately equal in area to that of the lake itself (Johnson and Harper, 1998). Various activities in and around the lake have been attributed as contributing to observed deterioration in the quality of the lake, including damage through abstraction, inputs of nutrient, pesticides and sediment, loss of the fringing papyrus 'buffer', and introduced species (Harper et al., 1990).

However, there is also increasing intensity of subsistence agriculture in the catchment, with common and widespread poor land management practices such as tillage to the banktop and riparian poaching by stock animals (Everard et al., 2002a; Everard and Harper 2002). Despite the considerable effort expended in identifying and managing pressures arising from the lake shore, there has been a pronounced lack of effort in characterizing the lake's catchment. This lack of knowledge of catchment characteristics and influences place at risk the efforts expended at the lake perimeter. Compared to Ugandan wetlands, controls on development within the Naivasha catchment are less advanced. Indeed, evidence of any intact extant wetlands is small, with site visits revealing cash and subsistence cropping on areas marked on the few maps available for the area denoting such features (such as Tourist Maps (K) Ltd 1996).

Everard et al. (2002a) have used the River Habitat Survey (RHS) method (Raven et al., 1998) to make a preliminary assessment of the structure of the wetland/riparian resources of the catchment in order to provide baseline information on the diversity and integrity of rivers. Flowering plant and bird diversity, as well as sediment morphology, have also been assessed (Everard et al., 2002b; Tarras-Wahlberg et al., 2002). Topography, arising from the underlying geology, creates a diversity of river systems draining into the lake (Figure 1.1) (Everard et al., 2002a). Some of these rivers are perennial and others ephemeral, and some have open confluences to the lake whilst others have blind channel endings. River channels vary from erosive rocky gorges and cascades to slack depositional channels, and run through landscapes as diverse as montane forest, open savannah and papyrus-dominated swampland. The overall catchment water balance appears critical to maintaining the unique character of the lake, with inputs of water to the lake dependent upon rainfall at high altitudes remote from the lake itself and often governed by microclimatic conditions (Gaudet, 1977; Gaudet and Melack, 1981; Everard et al., 2002a; Becht and Harper 2002). Riverine habitat in the lower reaches modifies sediment appreciably, altering its physical appearance and chemical reactivity and also its potential impacts on the lake (Tarras-Wahlberg et al., 2002). Papyrus swamp at river mouths to the north of the lake formerly served an important role in sedimentary deposition and the buffering of nutrient inputs to the lake (Gaudet, 1977), a role that now appears to be compromised by the incremental loss of swamp area (Kitaka, 2000 and 2002). Protection and,

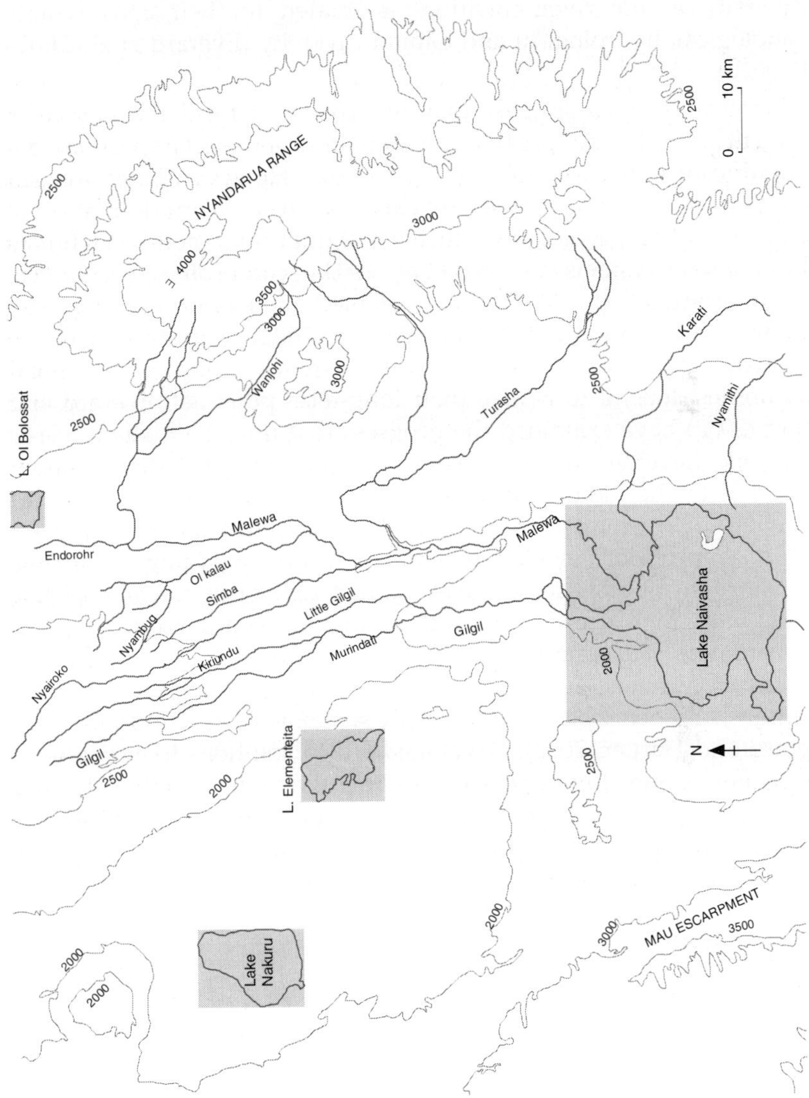

Fig 1.1 The topography and rivers of the Naivasha catchment (from Everard et al. 2002a)

ideally, restoration through appropriate use of this swampland is, therefore, critical to the continuing health of the lake (Zalewski and Harper, 2001). Parallel assessments of vascular plant and bird richness at RHS sites throughout the Naivasha catchment demonstrate that biodiversity in the river corridors is related to their considerable morphological, hydrological and habitat diversity (Everard et al., 2002a and 2002b).

Chemical studies suggest that the rivers of the Lake Naivasha catchment are still rather pristine and their anthropogenic impacts are not yet significant (Tarras-Wahlberg et al., 2002). However, habitat-based evaluation of river channels reveals localized degradation as a consequence of human activity which has significantly modified habitats within the river systems (Everard et al., 2002a), compromised ecological value (Everard et al., 2002b), and enabled increased sediment and pollutant loads entry to the lake (Kitaka, 2002). This threatens the supportive capacities of these river systems, upon which local communities depend to realize their long-term potential. Everard and Harper (2002) have examined the progress that must be made towards sustainable farming and both extensive and intensive development (Table 1.2) in the Lake Naivasha catchment, with river-edge habitat protection or restoration highlighted as a particularly important measure. Appropriate development and education regarding sustainable land management practices in the catchment, delivering catchment-scale benefits but which also create a self-interest for local landowners, will help promote sustainability, maintain relatively low inputs to agricultural systems, sustain traditional techniques adapted to local conditions, and hence be closer to a truly sustainable agricultural system (Everard and Harper, 2002). Developed-world solutions to agricultural and economic development, such as increasing input-intensity and monoculture growth in farming methods, should be treated cautiously, owing to the possibility of adverse ecological, social and economic consequences in an inappropriate environment.

HABITAT-BASED ASSESSMENT

Sustainable development has been identified as a priority to protect Ugandan wetlands (Everard et al., 1995) and the unique ecology of Lake Naivasha and its catchment (Everard and Harper, 2002). For both regions, the potential for loss of unique ecosystems, and declining social well being stemming from the loss of a supportive ecosystem functions with its consequences for regional economic activity and stability, is so great that the challenge of sustainable development must be grasped as a matter of urgency.

Table 1.2 Key sustainability challenges for intensive lakeside and extensive catchment-wide land use in the Lake Naivasha catchment, including some proposed approaches to solutions (from Everard and Harper, 2002)

Sustainability challenges for intensive horticulture	*Sustainability challenges for extensive subsistence land use*
Chemical Proceses • High inputs of fertilizers. • Energy intensity in agricultural practice. • Energy use in haulage and aviation fuel • Intensive use of pesticides. • Incautious reuse or disposal of plastics. • Use of persistent chemicals in processing.	*Chemical Processes* • Nutrient runoff to waterways. • Possible increases in fertiliser and energy use. • Possible increases pesticide use.
Physical Proceses • Conversion of important and rare habitat. • Fencing that disrupts migratory behaviour. • Excessive demands on surface and groundwater. • Reliance upon geothermal energy. • Degradation and erosion of soil.	*Physical Proceses* • Increasing habitat loss through population pressures. • Soil loss through erosion. • Loss of ecosystem services (particularly wetlands). • Spread of non-native weed species. • Threats from increased tillage.
Biological Processes • Loss of native crop species, varieties and practices. • Invasion by non-native weed species.	*Biological Processes* • Loss of natural crop pest predators.
Socio-economic Processes • Transfer of land ownership from endemic people. • Centralization of economic benefits to agri-businesses. • Disturbance of traditional cultures. • Loss of subsistence agriculture and cash cropping. • Inequity in economic rewards. • Inadequate protection from risks (i.e. pesticide use). • Marginalization of the poorest people to the poorest land • Luxury goods displace subsistence needs.	*Socio-economic Processes* • Loss of farmland and productivity for future use. • Loss of 'quality of life' through habitat loss. • Risks from increasing chemical and energy use.

Mechanisms discussed for addressing sustainability challenges

- *Recognition that the reputation of the region as a whole affects its competitiveness*, which should drive a more integrated approach to the problems facing the catchment as an integrated management unit by both those deriving the greatest economic gains and agencies interested in the development of the region.

(Table 1.2 Contd.)

(Table 1.2 Contd.)

- *Advancing a 'Sustainability Stewardship' scheme* that proactively addresses wider sustainability concerns in a transparent and audited manner, pre-empting the growing concerns of developed world markets. This is a particularly pressing problem for the export of water-intensive goods from arid regions, all dependent upon the use of large amounts of fossil fuel for transport.
- *Education throughout the catchment*, primarily with teachers and farmers, about improved land use practices effective local management based on 'enlightened self-interest' that benefits both the livelihoods of farmers and their families at the local scale yet has wider benefits across the Naivasha catchment.
- *Breaking the 'cycle of unsustainability'* that marginalizes people in the upper catchment from economic benefits, with consequent implications for lake quality through poor land use practices mediated via the catchment.

Sustainable utilization of sensitive and important wetlands is at the forefront of the challenge of delivering Wise Use (now agreed to be synonymous with sustainable development) as required under the Ramsar Convention (Ramsar 1971, 1988 and 1996). Degradation of the freshwater environment worldwide, and particularly the hydrological and ecological disconnection of river channels from the living landscapes of which they are a part, compromise the ecosystem functions upon which sustainable benefits depend (Everard, 1997a and 1997b; Boon et al., 2000). It is essential to address integrated sustainable development at the catchment scale rather than merely local management of ecosystems, sustaining the natural resources and ecosystem functions that ultimately support the social and economic needs of communities within catchments (Everard 1997a, 1999 and 2000, Calder, 1999; Everard and Powell, 2002). This approach relies upon an understanding of the behaviour of, and pressures upon, the catchment.

A number of 'forensic' methods are available to determine the historic extent of wetlands within catchments, prior to human activities that have degraded or eliminated them. Examples include methods applied in UK catchments by Hogan et al. (2000 and 2001) to map the functional importance of extant wetlands and the substantive extent of historic losses, as a guide to targeted advice to land managers on the sustainable use and recovery of catchment functioning. In the US, established wetland delineation procedures (Lyon, 1993), based on the three key features of hydrology, hydric soils and wetland-dependent vegetation, provide a model for defining the overall extent of wetland areas. The US methods have also been accepted in the Mediterranean area of Europe as the wetland delineation procedure underpinning the MEDWET programme (Wetlands International, 1996). Where databases are sparse in East Africa, a range of these relic and map-based features may be helpful in supporting an evaluation of the current and former wetland extent. Gleying, organic matter or other markers in sediment, anecdotal evidence including old farm and estate plans, some older maps, and potentially historic surveys and aerial photographs may provide evidence of the former extent of wetlands within tropical catchments. Habitat evaluation such as RHS, suitably modified to floodplain scale, can also provide evidence from 'break of slope' and other hydrogeomorphological features from which potential former wetland extent may be determined. Where habitat occurrence, extent and nature can be hindcasted, it may also be possible, at least in qualitative terms, to estimate their former contribution to catchment functioning and, consequently, the loss of 'carrying capacity' resulting from unsympathetic development such as drainage and tilling for short-term agricultural returns. The basic physical survey methods are already available, although considerable further development effort and ground-truthing would be required to convert them into dependable methodologies.

Methods for inferring function from morphology are conspicuously lacking at present, despite long-term acknowledgement of the need to better understand the functional value of wetlands (for example, Dugan, 1990; Everard, 1997a and 1997b). Further economic appraisal methods will also be necessary if the benefits of protection and/or restoration of wetlands and their multiple functions are to be translated into robust policies. Arguably, this need exists across Europe to fully implement the EU's Water Framework Directive, which provides at least some optimism that the tools necessary to underpin the supportive capacities of tropical wetlands may be developed within the relatively affluent developed world in the forseeable future. These measures will be important to deliver truly sustainable catchment land use, as also to target as priorities the most functionally-important 'hotspots' within the catchments. Methods developed for the SWAMP expert system may potentially be applied to degraded wetland areas in order to provide an assessment of the potentially sustainable and economically valuable uses for wetland that is not yet developed, or which is a priority for restoration. This linkage of catchment character to the pattern of human exploitation is central to sustainable development, protecting and/or restoring at least some of the beneficial ecosystem functions from which broader-scale social, economic and environmental benefits may flow.

A number of principles can be concluded from the application of habitat-based assessment in the two case studies referred to in this chapter.

i. Habitat-based assessment of wetland and river systems offers a helpful framework to determine ecosystem characteristics and processes. However, it is not automatically diagnostic of site- or catchment-scale health. Interpretation of habitat has to be undertaken cautiously, ensuring that monitoring and interpretation is attuned to the identification of catchment-scale processes—hydrological, erosion and sedimentation, ecological, etc.—yet is also sensitive to local conditions. Interpreted appropriately, a habitat can act as a surrogate measure for actual or potential ecosystem functioning.

ii. Interpretation within this broader context relies upon specific and appropriate protocols for data-gathering and analysis. This has been provided on a nationally attuned basis by SWAMP in Uganda and through the globally-applied RHS methodology in Kenya.

iii. As relatively static features reflecting the longer-term processes that have shaped them, the correct interpretation of habitat features may enable low-frequency monitoring and consequent management support. RHS and other related habitat assessment tools may, therefore, represent a cost-effective means to support low-intensity monitoring and interpretation of catchment health. Despite the high initial development costs of producing expert systems, SWAMP was also developed to increase the accessibility to expertise, and to provide training for local scientists, at reduced costs relative to routinely flying out experts from the developed world.

iv. Appropriate and sustainable development of wetlands can only be achieved from a thorough understanding of their characteristics and processes. This enables one to highlight problems 'upstream' that may need to be managed, as well as to identify the manner in which ecosystem functions can be recognized and utilized to support local needs through land use patterns that do not degrade natural processes.

v. Tools are required to assess the former extent, nature and functioning of wetland habitat within the catchments. This enables an assessment of lost 'carrying capacity', but also supports recommendations for sustainable patterns of development that may include restoration or protection of functionally critical habitat from which wider-scale benefits flow at catchment scale. Considerable development will be required to deliver robust policy, embodying continued social and economic well being.

vi. More information about, and effective outreach to, all sectors of society in catchments is essential in order to deliver awareness and practice of sustainable land use. This must simultaneously deliver local self-benefit as well as wider protection of ecosystem functions, and also promote greater inclusion of all people in strategic catchment planning (Everard, 2004). Sustainable uses must respect the culture of the local people and be tailored to address different local conditions, needs and traditions. Local people must be involved in finding solutions to their needs, with no presumption in favour of developed world solutions.

Tackling sustainable development in a concerted and serious manner is a mechanism by which the social development and economic benefits of regions of East Africa can be secured and developed in the longer term by protecting the ecosystem functions that support them. In doing so, it is essential to recognize that the needs of local people are integral to sustainable use, and that a 'museum piece' mentality towards ecosystem protection is not sustainable (Everard 1997a; Boon et al., 2000). Therefore, advice on potentially sustainable development practices has to be site-specific, though interpreted within the whole catchment system. One of the key strengths of both RHS and SWAMP is the capacity for interpretation of information, respectively, either from short stretches of the river channel with only small inputs of additional information about the floodplain and banktop, or from the characteristics of wetlands (or, in some cases, subsections of the larger wetlands). SWAMP's advice is intentionally location-specific, and may vary at two different sites on the same wetland, reflecting different habitat, biological, human, chemical or other factors. The studies reviewed in this chapter suggest that, if interpreted systemically, habitat-based information can be extremely helpful in assessing the processes operating in river and wetland systems (a more reliable approach than water quality where water is sometimes or frequently absent from the river channel). It may also provide a basis for advice on potentially sustainable forms of development tailored to local conditions, catchment-scale processes, biota and the needs of local people. This advice may apply not only to intact wetlands, but to the re-engineering of degraded or lost wetlands in order to restore their beneficial functions.

Both for Naivasha and Uganda, as also for aquatic resources throughout developing countries in the tropics, the consequences of continued non-sustainability are as significant as are the gains to be achieved from addressing sustainable development as a pressing and strategic challenge (Dugan, 1990; Denny, 1995; Everard et al., 1995; Everard and Harper, 2002). The majority of population growth and rise in material expectations are occurring in the developing world. It would be naïve to expect sustainable development automatically to occur in the best interests of ecosystems and their supportive functions. If we are to avoid the massive ecosystem destruction and loss of species and ecosystem functions that would otherwise accompany undirected development, it is incumbent upon the developed world, with its disproportionate share of money, power and scientific expertise, to transfer knowledge to the developing world. It is also a matter of self-interest, as all collectively benefit from, or conversely suffer from the loss of, ecosystem services such as climate control, carbon sequestration or purification processes that operate at the biospheric scale. If we wish to promote patterns of development in industrializing nations that follow a sustainable course, already-developed and affluent nations must be prepared to invest in knowledge transfer and the building of local expertise. Implementation of habitat-based assessment schemes may provide a robust and cost-effective means to support and monitor more sustainable development, especially when backed up with education to increase local capacity (Denny, 2001). Habitat-based assessments appear to offer potentially low-cost monitoring requirements, which may support the evaluation (or at least identification of the likelihood) of ecosystem functions.

Sustainable development of wetlands remains a pressing priority for East African catchments as a matter of long-term protection, not only for river basin ecosystems but also for the economic and social progress that depends upon them. This must be addressed at the catchment-scale to protect large-scale ecosystem functions, shaping the patterns of both intensive and subsistence land use. Long-term self-interest provides a compelling argument for sustainable development, the delivery of which will depend upon education and the spreading of good practice. However, achieving a transition from the current pattern of piecemeal exploitation will require leadership and long-term strategies, which, in turn, depend upon investment. Those deriving the greatest economic benefit from catchments in the developing world, including the beneficiaries along globalized supply chains remote from the region, have a vested interest (and also arguably a moral responsibility) to assist those closest to subsistence, thereby assuring overall sustainability of the catchments from which their profits ultimately derive (Everard and Harper 2002). Aid may well also be required to realize the goal of protection of long-term ecological, social and economic well being. Whilst the adverse consequences of continued unsustainable practice are highly significant for the region, and indeed the global biosphere, the benefits to be gained from shaping the course of future development are significant. Habitat-based monitoring of river corridors and wetlands may make an important and cost-effective contribution to determining appropriate management, protection or restoration, as an important contribution to sustainable development.

ACKNOWLEDGEMENTS

SWAMP was developed with the support of IBM UK. The original research in the Naivasha Catchment was supported by the Earthwatch Institute and assisted in the field by Earthwatch volunteers. It was made possible by a Kenya Government research permit to DMH, reference OP 13/001/12C/46. The authors are grateful to those organizations that enabled the research and development.

REFERENCES

Anon. 1999. 'The Lake Naivasha Management Plan'. Lake Naivasha Riparian Association, Naivasha, Kenya.

Becht R. and Harper, D.M. 2002. Towards an understanding of human impact upon the hydrology of Lake Naivasha, Kenya. Hydrobiologia, **488**, pp. 1-11.

Boon, P.J., Davies, B.R. and Petts, G.E. 2000. River conservation: A global imperative. In: P.J. Boon, B.R. Davies and G.E. Petts. (Eds). Global Perspectives on River Conservation: Science, Policy and Practice. John Wiley and Sons Ltd., Chichester, pp. xi-xvi.

Calder, I.R. 1999. The Blue Revolution: Land Use and Integrated Water Resources Management. Earthscan Publicatons Ltd, London.

Davies, B.R. and Day, J. 1998. Vanishing Waters. UCT Press, Cape Town.

Denny, P. (Ed.) 1985. The Ecology and Management of African Wetland Vegetation. Dr W. Junk Publishers, The Hague.

Denny, P. 1989. *Wetlands.* In: Strategic Resources Planning in Uganda. UNEP Report IX, UNEP, Nairobi, Kenya.

Denny, P. 1991. 'African Wetlands.' In: Finlayson, M. and Moser, M. (Eds), Wetlands. International Wetlands and Wildfowl Research Bureau, Slimbridge, pp. 115-148.

Denny, P. 1995. Benefits and priorities for wetland conservation: the case for national wetland conservation strategies. In: M. Cox, V. Straker and D. Taylor (Eds), Proceedings of the International Conference on Wetlands Archaeology and Nature Conservation. HMSO, London.

Denny, P. 2001. Research, capacity-building and empowerment for sustainable management of African wetland ecosystems. Hydrobiologia, **458**, pp. 21-31.

Denny, P. and Turyatunga, F. 1992. 'Ugandan Wetlands and their Management.' In: E. Maltby, P. Dugan and J.C. Lefeuvre (Eds) Conservation and Development: The Sustainable Use of Wetland Resources. Proceedings of the Third International Wetlands Conference, Rennes, France, 19th-23rd September 1988. IUCN Gland, Switzerland, pp. 77-84.

Dugan, P.J. 1990. Wetland Conservation: A Review of Current Issues and Required Action. IUCN, Gland, Switzerland. pp. 96.

Enniskillen. 2002. Introduction: the Lake Naivasha Management Plan – consensus-building to conserve an international gem. Hydrobiologia, **488**, ix-xii.

Everard, M. 1997a. 'Floodplain protection: Challenges for the next millennium.' In: R.G. Bailey, P.V. José and B.R. Sherwood (Eds), United Kingdom Floodplains, Westbury Academic and Scientific Publishing, West Yorkshire. pp. 477-483.

Everard, M. 1997b. Development of a British wetland strategy. Aquatic Conservation, **7**, pp. 223-238.

Everard, M. 1999. Towards sustainable development of still water resources. Hydrobiologia, **395/396**, pp. 29-38.

Everard, M. 2000. Aquatic ecology, economy and society: the place of aquatic ecology in the sustainability agenda. Freshwater Forum, **13**, pp. 31-46.

Everard, M. (2004). Investing in Sustainable Catchments. The Science of the Total Environment, **324/1-3**, pp. 1-24.

Everard, M. and Harper, D.M. 2002. Concluding remarks: Towards the sustainability of the Lake Naivasha Ramsar site and its catchment. Hydrobiologia, **488**, pp. 191-203.

Everard, M. and Powell, A. 2002. Rivers as living systems. *Aquatic Conservation,* 12, pp. 329-337.

Everard, M., Denny, P. and Croucher, C. 1995. SWAMP: A knowledge-based system for the dissemination of sustainable development expertise to the developing world. *Aquatic Conservation,* **5(4)**, pp. 261-275.

Everard, M., Vale, J.A., Harper, D.M. and Tarras-Wahlberg, H. 2002a. The physical attributes of the Lake Naivasha catchment rivers. *Hydrobiologia,* **488**, pp. 13-25.

Everard, M., Kuria, A., Maina, M., Vale, J.A. and Harper, D.M. 2002b. Aspects of the biodiversity of the rivers in the Lake Naivasha catchment. *Hydrobiologia,* **488**, pp. 43-55.

Everard, M., Powell, A. and Sweeting R.A. 2001. What Rio+10 must do for the freshwater environment. FBA News, **15 (Autumn 2001)**, pp. 1-4.

Gaudet, J.J. 1977. Natural drawdown on Lake Naivasha and the formation of papyrus swamps. *Aquatic Botany,* **3**, pp. 1-47.

Gaudet, J.J. and Melack, J.M. 1981. Major ion chemistry in a tropical African lake basin. *Freshwater Biology,* **11**, pp. 309-333.

Harper D.M., Mavuti, K.M. and Muchiri, S.M. 1990. Ecology and management of Lake Naivasha, Kenya, in relation to climatic change, alien species introduction and agricultural development. *Environmental Conservation,* **17**, pp. 328-336.

Henderson, I.G. and Harper D.M. (1992). Bird distribution and habitat structure on Lake Naivasha. African Journal of Ecology, **30**, pp. 223-232.

Hogan D, Maltby E, Blackwell M. 2000. Tamar 2000 Support Project: Wetlands Phase II Report. Wetlands Ecosystem Research Group, Royal Holloway Institute for Environmental Research, Egham, UK.

Hogan D, Maltby E, Blackwell M. 2001. Westcountry Rivers Project: Wetlands Report. Wetlands Ecosystem Research Group, Royal Holloway Institute for Environmental Research, Egham, UK.

IIED. 1994. Whose Eden? Empowering local communities to manage their wildlife resources. International Institute for Environment and Development Perspectives, **No. 13**, pp. 3-5.

Johnson, G.M. and Harper, D.M. 1998. Information for the sustainable management of shallow lakes: Lake Naivasha, Kenya. In: The Sustainable Management of Tropical Catchments. Eds. D.M. Harper and A.G. Brown. John Wiley & Sons, Chichester, pp. 259-274.

Lyon, J.G. 1993. Practical Handbook for Wetland Identification and Delineation. Lewis Publishers, Boca Raton, Florida.

Kitaka, N. 2000. Phosphorus Supply to a Shallow Tropical Lake and its Consequences—Lake Naivasha, Kenya. PhD Thesis, University of Leicester.

Kitaka, N., Harper, D.M., Mavuti, K.M. and Pacini, N. 2002. Chemical characteristics, with particular reference to phosphorus, of the rivers draining into Lake Naivasha, Kenya. Hydrobiologia, **488**, pp. 57-71.

Newson, M. 1994. *Hydrology and the River Environment*. Oxford University Press.

Ormerod, S. and Jüttner, I. 1998. 'Catchment sustainability and river biodiversity in Asia: a case study from Nepal.' In: *The Sustainable Management of Tropical Catchments*, Eds David Harper and Tony Brown. John Wiley and Sons Ltd, pp. 187-207.

Pacini, N. and Harper, D.M. 2000. River conservation in central and tropical Africa. In: P.J. Boon, B.R. Davies and G.E. Petts. (Eds). Global Perspectives on River Conservation: Science, Policy and Practice. John Wiley and Sons Ltd., Chichester. pp. 155-178.

Purseglove, J. 1988. Taming the Flood. Oxford University Press, Oxford.

Ramsar 1971. Convention on Wetlands of International Importance Especially as Waterfowl Habitat. 2/2/71, Ramsar, Iran.

Ramsar 1988. Proceedings of the Third Meeting of the Conference of the Contracting Parties. Ramsar Convention Bureau, IUCN, Gland, Switzerland. pp. 588.

Ramsar 1996. The Ramsar 25th Anniversary Statement. Resolution VI. 14, 6th Meeting of the Conference of the Contracting Parties, Brisbane, March 1996.

Raven, P.J., Holmes, N.T.H., Dawson, F.H., Fox, P.J.A., Everard, M., Fozzard I.R. and Rouen, K.J. 1998. River Habitat Quality: The Physical Character of Rivers and Streams in the UK and the Isle of Man. Environment Agency, Bristol.

Talling, J.F. and Lemoalle, J. 1998. Ecological Dynamics of Tropical Inland Waters. Cambridge University Press, Cambridge.

Tarras-Wahlberg, H., Everard, M. and Harper, D.M. 2002. Geochemical and physical characteristics of river and lake sediments at Naivasha, Kenya. Hydrobiologia, **488**, pp. 27-41.

Tourist Maps (K) Ltd. 1996. The Aberdares National Park and Environs. Tourist Maps (K) Ltd, Nairobi.

Wetlands International 1996. Mediterranean Wetland Inventory, Volume IV: Photointerpretation and Cartographic Protocols. Wetlands International, Delft.

Zalewski, M. and Harper, D.M. 2001. A 'sequential wetland' as a primary tool for the restoration of Lake Naivasha. In: Science and the Sustainable Management of Tropical Waters. (Eds) D.M. Harper and M. Zalewski, UNESCO IHP Technical Document No. 46. UNESCO, Paris.

CHAPTER

2

Seasonal Surfacewater Quality and Sediment Characteristics of the Urban Eutrophic Husainsagar Lake in Hyderabad, India

M. Vikram Reddy[1], and A. Vijaykumar[2]

[1]Salim Ali School of Ecology and Environmental Sciences, Pondicherry University, Kalapet, Pondicherry 605014, India

[2]Kedriya Vidyalaya, No. 2, Golconda, Hyderabad 500028, India

ABSTRACT

Analysis of the present state of the physico-chemical quality of the surfacewater and sediment at different sites in the littoral zone of the Husainsagar lake revealed that its water was green in colour, and at some places it was black, emanating a foul odour. The pH of the water was alkaline, and the electrical conductivity was high. The total hardness was also high, followed by chlorides, bicarbonates, sulphates, and aluminum content was low. The analysis of variance of spatio-temporal variations of different water quality characteristics—pH, electrical conductivity, total hardness, carbonates and bicarbonates, chlorides, sodium, iron, fluorides, boron, aluminium, silicates, dissolved oxygen, biochemical oxygen demand, ammonia nitrates, phosphates, potassium and sulphates showed significant variations. Further, Inductive Coupled Plasma Optical Emission Spectrometer (ICPOES) analyses of water and sediment of the littoral zone of the lake depicted the state of 22 elements, including the toxic heavy metals namely mercury, cadmium, lead and arsenic, the concentrations of which were much higher than those of the ISI, ICMR and WHO standards for drinking water. The concentration of many of these elements, including that of heavy metals, were significantly high at the site of the lake receiving the municipal raw sewage compared to other sites, clearly showing the municipal sewage to be the prime cause of pollution of the surfacewater of the lake. The textural composition of the surface sediment at different sites of the littoral zone of the lake showed its loamy nature. The spatio-temporal variations of different sediment characteristics also showed significant differences which also indicated that the site of the lake receiving the municipal sewage possessed higher concentrations of different elements and heavy metals compared to other sites of the lake. Human activities like Ganesh-idol immersion have increased the chemical pollution of the lake. Recurrence of fish kills were also recorded in the lake.

INTRODUCTION

Lakes and reservoirs, because of their resources, play a significant role in shaping the socio-economic conditions of the people inhabiting the area around these aquatic bodies. Moreover, the lakes in and around urban systems are known for lowering the ambient temperature. Despite their importance, many lakes are damaged by human activities and are severely contaminated by different types of wastes, according to the UN's 1997 Comprehensive Assessment of the Freshwater Resources of the World. Today, lakes all over the world—with probably an exception of Antarctica—suffer from ecological degradation, mainly cultural eutrophication, silting, etc. Monitoring of water quality such as physical, chemical and biological characteristics of these aquatic ecosystems is of paramount importance and a prerequisite in order to protect these significant water bodies from further degradation.

The concept of eutrophication, a well-known global phenomenon today, was first introduced by Weber (1907) to describe the enriched nutrient conditions determining the flora of German peat bogs. Later, Thienemann (1918) described deeper lakes as Oligotrophic and shallow lakes as Eutrophic, a term which was applied by Naumann (1919) to describe the nutrient-enriched conditions of Swedish aquatic ecosystems and their phytoplankton densities in relation to the concentration of phosphorus along with nitrogen and calcium, though he did not specify the ranges of nutrient concentrations for differentiating the types of lakes. Pearsall (1930) referred to the unevolved and evolved English lakes as Oligotrophic and Eutrophic, and was of the view that their eutrophic nature was primarily due to drainage basin erosion. Eutrophication, commonly known as the enriched state of inorganic nutrients particularly nitrogen and phosphorus of a water body, is manifested by an intense proliferation of algae and other aquatic plants or weeds, increasing the primary productivity, and their accumulation in water body in excess quantities. This accumulation can lead to detrimental changes in water quality and biological diversity and density of the water body (Vollenweider, 1968; National Academy of Science, 1969).

Natural eutrophication in original sense represented the natural ageing process of a lake. A lake receives allochthonous inflow of water from its surrounding drainage basin, along with eroded materials carried in the water from the land surface with various patterns of land use. The observed water quality and biological communities in an aquatic ecosystem, therefore, reflect the cumulative impact of all these allochthonous materials entering it. During the course of time, a lake slowly fills up with the suspended matter and other materials and ultimately becomes a marsh, which finally leads to the formation of a terrestrial system. The cultural eutrophication due to the discharge of raw or inadequately treated municipal sewage (major source of plant nutrients) into these water bodies, has increased several folds over the years, with the world's population increasing at a rate of 100 million per annum. This is the main environmental problem of lakes, particularly urban lakes, today.

There have been significant ecological consequences related to cultural eutrophication. As algal population die and sink to the bottom of a water body, their decay by bacteria can deplete the oxygen concentrations in bottom waters below the level required to support life, resulting in the mass killing of aquatic organisms, including fish. Various health hazards such as cholera, dysentery, typhoid and other water-borne diseases are aggravated by sewage discharge causing the cultural eutrophication of these water bodies, as they provide suitable habitats for the multiplication of causative organisms. The overall trophic responses of lakes and reservoirs to increased eutrophication has been explained by many earlier investigators (Stewart, 1968; Vollenweider, 1968; National Academy of Science, 1969; Lee, 1971; Welch, 1980; OECD, 1982).

Limnological studies on different temperate lake ecosystems are in replete (Wetzel, 1975, 2001). Investigations were carried out on the effects of cultural eutrophication on different limnological parameters of various lakes, Madison lakes (Sawyer et al., 1945), Lake Washington (Edmondson, 1961, 1970, 1991), Gull Lake, Lake Michigan (Moss, 1972), Round Lake, Minnesota (Shapiro & Weight, 1984), Lake Eire (Nicholls & Hopkins, 1993), and central European lakes (Thomas, 1969), and also on the changes in faunal and floral composition of natural waters in response to eutrophication (Morgan, 1970; Fanselius, 1977). Schindler (1975, 1990) probably was the first to carry out investigations on eutrophication in relation to recovery. There were a few excellent reviews on different aspects of structure and function of temperate lake ecosystems in relation to eutrophication, which gave an holistic understanding of their restoration and management (Hutchinson, 1975; Jorgensen, 1980; Ryding & Rast, 1989; Harper, 1992; Cooke, 1993).

In India, limnological studies on lakes, reservoirs and other lentic systems dates back to 1960s (Ganapati, 1960); George (1961) reported that of a pond ecosystem in Delhi. Later, such studies were undertaken on the reservoirs of south India, including that of Bhavanisagar reservoir (Sreenivasan, 1964, 1974) and Otteri Lake in Tamilnadu (Raghunathan et al., 2000), on Kuttiadi Lakc (Abbasi et al., 1996), Sasthamcotta Lake in Kerala (Prakasam & Joseph, 2000), the lakes of Ooty (Sivakumar et al., 2000), and Lake Powai in Mumbai (Maharashtra) in relation to pollution (Kohli et al. 1998). Zafar (1966, 1986) conducted limnological investigations on Husainsagar and other lakes in Hyderabad in Andhra Pradesh, followed by Munawar (1970) studying the limnology of certain ponds in the same region. Reddy & Prasad (1988) investigated the seasonal nutrient status of Banjara Lake, and Reddy (2001) on some water quality parameters of Osmansagar Lake. Limnological studies on lakes are relatively more in north India, such as on lakes in Kashmir (Kaul, 1977); Archar Lake (Blakhivetval, 1987), Dal Lake (Vass & Zutschi, 1978; Zutschi & Vass, 1978; Ara et al., 2003) and some of the forest lakes in Kashmir (Yousuf, 1979); Sukhna Lake at Chandigarh (Vasisht & Bhandal, 1979), Kumaran lakes (Sunder, 1996), Govindagarh Lake at Rewa (Mathew, 1975), Lake Fateh Sagar at Udaipur (Sharma & Durve, 1982), Lakhotia Lake (Khatri, 1984; 1986), Keetham Lake (Prakash, 1983), Nalkari Reservoir in Bihar (Sarkar, 1989), Hirakud Dam reservoir (Choudhury et al., 1979) in Orissa, lakes of Kolhapur, (Trivedy et al., 1983; Khatavkar et al., 1989), Nakana Lake (Maharashtra) (Patil & Nandan, 1989), lakes of Bhopal (Alexander 1976; Tanot & Bhatnagar, 1988; Tiwari et al., 1988; Valecha, 1985; Oomachan, 1988; Prakash , 1990; Adhlia et al., 1991; Dhote et al., 2001 and others) and Kapshi Lake (Fokmare & Musaddiq, 2002). The temporal variation in dissolved oxygen content was studied in Lakhotia Lake (Khatri, 1984, 1986). Water quality management of Talkatora Lake was studied by Gupta et al. (1999). Investigation on eutrophication of the lakes of Kashmir was first conducted about two and half decades back (Vass & Zutschi 1978; Zutschi & Vass, 1978). Keetham Lake at Agra (Prakash, 1983), some polluted tanks in Raipur (Unni, 1984), the Lake Pichhola at Udaipur (Sudan et al., 1984), and the Kalyanpur pond at Kanpur (Tripathi et al., 1987) indicated that studies on the structure and function of eutrophic lakes in India were meagre, which gradually increased in number. Never-the less, be more attention is needed for the holistic understanding of the eutrophication process in these lentic ecosystems.

Investigations on the physico-chemical characteristics of the lake sediment, which is an unseparable part of the lake, are about three decades old. Studies were conducted on the mineral composition of the bottom sediment in different basins of Dal Lake (Trisal, 1977), on some trace metals of surface sediment of Husainsagar (Prahalad & Seenayya, 1986) and that of lakes in Kashmir (Kango & Zutshi, 1988), and on the distribution of organic matter in the sediment of Vembanad Lake in Kerala (Murthy & Veerayya, 1972; and Anvar Batcha, 2001). The seasonal variations of the physico-chemical properties of the bottom sediment of Karwar Lake in Bihar and its nutrient status were studied about a decade back (Sinha et al., 1992; Sinha & Srivastava 1992; Singh & Jha, 2002). Further, the sediment of a typical tropical lake with reference to organic matter was investigated at Bhopal (Sinha et al., 1992). However, studies on the nutrient status of sediment of several eutrophic wetlands, including lakes in other countries, are far older (Tadajewaski, 1965; Rybak, 1969; Pieczynskai et al., 1975; Ryback and Sikarska, 1976; Forsberg, 1989).

Though ecological studies on different aspects of biotic and abiotic environment of the urban Lake Husainsagar have been initiated about four decades back (Munawar & Zafar 1964; Zafar, 1966), and later, followed by Hingorani et al., (1973-74), Prahalad & Seenayya (1986), Muley (1987), Ghosh & George (1989), Reddy & Vijayakumar (2001, 2002, 2003), sufficient information is still not available on the spatial and temporal variations of various surfacewater quality and sediment characteristics in relation to eutrophication of the lakes. The present study reports the state of spatio-temporal variations in the physico-chemical characteristics of the surfacewater and sediment of the littoral zone in relation to eutrophication, the effect of Ganesh-idol immersion, a very significant religious annual feature on some of the water quality characteristics and probable causes of recurrence of fish-kills in Husainsagar Lake.

LAKE AND ITS ENVIRONMENT

Husainsagar Lake is located at 15 North and 81 East and at 510 above msl, in the midst of the twin cities of Hyderabad and Secunderabad, the state capital of Andhra Pradesh in the south central India (Fig. 2.1). This lake has a surface area of 4.92 km^2, with a total volume of 9.44×10^6 m^3, maximum length of 32 km with store length of 14.2 km, and depth ranging from 1 to 12.5 m with a mean depth of 2.5 m (Prahalad & Seenayya, 1986).

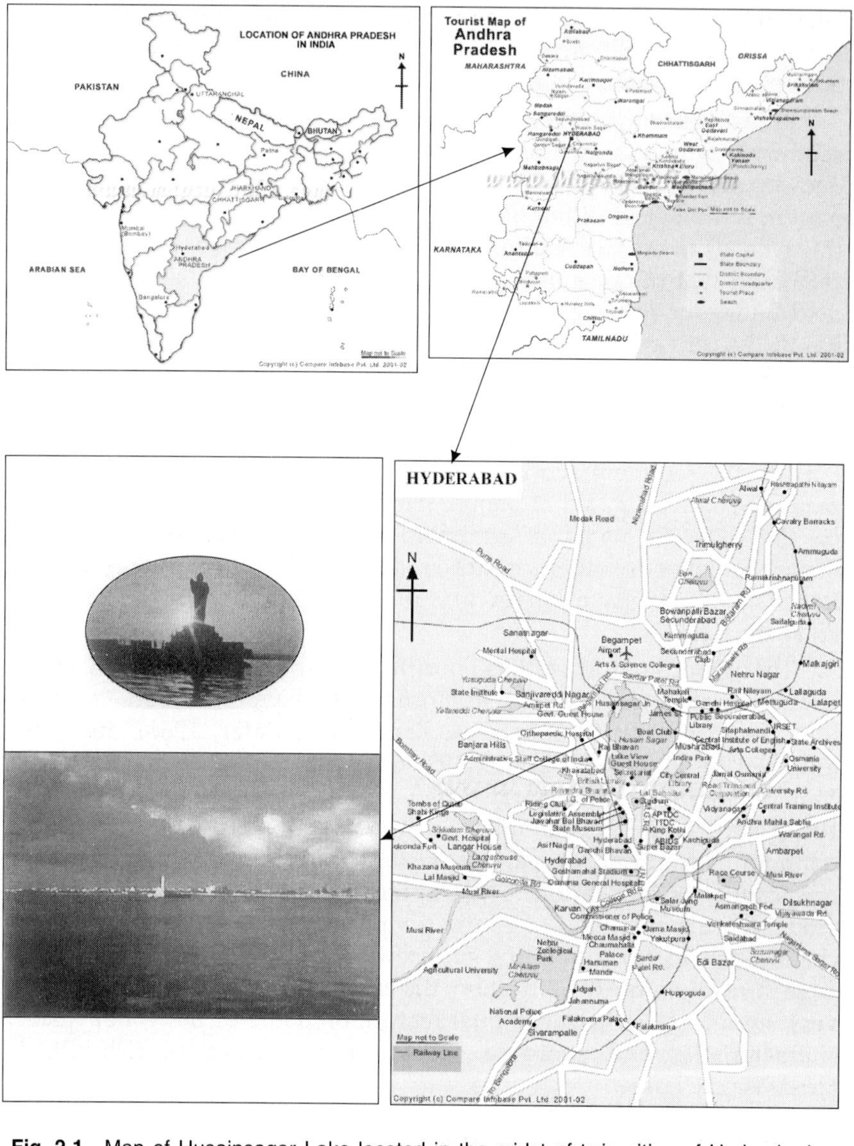

Fig. 2.1 Map of Husainsagar Lake located in the midst of twin cities of Hyderabad – Secunderabad, the state capital of Andhra Pradesh, India

History: Husainsagar Lake is said to have been constructed during 1550 to 1580 AD (the reign of Ibrahim Quli Qutub Shah, the fourth son of Sultan Quli Qutub Shah, the founder of the Qutub Shahi dynasty at Golconda). The construction and the supervision of the lake were entrusted to Hussain Shah Wali, the son-in-law of Ibrahim Quli Qutub Shah, and thus the name of the lake was given as Husainsagar. The Husainsagar, although commonly known as a lake, is a manmade reservoir built for serving the potable water needs of the people of Hyderabad (and Secunderabad). In its period of glory, the lake served as one of the primary sources of drinking water to the citizens of Hyderabad that developed into twin cities during 1894 to 1930 AD and later. The reservoir was constructed across Kalvaleru stream, a tributary of River Musi, which lost its importance as a source of portable water with the passage of time. It was a place of recreation for citizens as well as tourists for many generations. The lakewater was extensively used for gardening and washing activities. Even today, Husainsagar Lake is an ecological landmark on the cultural map of the state of Andhra Pradesh.

Geology: Underneath its water, the lake has the rock spread area around the "Gilbralter Rock Island", which is made up of residual jointed blocks of granite with a low dipping joint somewhere in the middle; that being hard, compact and massive, medium to course grained grey granite rock type. The surface of the rock is uneven. A colossus 16 m-tall monolithic statue of **Lord Buddha,** weighting 350 tonnes, is placed on the rock of Gilbralter in the middle of the lake, which reminds the visitor the Buddhist heritage of the State.

Inflow: The lake remained in an advanced stage of eutrophication with permanent algal blooms, poor water quality and with green colour as large quantities of municipal sewage (both domestic and industrial wastewaters) continued to pollute the lake for the past five to six decades. The lake received its inflow in the form of untreated raw-sewage and industrial effluents of about 34 million cubic metres/annum mainly from Picketnallah, Kukatpally nallah, Banjara nallah and Balkapur channel till date. The natural inflow (28 to 33 MCM/annum) into the lake is from a total catchment area of about 240 km^2. The average runoff into the lake is between 28 to 33 MCM. In summer, the situation worsens and the lake virtually becomes a cesspool and a breeding site for mosquitoes and other aquatic and semi-aquatic vectors.

Today, the human activities on the lake surroundings have affected both the quality and quantity of its water. Continuous washing of dirty clothes, vehicles and buffaloes and human defecation on its foreshore and in the littoral zone, runoff from catchments containing municipal sewage entering into the lake have all led to the excessive loading of external nutrients, enhancing the already advanced stage of its eutrophication.

This has caused over-growth of aquatic vegetation, such as algae, water hyacinth, and Ipomea on its surfacewater and in the littoral zone, which has further depleted the oxygen content in the lake water, affecting the aquatic organisms, including fishes.

During a long-term study starting from 1988 to 1991 and September 1998 to May 2000, the surfacewater quality and sediment characteristics were monitored at four sampling sites (stations) in the littoral zone of the lake, as this zone is the one most often damaged by human activities (Davis & Hirji, 2003). The sampling sites were selected on the basis of the alochthonous nutrient input into the lake, and the human disturbance in its foreshore and littoral zone.

Station–I (near Hyderabad Boat Club) located in between the Police Station and Hyderabad Boat Club on the bank of Husainsagar Lake, was mainly characterized by severe boating activities, with spilled oil on the surfacewater. This site was used by the boat club and was relatively dotted with slightly more vegetation, particularly trees on the bank, shedding their leaves into the littoral zone.

Station–II (near Secunderabad Boat Club) located on the Secunderabad area of the lake, was the protected zone of the military cantonment rest house. It received wasted from the kitchen and dining hall of the rest house. This site has good aquatic vegetation in the littoral zone, with even a few trees on its bank.

Station–III (near Sanjeevayya Park) mainly characterized by severe manual washing activities using detergent (containing phosophates), was the "Dhobi Ghats" area. Apart from clothes, vehicles and domesticated cattles and buffaloes were also washed in the lakewater, adding not only the organic pollution but also phosphorus, leading to eutrophication of the lake. The cattle waded into the water and remain there for a long time, particularly during summer. Besides, people living in nearby slums used this shore of the lake for open defecation, which gradually entered the lake. The water was usually dark green in colour with algal blooms and *Eichhornia crassipes* Solms afloat.

Station-IV (near sewage inlet passing through the Necklace Road) is a reclaimed area of the lake. This area of the lake earlier used to receive raw municipal sewage directly from the catchments of Banjara hills, Begumpet and Raj Bhavan areas of the city. It was reclaimed, filled with soil brought from outside and the necklace road was constructed, which prevented the direct inflow of the municipal sewage and storm runoff into the lake. Even today, a nallah passes underneath the road, dumping black municipal sewage into the lake. Moreover, the site is nearer to the inlet point of Kukatpally nallah, bringing the industrial effluents from Kukatpally and Balanagar industrial areas, and domestic sewage from the nearby catchment areas of the city. The lakewater at this site was usually black in colour, emanating a foul odour. The inflow of municipal sewage has caused nutrient enrichment and eutrophication of the lake. Moreover, thousands of Ganesh and Durga idols of varying sizes were immersed every year in the lake, creating a serious threat to its water quality.

Climatic Conditions: The lake experiences three seasons during the annual cycle: the winter extending from November to January, summer from February to May and monsoon from the first week of June to October. The average diurnal maximum temperature ranged from 28.7 to 39.9°C and the minimum temperature ranged from 12.8 to 25.2°C. The average of maximum and minimum temperatures were 32.6 to 20.2°C, respectively. The relative humidity ranged from 47 to 85% in the morning and from 23.0 to 75.0% in the evening.

The total monthly rainfall ranged from zero to 240 cm, and there was normally no rainfall during December to March, the total monthly rainfall being high during the period from June to October.

PHYSICO-CHEMICAL CHARACTERISTICS OF THE SURFACEWATER

The physico-chemical characteristics of lake ecosystems are influenced by allochthonous inputs of various types of nutrients and chemical pollutants entering through municipal sewage, industrial effluents and agricultural runoff (Davis & Hirji, 2003). These characteristics influence the biotic components of the ecosystem and control their diversity and distribution in time and space. The physical and chemical characteristics exert their influence both individually and collectively, and their interaction brings about a conducive environment that ultimately conditions the origin, development and finally, succession of the biotic communities, the "biodiversity signatures" of the ecosystem. Further, the dynamic biotic communities, in turn, continuously interact and alter the abiotic environment, particularly in an aquatic ecosystem, leading to a constant interaction between the biotic and abiotic components. Each ecosystem on the biosphere has its characteristic abiotic and biotic features, and their holistic understanding is essential for its effective management. Some of these aspects of the Husainsagar urban eutrophic lake are presented in this chapter.

The Physical Characteristics

The colour of water in the littoral zone of Husainsagar Lake at different sites across the seasons was green because of the blue-green algal blooms, particularly of *Microcystis* sp. (Ghosh & George, 1989) and at one or two sites, it was black, clearly indicating that the water of the lake is totally polluted; such perturbation in the water quality is further supported by a fishy stink. Similarly, the colour of water of Lake Talkatora was conspicuously green due to excessive algal growth. Besides, in some parts of the lake the colour was black due to septic conditions (Gupta et al., 1999). The colour of a lakewater has a significant relationship with its seasonal oxygen levels (Crisman et al., 1998).

Temperature is one of the most important physical factors of the environment, as it has a profound influence on the survival and distribution of its biotic communities, particularly planktonic organisms in aquatic ecosystems, which constituted the very base of the energy cycle. Water temperature is the most apparent factor affecting the seasonal cycle and abundance of aquatic organisms in aquatic ecosystems (Welch, 1952; Hussainy, 1967; Cairns, 1970; Hutchinson, 1975; Mathew, 1975; Kaushik et al., 1990). There was a wide spatio-temporal variation in water temperature ranging between 19 to 33°C in different sites, being lower during winter and higher during summer and early part of the rainy seasons in the Husainsagar Lake. Occasionally, high temperature of more than 33°C was recorded at some sites of the lake, which could be due to the discharge of hot effluents from industrial zones (Morrissy, 1971).

Earlier investigators have observed the water temperature fluctuating in different lentic systems, ranged from 17.0 to 33.0°C in the Husainsagar (Ghosh & George, 1989), 21 to 30°C in Hirakud Dam Reservoir (Choudhury et al., 1979), 23 to 35°C in Kalamba Lake (Khatavkar et al., 1989), 16.0 to 35.0°C in the eutrophic Keetham Lake (Prakash, 1983), 27.8 to 31.2°C in Lake Sasthamcotta (Prakash & Joseph, 2000), 19.3 to 23.1°C in Kumaon Himalayan Lakes (Sunder, 1996), 5.4 to 29.4°C (Trisal, 1987) and 06.0 to 28.0°C in the Dal Lake (Ara et al., 2003). Interestingly, Rosas et al. (1985) reported that the temperature being homogeneous throughout a tropical lake in Mexico, the lowest temperature (15°C) was recorded in January, while the highest (23°C) was registered in June.

pH, the hydrogen ion concentration, is another important abiotic parameter that affects the diversity and distribution of species in any aquatic ecosystem. It also controls the distribution of flora and fauna. pH ranging between 6.7 to 8.4 is suitable while values below 5.0 or above 8.8 are detrimental (Ellis, 1937). pH ranging between 6.5 and 8.5 is best for drinking water (Robert et al. 1940). Waters with pH less than 4 are acidic and much less productive than alkaline waters and the aquatic organisms in such waters are often diseased (Alikunhi, 1957). While more acidic conditions are detrimental, alkaline conditions support productivity and diversity (Todd, 1970; William, 1987). The pH of a typical eutrophic lake ranges from 7.7 to 9.6 (Wetzel, 1975). The pH of the Husainsagar lakewater was in the above range and alkaline, ranging from 7.9 to 8.2 across the seasons with an average of 8.2 (Fig. 2.2), displaying not only significant variation in between the sites of the lake but also across the seasons (Table 2.1). It was also within the desirable limits of the Bureau of Indian Standards (ISI, 1982) and the Indian Council of Medical Research (ICMR 1975) standards for drinking water. This water became more alkaline (8.1) during post-monsoon, which gradually decreased with the onset of summer, reaching 7.0 (Ghosh & George, 1989). It exhibited considerable variations during monsoons, probably due to continuous water movement and turbulation, and inflow of rainwater. The pH of the lakewater showed significant correlation with carbonates and bicarbonates, as its changes in water is governed by the amount of free CO_2, CO_3, and HCO_3 in summer. Besides, Carbon dioxide being soluble in lakewater, forms carbonic acid, which dissociates and raises the concentration of hydrogen ions (lowering the pH) (William, 1987).

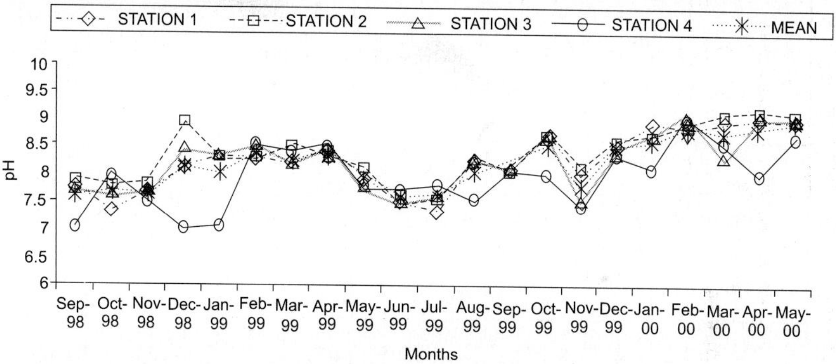

Fig. 2.2 Spatio-temporal variations in pH of surfacewater in the littoral zone of the Husainsagar Lake

The pH range varies in different lakes and reservoir ecosystems mostly because of the alkalinity in eutrophic lakes: 7.3 to 8.4 in Lake Nainital (Das, 1978), 7.3 to 8.5 in Hirakud Reservoir (Choudhury et al., 1979), 7.4 to 8.4 in Matyatal (Panna) (Nayak et al., 1982), 7.2 to 7.6 in Keetham Lake (Prakash, 1983), 7.5 to 8.1 in Badka Reservoir (Verma & Munshi, 1983), above 8.5 in Ranital (Jabalpur) (Patil et al., 1983), 7.8 in Lake Pichola (Sudan et al., 1984), 7.5 to 8.9 lakes at Kolhapur (Khatavkar et al., 1989), 9.5 to 9.8 in Talkatora Lake (Gupta et al., 1999), 7.4 to 9.8 in Bhopal Lake (Oommachan, 1988), 7.9 to 8.2 in Kumoan Himalayan lakes (Sunder, 1994), 8.3 to 8.4 (Trisal, 1987) and 6.7 to 7.8 in the Dal Lake (Ara et al., 2003). It was slightly acidic (6.9) in Sasthamcotta Lake (Prakasam & Joseph, 2000). The lakewater in arid conditions was alkaline, except

Table 2.1 Analysis of variance of different water quality characteristics at four sites in the littoral zone of Husainsagar Lake during the study period from September 1998 to May 2000

Parameter	*Mean*	*SEM*	*CV % coefficient of Variation %*	*F-test (stations)*	*F-test (months)*	*EMSS*	*CD*	*Change (%) during the period end-start*
pH	8.914	0.149	3.63%	6.95**	9.42**	0.088	0.421	1.372
Electrical Conductivity	1140.667	65.622	11.51%	NS	2.82**	17225.07	185.607	196
Calcium	63.44	4.092	12.90%	8.4**	8.64**	66.965	11.573	13.25
Magnesium	40.81	3.152	15.45%	NS	10.21**	39.741	8.915	6.5
Carbonate	25.599	7.576	59.19%	6.37**	7.24**	229.599	21.429	50.5
Bicarbonate	167.821	13.962	16.64%	6.96**	5.42**	779.721	39.49	-24
Nitrate	8.145	2.525	62.00%	NS	4.05**	25.497	7.141	-15.514
Phosphate	3.541	0.644	36.35%	NS	8.18**	1.657	1.82	2.55
Sulphate	116.024	10.384	17.90%	5.95*	5.79**	431.277	29.369	-34.375
Sodium	54.56	9.718	35.62%	NS	12.06**	377.748	27.486	-47.75
Potassium	17.28	2.363	27.36%	NS	2.24**	22.343	6.685	4.15
Iron	0.144	0.061	83.86%	NS	3.58**	0.015	0.171	-7.03
Total Hardness	320.512	8.796	5.49%	5.52**	10.03**	309.5	24.88	-5.5
Fluoride	1.012	0.067	13.33%	5.5**	39.06**	0.018	0.191	0.505
Boron	0.438	0.298	136.03%	5.62**	3.27**	0.355	0.843	-1.5
Ammonia	1.812	1.206	133.05%	5.33**	3.9**	5.816	3.41	0
Aluminium	0.027	0.043	321.67%	NS	2.58**	0.008	0.123	0.013
Silicon	14.994	2.726	36.36%	NS	NS	29.721	NS	-4.495
Chloride	235.997	14.971	12.69%	2.92**	19.49**	896.483	42.343	151.385
Biological Oxygen Demand	15.748	3.315	42.10%	204.36**	NS	43.952	NS	12.575
Dissolved Oxygen	4.003	0.525	26.25%	60.43**	5.27**	1.104	1.486	-1.298

SEM: Standard Error of Means; EMSS: Error Mean Sum of Squares; CV: Coefficient of Variation; CD: Critical Difference Level of significance **$P<0.05$; *$P<0.01$.

during dry spells and winter, when the concentrations of free carbon-dioxide lowered the pH to acidic level at the bottom (Khatri, 1986).

Electrical Conductivity (EC) of water is related to the total dissolved salts and reflects its nutrient status. High values of electrical conductivity indicate a high nutrient load in the water (Rodhe, 1969; Otsuki and Wetzel, 1974). Specific conductance of water, the numerical expression of its ability to carry an electric current, is an indicator of ionic composition and any alteration in it reflects a corresponding change in ionic concentration in a proportional manner.

The seasonal average electrical conductivity of the water in the littoral zone of the Husainsagar Lake ranged between 1135 to 1159 μmhos/cm among various sites (Fig. 2.3) and was significantly different spatially (Table 2.1), the mean average EC being 1141 μmhos/cm. However, the EC varied significantly across the seasons, being high in summer and decreasing in monsoon and winter, successively. This

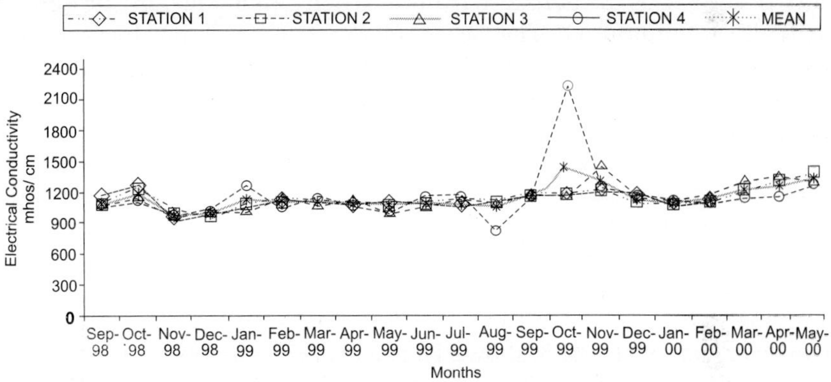

Fig. 2.3 Spatio-temporal variations in electrical conductivity of surfacewater in the littoral zone of the Husainsagar Lake

phenomenon could be due to the summer inflow into the lake, most of which was domestic sewage and industrial effluents. In other seasons, there was greater waterflow from non-polluted zones and the dilution effect resulted in the low values of conductance. It showed a positive correlation with a temporal variation in the potassium content at most of the sites, and with that of aluminium in one or two sites, and negative correction with sulphate in most of the sites, and with phosphates, bicarbonates, total hardness, ammonia, sodium, silicate and chloride at one or two sites of the lake. The annual average electrical conductivity was 52.81 $\pm$ 9.20 μmhos in Lake Sasthamcotta, which was low probably due to scarce amount of dissolved salt (Prakasam & Joseph, 2000). It was

436.6 µmhos/cm in the water of the eutrophic Lake Pichola (Sudan et al., 1984) and 1040 to 2040 µmhos/cm in the Talkatora Lake (Gupta et al., 1999). The EC ranged from 361 to 398 µmhos/cm in Dal lakewater (Trisal, 1987). However, it ranged from 0.11 to 0.29 µmhos/cm in the unpolluted Hirakud Dam Reservoir (Choudhury et al., 1979) and 182.2 to 319.1 µmhos/cm even in the unpolluted water of Kumaon Himalayan lakes (Sunder, 1996).

Chemical Characteristics

Among the chemical characteristics of the surfacewater of the Husainsagar Lake, the average concentration of magnesium, that of total hardness and fluoride were higher than those of the desirable limits of BIS (ISI, 1982) and ICMR (1975) standards for drinking water.

The Total Hardness (TH) is the characteristics of water representing the total concentration Ca and Mg ions expressed as mg of $CaCo_3$/litre and is known as carbonate hardness when other ions are present in insignificant amounts. The seasonal average total hardness of the water in the littoral zone of the Husainsagar Lake ranged from 329 to 310 mg/l among the sites (Fig. 2.4), the mean average total hardness being 320.51 mg/l, which is higher than the desirable limits of BIS (ISI, 1982) and ICMR (1975) standards for drinking water, as stated earlier. Its concentration showed a significant variation among different sites of the lake as well as across the seasons (Table 2.1). Its temporal variation showed a significant positive correlation with that of fluoride and ammonia, and a negative correlation with that of dissolved oxygen and with Biochemical Oxygen Demand (BOD) at one or two sites of the lake. However, Das et al. (2003) showed a negative correlation between the TH and fluoride content. The TH ranged from 57.5 to 80 mg/l in the Hirakud Dam Reservoir (Choudury et al., 1979), 61 to 150 mg/l in the eutrophic Keetham lakewater (Prakash, 1983) and an average of 80.3 mg/l in Kapshi lakewater (Fokmare & Musaddiq, 2002). Total hardness is often employed as an indicator of water quality, which depends on the concentration of Co_3 2– and H_{co3}^- salts of calcium and magnesium (temporary hardness) or sulphate, chloride or other anions of mineral acids (permanent hardness).

Calcium is a major chemical precipitate in lake ecosystems. Calcium and magnesium are the principal cations imparting hardness. Their ecological significance in the growth and population dynamics of freshwater organisms is well known. The higher levels of calcium and magnesium are mostly due to sewage or organic pollution (Beeton, 1969; Munawar, 1970). Organic pollution is known to play a potential role in metabolism and growth of aquatic organisms. The seasonal average

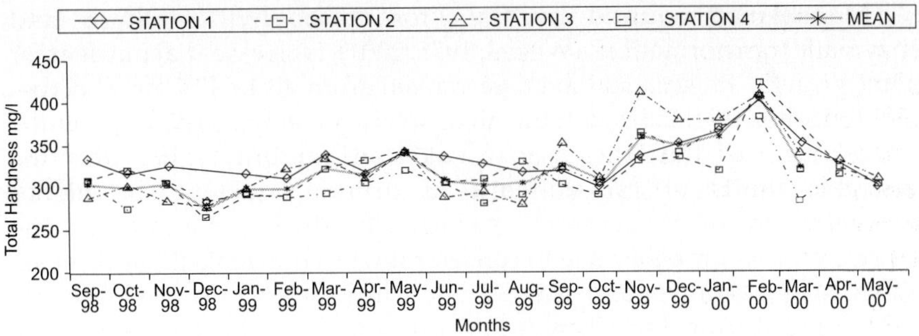

Fig. 2.4 Spatio-temporal variations in total hardness of surfacewater in the littoral zone of the Husainsagar Lake

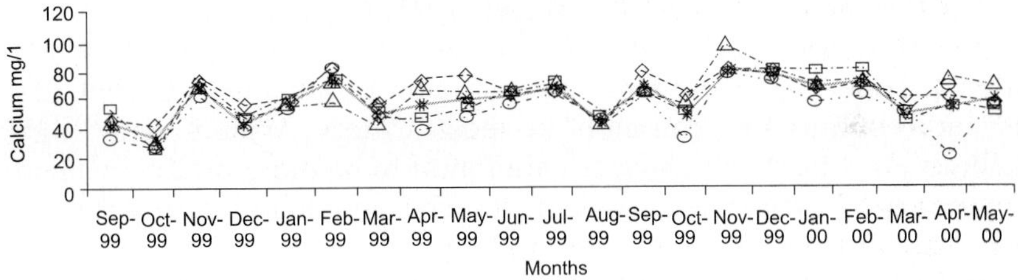

Fig. 2.5 Spatio-temporal variations in calcium of surfacewater in the littoral zone of the Husainsagar Lake

concentration of calcium in the water in the littoral zone of the Husainsagar Lake ranged between 69 to 56.4 mg/l (Fig. 2.5), its mean average being 63.4 mg/l. This level was below the desirable limits of ISI and ICMR standards of drinking water. The concentrations of calcium in water at various sites of Husainsagar Lake was significantly different and showed a significant variation across the seasons (Table 2.1). The temporal variation in the concentration calcium in the Husainsagar lakewater showed a significant positive correlation with carbonates, total hardness, ammonia and aluminium, and a negative correlation with that of magnesium, silicon and dissolved oxygen at some of its sites. The positive correlation was probably due to the addition of detergents into the lakewater through washing activities. Any value of calcium above 25 mg/l indicates richness of calcium in water (USEPA, 1975) and its eutrophic nature. As per this definition, the water of the Husainsagar Lake is rich in calcium and can be identified as eutrophic in nature. It was 22.5 mg/l in the eutrophic Lake Pichola (Sudan et al., 1984), ranged from 18 to 39 mg/l in Dal Lake (Trisal, 1987), 24.8 to 29.6 mg/l in Kalamba Lake (Khatavkar et al., 1989), and averaged 64.7 mg/l in Kapshi Lake (Fokmare & Musaddiq, 2002). It was very less, 0.0 to 4.8 mg/l in Kuttaidi Lake (Abbasi et al., 1996).

Magnesium required by flora for chlorophyll synthesis and enzymatic transformations (Wetzel, 1975, 2001), is an essential nutrient for plant growth. Its seasonal average ranged from 40 to 42.7 mg/l across different sites (Fig. 2.6), with the mean average beings 40.81 mg/l in the littoral zone of the Husainsagar Lake, an amount higher than the desirable limits of ISI standard of drinking water. The higher concentration of magnesium, particularly during summer in the lakewater, was probably due to concentration of nutrients (Sharma et al., 1978). Magnesium content was 9.2 mg/l in Lake Pichola (Sudan et al., 1984), ranged from 8 to 15 mg/l in the waters of Dal Lake (Trisal, 1987), 2.7 to 23.6 mg/l in Kalamba Lake (Khatavkar et al., 1989), averaged 25.6 mg/l in Kepshi Lake (Fokmare & Musaddiq, 2002) and 0.24 to 2.2 mg/l in Kuttiadi Lake (Abbasi et al., 1996).

The Carbonates (Co_3 2-), Bicarbonates (HCO^-_3) and hydroxide (OH^-) ion concentrations define alkalinity of water and all the three, when added together, form the total concentration of alkalinity. Carbonates and bicarbonates are common in most waters. Waters with high alkalinity, hardness and chlorides are found to be more productive and support a rich biodiversity (Moyle, 1946). The amount of bicarbonate was seen more during the rainy season (August) because of the inflow of mineral salts along with water, leading to an increase in the total alkalinity, which ranged from 60 to 88 mg/l in the reservoir water (Choudhury et al., 1979). These salts, after rains, are either carried out either when water was released from the reservoir or some mineral salts settled down at the bottom sediment. The total alkalinity of lakewater remaind above 100 ppm in higher productive water bodies (Alikunhi, 1957).

The south Scottish lakes were classified on the basis of alkalinity into three categories: as nutrient poor with the range of 1-15 ppm; moderately nutrient rich (16-60 ppm); and nutrient rich (>60 ppm) (Spence, 1964). The seasonal average amount of carbonates of the water in the littoral zone of the Husainsagar Lake ranged between 16.4 to 33.6 mg/l (Fig. 2.7), its mean average being 25.59 mg/l; while the average bicarbonates ranged between 151 to 190 mg/l, its mean average concentration being 167.82 mg/l (Fig. 2.8). The concentration of carbonates and bicarbonates of water of the lake differed significantly among different sites and also showed significant variations across the seasons (Table 2.1). The carbonates and bicarbonates showed a significant correlation with pH, and free carbon dioxide (Nirmal Kumari, 1984). However, the carbonates and bicarbonates in the eutrophic Lake Pichola were 1.6 and 168 mg/l, respectively (Sudan et al., 1984). The total alkalinity ranged from 10.1 to 14.55 mg/l in Lake Sasthamcotta (Prakasam & Joseph, 2001), and was high ranging from 110 and 140 mg/l in Lake Malpur (Qadri & Yousuf,

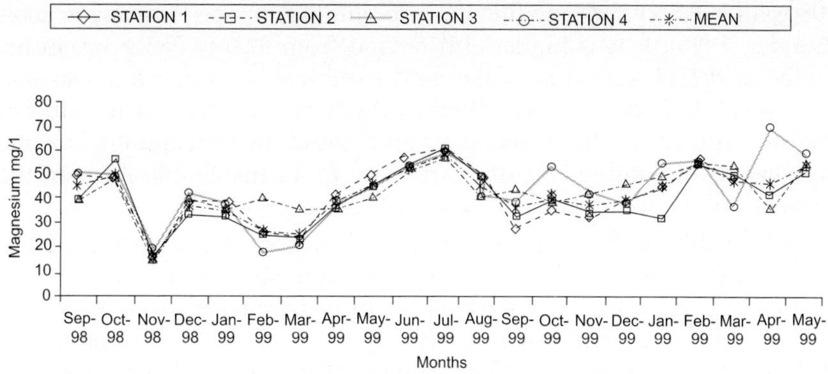

Fig. 2.6 Spatio-temporal variations in magnesium of surfacewater in the littoral zone of the Husainsagar Lake

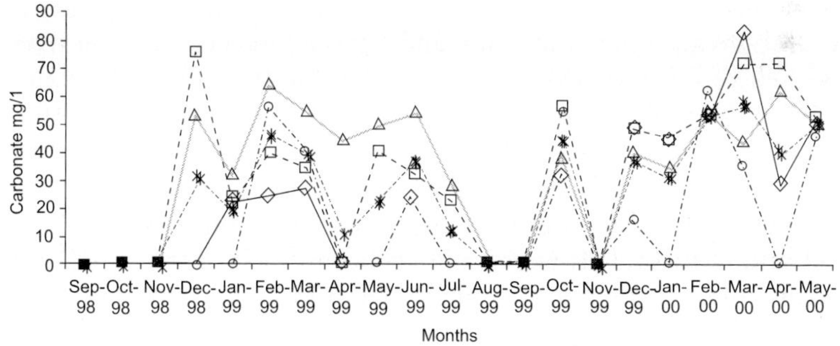

Fig. 2.7 Spatio-temporal variations in carbonates of surfacewater in the littoral zone of the Husainsagar Lake

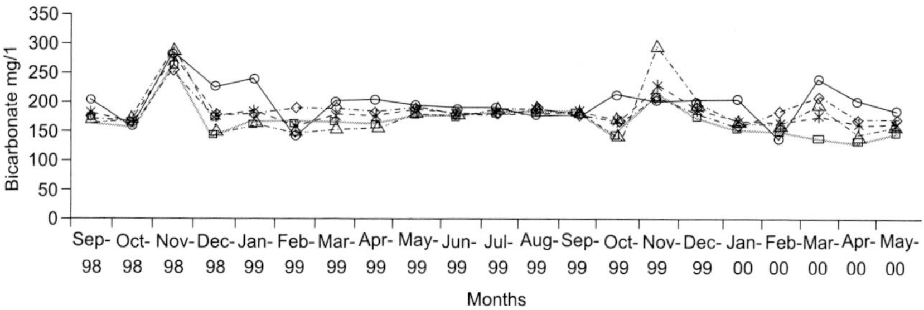

Fig. 2.8 Spatio-temporal variations in bicarbonates of surfacewater in the littoral zone of the Husainsagar Lake

1980), and 96.5 to 119.8 mg/l in the unpolluted Kumaon Himalayan lakes (Sunder, 1996). It was higher and ranged from 92.0 to 385 ppm, in the water in the Husainsagar Lake, and exhibited well-marked seasonal variations (Ghosh & George, 1989), which may be related to nutrient concentrations in the lakewater, and could be attributable to the conversion of soluble calcium carbonate to an insoluable form in the presence of carbon dioxide (Otsuki & Wetzel, 1974).

Chlorides as chloride ions (Cl^-), are major anions in water and wastewater. Chlorides determine the total salinity of water. The ecological significance of chlorides lies in their potential to regulate the salinity of water and exert a consequent osmotic stress on biotic communitics. The relatively low salinity of freshwater has a profound influence on the distribution of biota (Srivastava, 1988; Das, 1989). The chloride concentration is higher in municipal wastewater because of human diet being rich in NaCl. The higher concentration of chlorides of water is considered to be an indicator of pollution due to higher organic waste and sewage mixing in water and it has a direct connection with the levels of eutrophication (Munawar, 1970; Sinha, 1986; Das 1989; Gupta et al., 1999). Most of its increase in Lake Erie was due to industrial effluents and municipal waste (Ownby & Kee, 1967).

Freshwater normally contains about 8.3 mg/l of chlorides; the value would be much more higher in waters when polluted by domestic sewage and industrial effluents. High concentration of chloride indicates large amount of organic matter of animal origin and domestic wastes in the water, which is suggestive of eutrophication (Sinha, 1986). Further, it has also been reported that high chloride content of water of any lake with an annual average of 83.7 mg^{-1} is an indicator of organic pollution (Todd, 1970; Vass & Zutshi, 1983; Gopal & Durve, 1989). The seasonal average chloride content of the water in the littoral zone of the Husainsagar Lake ranged between 219.7 to 245 mg/l, its mean average concentration being 236.0 mg/l (Fig. 2.9). Its concentration in the lakewater was higher than that of the desirable limits of ISI and ICMR standards for drinking water (200 mg/l). Higher values of chlorides in the summer could be due to higher temperature associated with evaporative concentration, and the rate of decomposition. Its low values, during monsoon, was probably due to dilution of water. The observed values of the lakewater indicated heavy organic pollution in the littoral zone, particularly at the site of the lake receiving municipal sewage directly. The variation in its concentration in the lakewater were significantly different in between the sites and also across the seasons (Fig. 2.9) (Table 2.1). Chlorides and other parameters such as high values of nitrates, DOM, TDS and bicarbonates of the water of Husainsagar Lake are indicative of its high nutrient load.

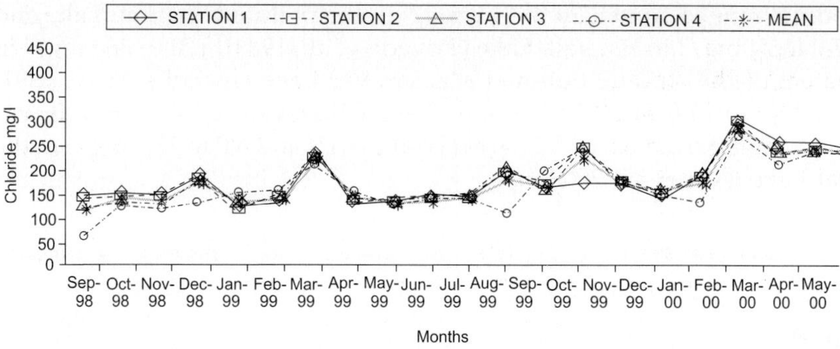

Fig. 2.9 Spatio-temporal variations in chlorides of surfacewater in the littoral zone of the Husainsagar Lake

The chloride content of water in Hirakud Dam Reservoir ranged from 15 to 25 mg/l (Choudury et al., 1979) and from 21.3 to 82.4 mg/l and 39.8 to 82.3 mg/l in the eutrophic lakes Shambu and Rankala, respectively (Trivedy et al. 1983). It was 28.0 mg/l in Lake Pichola (Sudan et al., 1984) and ranged from 405 to 690 mg/l in Talkatora Lake (Gupta et al., 1999). Chloride content of Kuttiddi Lake was very low, ranging from 4.5 to 11.8 mg/l (Abbasi et al., 1996) and 9.7 to 14.9 in the water of the unpolluted Kumaon Himalayan lakes (Sunder, 1996). However, it ranged from 47 to 90 mg/l in eutrophic Keetham lakewater (Prakash, 1983), 15 to 38 mg/l in Bhopal Lake (Oommachan & Belsare, 1986), 7.4 to 25.5 mg/l in Kalamba lakewater (Khatavkar et al., 1989) and 17.5 to 36.5 mg/l in the Dal Lake (Ara et al., 2003). The presence of higher amount of chloride influenced the amount of dissolved oxygen, which may adversely affect the aquatic organisms (Datta Roy et al., 1984).

Sodium concentration in the receiving waters varies according to the sewage, industrial effluents, and the use of sodium compounds for corrosion control and water-softening processes. Its concentration in surfacewater may be less than 1 mg/l or more than 300 mg/l, depending upon the geographical area. Lakes containing a high concentration of sodium sulphate are called bitter lakes while those containing sodium carbonate are called alkali lakes. High concentration of sodium is the result of pollution (Beeton, 1969). The seasonal average sodium content of the water in the littoral zone of the Husainsagar Lake ranged between 53.7 to 55.3 mg/l, its mean average concentration being 54.56 mg/l. Its concentration in the water showed a significant temporal variation across the seasons (Fig. 2.10) and a significant positive correlation with that of potassium, ammonia and total hardness, and negative correlation with fluoride at different sites of the lake. However, Das et al. (2003) showed a positive correction between N^+ and fluoride content. The amount of sodium ranged from 22.0 to 64.0 mg/l in the polluted Shambu Lake and 27.0 to 74 mg/l in Rankala Lake (Trivedy et al., 1983), 170 to 380 mg/l in waters of the sewage-polluted Mansarovar Lake (Adholia et al., 1991) and 14.0 to 31.0 mg/l in Kalamba Lake (Khatavkar et al., 1989). It was low, ranging from 3.6 to 7.2 mg/l (Trisal, 1987) and 6.0 to 13.6 mg/l in the Dal Lake (Ara et al., 2003).

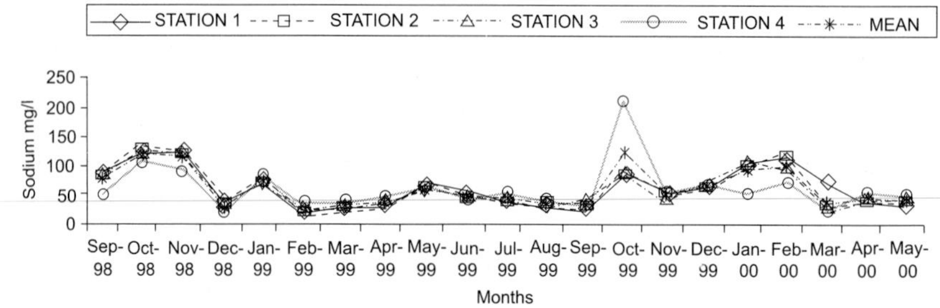

Fig. 2.10 Spatio-temporal variations in sodium of surfacewater in the littoral zone of the Husainsagar Lake

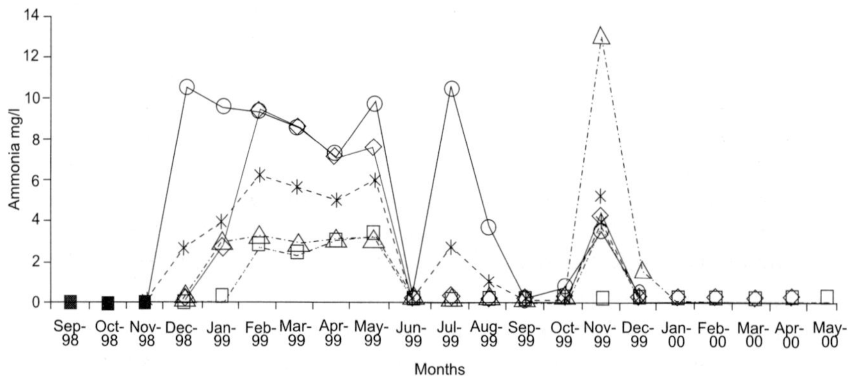

Fig. 2.11 Spatio-temporal variations in ammonia of surfacewater in the littoral zone of the Husainsagar Lake

Ammonia is produced by the microbiological activity of organic nitrogenous matter (Kuznesov, 1968). Higher concentration of ammonia, a product of ammonification of the organic matter, indicates organic pollution of the water (Rybak & Sikarska, 1976). The seasonal average ammonia concentration of the water in the littoral zone of the Husainsagar Lake varied widely, ranging between 0.56 to 3.4 mg/l (Fig. 2.11), its mean average concentration being 1.8 mg/l. Its concentration was significantly different not only in between different sites of the lake but also across the seasons (Table 2.1). The observed decrease in ammonia in some of the months may be attributable to its utilization by phytoplanktons (Reddy & Prasad, 1988). Its temporal variation showed a significant positive correlation with that of BOD, and aluminium at a few sites. Ammonia-N can range from 0 to 5 mg/l in unpolluted surfacewater (Wetzel, 1975). The ammonia ranged from 63.0 to 404.6 μg/l in the water of Banjara Lake (Reddy & Prasad, 1988), 0.14 to 1.92 mg/l in Kalamba Lake (Khatavkar et al., 1989), and 250.0 to 502.5 μg/l in the Dal lakewater (Ara et al., 2003). Its higher concentration in water polluted by sewage or some kind of industrial waste containing either organic nitrogen or directly ammonia was toxic to fish (Varshney, 1985).

Nitrate (No_3), one of the most important nutrients in an aquatic system, enters the lake through domestic sewage, runoff with residual fertilizers, and industrial effluents (Hasler, 1947; Pieczynskai et al., 1975; Wetzel, 1975; Alexander, 1976; Sharma & Durve, 1982). The inflow of sewage and runoff increase the concentration of NO_3. Nitrogen and its compounds form a complex system in lakes, appearing as free nitrogen in solution, organic compounds, ammonia, nitrate and nitrite (Todd, 1970; Wetzel, 1975). Oxidation of nitrogenous matter leads to the formation of nitrates, through an intermediate formation of ammonia and nitrite. A nitrate-nitrogen is the oxidized state of the element in water, and its cycle operates in many shallow eutrophic lakes (Holden & Cairns, 1979). Nitrates stimulate nutrient-linked eutrophication and algal blooms, and are heavily consumed in the upper strata of lakes during the periods of high productivity of phytoplanktons (Stewart, 1968). Though nitrate is essential for enhancing productivity, its concentration of more than 50 mg/l is undesirable in water used for domestic purpose, as per the European Health Standards for drinking water, and higher than 45 mg/l is unacceptable in the United States of America (Mason, 1996). The world average of nitrates in unpolluted freshwater is 0.3 mg/l (Reid & Wood, 1976). The unpolluted tropical waters are generally deficient in nitrate levels (Ganapati, 1960). Nitrate-nitrogen level of 0.15 mg/l was suggested as the critical level, beyond which a water body may indicate eutrophication (Sawyer et al. 1945). This level was 0.4 mg/l in the water of Badua Reservoir (Verma & Munshi, 1983); in eutrophic Lake Pichola it was 0.16 mg/l (Sudan et al., 1984). It ranged from 0.56 to 3.76 mg/l in the water of Banjara Lake (Reddy & Prasad, 1988), 1.65 to 6.18 mg/l in Shambu Lake and 1.15 to 18.5 mg/l Rankala Lake polluted with sewage (Trivedy et al., 1983), 0.53 to 16.3mg/l in Kalamba Lake because of runoff from agricultural catchment area (Khatavkar et al., 1989), and was 0.67 mg/l in Kapshi lakewater (Fokmare & Musaddiq, 2002). It was high, 12 to 16 mg/l in Talkatora Lake (Gupta et al., 1999), which was because of inflow of municipal sewage. The seasonal average nitrate of the water in the littoral zone of the Husainsagar Lake ranged between 7.5 to 9.2 mg/1 (Fig. 2.12), its mean average concentration being 8.15 mg/l, which is in a high range, but within the desirable limits of ISI and ICMR standards for drinking water. The inorganic N was 3.2 mg/l in Husainsagar a decade back (Zafar, 1986). Though the variation in nitrate concentration of the water was not significantly different among the sites of the lake, it showed a significant variation across the seasons (Table 2.1). Its increase during the summer (Fig. 2.12) was probably due to the runoff inflown into the lake due to pre-monsoon showers, along with evaporation of water in the lake (Trivedy et al., 1983; Geldermalsen, 1985; Das, 1989). Moreover, the runoff during rainy season also contributes to the nitrate-nitrogen level of lakes and reservoirs (Badge & Verma, 1985).

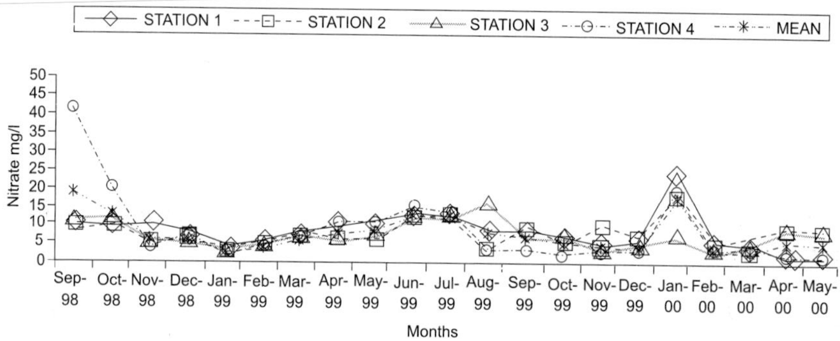

Fig. 2.12 Spatio-temporal variations in nitrates of surfacewater in the littoral zone of the Husainsagar Lake

Nitrate is often identified as a regulating element in situations where abatement is necessary to control eutrophication (Stewart, 1968; Edmondson, 1970; Varshney, 1985; Das, 1989). The determination of the quality of nitrogen is highly significant to measure the status of pollution, as it gives a picture of relative easy availability of decomposable organic matter. A number of studies on lentic systems have shown that nitrate is abundant during winter and spring preceded by heavy rains in the catchment (Alexander, 1976; Dakshini & Soni, 1979; Das, 1989). Nitrate nitrogen showed high positive correlation with depleted dissolved oxygen in Lake Erie (Burns & Ross, 1971), and fluctuated inversely with the dissolved oxygen (Munawar, 1970).

Phosphate is a major limiting nutrient of any aquatic ecosystem, particularly in small lakes (Schindler, 1975, 1990). Orthophosphate and other organic phosphates are the most important phosphorus compounds in a lake, and all inorganic phosphate is usually considered as PO_4. Phosphates and nitrates are heavily consumed in the upper strata of the lakes during periods of high productivity of phytoplanktons (Hussainy, 1967). The nitrate and phosphate ratio (N:P) indicates the degree of pollution caused by sewage. It can also indicate the recovery rate after the diversion of sewage entering the water body (Edmondson, 1970). Increased concentrations occur in deeper portions due to the decaying process of biological material, and also regeneration and release from the sediment, especially during anoxic conditions in shallow lakes (Burns & Ross, 1971; Golterman, 1975; de Anda, et al., 2000). Phosphates are present in small quantities in natural water, unless the water body is contaminated by domestic sewage or agricultural runoff (Edmondson, 1970; Hutchinson, 1975; Wetzel, 1975; William, 1987). Phosphate is essential for growth, being a major constituent in ATP and NADP, and acts as a limiting factor for the productivity of an ecosystem. Often, phosphates entering the receiving water bodies from raw wastewater, agricultural drainage (as these are used as fertilizers) or certain industrial waste, stimulate primary producers of nuisance value. These are the major ingredients of most detergents (sodium tripolyphosphates), industrial effluents and domestic sewage, and favour luxuriant growth of algae, causing their blooms. This extensive algal growth consumes most of the available dissolved oxygen from the water, leading to its depletion—being detrimental to the growth of other organisms, and such algae produces a foul smell upon decay (Varshney, 1985; Das, 1989; Wetzel, 1975, 2001).

The seasonal average phosphate concentration of the water at different sites in the littoral zone of the Husainsagar Lake ranged between 3.4 to 3.8 mg/l (Fig. 2.13), its mean average concentration being 3.54 mg/l, which is many times higher than that of the ICMR (1975)

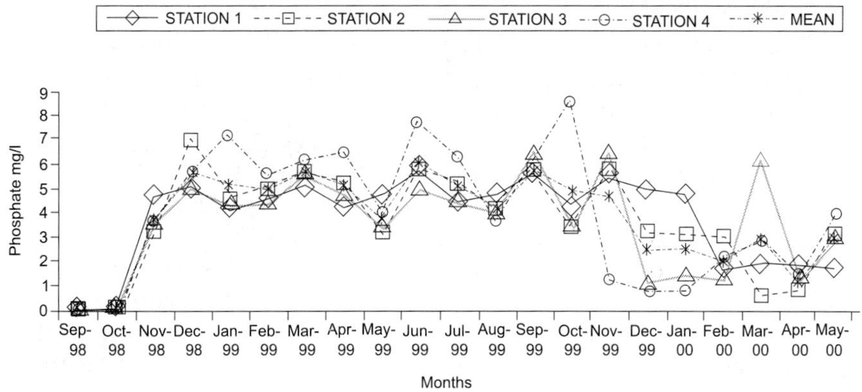

Fig. 2.13 Spatio-temporal variations in phosphates of surfacewater in the littoral zone of the Husainsagar Lake

standards for drinking water and more than sufficient to cause algal blooms. Its concentration in the lakewater varied significantly across the seasons (Table 2.1). The temporal variation of phosphate in lakewater showed positive correlation with that of iron, ammonia and BOD, and negative correlation with that of sodium and potassium at some sites of the lake. It was only 0.43 mg/l in the lakewater about a decade back (Zafar, 1986); it ranged from 1.3 to 3.5 mg/l in eutrophic Keetham lakewater (Prakash, 1983) and 0.4 to 1.6 mg/l in the upper lake and 0.2 to 0.75 mg/l in the lower lake of Bhopal, due to municipal sewage inputs (Velacha, 1985). The yearly average phosphate concentration was as low as 0.12 mg/l in Badua Reservoir (Verma & Munshi, 1993), and 0.38 mg/l in Lake Pichola (Sudan et al., 1984). It ranged from traces to 0.93 in Lake Kalamba (Khatavkar et al., 1989), 0.43 ± 0.2 mg/l in Lake Sasthamcotta (Prakasam & Joseph, 2000), 0.14 to 0.36 mg/l in Shambu Lake and 0.11 to 0.44 in Rankal Lake (Trivedy et al., 1983), 0.11 to 0.27 mg/l in the surfacewater of upper lake of Bhopal (Prakash, 1990), 0.16 to 0.8 mg/l in water of Talkatora (Gupta et al., 1999) and 0.28 to 0.32 in Kapshi Lake (Fokmare & Musaddiq, 2002). The concentration of phosphorus in the water should be less than 0.05 mg/l for controlling algal growth, and any amount of it in excess of 0.03 ppm is an indicator of pollution and is sufficient to cause algal blooms (Sawyer et al., 1945; Wetzel, 2001). Higher value of phosphate in the water of the Husainsagar lake was most probably due to the inflow of allochthonous nutrients in the form of municipal sewage, leading to further eutrophication of the lake (Hasler, 1947; Hutchinson, 1967; Beeton, 1965; Lee, 1971; Pieczynskai et al., 1975; Jonason, 1984; Gopal & Durve, 1989). A number of investigators have reported higher values of phosphate in different water bodies receiving domestic sewage (National Academy of Science, 1969; Pieczynskai et al., 1975; OECD, 1982).

Potassium released during the process of decomposition, is essential for photosynthetic activity. Its seasonal average concentration in the surfacewater in the littoral zone of the Husainsagar Lake ranged in between 16.4 to 19.2 mg/l, the mean average concentration being 17.3 mg/l. Its concentration in the water showed a significant temporal variation across the seasons (Fig. 2.14) (Table 2.1), which showed a significant positive correlation with that of chloride concentration, total hardness and aluminium contents of water at different sites of the lake. It ranged from 0.2 to 3.0 mg/l in Kalamba lakewater (Khatavkar et al., 1989), 0.2 to 1.66 mg/l (Trisal, 1987) and 3.0 to 6.5 mg/l in Dal Lake (Ara et al., 2003), 30 to 160 mg/l in eutrophic Shambu Lake and 3.0 to 9.0 mg/l in Rankala Lake (Trivedy et al., 1983) and 25.0 to 200.0 mg/l in the water of Mansarovar Lake receiving a sewage inflow (Adholia et al., 1991).

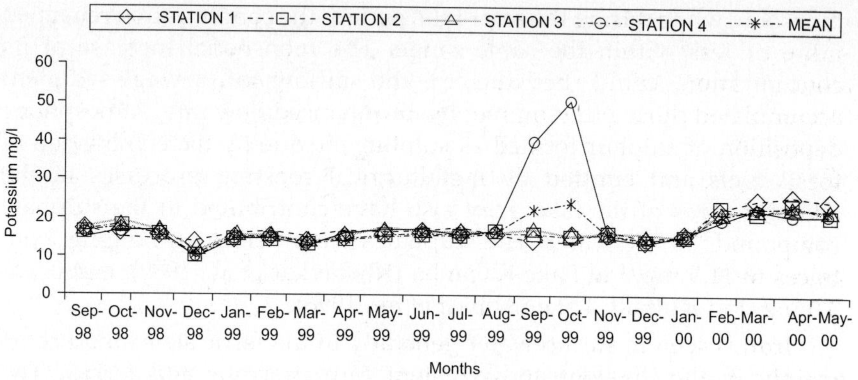

Fig. 2.14 Spatio-temporal variations in potassium of surfacewater in the littoral zone of the Husainsagar Lake

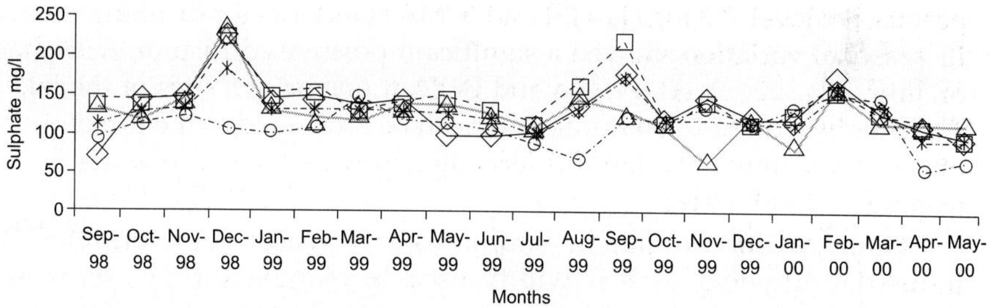

Fig. 2.15 Spatio-temporal variations in sulphate of surfacewater in the littoral zone of the Husainsagar Lake

Sulphate (So_4^{2}) ions usually occur as principal ions in lakewaters, and are derived form sedimenting rocks. Sulphur enters into the biomass largely as sulphates (So_4^{2-}) (Kuznesov, 1968; Harper, 1992; Mason, 1996). Heavy pollution due to sewage and unpleasant industrial effluents generate obnoxious odours as sulphates undergo reduction into sulphide and escape as H_2S gas, a stage that is popularly identified with the "dealth" of a lake (Welch, 1980). The seasonal average sulphate content of the water in the littoral zone of the Husainsagar Lake ranged between 101.5 to 119 mg/l, its average concentration in the water being 118.02 mg/l, below the maximum desirable limit (150 mg/l) of ISI and ICMR standards of drinking water. Its concentration in the water at one or two sites during winter was higher than that of the desirable limits of ISI and ICMR standards. Its concentration though did not show much spatio-temporal variation in the lakewater (Fig. 2.15), its temporal variation, however, showed a negative correlation with that of potassium and total hardness at some sites of the lake. Its concentration was high during monsoon, while during the rest of the period, the concentration remained more or less within the same range. The monsoonal increase of its concentration could be due to the inflow of sewage-sediment accumulated during the summer by an enhanced flow rate. Atmospheric deposition of sulphur formed as sulphur dioxide by the combustion of fossil fuels and emitted by metallurgical roasting processes in the adjacent areas of the lake, may also have contributed to the sulphate compounds of the water. The sulphates were very low, ranging from traces to 11.5 mg/l in Lake Kalamba (Khatavkar et al., 1989), and 0.0 to 3.0 mg/l in Kuttaidi Lake (Abbasi et al., 1996).

Iron is seen in surfacewater generally in the ferric state, and occurs mainly in the divalent and trivalent form (ferrous and ferric). The seasonal average iron concentration in the surfacewater in the littoral zone of Husainsagar Lake ranged between 0.12 to 0.21 mg/l (Fig. 2.16), its mean average concentration being 0.144 mg/l. It was lower than the permissive level, 0.3 mg/l of ISI and ICMR standards for drinking water. Its seasonal variation showed a significant positive correlation with that of ammonia, dissolved oxygen and BOD at one or two sites of the lake. The presence of iron in natural water can be attributed to the dissolution of rocks and minerals, landfill leachates, sewage or waste water from iron-related industries.

Fluoride concentration in natural water increases due to addition of industrial effluents, as it is widdy used in manufacturing processes. Prolonged use of alkaline drinking water containing a high concentration of fluoride may cause dental and skeletal fluorosis (disfigurement of the teeth and bones), both in human beings and domestic animals (William, 1987). The World Health Organization has setup the fluoride limit to 1.5 mg/l in order to prevent dental cavities. The Bureau of Indian Standards has laid down the Indian Standards as 1.0 mg/l as the maximum permissible limit for drinking water. The seasonal fluoride content of the water in the littoral zone of the lake ranged between 0.99 to 1.1 mg/l (Fig. 2.17), its mean average concentration being 1.01 mg/l. Its concentration in the lakewater showed significant variation among different sites as well as temporal variation across the seasons (Table 2.1). Its concentration in some of the months, particularly in February, at all the sites was higher than the desirable limits of ISI (0.6 to 1.2 mg/l), ICMR (1.0 to 1.5 mg/l) and WHO (1.5 mg/l) standards for drinking water. Hammer (1985) reported fluoride content of less than 0.5 mg/l in the surfacewaters.

Boron through an essential micronutrient, is dangerous for plant growth if its concentration in irrigated water exceeds 1 mg/l. The seasonal average boron concentration of the water in the littoral zone of the Husainsagar Lake ranges between 0.14 to 0.72 mg/l (Fig. 2.18), its

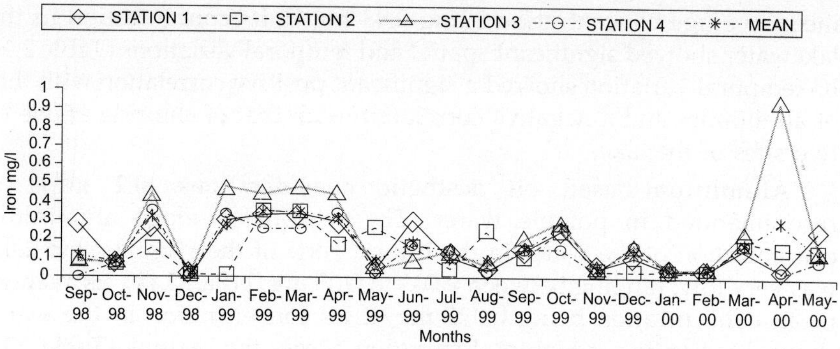

Fig. 2.16 Spatio-temporal variations in iron of surfacewater in the littoral zone of the Husainsagar Lake

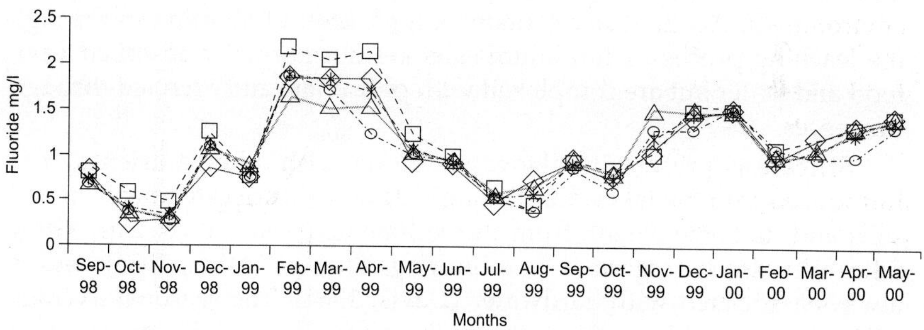

Fig. 2.17 Spatio-temporal variations in fluoride of surfacewater in the littoral zone of the Husainsagar Lake

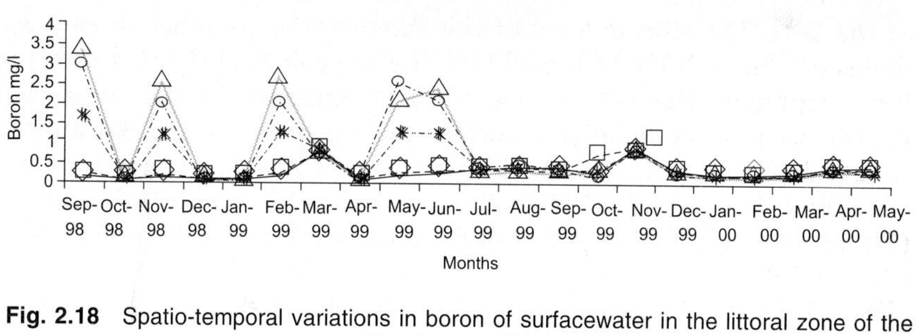

Fig. 2.18 Spatio-temporal variations in boron of surfacewater in the littoral zone of the Husainsagar Lake

mean average concentration being 0.43 mg/l. Its concentration in the lakewater showed significant spatial and temporal variations (Table 2.1). Its temporal variation showed a significant positive correlation with that of aluminium and a negative correlation with that of chloride at one or two sites of the lake.

Aluminium based on aesthetic considerations, 0.2 mg/l is recommended in potable water. The seasonal average aluminium concentration of the water in the littoral zone of the Husainsagar Lake was very low, ranging between 0.003 to 0.05 mg/l (Fig. 2.19), its average mean concentration being 0.027 mg/l. Its concentration in the water showed a significant temporal variation across the seasons (Table 2.1), and a significant negative correlation with that of dissolved oxygen at one or two sites of the lake. Industrial waste, erosion, and leaching of minerals and soils are the main pathways of aluminium in the aquatic environment. Acidic water contains a high level of aluminium through the leaching process. Aluminium salts are not normally absorbed from food and water but are complexed with phosphate and excreted through the faeces.

Silicate is present in lakewaters along with other nutrients. It is introduced into the lakewater through effluents from catchment-eroded soils and, to some extent, from the sediments (Atkin, 1976; Das, 1989). Silica can also be regarded as a limiting nutrient (Atkin, 1976), and is always associated with hardwater (Davis, 1964). The seasonal average silicate concentration of the water in the littoral zone of the Husainsagar Lake ranged between 14.3 to 16.2 mg/l, its mean average concentration being 14.51 mg/l. Its temporal variation showed significant negative correlation with that of dissolved oxygen and chloride at one or two sites of the lake. The silicate level of the lakewater was higher during the monsoon (Fig. 2.20), which could be due to geological characteristics of the catchment. The weathering of rocks/minerals in the catchment mainly account for dissolved silica in the receiving waters. Silica often ranges between 1 to 10 mg/l in natural water. Its content in the surfacewater of Hirakud Dam Reservoir ranged from 10 to 20 mg/l (Choudury et al., 1979) and that of Badua Reservoir was about 25.0 mg/l (Verma & Munshi, 1983). It was very low, ranging from 1.5 to 3.6 mg/l (Trisal, 1987) and 1.1 to 2.7 mg/l in Dal Lake (Ara et al., 2003). Its decrease in water is attributable to the blooming of diatoms (Reddy & Prasad, 1988). The production of diatom blooms is a major process for reducing silicate concentrations in water.

Dissolved Oxygen (DO) is one of the most important abiotic factors influencing life in an aquatic ecosystem. It is essential for the metabolism of all the aquatic organisms inhabiting the water body (Beeton, 1965, 1969; Wetzel, 1975, 2001), and probably, is the most common

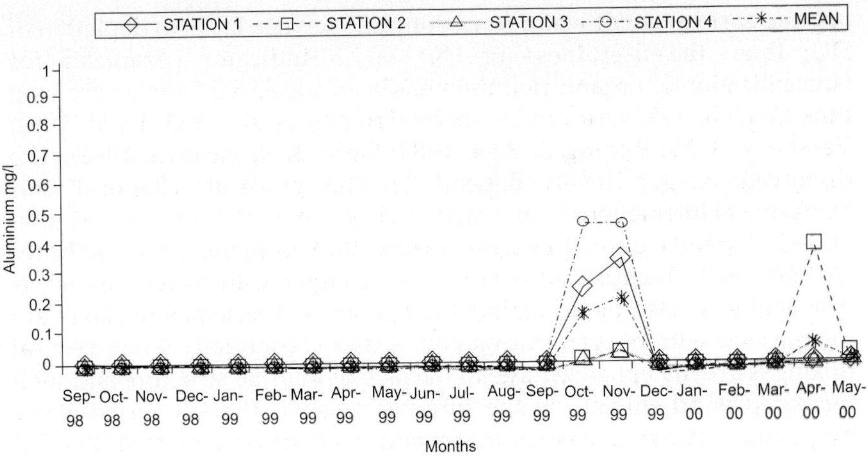

Fig. 2.19 Spatio-temporal variations in aluminium of surfacewater in the littoral zone of the Husainsagar Lake

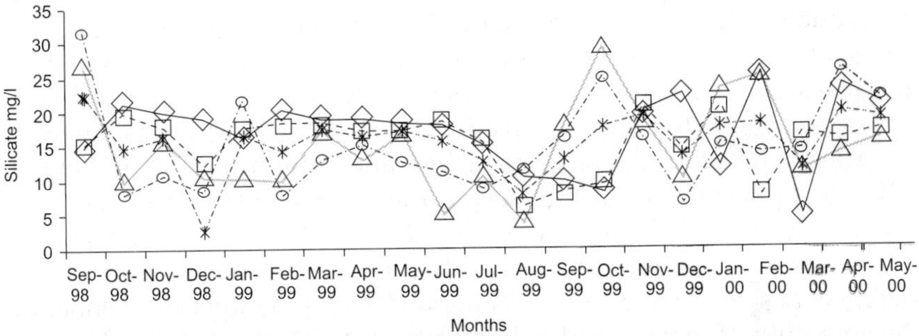

Fig. 2.20 Spatio-temporal variations in silicate of surfacewater in the littoral zone of the Husainsagar Lake

measurement of biological significance. The two main sources of DO in water are: (i) diffusion from air; and (ii) photosynthetic activity within the water. The process of biological production (photosynthesis) releases oxygen into the water, while biological decay (decomposition) utilizes the same. Various chemical reactions within the lake ecosystem also affect the concentration of DO.

The concentration of DO indicates the water quality. Depletion of DO, perhaps, is the most critical manifestation of eutrophication and effects of low levels of DO on aquatic ecosystem is a subject of extensive research (Alexander, 1976; Subramanian et al., 1988; Gopal & Durve, 1989; Das, 1989; Harper, 1992; Reddy & Rao, 1995; Mason 1996). The organic enrichment of water from domestic sewage leads to depletion of DO; thus, the usefulness of DO as an indicator parameter for eutrophication or organic pollution has been suggested from a very long time (Welch, 1952; Alexander, 1976; Trivedy et al., 1983; Unni, 1984; Varshney, 1985; Ryding & Rast, 1989; Sinha & Srivastava, 1992). The dissolved oxygen levels depend on the physical, chemical and biochemical interactions in an aquatic ecosystem. The temporal variation of DO depends upon the temperature that influences its solubility (APHA, 1989). The diffusion of oxygen through the air-water interface is principally affected by the ambient temperature. Consequently, lakes are replenished with oxygen during cold seasons, especially when vertical mixing is greatly enhanced due to the lack of thermal structure and with increased wind stirring. In the summer seasons, though surfacewater may remain more or less saturated and even super saturated, the DO concentrations are lower beneath the surface, which could be due to higher water temperature and oxygen consumption as a result of biological decay, causing its depletion. Its depletion also occurs near the bottom due to the processes at the mud-water interface, which is still inadequately understood, particularly in tropical eutrophic lakes. In monsoon, the DO is highly influenced by rainfall, with the circulation and mixing of water because of its flow pattern apart from the pollution stress being diluted. Higher contents of DO during monsoon were recorded when the temperature is low and minimum DO in summer when temperature was high (Alexander, 1976; Das, 1989; Sinha & Srivastava, 1992; Reddy & Rao, 1995). The high DO values are related to a higher rate of photosynthesis (Hussainy, 1975; Hutchinson, 1975; Balkhi et al., 1987; Srivastava, 1988) and the lower temperature enhances the solubility of oxygen and its retention in a dissolved state. However, the turbidity of water affects the penetration of light that, in turn, influences the liberation of oxygen during the process of photosynthesis.

The lower oxygen content of the water is a serious limiting factor. and low dissolved oxygen conditions create unfavourable environmental conditions for living organisms (Kaul, 1977; Das, 1989; Sarkar, 1989). One of the probable reasons for the mortality of fish in lentic systems may be the low dissolved oxygen content, since the lowest value of oxygen to maintain fish fauna is 5 mg/l and the critical value is 3 mg/1, and this range of DO is below the range of normal requirement of the fish (Alikunhi, 1957). The World Health Organization has recommended DO of 5+ mg/l for all domestic purpose. The average seasonal dissolved oxygen of the water in the littoral zone of the Husainsagar Lake ranged between 1.7 to 5.9 mg/l (Fig. 2.21), its mean average concentration being 4 mg/l. Moreover, the DO concentration in the water varied significantly in between different sites and across the seasons (Table 2.1). Therefore,

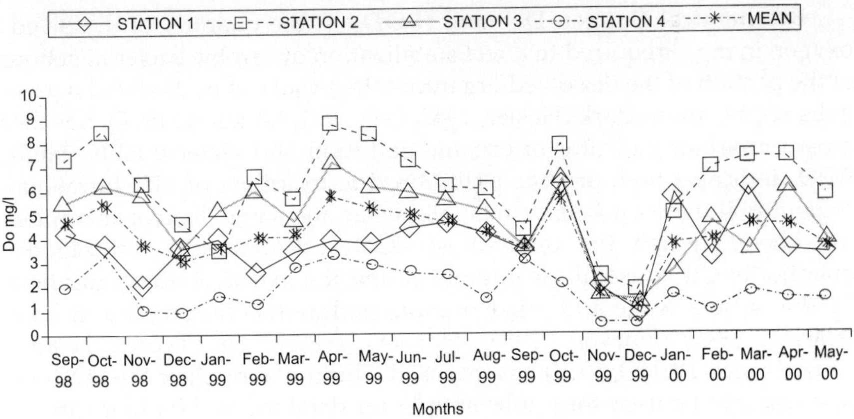

Fig. 2.21 Spatio-temporal variations in dissolved oxygen of surfacewater in the littoral zone of the Husainsagar Lake

based on the DO level of water, the lake can be classified as eutrophic and is mainly because of allochthonous input of municipal sewage into the lake. Earlier reports also recorded low DO levels in the Husainsagar Lake, because of direct inflow of municipal sewage into the lake. The monthly average values ranged from as low as 0.4 to 13.2 ppm, the latter value being recorded during the monsoon, which is attributable to the agitation of water by wind and rain and the subsequent diffusion of aerial O_2 in the water; it was low in winter and summer (Ghosh & George, 1989).

The dissolved oxygen was generally low during pre-monsoon and post-monsoon seasons and moderately high during monsoon period with the mean values of 4.02, 4.09 and 4.36 mg/l, respectively. It ranged between 2.76 and 5.29 mg/l during the three seasons. Higher dissolved oxygen content 14.5 ppm was recorded during the rainy season when there was plenty of aquatic vegetation covering the water surface, while relatively lower DO (8.4 ppm) was recorded during the summer (Gahtori et al. 1981). It was 11.3 mg/l in the unpolluted Kapshi Lake (Fokmare & Mussaddiq, 2002), and ranged from 6.8 to 12.0 mg/l in the Hirakud Dam Reservoir (Choudury et al., 1979), 9.2 to 10.6 mg/l in winter in the unpolluted Kumaon Himalayan lakes (Sunder, 1996) and 7.7 to 9.8 mg/l (Trisal, 1987) and 9.4 to 12.41 mg/l in winter (Ara et al., 2003) in Dal Lake—the former value was due to an increased temperature in August. The dissolved oxygen content of the water was little or low, 3.5 mg/l most of the time in the eutrophic tanks (Patil et al., 1983), and ranged from 3.3 to 6.4 mg/l in Keetham Lake (Prakash, 1983), 5.6 to 8.9 mg/l in upper lake of Bhopal (Tanot & Bhatnagar, 1988), and 6.0 to 10.2 mg/l in lower lakewater (Tiwari et al. 1988).

Biochemical Oxygen Demand (BOD_5) is the quantity of dissolved oxygen in mg/l required to affect stabilization of aerobic bacterial action of the portion of the dissolved organic matter, that can be oxidized in five days at 20°C in the dark (Hasler, 1947; Lee, 1971; Londnev, 1976). It is the most important indicator of organic pollution and water quality. High BOD_5 indicates high organic pollution due to inflow of allochthonous organic pollutants such as municipal sewage and a high load of anaerobic microbes. Though the method of BOD_5 estimation is not strictly quantitative, it is useful for understanding the extent of eutrophication by domestic sewage and other organic pollutants entering into a lake (Hasler, 1947; Jonason, 1984; Harper, 1992;). The World Health Organization (WHO, 1982) has imposed a limit of 6 mg/l of the BOD_5 of raw water to be used for public supply for drinking and bathing, and it should not exceed the limit of 3 mg/l in water, according to the ISI drinking water standards.

The seasonal average BOD_5 of the water in the littoral zone of the Husainsagar urban lake ranged between 13.56 to 19.42 mg/l (Fig. 2.22), the mean average being 15.74 mg/l. It showed the improved state of the lakewater as compared to the earlier state (author's unpublished data). The BOD_5 of water was high at the site of the lake, directly receiving the inflow of raw municipal sewage (Pieczynskai et al. 1975). Temperature, DO and BOD_5 form a complex and any alteration in one of these has a profound influence on the other. The BOD in the water of eutrophic Lake Pichola was 15.4 mg/l (Sudan et al., 1984), and was very high, ranging from 201 to 509 mg/l during 1977 to 1980 in Keetham lakewater (Prakash, 1983). It ranged from 12 to 360 mg/l in eutrophic Shambu Lake and 120 to 340 mg/l in Rankala Lake (Trivedy et al., 1983), 13.5 to 34.5 mg/l in surfacewater of the site receiving sewage in upper lake of Bhopal (Tiwari et al., 1988), 9.0 to 112 mg/l in surfacewater of Lake Mansarovar receiving sewage (Adholia et al., 1991), and 114 to 170 mg/l in Talkatora

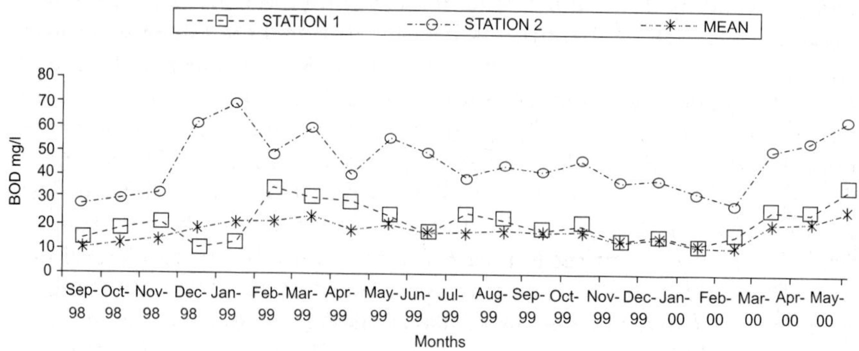

Fig. 2.22 Spatio-temporal variations in biochemical oxygen demand of surfacewater in the littoral zone of the Husainsagar Lake

Lake (Gupta et al., 1999), which was mainly due to allochthonous nutrient inputs. However, it ranged from 1.8 to 4.5 mg/l in the unpolluted surfacewater of Hirakud Dam Reservoir (Choudhury et al., 1979), 1 to 6.8 mg/l with an average of 2.7 mg/l in Osmansagar Lake (Reddy, 2001), and was 17.6 mg/l in the Kaphsi Lake (Fokmare & Musaddiq, 2002). These findings clearly showed an increase in BOD_5 in the water of unpolluted oligo- or meso-trophic to eutrophic state of lakes.

Analysis of the surfacewater of the littoral zone of the lake using *Inductive coupled plasma optical Emission spectrometer (ICPOES)* revealed that among the various elements, the average concentration of calcium was high (25.14 mg/l), whereas that of copper was low (0.067 mg/l). Further, it was noticed that aluminium was nil in the water. Among the metals present in the water, their average concentration was in the following descending order: Ca > Mg (7.78 mg/l) > K (0.36 mg/l) > Li (0.34 mg/l) > Pb (0.29 mg/l), the latter being absent in the water at one or two sites of the lake. Among the non-metals, their average concentrations were in the following descending order: S (17.45 mg/l) > Si (3.59 mg/l) > Se (1.02 mg/l) > P (0.8 mg/l) > B (0.26 mg/l) > As (0.12 mg/l). Interestingly, the concentrations of B, Si, P, S and Se of water were higher at the site of the lake which received the direct inflow of municipal sewage (Vijayakumar & Reddy, 2001). Among transitional metals of the water, the average mercury (Hg) concentration (0.69 mg/l) was higher, and that of others was in the following descending order: Zn (0.34 mg/l) > Mn (0.31 mg/l) > Ni (0.27 mg/l) > Cd (0.26 mg/l) > Fe (0.2 mg/l) > Cr (0.18 mg/l) > Co (0.11 mg/l) > Mo (0.1 mg/l) > Cu. It is again interesting to note that the concentrations of Mo, Mn, Fe, Ni, Zn, and Hg were higher at the site of the lake, which received the direct inflow of municipal sewage (Vijaykumar & Reddy, 2001). An earlier investigation reported 10.0, 1800.0 and 72.3 μg/l concentration of Hg, Mn and Cr, respectively, in the water of Husainsagar Lake (Prahalad & Seenayya, 1986). The water of Marlimund Lake and Ooty Lake contained 4.7 and 3.0 μg/l Pb, 11.6 and 7.9 μg/l Zn, 0.2 and 1.1 mμ/l Cd, 571 and 6342 μg/l Fe, 2.7 and 6.8 μg/l Cr and 1.3 and 4.4 μg/l Cu, respectively (Sivakumar et al. 2000).

Further, among the metals, non-metals and transitional metals, the average concentration of Cu and P was slightly higher, that of Pb and As was more than twice, Mn and Cr was more than three times, Cd was more than 25 times, Se was more than 100 times and Hg was 690 times higher than those of the desirable limits of BIS (ISI, 1982) and ICMR (1975) standards for drinking water. Such higher concentrations of these persistent chemicals in surfacewater of the lake are harmful (Vijaykumar & Reddy, 2001).

Physico-Chemical characteristics of the surface sediment

The sediment is an inseparable part of any aquatic ecosystem. It is considered to be an important sink for various elements, including trace metals and the chief storage of nutrients. It is a complex heterogeneous system, made up of organic matter, various nutrients and micro- and macroorganisms. The gradual increase of sediment thickness through a period of time may threaten the very existence of the lake. The bottom morphology of the lake can be greatly influenced by deposition of the sediment carried by inflown water.

The **textural composition** of the sediment varies from lake to lake. At different sites of the Husainsagar Lake, it is loamy, comprising average sand content of 79%, silt 6% and clay 15%, indicating its loamy nature (Murthy & Veerayya, 1972). The sediment was mainly loamy and sandy (65 to 89%) at some sites of lower lake in Bhopal (Oommachan & Belsare, 1986), silt-loam in the Dal Lake (Trisal, 1987) and loamy in texture in Kawar Lake (Singh & Jha, 2002).

The **colour** of the sediment at different sites of the lake ranged from dark brown to black, indicating the presence of considerable higher quantity of organic matter (humus) received from the inflowing sewage (OECD, 1982).

The **temperature** affects the biotic potentiality of not only the water but also the sediment and also controls the rate of decomposition and mineralization in the sediment. At different sites of the sediment in the littoral zone of the lake, it ranged from 18 to 32°C, the former during winter and the latter during summer, and its temporal variation was influenced by that of the water.

The seasonal average **pH** of the sediment in the littoral zone of Husainsagar Lake was alkaline and showed a slight variation between different sites ranging between 7.1 to 7.3 with an average of 7.23 (Fig. 2.23). It ranged from 6.6 to 7.2 in the sediment of Dal Lake (Trisal, 1977) and was slightly acidic in the sediment of Kawar Lake (Singh & Jha, 2002), which may be attributable to excessive decomposition of organic matter (Payne, 1986).

The seasonal average **Electrical Conductivity** of the sediment at different sites in the littoral zone of the lake ranged between 629 to 1386 μmhos/cm (Fig. 2.24) with an average of 1003.57 μmhos/ cm. It was high at the site of the lake receiving the direct inflow of sewage with suspended particulate organic matter (Richardson et al., 1978). There was significant variations in EC of the sediment in between the sites (Table 2.2). In general, it was high during summer and low in winter. Its temporal variation showed significant positive correlation with that of phosphorus, potassium and nitrogen and percentage organic carbon.

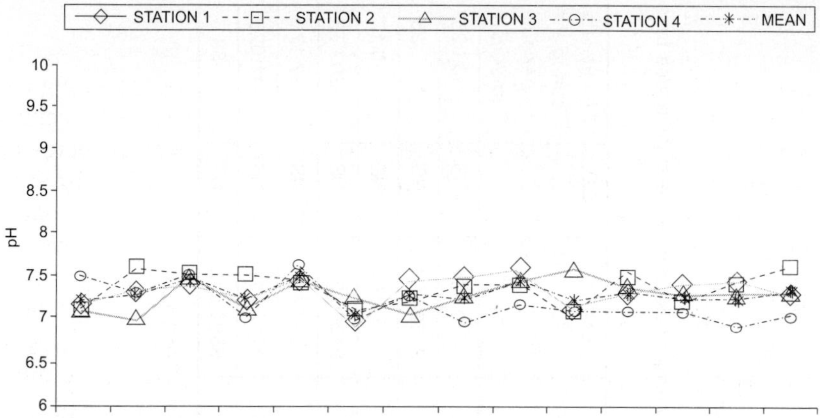

Fig. 2.23 Spatio-temporal variations in pH of surface sediment in the littoral zone of the Husainsagar Lake

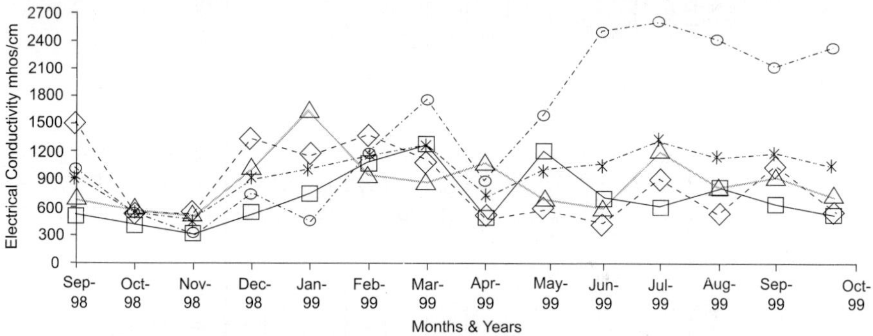

Fig. 2.24 Spatio-temporal variations in electrical conductivity of surface sediment in the littoral zone of the Husainsagar Lake

The seasonal average percentage of **organic carbon** of the sediment of different sites in the littoral zone of the Husainsagar Lake varied between 0.41 to 0.94%, with an average of 0.67%. It was high at the site of the lake receiving the direct inflow of municipal sewage and was significantly different from that of the other sites (Fig. 2.25) and also varied significantly across the seasons too (Table 2.2). Its temporal variation showed a significant positive correlation with that of nitrogen, potassium and phosphorus of the sediment at one or two sites of the lake (Murthy & Veerayya, 1972; Anvar Batcha, 2001). The sediment, based on the amount of organic matter present, is classified into two types: mineral sediment possessing less than 20-35% organic matter; and organic sediment containing more than 20 to 35% organic matter (Richardson et al.

Table 2.2 Analysis of variance of different characteristics in the sediment of littoral zone of Husainsagar Lake during the study period from September 1998 to October 1999

Parameter	*Mean*	*SEM*	*CV % coefficient of Variation %*	*F-test (stations)*	*F-test (months)*	*EMSS*	*CD*	*Change (%) during the period end-start*
pH	7.232	0.092	2.53%	NS	NS	0.034	NS	-0.025
Electrical Conductivity	901.786	248.187	55.04%	6.28**	NS	246387.4	NS	-25
Organic Carbon	0.669	0.163	48.85%	NS	2.15*	0.107	NS	-0.12
Available Nitrogen	225.816	73.086	64.73%	4.01*	NS	21366.31	NS	-207.75
Nitrate	182.946	6.85	7.49%	2.94*	NS	187.663	NS	27.175
Phosphate	61.061	17.485	57.27%	10.53**	NS	1222.916	NS	-11.025
Potassium	275.9	46.348	33.60%	NS	NS	8592.512	NS	-68.6

SEM: Standard Error of Means; EMSS: Error Mean Sum of Squares; CV: Coefficient of Variation; CD: Critical Difference
Level of significance $^{**}P<0.05$; $P<0.01$.

1978). This level ranged between 9.75 to 18.56% in the sediment of Dal Lake (Trisal, 1977).

Nitrogen is the most important limiting nutrient in an aquatic ecosystem (Carney et al., 1993). The seasonal average available nitrogen of the sediment at different sites of the littoral zone of Husainsagar Lake ranged between 163 to 338.8 kg/h., with an average of 225.8 kg/h. (Fig. 2.26). Its concentration is significantly different in between sites (Table 2.2) and was higher during some months at the site receiving the inflow of municipal sewage as compared to other sites of the lake, which

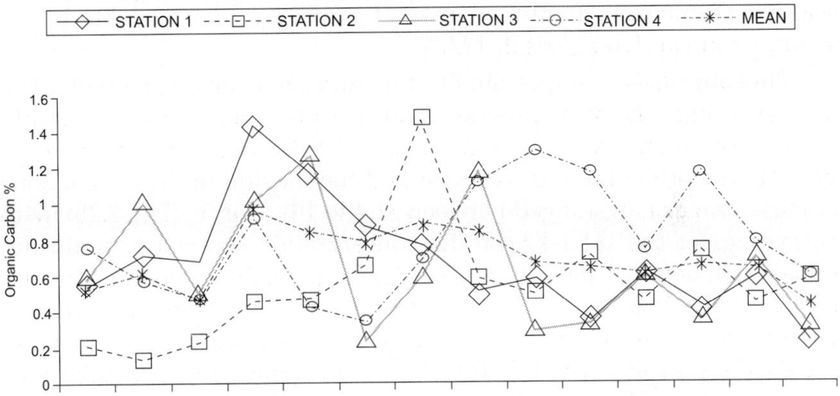

Fig. 2.25 Spatio-temporal variations in organic carbon of surface sediment in the littoral zone of the Husainsagar Lake

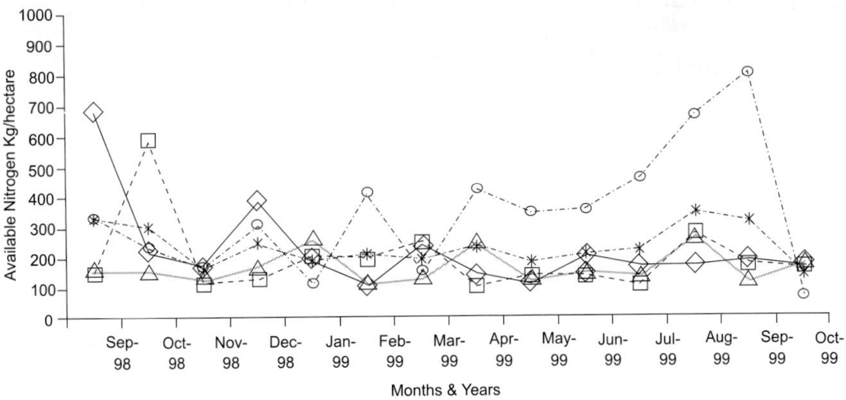

Fig. 2.26 Spatio-temporal variations in available nitrogen of surface sediment in the littoral zone of the Husainsagar Lake

was probably due to decomposition of organic matter and nitrogen mineralization. Its temporal variation showed a significant positive correlation with that of potassium, phosphorus and sodium at some of the sites of the lake (National Academy of Science, 1969).

The seasonal average **Nitrate** content of the sediment at different sites of the littoral zone of Husainsagar Lake ranged between 177.45 to 191.97 kg/h. with an average of 182.9 kg/h. (Fig. 2.27). Its concentration showed a significant variation in between different sites, being higher at the site that received the direct inflow of the sewage into the lake. However, it did not show any significant temporal variation across seasons (Table 2.2). It ranged from 1.0 to 22% (dry weight) in the sediment in Dal Lake (Trisal, 1977).

Phosphorus is an important limiting nutrient in any water body. The biological and chemical processes release phosphorus from a highly organic sediment (Mantigny & Prairie, 1993). The seasonal average phosphorus content of the sediment at different sites of the littoral zone of Husainsagar Lake ranged between 31.4 to 101.3 kg/h. (Fig. 2.28), with an average of the 161.1 kg/ha. Its concentration showed a significant variation in between the sites (Table 2.2), being higher at the site directly receiving the inflow of municipal sewage, which is probably due to decomposition of organic matter in the sediment. Its gradual increase across the season was probably due to the inflowing allochthonous organic matter. However, its variation was not significantly different across the seasons (Table 2.2). Its temporal variation showed a significant positive correlation with that of potassium at a site or two in the lake. It ranged between 0.08 to 0.29% dry weight in the sediment of Dal Lake (Trisal, 1977). Macrophytes act as sinks for both N and P, and take upto

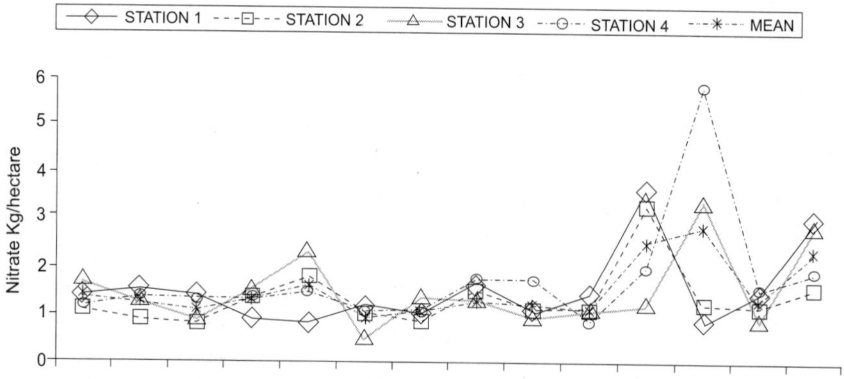

Fig. 2.27 Spatio-temporal variations in nitrate of surface sediment in the littoral zone of the Husainsagar Lake

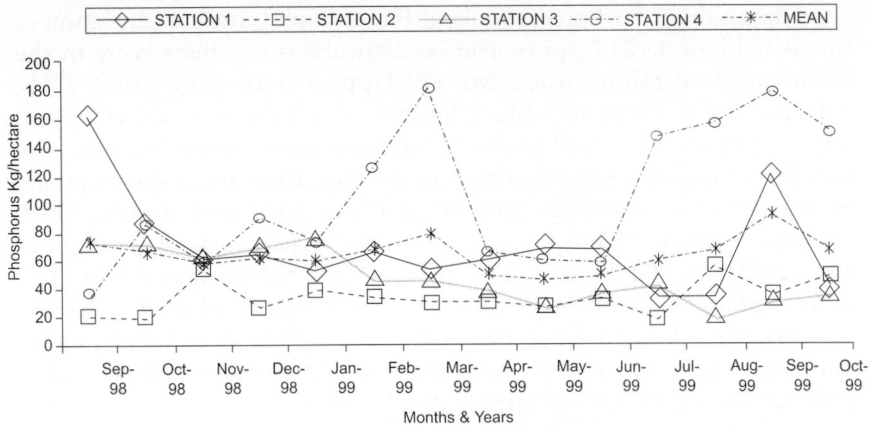

Fig. 2.28 Spatio-temporal variations in phosphorus of surface sediment in the littoral zone of the Husainsagar Lake

60% from the sediment during their growth phase (Donk et al. 1993). These nutrients are returned into the water body after their death.

The seasonal average **potassium** of the sediment at different sites in the littoral zone of the lake ranged between 248 to 302.8 kg/h., with an average of 275.9 kg/ha. It is readily used by submerged plants for growth. Its temporal variation showed a significant negative correlation with that of nitrate of the sediment at one or two sites of the lake. It ranged from 0.64 to 1.3% (dry weight) in the sediment of Dal Lake (Trisal, 1977).

Analysis of the surface sediment of the littoral zone of the Husainsagar Lake using *ICPOES*, revealed that among the analysed elements, calcium content was high (859.2 ppm), cobalt was low (0.58 ppm/l), and aluminium was absent. Among the metals of the sediment of the lake, their concentration showed a descending order, that is as follows: Ca > Mg (76.1 ppm) > K (20.7 ppm) > Li (1.5 ppm) > Pb (12.7 ppm). Interestingly, all these metals except Pb were high in concentration at the site receiving the direct inflow of raw sewage into the lake (Vijayakumar & Reddy, 2001). The average concentration of Pb in the sediment of Dal Lake was 26 $\pm$ 2.7 μg/g (Kango & Zutshi, 1988). The higher concentration of Pb in the Dal Lake sediment is attributable to the input from the carpet dye industry into the lake.

Among non-metals of the sediment, their concentration showed a descending order that is as follows: S (493.8 ppm) > Se (16.9 ppm) > Si (4.5 ppm) > P (3.9 ppm) > As (2.8 ppm) > B (1.4 ppm). It showed a similar trend to that of metals at the site that received the inflow of municipal sewage into the lake, due to which the concentrations of all these non-metals were high (Vijayakumar & Reddy, 2001).

Among the transitional metals of the sediment, the concentration of zinc was highest (25.7 ppm). The concentrations of others were in the following descending order: Mn (17.3 ppm) > Fe (13.9 ppm) > Hg (5.1 ppm) > Cu (3.9 ppm) > Mo (2.5 ppm) > Ni (1.8 ppm) > Cd (1.2 ppm) > Cr (0.9 ppm) > Co (0.57 ppm). These transitional metals, namely Mo, Mn, Fe, Ni, and Zn, were also high in concentration at the site receiving the direct inflow of sewage into the lake (Vijayakumar & Reddy, 2001). Earlier, Prahalad & Seenayya (1986) reported 912.0 mg/kg dry weight of Mn, 12400.0 mg/kg dry weight of Hg and 120.6 mg/kg dry weight of Cr in the surface sediment of the lake. The concentration of Zn, Cu, Ni and Co were 184 $\pm$ 16.2, 61 $\pm$ 6.6, 10 $\pm$ 0.95, 8 $\pm$ 1.3 μg/g, respectively, in the sediment of Dal Lake, and the concentration of Zn was much higher at the polluted site of the lake (Kango & Zutshi, 1988).

Physico-chemical characteristics of surface water in relation to that of sediment of the lake

Changes in the physico-chemical characteristics of surfacewater affect those of the sediments in an aquatic ecosystem, which was probably because of the gradual deposition of suspended sediment of the water. The temporal variation in EC of surface water showed a positive correlation with that of nitrogen of the sediment and that of potassium at some of the sites of the lake. Such a variation in nitrate content of water showed a significant positive correlation with that of potassium of the sediment and a negative correlation with that of phosphorus of the sediment at some of the sites of the lake (Singh & Jha, 2002). Such variations of phosphate content of water, again, showed a significant positive correlation with that of organic carbon and EC of the sediment and significant negative correlation with that of nitrogen and phosphorus of the lake sediment. The temporal variation in sodium in the surfacewater showed a significant positive correction with pH and phosphorus of the sediment and a significant negative correlation with that of organic carbon of the sediment of the lake; while that of potassium of water showed a significant positive correlation with that of EC and phosphorus of the sediment, and a negative correlation with organic carbon and pH of sediment of the lake. It indicates that the physico-chemical variables of surfacewater have a profound influence on that of sediment.

Effect of Ganesh-idol Immersion on Water Quality

Immersion of thousands of idols of Lord Ganesh in the Husainsagar Lake on the Ananta Chaturdasi day, falling in the month of either August or September, is an annual phenomenon. These idols are made using plaster of paris, clay, white cement, paper or paper-powder, chalk powder, fevicol, soap and clothes, gunny bags, iron mesh or even coconut fibres, supported by small iron rods or wooden planks and are coloured with different types of paints such as varnish, french polish mixing turpentine oil or kerosene, and various water colours. When the idols are immersed in the lakewater, these coloured chemicals dissolve slowly and become pollutants, leading to a significant alteration in the water quality, which in turn, affect the biotic component of the lake. Significant increases in concentration of calcium, phosphorus, potassium, silicon, magnesium, mercury, lead, arsenic, iron and molybdenum in the lakewater were observed after the Ganesh idol immersion, which is a serious problem leading to increase in the chemical pollution of the lake (Reddy & Vijaykumar, 2001). Important parameters such as turbidity, total alkalinity and total hardness have increased in water of lower and upper lakes of Bhopal after the immersion of idols of Ganesh and Durga, as compared to that of the water prior to idol immersion, which was attributed to the addition of biodegradable and non-biogradable substances into the lake due to the immersion (Dhote et al. 2001).

The arsenic, a noxious element, has increased significantly eightfolds in the lakewater after the idol immersion, as compared to its ISI and ICMR standards for drinking water, the highest desirable limits in water being 0.05 mg/l. The concentrations of lead and mercury, potentially obnoxious persistent heavy metals, have also increased manifold in the lakewater due to the system of idol immersion (Reddy & Vijaykumar, 2001), compared to their highest desirable limits of ISI and ICMR standards for drinking water. The concentration of lead in the lakewater has increased more than fourfolds after idol immersion, though its average concentration in the water was less than that of the idol immersion sites in the lake. The concentration of mercury, the causative chemical pollutant of the famous **Minamata** disease in Japan, was higher compared to the other trace elements in the lakewater, and it has increased to more than five to six hundred folds in the water, including that of the immersion sites, as compared to the highest desirable limits of ISI and ICMR standards for drinking water. After the immersion of the idols, its concentration has increased further to more than seven hundred fifty folds in the lakewater (Reddy & Vijayakumar, 2001).

Heavy metals are known to be persistent pollutants in the aquatic environment, and gradually accumulate and magnify through the process known as "Bioaccumulation and Biomagnification", while moving up in the food chain. Thus, lead and mercury may magnify in their concentrations at different trophic levels including fishes and birds inhabiting the lake, which finally reach the human body. The local fisherman catch fish from the lake to earn their livelihood and sell them in the fish markets of the twin cities. People inadvertently buy these fishes and consume the same; thereby getting exposed to these toxic metals by a daily intake of small quantities for a long period of time and are then afflicted from various serious diseases with new symptoms that are strange to medical literature. Organic compounds of mercury, including methyl mercury, on entering the human body, is known to concentrate in the brain and destroy the brain cells that control the central nervous system. Thus, the people consuming contaminated fish caught from the lake over a period of time may become afflicted with mercury poisoning.

Interestingly, the concentration of some of the other elements belonging to transitional metals—chromium, cobalt, nickel, copper, zinc and cadmium, to metals—lithium and to non-metals—boron and selenium have vanished from the immersion sites of the lake (Reddy & Vijaykumar, 2003). Vanishing of these chemicals, particularly Se and Cd, that are toxic to human beings and present in the lakewater in concentration more than 20 to 26 times higher and Cr and Mn being present three times higher than those of maximum or highest desirable limit of ISI, ICMR and WHO standards for drinking water, is quite intriguing. The concentration of manganese, potassium, phosphorus and sulphur dropped off considerably from the surfacewater following idol immersion. The disappearance of all these elements following the idol immersion can be attributable to the adsorption processes of the dissolved metal ions by the materials, particularly the coloured hydrous metal oxides used in paintings, varnishes and water colours on the idols, the resultant having settled down on the sediment of the lake (Reddy & Vijaykumar, 2003). Such vanishing of the elements from the lakewater may also be due to the formation of some insoluble salts and metal ligand complexes that have settled down on the lake sediment (Cai et al., 1988).

The findings on the vanishing of the elements from the lake surfacewater following the Ganesh-idol immersion though plead for a sort of ameliorating effect of idol-immersion on the water quality, the significant increase in concentration of other toxic heavy elements in the lakewater is detrimental to its water quality.

Fish kills

Guidelines to classify fish kills in different water bodies based on the numerical abundance of dead specimens in a specific water body (APHA, 1989) suggests that one to 100 dead or dying fishes confined to a small area or stretch of stream are called "minor kills", while 100 to 1000 dead or dying fishes of various species in one to two km of stream or equivalent area of a lake are called as "moderate kills", whereas 1000 or more dead or dying fishes of many species in a stretch of stream upto 16 km or more or equivalent area of a lake are called "major fish kills". Though exact figures on the number of individuals of different species of fishes killed in Husainsagar Lake are not available as it was a recurring feature, our observations at some of the sites of the lake soon after the occurrence of "fish kill" as well as from the repeated press reports showed that several thousands of fishes have been killed and found floating on the water surface of the lake. As per the APHA guidelines, these fish kills could be classified as "moderate" or "major kills". Earlier, Hingorani et al. (1973-74) reported such mass mortality of different species of fishes, viz., *Etroplus suratensis, Callichrous pabda, Barbus* sp. and *Mystus* sp. in the lake and attributed the same to asphyxiation, as the dead fishes showed bulged abdomen, pale gills and protruding eyes. Such mass killing of fish is recurring almost every year. The causative factors of the fish kills may probably be attributable to municipal or industrial wastes inflown into the littoral zone of the lake, leading to the severe depletion of the dissolved oxygen. The analysis of water in the fish-killed areas of the lake showed that it contained residues of strong acids, alkalies, pesticides and heavy oils, etc. and the effluent discharge of IDPL company (Hingorani et al., 1973-74), which may be the probable cause for such fish kills.

Industrial pollution and raw municipal sewage have degraded the water quality in the lake, and probably also caused the fish kills in the lake (Muley, 1987). There were many industries manufacturing a variety of products ranging from drugs, dyes, chemicals, paints to heavy machinery tools, etc., releasing the effluents to the Kukatpally "K-main" nallah, the main feeding channel of the lake. Most of these industries released their treated and untreated toxic effluents in the vicinity of the feeder channel, which ultimately flowed into the lake, causing severe degradation of its water quality. All these factors together might have a synergistic effect, causing the fish kills in the lake.

Several theories and hypotheses on the fish kills have been put forward in the form of unpublished reports (Muley, 1987). One of the common observation expressed in the fish kills was due to the sudden depletion of dissolved oxygen in the lake, which might have induced migration of the fish species that need more oxygen for survival and, during the course, their gills might have clogged due to algal blooms on the surface, leading to mass mortality. Moreover, with regard to depletion of oxygen in the surfacewaters of the lake, several explanations have been put forward. According to a few, the influx of organic matter and sewage brought by the storm water as well as sudden change in the photosynthetic activity followed by high community respiration might have lead to depletion of dissolved oxygen in the surfacewaters. These fluctuations, associated with critical values of certain toxic elements like of ammonia (NH_4) and or hydrogen sulphide (H_2S), might have lead to "fish kills" in the lake. Critical levels of dissolved oxygen in the water of the lake had prevailed on many occasions, which might have caused the mass mortality of fish.

Highly toxic elements like heavy metals are concentrated in the surfacewaters of the lake as well as in its sediment due to the inflown industrial effluents (Vijayakumar & Reddy, 2001; Reddy & Vijaykumar, 2003) and sewage pollution in the lake (Reddy & Vijaykumar, 2002). These toxic elements concentrated during the dry season and, together with the influx of water along with municipal sewage followed by pre-monsoon heavy showers entering the lake, may have caused severe depletion of the dissolved oxygen, increasing the eutrophication of the lake (Reddy & Vijaykumar, 2002), and this finally may have resulted in the mass mortality of fishes.

Nevertheless, what is required is a robust system of monitoring the water quality and sediment characteristics associated with "Water Quality Criteria" of the lake, in relation to the anthropogenic activities and land-use taking place in the catchments, and environmental education and awareness, in order to prevent the nutrient enrichment in the lake. This is the primary step in preventing eutrophication, and is essential for efficient management and restoration of the lake ecosystem. It will also give the lake authorities an early warning to take up timely action against pollution.

REFERENCES

Abbasi, S.A., Bhatia, K.K., Kunhi, A.V.M. & Soni, R. 1996. Studies on the limnology of Kuttiadi Lake (North Kerala). Ecol. Env. & Cons., 2: 17-27.

Adholia, U.N., Shrivastava, V., Chukraborty, A., Vyas, A. & Jacob, M. 1991. Hydrography of Mansarovar Reservoir Bhopal. Natl. Symp. on Planning for Environ. Sust. Develop., New Delhi, p. 21.

Alexander, K.C. 1976. Limnological studies on the upper and lower lakes of Bhopal. Ph.D. Thesis, Vikram Univ, Ujjain.

Alikunhi, A. 1957. Fish culture in India. F.M. Bull. India Coun. Agrl. Res., 20: pp. 1444.

Anvar Batcha, S.M. 2001. Studies on distribution of organic matter content in the sediment of north Vembanad Lake. Journal of Environment & Pollution 8(1): 67-70.

APHA. 1989. Standard methods for examination of water and waste water. 17th Edition. American Public Health Association, IMC., New York, U.S.A.

Ara, S., Khan, M.A. & Zargar, M.Y. 2003. Physico-chemical characteristics of Dal lake water, Kashmir Valley, India. Indian Journal of Environment & Ecoplan, 7(1): 47-50.

Atkin, W.R.G. 1976. Seasonal changes in the silica content of natural waters in relation to phytoplanktons. J. Mar. Biol. Asson., 14(1): 89.

Badge, U.S. & Verma, A.K. 1985. Physico-chemical characteristics of water of J.N. lake at New Delhi. Indian J. Ecol., 12(1): 151-156.

Balkhi, M.H., Yosuf, A.R. & Qadri, M.Y. 1987. Hydrobiology of Anchar Lake, Kashmir. Comp. Physiol. Ecol. 12(3): 131-139.

Beeton, A.M. 1965. Eutrophication of the St. Lawrence great lakes. Limnol. & Oceanogr., 10: 240.

Beeton, A.M. 1969. Changes in the environment and biota of the Great Lakes, In: Eutrophication Causes, Consequence and Correctives. pp 150-187, Natl. Acad. Science, Washington, D.C.

Burns, N.M. & Ross, C. 1971. Nutrient relationships in a stratified eutrophic lake. Proc. 14th Conf. Great Lakes Res. 749-760.

Cai, W.C. & Hu, B.L. 1988. The effects of sewage discharge on the adsorption of heavy metal ions by the small particles of Yangtse River estuarine (sediment). Mar. Sci., 1: 20-23.

Cairns, J. Jr. 1970. Ecological management of problems caused by heated waste water discharge into the aquatic environment. Water Resources Bull., 6: 668-678.

Carney, H.J., Bimford, M.W., Martin, B.R. & Goldman, C.R. 1993. Nitrogen and phosphorus dynamics and retention in ecotones of Lake Titicaca, Bolivia, Peru. Hydrobiologia, 251: 39-47.

Cooke, G.D. 1993. Restoration and Management of Lakes and Reservoirs Lewis, Boce Raton.

Choudhury, N.K., Pradhan, P.K. & M.C. Dash 1979. Some physico-chemical factors and phytoplanktons of Hirakud Dam resevoir. Geobios, 6: 104-106.

Crisman, T.L., Chapman, C.A. & Chapman, L.J. 1998. Predictors of seasonal oxygen levels in small Florida lakes: The importance of colour. Hydrobiologia, 368: 149-155.

Dakshini, K.M.M. & Soni J.K. 1979. Water quality of sewage drains entering Jamuna in Delhi. Indian J. Environ. Hlth. 21(4): 354-361.

Das, B., Talukder, J., Sarma, S., Gohair, B., Dutta, R.K. Das, H.B. & Das, S.C. 2003. Fluoride and other inorganic constituents in groundwater of Guwahati, Assam, India. Curr. Sci, 85: 657-661.

Das, S.M. 1978. High pollution in lake Nainital, U.P. as evidenced by biological indicators. Science and Culture, 44(5): 236-237.

Das, S.M. 1989. Handbook Of Limnology and Water Pollution with Practical Methodology. South Asian Publishers, New Delhi, pp. 174.

Datta Roy, S., Maity, B.R. Dutt A. & Banerjee, R.D. 1984. Pollution load of paper pulp mill effluents. Environ. & Ecol., 2: 29-34.

Davis, S.N. 1964. Silica in streams and groundwaters. Am J. Sci., 262: 870-92.

Davis, R. & Hirji, R. (Series Editors) 2003. Water Resources and Environment. Technical Note G.2. Lake Management, World Bank, Washington, D.C., pp. 28.

de Anda, J., Shear, H., Maniak, U. & Riedel, G. 2000. Phosphorus balance in Lake Chapala (Mexico). J. Great Lakes Res. 26: 129-140.

Dhote, S., Varughese, B. & Misra, S.M. 2001 Impact of idol immersion on water quality of twin lakes of Bhopal. Indian J. Environ. Prot., 21: 998-1005.

Donk, E.V., Gulati, R.D., Iedema, A. & Meulemans, J.T. 1993. Macrophyte-related shifts in the nitrogen and phosphorus contents of the different trophic levels in a biomanupulated shallow lake. Hydrobiologia, 251: 19-26.

Edmondson, W.T. 1961. Changes in Lake Washington following an increase in the nutrient income. Verh. Int. Ver. Limnol., 14: 187-175.

Edmondson, W.T. 1970. Phosphorus, nitrogen and algae in Lake Washington after diversion of sewage. Science, 169: 690-691.

Edmondson, W.T. 1991. The Uses of Ecology: Lake Washington and Beyond Univ. Washington press, Seattle.

Ellis, M.M. 1937. Detection and measurement of stream pollution. Bull. U.S.B.U. Fish, 48(22): 365-437.

Fanselius, S.H. 1977. The eutrophicating effect of organic matter and nutrient elements on natural waters. Proc. FAO/SIDA Seminar, pp. 1032-1041.

Fokmare, A.K. & Musaddiq, M. 2002. A study of physico-chemical characteristics of Kapshi Lake and Purna river waters in Akola district of Maharastra (India). Nature, Environment and Pollution Tech., 1: 261-263.

Forsberg, C. 1989. Importance of sediments in understanding nutrient cycling in lakes. Hydrobiologia, 176/177, 263-277.

Gahtori, O.D., Chaturvedi, L.D. & Joshi, B.D. 1981. Seasonal variations in a sewage-fed ecosystem. Comp. Physiol. Ecol., 6(3): 213-216.

Ganapati, S.V. 1960. Ecology of Tropical Waters. In: D. Raghavan & P. Kachroo (Eds.). Proc. Symp. Algology, ICAR, New Delhi, pp. 204-218.

Geldermalsen, L.A.V. 1985. Evaluation of seasonal nutrient dynamics in the Oosterschelde (SW Netherlands). Neth. J. Sea Res., 19: 207-216.

George, M.G. 1961. Diurnal variation in two shallow ponds in Delhi. Curr. Sci., 30: 265-269.

Ghosh A. & George J.P. 1989. Studies on the abiotic factors and zooplankton in a polluted urban reservoir Husainsagar, Hyderabad: Impact on water quality and embryonic development of Fishes. Indian J. Environ. Hlth., 31(1): 49-59.

Golterman, H.L. 1975. Limnology. Elsevier Sci. Publ. Co., Amsterdam.

Gopal, R.N. & Durve, V.S. 1989. Cultural eutorphication of the Lake Rangasagar, Udaipur, Rajastan, India. J. Envi. Biol., 10(2): 127-134.

Gupta, A.B., Jain, R. & Gupta, K. 1999. Water quality management for the Talkatora Lake, Jaipur—A case study. Water Sci. Tech., 40: 29-33.

Hammer, M.J. 1985. Water and Wastewater Technology. S.I. Version, John Willey & Sons, Inc., New York.

Harper, D. 1992. Eutrophication of Freshwater. Chapman and Hall, London.

Hasler, A.D. 1947. Eutrophication of lakes by domestic drainage. Ecology 28: 288-295.

Hingorani, H.G., Rao, K.M. & Ramachandran, R. 1973-74. An attempt to study pollution and fish mortality in Husainsagar reservoir in Hyderabad. J. Indian Fish. Assoc., 3 & 4: pp. 4.

Holden, A.V. & Caines, L.A. 1974. Nutrient Chemistry in Lake Leven, Kinrose. Proc. Royal Society of Edinberg, 74: 101-212.

Hussainy, S.V. 1967. Studies on the limnological and primary production of a tropical lake. Hydrobiol., 30 (3-4): 335-352.

Hutchinson, G.E. 1975. A treatise on limnology. Introduction to lake biology and limnoplankton. 2nd Edn., John Wiley and Sons, Inc., New York, pp. 1115.

ICMR 1975. Manual of standards of quality for drinking water supplies. Indian Council of Medical Research (ICMR), New Delhi, Special Report No. 44. 27.

ISI 1982. Indian standard specifications for drinking water. DOC, DDC, Indian Standard Institute. New Delhi.

Jonason, P.M. 1984. The Ecosystem of eutrophic Lake Esrom. Lakes and Reservoirs, Elsevier Sci. Publi, B.V. Amsterdam.

Jorgensen, S.E. 1980. Lake management, Pergaman, Oxford Kalff, J. 2002. Limnology: Inland Water Ecosystems, Prentice Hall Inc., New Jersey.

Kango, R.A. & Zutshi, D.P. 1988. Trace metal content of some Kashmir Himalayan lake sediments. Geogios, 15: 280-281.

Kaul, V. 1977. Limnological survey of Kashmir lakes with references to trophic status and conservation. Interm. J. Ecol. & Environ. Sc., 3: 29-44.

Kaushik, S., M.N. Saxena, S. Sharma & Saksene, D.N. 1990. Abundance of insects in relation to physico-chemical characteristics of pond water at Gwalior (M.P). Pro. Natl. Adad. Sci., India, 60 (B): 103-106.

Khatavkar, S.D., Kulkarni, A.Y. & Goel, P.K. 1989. Limnological study on two lentic freshwater bodies at Kolhapur with special reference to pollution. IJEP., 9: 198-203.

Khatri, T.C. 1984. Seasonal variation in the ecosystem of the Lakhotia Lake. Proc. Indian Acad. Sci., (Sec. B), 93 (7): 897-702.

Khatri, T.C. 1986. Limnological studies of a tropical lake of western Rajasthan. J. Envin. Sci., 2(2): 77-85.

Kohli, M.P.S., Dube, K., Prakash, C., Chaturvedi, C.S. & Jadhar, R.J. 1998. Observations on some hydrobiological features of Lake Powai near Mumbai (Maharastra). J. Inland Fish. Soc. India, 30: 37-49.

Kuznesov, S.I. 1968. Recent study on the role of microorganisms in the cycling of substances in lakes. Limnol. & Oceanogr. 13: 211-224.

Lee, G.F. 1971. Eutrophication. In: Encyclopedia of Chemical Technology, Suppl. Vol., John Wiley & Sons. Inc. New York, pp. 315-38.

Londnev, L. 1976. Eutrophication of lakes, causes, effects and means for control, with emphasis on lake rehabilation. Report prepared for World Health Organization. Regional office for Europe Copenhagen, Denmark. p. 78.

Mantigny, C. & Prairie, Y. 1993. The relative importance of biological and biochemical processes in the release of phosphorus from a highly organic sediment. Hydrobiologia, 253: 141-150.

Mason, C.F. 1996 (3rd Ed.). Biology of Freshwater Pollution, Longman, Essex, pp. 356.

Mathew, P.M. 1975. Limnology and productivity of Govindgarh Lake. Rewa M.P. J. Inland Fish. Soc., India, 7: 16-24.

Morgan, N.C. 1970. Changes in the fauna and flora of a nutrient-enriched lake. Hydrobiologia, 35: 544-553.

Morrissy, N.M. 1971. Temperature relationships in small water bodies of freshwater with special reference to trout streams in south Australia. Bull. Aust. Soc. Limnol., 4: 8-20.

Moss, B. 1972. Studies on Gull lake, Michigan. II. Eutrophication—evidence and Prognosis. Freshwater Biol. 2: 304-320.

Moyle, J.B. 1946. Some indices of lake productivity. Trans. Am. Fish Soc., 76: 322-334.

Muley. E.V. 1987. Investigations on fish kills in Lake Husainsagar, Hyderabad. Project report, Municipal Corporation of Hyderabad, A.P.

Munawar, M. 1970. Limnological studies on freshwater pond of Hyderabad, India II. Biocenose. Hydrobiol., 36(1): 105-128.

Munawar, M. & Zafar, A.R. 1964. A preliminary study of the vertical movement of *Eudorina elegans* and *Trinema linean* during a bloom caused by them. Hydrobiologia, 29: 140-148.

Murthy, P.S.N. & Veerayya, M. 1972. Studies on sediments of Vembanad Lake, Kerala Part I: Distribution of organic matter. Indian J. Mar. Sci., 1: 45-51.

National Academy of Science, 1969. Eutrophication: Causes, Conseqeunces and Correctives. Symposium Proceedings, National Academy of Sciences, Washington, D.C., pp. 661.

Naumann, E. 1919. Nagra sympunkter andaende limnoplanktons ocekologic madsarskild hynsyn tool fytoplankton. Svensk. Botanisk Tidskrift, 13: 129-163.

Nayak., T.R., Iyer, S.J. & Jaish, S.C. 1982. Seasonal variations of rotifers and certain physico-chemical factors of Matyatal, Panna (M.P), India. Comp Phy. Ecol., 7(3): 165-169.

Nicholls, K.H. & Hopkins, G.J. 1993. Recent charges in Lakes Erie (North shore) Phytoplankton: Cumulative impact of Phosphorus loading reductions and the zebra mussel introduction. J. Great Lakes Res., 19: 637-647.

Nirmal Kumari, J. 1984. Studies on certain biochemical aspects of hydrobiology. Ph.D. Thesis. Osmania University, Hyderabad, A.P.

OECD (Organization for Economic Cooperation and Development), 1982. Eutrophication of waters. Monitoring, Assessment and Control. Final Report. OECD Cooperative Programme on Monitoring of Inland Waters (Eutrophication Control), Environmental Directorate, OECD, Paris, pp. 154.

Oommachan, L. 1988. Macrobenthos of lower lake, Bhopal. Proc. Nat. Symp. Past, Present and Future of Bhopal lakes, pp. 27-31.

Oommachan, L. & Belsare, D.K. 1986. Bottom sediments and bathymetric distribution of oligochaetes in the lower lake of Bhopal. J. Hydrobiol. 24: 57-62.

Otsuki, A. & Wetzel, R.G. 1974. Calcium and total alkalinity budget and calcium carbonate precipitation of hardwater lake. Hydrobiol., 73: 14-30.

Ownbey, C.R. & Kee, D.A. 1967. Chlorides in Lake Erie. Proc. Conf. Great Lakes Res., 10: 382-389.

Patil, S.G., Singh, D.P. & Harshey D.K. 1983. Ranital Jabalpur: A sewage polluted water body, as evidenced by chemical and biological indicators of pollution. J. Environ. Biol. 4(2): 43-49.

Patil, S.S. & Nandan, S.N. 1984. Limnological Studies on Nakana Lake of Dhule (Maharashtra). Proc. of Natl. Symp. Ecol. Environ. Impact and Organism response, pp. 91-94.

Payne, A.I. 1986. The ecology of tropical lakes and rivers. John Wiley & sons, New York, pp. 301.

Pearsall, W.H. 1930. Phytoplankton in the English Lakes-I. The proportions in the water of some dissolved substances of biological importance. J. Ecol., 18: 306-320.

Pieczynskai, E., Sikorsia U. & Ozimek, T. 1975. The influence of domestic sewage on the littoral zones of lakes. Pol. Arch. Hydrobiol., 22: 141-156.

Prahalad, A.K. & Seenayya, G. 1986. In situ compartmentation and biomagnifications of Copper and Cadmium in industrially polluted Husainsagar Lake, Hyderabad, India. Arch. Environ, contain. Toxicol. 15: 417-425.

Prakasam, V.R. & Joseph, M.L., 2000. Water quality of Sasthamcotta Lake, Kerala (India) in relation to primary productivity and pollution from anthropogenic sources. J. Environ. Biol., 21: 305-307.

Prakash, C. 1983. Water quality of Keetham Lake (Soor Sarovar). J. Environ. Biol., 4: 193-200.

Prakash, D.Y. 1990. Primary productivity and energy flow in upper lake, Bhopal. Indian J. Environ. Health, 32: 132-139.

Qadri, M.Y., & Yousuf. A.R. 1980. Limnological studies on Lake Malpur Sar. Biotope. Geobios 7: 117-119.

Raghunathan, M.G., Mahalingam, S.F Vanithadevi, K. 2000. A study of Physico-chemical characteristics of Otteri Lake and Pollar river in Vellore town (T.N.), India. J. Aqua. Biol., 15: 56-58.

Reddy, G.N. & Prasad, M.N.V. 1988. Seasonal nutrient status of Banjara Lake, Hyderabad. Proc. Indian Acad. Sci. (Plant Sci.), 98: 515-522.

Reddy, M.V. 2001. Some water quality parameters in relation to benthic macroinvertebrate bioindicators of littoral zone of an Oligotrophic lake nearby Hydrabad city (India). In: Partnerships for Sustainable Life in Lake Environments: Making Global Freshwater Mandates Work. Proceedings, 9th Intern. Conference on Conservation and Management of Lakes, Biwako, 2001, pp. 4.

Reddy, M.V. & Rao, B.M. 1998. Biomonitoring of Water Quality of an urban reservoir using benthic macroinvertebrates In: M.V. Reddy & V.B. Rao (eds) Environmental Monitoring and Management, SDLCE, K.U., Warangal, pp. 221-263.

Reddy, M.V. & Vijaykumar, A. 2001. Effects of Ganesh-idol immersion on some water quality parameters of Husainsagar Lake. Current Science, 81: 1412-1413.

Reddy, M.V. & Vijaykumar, A. 2002. Municipal sewage, the major cause of pollution in Husaisagar lake. In: B. Venket Rao (Ed. In-chief), Hydrology and Watershed Management with focal theme on water quality and conservation for sustainable Development. Cent. Water Resources, J.N.T.U., Hydrabad, India , pp. 632-639.

Reddy, M.V. & Vijaykumar, A. 2003. Disappearance of some toxic elements in the surfacewater of Husainsagar Lake following Ganesh-idol immersion. Indian J. Environ. Hlth, 45(1): 161-162.

Reid, G.K. & Wood, R.D. 1976. Ecology of Inland Waters and Estuaries. Reinhold Publ. Corp., New York.

Richardson, C.J., D.L. Tilton, J.A. Kadlac, J.P.M. Chamic, & Wentz, W.A. 1978. Nutrient dynamics of northern wetland ecosystems. In: R.E. Goodt, G.F. Whigham (eds), Freshwater Wetlands. Ecological Process and Management Potential, Academic Press, New York. pp. 217-241.

Robert, S.C., J. Gridley, & William, M. 1940. Chemical methods for the study of river pollution. H.M.S.O. London., Fish. Invest. Ser. I.

Rodhe, W. 1969. Crystalisation of eutrophication concepts in northern Europe. In: Eutrophication causes, consequences and correctives, Nat. Acad. Sci. Washington Nat. Res. Council. Publ., pp. 50-64.

Rosas, I., Mazari, M, Saavedra, J, & Baez, A.P. 1985. Benthic organisms as indicators of water quality in lake Patzcuaro, Mexico. Water Air Soil Pollution 25: 401-414.

Rybak, J.I. 1969. Bottom sediments of the lakes of various trophic type. Ekol. Pol. 17: 611-662.

Rybak, J.I. & Sikarska, V. 1976. Environment. In: Selected problems of lakes littoral ecology (Ed.) E. Pieczynska, Univ. of Warsaw, Warsaw.

Ryding, S. & Rast, W. 1989. The control of eutrophication of Lakes and Reservoirs. UNESCO, Paris.

Sarkar, S.K. 1989. Phosphorus and aquatic ecosystems. Indian Rev. Life Sci., 9: 227-251.

Sawyer, C.N., Lackey, T & Lenz. R.T. 1945. Investigations of odour nuisance occurring in the Madison lakes particularly Monosa, Waubesa, and Kegonsa. Report Governor's Committee, Madison, Wisconsin, 92-100.

Schindler, D.W. 1975. Whole lake eutrophication experiments with phosphorus, nitrogen and carbon. Verh. Internat. Verein. Limnol. 19: 389-400.

Schindler, D.W. 1990. Experimental perturbations of whole lake as test of hypothesis concerning ecosystem structure and function. Oikos, 57: 25-41.

Sharma, K.P., Goel, P.K., & Gopal, B. 1978. Limnological studies of polluted freshwater. I. Physico-chemical characteristics. Int. J. Ecol. Environ, Sci., 4: 89-105.

Sharma, L.L & Durve, V.S. 1982. Limnology vis-à-vis Conservational need of Lake Fateh Sagar, Udaipur. Poll. Res. 9 (1-2): 25-29.

Shapiro, J. & Weight, D.I. 1984. Lake restoration by manipulation: Round Lake, Minnesota, the first two years. Freshwat. Biol., 14: 371-383.

Singh, B.K. & Jha, N.N. 2002. Nutrient status of sediments of Karwar wetland in Bihar. Nat. Environ. & Poll. Tech., 1: 107-114.

Sinha, A.K. Singh, D.K., Baruah, A. & Sharma, U.P. 1992. Seasonal variation of physico-chemical properties of bottom sediment of Kawar Lake, Begusari (Bihar). J. Freshwater Biol., 4(4): 249-254.

Sinha, A.K., & Srivastava, A.K. 1992. Impact of industrialisation and urbanization on water quality of lakes at Rae Bareli (India)—A case study. Proc. National Symp. on Environment, 25-26.

Sinha, M.P. 1986. Limnobiotic study on tropic status of a polluted freshwater reservoir of coalfield area. Poll. Res., 15: 13-18.

Sivakumar, R., Mohanraj, R. & Azeez, P.A. 2000. Physico-chemical analysis of water sources of Ooty, south India. Poll. Res. 19: 143-146.

Spence, D.H.N. 1964. The macrophyte vegetation of lakes, swamps, and associated ferns. In: J.H. Burnett (ed.). The Vegetation of Scotland, Oliver and Boyd, Edinburgh, pp. 306-425.

Sreenivasan, A. 1964. Limnological studies and fish yield in three upland lakes of Madras state, India. Limnol. & Oceanogr., 9(4): 564-575.

Sreenivasan, A. 1974. Limnological factors of a tropical impoundment Bhavanisagar Reservoir, Tamil Nadu, India. Int. Revue. Ges. Hydrobiol., 59: 327-342.

Srivastava, S.K. 1988. Factors affecting plankton population in a tropical pond. A statistical approach. J. Environ. Biol., 9(4): 401-408.

Stewart, W.D.P. 1968. Nitrogen input into the aquatic system. In: D.F. Jackson, (ed.), Algae, Man and the Environment. Syracuse Univ. Press. Syracuse. New York, pp. 53-72.

Subramanian, M.A., Vishwanathan, S. & Varadraj, G. 1988. A report on physico-chemical parameters of sewage and industrial effluents in and around Erode, Tamil Nadu. Indian Zoologist, 12 (1&2): 185-87.

Sudan, M., Sharma, L.L. & Durve, V.S. 1984. Eutrophication of the Lake Pichola in Udaipur, Rajastan. Poll. Res., 3: 39-44.

Sunder, S. 1996. Composition, distribution and seasonality of benthic invertebrates in two important Kumaon Himalayan lakes. Indian J. Fish; 43: 187-194.

Tadajewaski, A. 1966. Bottom sediments in different limnetic zones of a eutrophic lake. Ekol. Pol., 414: 321-341.

Tanot, P. & Bhatnagar, G.P. 1988. Limnobiological studies of upper lake Bhopal. Proc. Natt. Symp. Past, Present and Future of Bhopal lakes, pp. 41-44.

Thienemann, A. 1918. Untersuchungen uber die Beziehungen Zwischen der Sauerstoffgen haltdes wassers und der Zusammensetzug der fauna in norddeutschen seem. Arch. Hydrobiol. 12: 1-65.

Thomas, E.A. 1969. The process of eutrophication in central European lakes. Eutrophication, causes, consequences, correctives. National Academy of Sciences, Washington, D.C., pp. 29-49.

Tiwari, R.K, Saxena, A.K. F., Kulshrestja, S.K. 1988. Evaluation of certain aquatic worms as indicators of water quality in Lower lake Bhopal. Prog. Natt. Symp. Past, Present and Future of Bhopal Lakes, pp. 67-76.

Todd, D.K. 1970. Water Encyclopedia, Water Information Centre, Water Res. Bldg., Manhasset, Isleport Washington, New York.

Tripathi, A.K., Pandey, S.N. & Tiwari, R.K. 1987. Eutrophication study of Kalyanlpur Pond. Kanpur (India). Proc. Natl. Acad. Sci. India, 57 B (111): 279-283.

Trisal, C.L. 1977. Studies on Primary Production in some Kashmir lakes. Ph.D. thesis, Univ. Kashmir, Srinagar, India.

Trisal, C.L. 1987. Ecology and Conservation of Dal Lake, Kashmir. Water Resources Develop., 3: 44-54.

Trivedy, R.K., Garud, J.M. & Goel, P.K. 1983. Limnology of five lakes in Kohlapur with reference to human activity. Environment & Ecology, 1(1): 41-43.

Unni, K.S. 1984. Limnology of a sewage polluted tank in central India. Int. Revue. Ges. Hydrobiol. 9: 553-565.

U.S.E.P.A. 1975. United States Environmental Protection Agency Quality Criteria for Water. Train (ed.), Caste House Publ. Ltd. Great Britain.

Varshney, C.K. 1985. Water Pollution and Management, Wiley Eastern Limited, New Delhi, pp. 241.

Vasisht, H.S. & Bandal, R.S. 1979. Seasonal variation of benthic fauna in some north Indian lakes and Ponds. Indian J. Ecol., 6: 33-37.

Vass, K.K. & Zutschi, D.P. 1978. Limnological studies on Dal lake: Morphometry and physical features. J. Inland. Fish. Soc., Indian, 11: 12-21.

Velacha, V. 1985. Ecology of Phytoplanktons in Lower Lake of Bhopal. Ph.D thesis, Bhopal Univ., Bhopal.

Verma P.K. & Munshi, J.D. 1983. Lemnology of Badka reservoir of Bhagalpur, Bihar. Proc. Ind. Natl. Sci. Acad. 349: 598-609 (in CD).

Vijaykumar, A. & Reddy, M.V. 2001. Elemental contamination of water and sediment of Husainsagar Lake in Hyderabad (Andhra Pradesh), with particular reference to metals. Proc. Intern. Conf. Indust. Pollution and Control. Tech. (ICIPACT), JNTU, Hyderabad, pp. 6.

Vollenweider, R.A. 1968. Scientific fundamentals of the eutrophication of lakes and flowing waters with particular reference to nitrogen and phosphorus as factors in eutrophication. Organisation for Economic Cooperation and Development, Paris, DAS/CSI/68.27. pp. 192, Annex. pp. 21, Bibliography pp. 61.

Weber, C.A. 1907. Aufbau and Vegetation der Moore Norddeut Schlands. Beibl. Bot. Johrb. 90: 1954 (Suppl. to Bot. Johrb. 40).

Welch, E.B. 1980. Ecological effects of wastewater. Cambridge University Press, Cambridge, U.K. pp. 337.

Welch, P.S. 1952. Limnology. McGraw Hill Book Co. 2nd Ed., New York.

Wetzel, R.G. 1975. Limnology. W.B. Saunders Company, Philadelphia, pp. 743.

Wetzel. R.G. (ed.) 2001. Limnology. W.B. Saunders Company, Philadelphia, pp. 743.

WHO 1982. International standards for drinking water. World Health Organisation, Geneva.

William, M.L. Jr. 1987. Tropical Limnology. Ann. Rev. Ecol. Syst., 18: 159-184.

Yousuf, A.R. 1979. Studies on the limnology and fisheries of Lake Manasbal, Kashmir. Ph.D. Thesis Univ. Kashmir.

Zafar, A.R. 1966. Limnology of the Husainsagar Lake, Hyderabad, India. Phykos, 115-126.

Zafar, A.R. 1986. Seasonality of phytoplankton in some south Indian lakes. Hydrobilogia, 138: 177-187.

Zutschi, D.P. & Vass, K.K. 1978. Limnological studies on Dal Lake—Chemical features. Indian J. Ecol., 5(1): 90-97.

CHAPTER

3

Phosphorus and Eutrophication in a Subtropical Lake Basin Lake Chapala–Mexico

Harvey Shear[1] and José de Anda[2]

[1] Environment Canada-Ontario Region, 4905 Dufferin Street, Toronto, Canada M3H 5T4.

[2] Centro de Investigación y Asistencia en Tecnología y Diseño del Estado de Jalisco, A. C., Normalistas 800 S. H., Guadalajara, Jalisco, México, C. P. 44270.

ABSTRACT

The world is experiencing a significant growth in human populations, and freshwater is fast becoming a scarce resource. Lakes represent a major source of such freshwater, but are coming under increasing pressure through exploitation for agriculture, industry, drinking water and hydroelectricity generation. This is particularly the case in tropical countries. This chapter presents a case study, Lake Chapala, the largest lake in Mexico. The primary tributary of Lake Chapala is the Lerma River. Much of the flow of this tributary is used in a non-conservative fashion upstream for agricultural and industrial purposes. Large quantities of domestic, agricultural, and industrial sewage from the entire Lerma-Chapala basin still flow untreated into the watershed and eventually to the lake, resulting in excessive inputs of phosphorus (P). High nutrient concentrations in the lake have led to degraded water quality, resulting in growths of floating aquatic vegetation of bulrush and water hyacinth, as well as blue-green algae. Spatial and temporal P distributions in the lake during the past 24 years have been discussed in relation to the hydrologic changes. A method based on the bathymetric features of the lake estimates the amount of total P in the lake. It has been well established in the case of temperate lakes that the most effective long-term measure for the control of eutrophication in a water body is a reduction in the input of external nutrients. In tropical shallow lakes, internal recycling of phosphorus from sediments represents a significant long-term input. Determining the nutrient balance of any lake or reservoir constitutes a very important tool to understand its chemical and biological processes. Equally, the knowledge of existing relationships between hydrological variables, nutrient concentrations, and primary productivity can help in determining the best way of restoration where eutrophication problems are involved. These relationships are discussed in this chapter.

BACKGROUND

The goals of the work described in this chapter are to describe the phosphorus concentrations and loading trends for Lake Chapala during the period 1974 to 1998. Smith et al. (1999) discussed the general features of the impact of humans on biological communities, as well as on the earth's biogeochemical cycle. Human activities, mainly through the use of phosphorus-containing fertilizers, and discharge of sewage (treated or untreated) containing biologically available phosphorus, have altered the nutrient status of many of the earth's water bodies.

Rast and Thornton (1996) gave a comprehensive overview of the current research and control options for temperate and tropical lakes suffering from the effects of cultural eutrophication. They also describe the evolution of the concepts of eutrophication and the term "eutrophic, mesotrophic and oligotrophic" to describe the nutrient status of lakes. The OECD (1982) has defined eutrophication as "the nutrient enrichment of waters which results in the stimulation of an array of symptomatic changes, among which are increased production of algae and macrophytes, deterioration of water quality and other symptomatic changes that are found to be undesirable and interfere with water uses". All lakes undergo a natural process of eutrophication, whereby nutrients such as nitrogen and phosphorus from terrestrial sources enter the lake and, over centuries or millennia, contribute to the natural ageing of a lake. In the presence of human inputs of these nutrients, especially phosphorus, this process is accelerated.

In the 1960s, there was considerable scientific debate about the causes of eutrophication focussing on nitrogen, phosphorus and carbon as the causative agents of eutrophication. Nitrogen and carbon were ruled out as contributors to eutrophication because these elements are abundant in the atmosphere and are not easily controlled. The definitive work of Schindler and coworkers at the Experimental Lakes Area in Northwestern Ontario, Canada in the late 1960s early 1970s (Schindler and Fee, 1974) showed—through whole lake enrichment experiments—that phosphorus was the nutrient controlling eutrophication.

Phosphorus can enter a lake from two sources—point sources such as sewage treatment plant effluents, and non-point sources, such as agricultural runoff. The strategies to control phosphorus inputs from external sources differ, depending on the source, but in either case, it is generally accepted that control of external sources is the most effective means of controlling eutrophication in the long term (Ryding and Rast, 1989).

The importance of Lake Chapala (see Fig. 3.1) is that it incorporates all possible beneficial uses such as recreation, agriculture, water supply, fishing, aquatic sports, etc. Unfortunately, large quantities of domestic, agricultural, and industrial sewage from the entire Lerma-Chapala basin still flow untreated into the watershed (Hansen et al., 1995) eventually reaching the lake. In addition, erosion in the hills surrounding the lake as a result of deforestation and change in land uses constitute the main non-point nutrient source to the lake (de Anda et al., 2000).

The existing in-basin nutrient concentrations promote the growth of floating aquatic vegetation known as water hyacinth (*Eichhornia crassipes*) and bulrush ("tule") (*Typha latifolia*), that have bloomed in the Lerma River, Lake Chapala, and Santiago River (the primary natural outflow of the lake) in recent years. Growths of these weeds have, in the past, covered littoral areas, and have adversely impacted commercial and recreational fishing, waterfowl feeding, boating and swimming, as well as the aesthetic beauty of the lake (de Anda et al., 1998). For example, in 1993, aquatic vegetation covered more than 135 km^2 representing about 13% of the entire surface of the lake (INEGI, 1995). Most of the water hyacinth present in the lake grows in the Lerma River, and is then carried to the lake during the rainy season. Recent clean up measures have physically removed a significant part of the water hyacinths from Lerma River, preventing most of the plants from entering the lake.

More recently, there have been blooms of a blue-green algal species identified as *Anabaena flos-aquae*, a cyanobacterium species that could cause taste and odour problems in the potable water supply for Guadalajara (CNA, 1998; CNA personal communication).

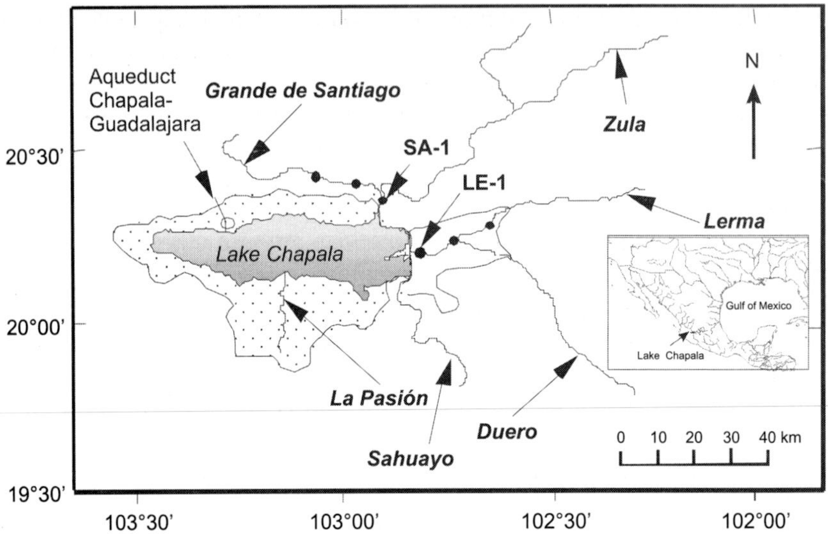

Fig. 3.1 Lake Chapala and its surroundings

Mexican standards allow municipal sewage discharges of up to 5 mg l^{-1} for total P to lakes and reservoirs (SEMARNAP, 1996). In comparison, the Canada-US Great Lakes Water Quality Agreement stipulates a maximum discharge for total P concentrations of 0.5-1.0 mg l^{-1} for sewage treatment plants discharging > 4.5×10^6 l day^{-1} (Governments of Canada and of the United States of America, 1987).

The United States Environmental Protection Agency (USEPA) water quality criteria state that to control algal growth, phosphates should not exceed 0.05 mg l^{-1} if streams discharge into lakes or reservoirs, 0.025 mg l^{-1} within a lake or reservoir, and 0.10 mg l^{-1} in streams or flowing waters not discharging into lakes or reservoirs (USEPA, 1999). Surface waters that are maintained at 0.01 to 0.03 mg l^{-1} of total P tend to remain uncontaminated by algal blooms.

Because of the high Mexican standards, the wastewater treatment plants along the Lerma River and along the shoreline of the lake are not required to have tertiary treatment to eliminate phosphorus discharges to these water bodies. Furthermore, detergents containing phosphates and several mixtures of NPK (nitrogen, phosphorus, potassium) fertilizers are permitted in Mexico without any restrictions on their use. In addition, the main land uses along the shoreline of the lake, and the part of the Lerma River close to the lake are agricultural and livestock based. These activities contribute high phosphorus loads through direct runoff, principally in the rainy season.

The methods used to produce the information presented in this chapter are described in de Anda et al. (1998, 2001), and de Anda and Shear (2001). The stations at which samples were collected are shown in Fig. 3.2.

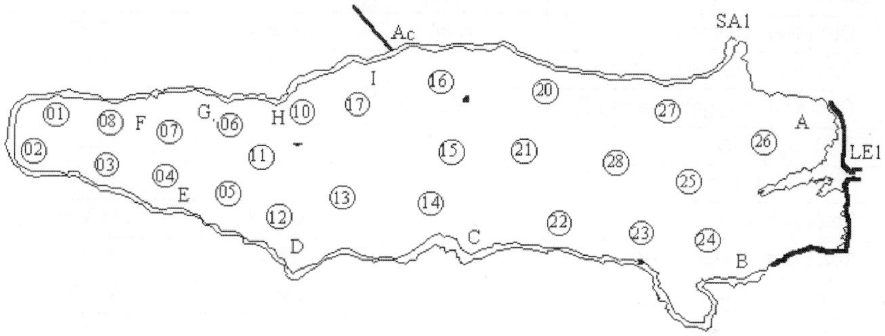

Fig. 3.2 Distribution of sampling points in Lake Chapala

HYDROLOGIC VARIABLES

The hydrology of Lake Chapala has been described by de Anda et al. (1998) and Aparicio (2001). The volume, surface area and depth of Lake Chapala have undergone significant variations over time, some of them caused by the climate, and others by the exploitation of the upstream water resource as discussed further in this chapter. Table 3.1 summarizes some of these variations.

Table 3.1 Relevant hydrologic values in Lake Chapala

Values	*Volume (Mm^3)*	*Surface area (km^2)*	*Mean depth (m)*	*Date*
Minimum	940.37	658.93	1.43	July 1955
Mean	5341.78	1,071.17	4.98	1934 to 1998
Maximum	9576.29	1,160.08	8.25	October 1935

PHYSICAL AND CHEMICAL VARIABLES

Long-and short-term variations of some physical and chemical parameters of Lake Chapala have been studied in previous works by Limón et al. (1989) and Limón and Lind (1990). Water quality parameters for the past three decades are reported in Table 3.2. These data show that the dissolved oxygen in the lake reaches values close to 100% of saturation at the water surface. The E-W and W-E wind velocities reach

Table 3.2 Average water quality parameters of Lake Chapala reported for three periods*

Parameter (mg l^{-1})	*1974-79*	*1980-89*	*1990-98*
Temperature (°C)	22.05	22.08	21.78
pH	8.56	8.66	8.54
Dissolved Oxygen	6.84	6.87	7.13
Biological Oxygen Demand	1.53	1.73	2.01
Chemical Oxygen Demand	17.21	24.27	34.97
Total alkalinity	186.85	247.44	293.25
Total hardness	147.79	190.66	218.16
Total solids	374.21	617.8	748.26
Total suspended solids	22.44	34.57	47.09
Total dissolved solids	351.77	583.23	698.97
Organic nitrogen	0.96	-	1.35
Nitrate	0.51	0.33	0.24
Ammonia	0.14	0.12	0.43
Total phosphorus	0.40	0.62	0.56
Phosphates	0.17	0.40	0.37

* (CNA, 1998)

values above 9 km h^{-1} and, considering the shallow characteristics of the lake, provide enough energy to maintain the waters well mixed (Simons, 1984). The lakewater is moderately hard and consistently alkaline. Secchi disk and turbidity values show that the lake is highly turbid, limiting the primary productivity (Lind et al., 1992a and 1992b). This turbidity is caused by: (1) sediment resuspension in the lake; (2) input of significant amounts of suspended solids from the Lerma River; and (3) erosion of the hills surrounding the lake. Chemical oxygen demand, total solids, total suspended solids, total dissolved solids, hardness and alkalinity have all increased significantly during the period of study. While nitrate has decreased over the period of study, organic nitrogen, ammonia, total phosphorus, phosphates have all increased in concentration.

One of the major environmental disturbances to Lake Chapala is the disruption of its hydrologic regime. Figures 3.3 and 3.4 show the annual average volume and depth of Lake Chapala and rainfall and flow into Lake Chapala for a 60+ year period. These data are based on measurements of the CNA. These figures show a clear relationship between rainfall/inflow and lake depth for much of the recorded period, particularly in the middle of the 1950s when Mexico experienced a severe, prolonged drought. Figures 3.3 and 3.4 also show that the

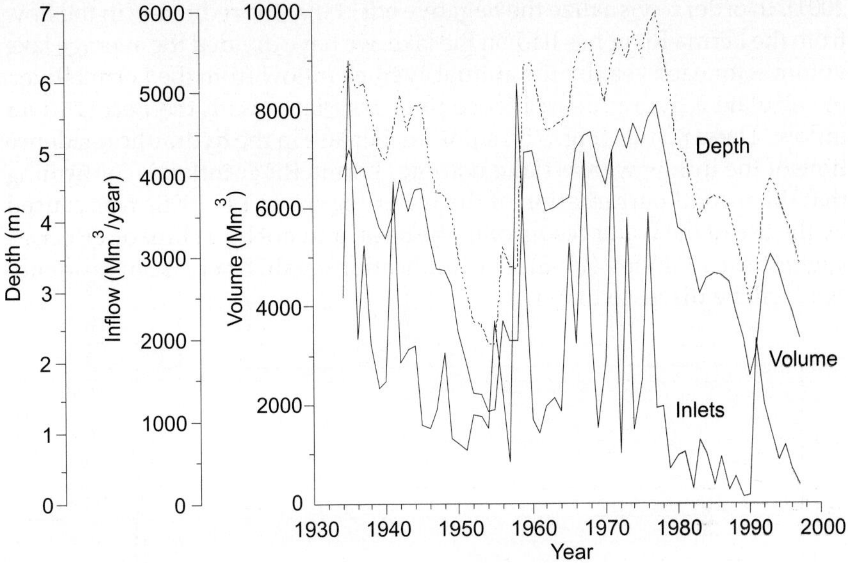

Fig. 3.3 Annual average volume and depth of Lake Chapala

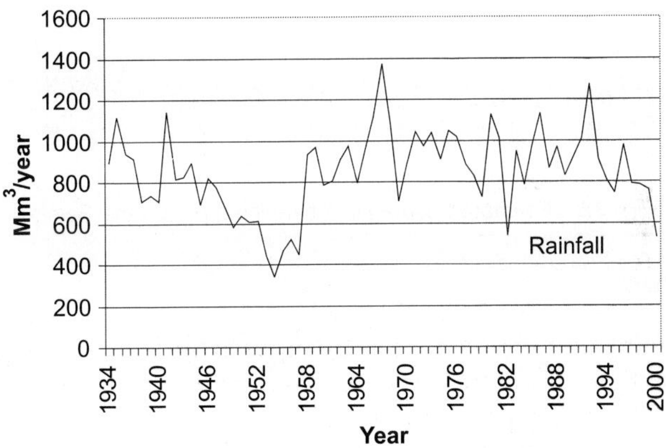

Fig. 3.4 Rainfall and flow into Lake Chapala

hydrologic equilibrium of the system was severely perturbed at the beginning of the 1990s. The correlation to rainfall patterns, however, is not apparent. This second perturbation was caused by excessive water withdrawals in the Lerma River basin, as discussed by de Anda et al. (1998).

Since the late 1970s, the Lerma River inflow to Lake Chapala has been drastically reduced due to the over exploitation upstream (Aparicio,

2001). In order to visualize the negative effect that the reduction in the flow from the Lerma River has had on the lake, we have divided the average lake volumes for each year by the annual average inflow from the Lerma River, to calculate a hydraulic residence time associated with the Lerma River inflow. These results (Fig. 3.5.) show an increase in the hydraulic residence time of the inflow waters (lake volume/Lerma River inflow), confirming that the second perturbation of the water regime in the 1990s was caused by the use of the water resource in the basin and not by a climatic effect, as suggested by Filonov (1998). The implications of this increase in residence time will be discussed later.

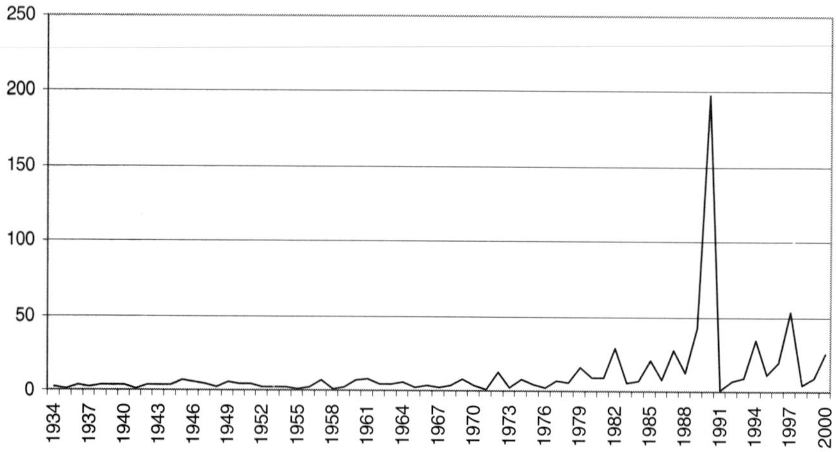

Fig. 3.5 Estimated hydraulic residence time (Lerma River)

Phosphorus

The only significant form of inorganic P input to the lake is orthophosphate (PO_4^{-3}). More than 90% of the P is bound as organic phosphates and cellular constituents in living particulate matter of seston, or associated with organic materials (Wetzel, 1975). Figure 3.6 shows the variation in these amounts in the lake during the period 1974-1998.

Concentrations as shown in Fig. 3.7 were estimated by dividing the total amount by the volume of the lake. It is observed that there is no clear long-term trend for P concentrations in the lake, but there are some periods where the P concentrations increase significantly (Fig. 3.7). The high levels of total P reached in 1976 were probably due to very high levels of the lakewater, which in this year, flooded highly fertile agricultural lands around the lake. After this event, phosphorus probably

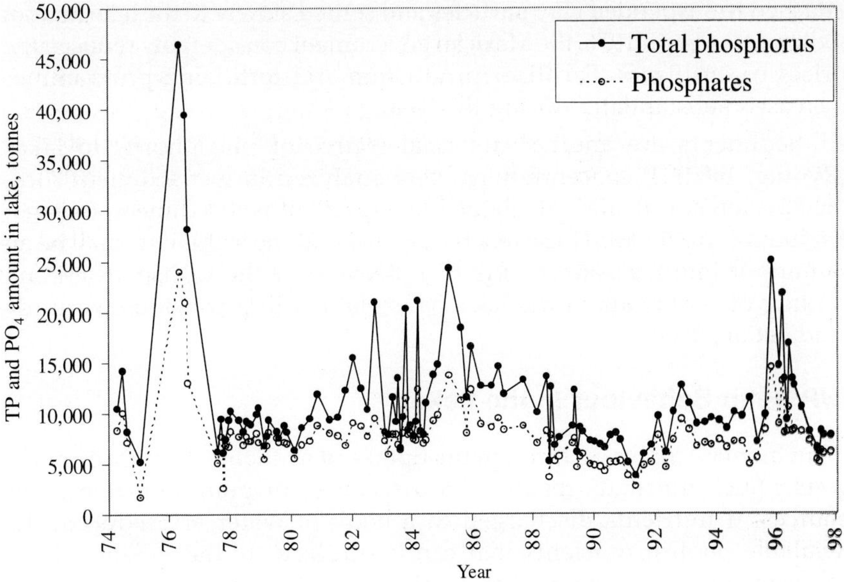

Fig. 3.6 Amount of total P and PO_4^{-3} in the lake

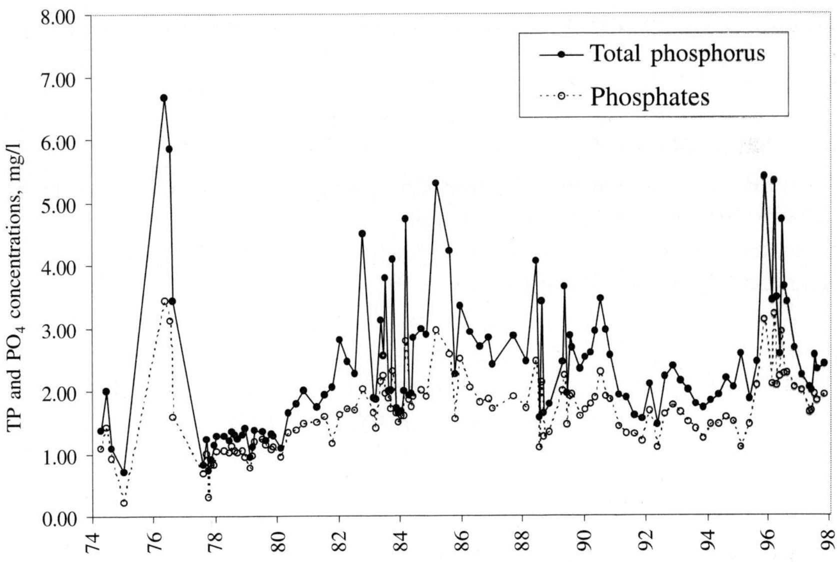

Fig. 3.7 Concentration of total P and PO_4^{-3} in the lake

attached to suspended clay particles and settled slowly to the lake bottom sediments. Also in 1976, the Mexican government considerably reduced the prices of fertilizers. Fertilizer production and fertilizing programmes increased substantially during this year.

Sediments are another potential source of phosphorus to lakes (Ryding, 1982). P concentrations were analyzed in wet sediment cores (80-82% moisture), and are about 200 mg kg^{-1} of wet sediment, (Solana-Espinosa et al., 1998). These results suggest that the sediments may be an important internal source/sink of P because of the resuspension and settling of sediments in the lake, as explained in a previous work (de Anda et al., 2000).

N/P Ratio Behaviour in the Lake

Plant biomass increases through the uptake of available P and N from the water. Such nutrients result from discharges of point and non-point sources. If nutrients discharged to a body of water are reduced, the available pool of nutrients that can be utilized by the plants will be reduced and, in general, the total plant biomass will be reduced (Thomann and Mueller, 1987).

For lakes with inputs of non-point source nutrients, P usually controls growth. Lakes that may have significant point source inputs may be N limited, but can also be made P limited by removal of P at sewage treatment plants. The N/P ratio is, therefore, a useful measure for understanding the relationship between N, P, and biomass. As with all such procedures, however, care must be taken in its interpretation and use (Thomann and Mueller, 1987).

While the focus of this chapter is on phosphorus, we have calculated the concentrations of various nitrogen compounds in the lake as well as subsequently, the N to P ratios (de Anda and Shear, 2001). The trend of this ratio has been illustrated in Fig. 3.8.

An increasing N/P value of about 2.0 to 2.5 is observed for the entire monitoring period. N/P ratios below 4 may stimulate the growth of noxious blue-green algae in freshwater (Thomann and Mueller, 1987). These algae can obtain nitrogen for growth by fixing the atmospheric nitrogen. Such algae tend to form unsightly surface mats which can result in serious impairment of water uses.

The N/P is, however, not the entire story of eutrophication. In Lake Chapala, the high amount of suspended sediments limits phytoplankton production processes (Lind et al., 1992a, 1992b; Davalos-Lind, 1993). There is no agreement between the empirical models that relate biomass production or trophic status to nutrient concentrations. Such models

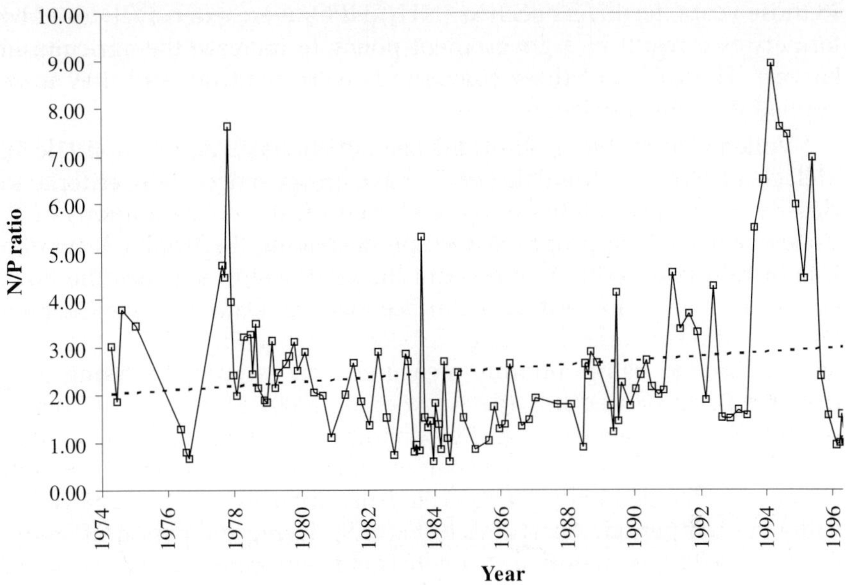

Fig. 3.8 N/P ratio for Lake Chapala

assume that the phytoplankton growth and biomass are controlled by phosphorus and nitrogen availability (Lind et al. 1993). In the case of Lake Chapala, light limits the phytoplankton production. A similar situation is seen in another Mexican lake, Lake Patzcuaro (Chacon, 1993). The high nutrient concentrations do, however, result in nuisance growths of aquatic vegetation (water hyacinth, *Eichhornia crassipes*) and bulrush ("tule") (*Typha latifolia*), including the growth of this vegetation along the Lerma River (INEGI, 1995), and may have led to the death of fisheries in the Lerma River basin.

In conclusion, we know that the Lerma River basin is home to about 10 million Mexicans, and is the industrial and agricultural heartland of the country. This high population, coupled with intensive agriculture, has led to severe pollution problems in the basin, as well as downstream in Lake Chapala. Much of the sewage and agricultural runoff is untreated, resulting in highly available forms of phosphorus entering the aquatic ecosystem. Approximately, 70% of the total P occurs as PO_4^{3-}. The source is mainly anthropogenic, although some may also originate from the decomposition of organic matter.

Because of the intensive agriculture, the Lerma River basin has a high use of fertilizers and, consequently, high levels of nutrient pollution. From 1976 to 1977, phosphates reached high concentrations in the lake.

In those years, fertilizers such as $(NH_4)_2HPO_4$ were extensively used by farmers as a result of a government policy to increase the agricultural harvest. The prices of these compounds were reduced, and they were applied without restriction.

Vollenweider (1975), Rast and Lee (1978), Salas and Limón (1985), Melack (1992) and Maniak (1997) have proposed certain criteria to classify the trophic status of a lake based on the concentrations of P. Vollenweider (1975) proposed a graph describing the trophic level of a lake or reservoir taking into account the relationship between the flow, lake surface area, and annual P surface loading. Data for Lake Chapala for three different time periods have been shown in Fig. 3.9. These values were estimated using monthly measured and estimated loading data from the Lerma River only (de Anda et al., 2000). Subsequently, yearly surface PO_4 loads were obtained by dividing the total estimated monthly loads by the surface area of the lake, and then summing over a 12-month period. Each circle in Fig. 3.9 was calculated by using the average yearly value in each period. According to Fig. 3.9, during the period of study, Lake Chapala has maintained a consistent eutrophic status. As noted earlier, however, the manifestation of the eutrophic state is in extensive

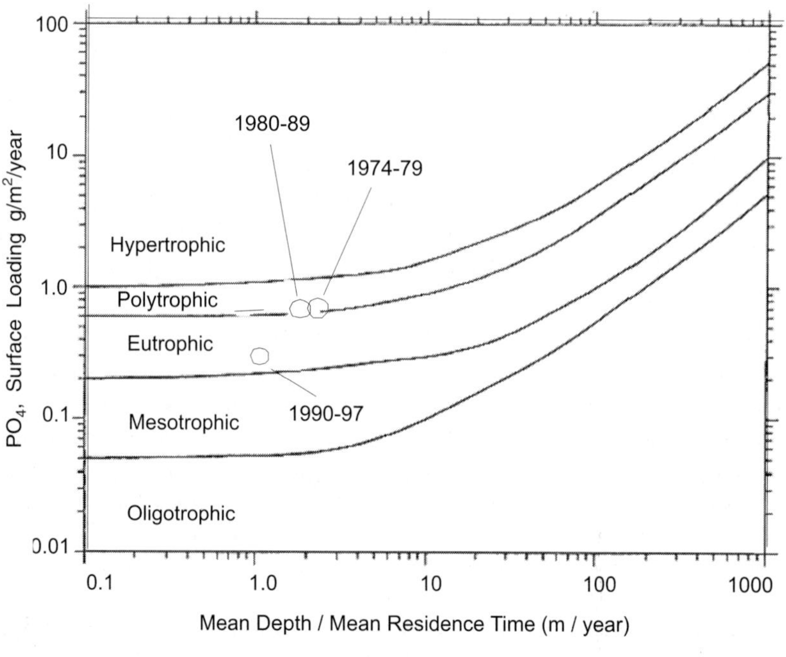

Fig. 3.9 Trophic status of Lake Chapala

growths of rooted and floating aquatic vegetation, and not in excessive populations of phytoplankton. This is due to light limitation caused by sediments constantly kept in suspension by wind-driven forces over the lake.

REMEDIATION OF EUTROPHICATION

The overuse of water resources and the excessive inputs of nutrients such as phosphorus, resulting in the uncontrolled growth of nuisance aquatic vegetation, plague many tropical lakes. It has been well documented (Rast and Thornton, 1996) that the easiest, cheapest and most effective means of controlling the nutrient enrichment in a temperate lake is to control the phosphorus input. Many examples of successful eutrophication control exist, such as in the Great Lakes (Governments of Canada and the United States, 2001), Lake Washington (Edmondson and Lehman, 1981) Lake Zurich (Sas et al., 1989) and Lake Norrviken in Sweden (Ahlgren, 1978). In contrast, eutrophication control based solely on external controls has failed where there has been extensive internal loading of phosphorus from nutrient-rich sediments, such as in Lake Shagawa (Larsen et al., 1981; Chapra and Canale, 1991) or Lake Trehorningen (Ryding, 1982). In the case of Lake Trehorningen, the solution was to dredge the contaminated sediments. When this was done, the lake recovered (Ryding, 1982). In the case of Lake Shagawa, the internal loading is so high that the lake is predicted to take >100 years to recover (Chapra and Canale, 1991).

Lake Chapala is in a similar situation as these latter two lakes, having a net accumulation of total P in the sediments and a net internal loading of total P of 615-1150 metric tonnes/year. There are several options for control of the nutrient situation in Lake Chapala.

Do nothing

This option is obviously of no initial cost, but does not take into account the economic losses that will accrue when the lake degrades even further. There will be further loss of fishing opportunities (either sport or commercial); degradation of nearshore areas with extensive growths of tule; and degradation of raw water quality for drinking water.

Control of Nutrients Upstream

The control of nutrients in the Lerma River Basin is a major undertaking. For non-point sources, one needs to consider the extensive agricultural operations. Over 66 million dairy and beef cattle, pigs and poultry comprise the livestock of the basin, with some 3.2 million hectares of land devoted to agriculture of a total land area of 4.8 million hectares (Leon et al., 2001). Much of the agricultural wastes are discharged untreated into the Lerma River. Treatment of farm waste from livestock operations is feasible, but requires treatment at the farm site, as opposed to some centralized system of collection of farm waste. In the case of polluted runoff from vegetables or other row crops, the problem lies in the quantities of chemical fertilizers applied and the flood irrigation techniques used in the Lerma Basin. This results in significant non-point sources of phosphorus to the river. Techniques for control of agricultural waste have been well documented (Brenner and Brenner, 1998).

For point sources, one needs to determine the status of each sewage treatment plant (STP) in the basin, and also determine how many people are not on any system. Next, one has to determine the effectiveness of each STP, in terms of nutrient removal, and then develop a management plan for each plant to reduce the phosphorus discharged to the river. Phosphorus removal involves chemical precipitation, resulting in the production of a chemical sludge which must be disposed of. If the sludge is not contaminated with metals or toxic organic chemicals, it can be applied on land as an excellent fertilizer. If the sludge is deemed to be toxic, it must be disposed of in an appropriate facility (MOE, 1996). There are significant financial and personnel costs involved with P removal from sewage treatment plants. For example, in the Great Lakes, the costs to build/upgrade STPs to allow for P removal was some $10B for Canada and the USA (McGucken, 2000) and the requirement for hundreds of highly trained staff to operate the plants efficiently and effectively. Realistically, for the Lerma Basin, this option may only be viable for the largest plants.

P Removal from Detergents

One of the additional approaches taken in the Great Lakes was the removal of phosphorus from laundry detergents. Phosphorus had been added to synthetic laundry detergents since their introduction in the 1940s. Phosphorus acted as a softening agent to allow the detergent to clean better. With the signing of the Great Lakes Water Quality Agreement in 1972 (Governments of Canada and the United States, 1972), removal of P from detergents became an active part of the Great Lakes agenda. By 1990, all jurisdictions in the Great Lakes Basin had removed all or most of the P from laundry detergent formulations. (McGucken, 2000) This is a relatively low cost method of removing a biologically active form of P.

Reduced Fertilizer Usage

Since the 1940s, the use of chemical fertilizers has increased significantly throughout the world, but it has also brought associated environmental problems (Kellogg et al., 1992). In Mexico, fertilizers are used extensively on row crops to ensure several good harvests a year. Some of the fertilizers applied to the land are not taken up by plants, but find their way to the nearest water course, leading to the high P loadings, as can be seen in the Lerma River. A strategy to reduce P loadings to the Lerma River would couple the effective management of the land through prevention of erosion, with a reduction in the use of chemical fertilizers. This is a relatively low cost option, although one needs to conduct an economic analysis to determine what, if any, loss of income would result from such a strategy.

Dredging the Lake Sediments

One final option to reduce phosphorus loading would be to dredge the nutrient-rich sediments of the lake. This method has the advantage of permanently removing this source of internal loading. The negative aspect of this method is the cost to dredge a lake the size of Lake Chapala (77 x 15 km). It may, however, be feasible and cost effective to dredge certain areas of the lake that are known to contain high PO_4 concentrations. The east end of the lake has the highest PO_4 concentrations (de Anda et al., 2001), so it might be logical to dredge in this area.

ACKNOWLEDGMENTS

This research was supported by following sponsors: Universidad Nacional Autónoma de México (UNAM). Programa de Apoyo a las Divisiones de Estudios de Posgrado 1997 (PADEP). Apoyo a proyectos de investigación. Tesis doctorales. Instituto Mexicano de Tecnología del Agua. Ing. J. Jesús Amezcua Cerda. Comisión Nacional del Agua (CNA). Centro de Estudios Limnológicos (CEL). Mexico. Environment Canada-Ontario Region, Toronto, Canada. Deutscher Akademischer Austauschdienst (DAAD). Germany. Prof. Dr Ing. Ulrich Maniak and co-workers. Technische Universität Braunschweig. Leichtweiss-Institut für Wasserbau, Abteilung Hydrologie und Wasserwirtschaft. Germany.

REFERENCES

Ahlgren, I. 1978. Response of Lake Norrviken to reduced nutrient loading. Verh. Int. Verein. Limnol. 20, 846-850.

Aparicio, J. 2001. Hydrology of the Lerma-Chapala watershed:. In: The Lerma-Chapala Watershed, Evaluation and Management. (Eds) A. Hansen and M. van Afferden. Kluwer Academic/Plenum Publishers. New York. pp. 385.

Brenner, Fred J. and Brenner, Elaine K. 1998. A watershed approach to agricultural nonpoint source pollution abatement. In: Watershed Management—Practices, Policies and Coordination. (Ed.) Robert J. Reimold. McGraw-Hill. pp. 391.

Chacón, A.T. 1993. Lake Patzcuaro, Mexico: Watershed and water quality deterioration in a tropical high altitude latin american lake. Lake and Reserv. Manage. 8(1) 37-47.

Chapra, S.C. and Canale, R.P. 1991. Long-term phenomenonological model of phosphorus and oxygen for stratified lakes. Wat. Res. 25, 707-715.

CNA (National Water Commission). 1981. Analytical results and vertical distribution in Lake Chapala (Oct. 1977-Oct. 1981). Gerencia Regional Lerma-Santiago. Centro de Estudios Limnológicos. Guadalajara, Jalisco. MEXICO. Unpublished.

CNA (National Water Commission). 1998. Water quality data in Lake Chapala. Gerencia Regional Lerma-Santiago. Centro de Estudios Limnológicos. Guadalajara, Jalisco. MEXICO. Unpublished.

CNA (National Water National). 1999. Diagnóstico hidráulico y lineamientos estratégicos en la Cuenca Lerma-Chapala. Presentación ante el Consejo de Cuenca. Informe Técnico, Abril 20 de 1999. Gerencia Regional Lerma-Santiago. Guadalajara, Jalisco. MEXICO. Unpublished.

Dávalos-Lind, L. O, Lind, O. T. 1993. The changing state of limnology in México: Lake Chapala as an example. Verh. Internat. Verein. Limnol. 25, 427-430.

de Anda, J., Quiñones-Cisneros, S., French, R., Guzmán, M. 1998. Hydrologic balance of Lake Chapala (Mexico). Journal of the American Water Resources Association. 34(6), 1319-1331.

de Anda, J., Shear, H., Maniak, U., Riedel, G. 2000. Phosphorus balance in Lake Chapala (Mexico). J. Great Lakes Res. 26(2): 129-140.

de Anda, J. and H. Shear. 2001. Nutrients and eutrophication in Lake Chapala. In: The Lerma-Chapala Watershed—Evaluation and Management. (Eds.) A. Hansen and M. van Afferden. Kluwer Academic/Plenum Publishers. New York. pp. 385.

de Anda, J., Shear, H., Maniak, U., Riedel, G. 2001. Phosphates in Lake Chapala (Mexico). Journal of Lakes and Reservoirs: Research and Management. **6:** 313-321.

de Anda, J., Shear, H., Maniak; U., and Zarate-del Valle, P. 2004. Solids Distribution in Lake Chapala (Mexico). Journal of the American Water Resources Association. 40(1): 97-109.

de León-Mojarro, Benjamín, Medina-Mendoza Raúl and González-Casillas Arturo 2001. Natural Resources in the Lerma-Chapala Basin. In: The Lerma-Chapala Watershed—Evaluation and Management. (Eds) A. Hansen and M. van Afferden. Kluwer Academic/Plenum Publishers. New York. pp. 385.

Edmondson, W. T. and Lehman J. R. 1981. The effects of changes in the nutrient income on the condition of Lake Washington. Limnol. Oceanogr. 26, 1-29.

Filonov, A. E. 1998. Will Lake Chapala die? Can it be saved? Article in the Magazine "Teorema", September-November 1998. México.

Governments of Canada and of the United States of America. 1972. Great Lakes Water Quality Agreement. Ottawa and Washington.

Governments of Canada and of the United States of America. 1987. Great Lakes Water Quality Agreement. Ottawa and Washington.

Governments of Canada and of the United States of America. 2001. State of the Great Lakes 2001. Toronto and Chicago. pp. 82.

Hansen, A. M., León, A. and Bravo, L. 1995. Sources of contamination and enrichment of metals in the sediments of the Lerma-Chapala basin. Ingeniería Hidráulica en México, 10(3), 55-69.

Kellogg, Robert L., Maizel Margaret S., and W. Goss Don 1992. Agricultural chemical use and ground water quality: Where are the potential problem areas? Report of the U.S. Dept. Agr., Soil Cons. Serv., Econ. Res. Serv., Cooperative State Res. Serv., National Center for Resource Innovations.

Larsen, D.P., Schultz, D.W. and Malueg, K.W. 1981. Summer internal phosphorus supplies in Shagawa Lake, Minnesota. Limnol. Oceanog. 26, 740-753.

León-Vizcaíno, L. F., Martínez-Austria, P., Aldama-Rodríguez, A. A., 1994. Advances in the integrated management of the Lerma-Chapala basin. First trinational workshop on water, development and environment. Toluca, Estado de México. Mexican Institute of Water Technology, pp. 43-55.

Limón, J. G., Lind, O. T., Vodopich, D. S., Doyle, R., and Trotter, B. G. 1989. Long- and short-term variation in the physical and chemical limnology of a large, shallow, turbid tropical lake (Lake Chapala, Mexico). Arch. Hydrobiology. 83, 57-81.

Limón, J. G. and Lind, O. T. 1990. The management of Lake Chapala (Mexico): Considerations after significant changes in the water regime. Lake and Reservoir Management, 6(1), 61-70.

Lind, O. T., Doyle, R., Darrell, S. Vodopich, Trotter, B. G., Glass, J., Dávalos-Lind, L. O., Limón, J. G. 1992a. Lake Chapala: Factors controlling phytoplankton production. Ingeniería Hidráulica en México. 7 (2/3), 17-28.

Lind, O. T., Doyle, R., Vodopich, D. S., Trotter, B. G., Limón, G., Dávalos-Lind, L. 1992b. Clay turbidity: Regulation of phytoplankton production in a large, nutrient-rich tropical lake. Limnol. Oceanogr., 37(3), 549-565.

Lind, O. T., Terrell, T. T., Kimmel, B. L. 1993. Problems in reservoir trophic-state classification and implications for reservoir management. In: M. Straskaba, J. G. Tundisi & A. Duncan (Eds), Comparative Reservoir Limnology and Water Quality Management, Kluwer Academic Publishers, the Netherlands. pp. 57-67.

Maniak, U. 1997. Hydrology and Water Management. Introduction for Engineers. 4th Edition. Springer-Verlag, Berlin Heidelberg New York.

Melack, J. M. 1992. Eutrophication and water quality in lakes in tropical alluvial plains. Ingeniería Hidráulica en México, 7 (2/3): 142-147.

McGucken, W. 2000. Lake Erie rehabilitated: controlling cultural eutrophication, 1960s-1990s. University of Akron Press. pp. 318.

MOE. 1996. Ontario Ministry of the Environment and Ontario Ministry of Agriculture Food and Rural Affairs. Guidelines for the utilization of biosolids and other wastes on agricultural lands. Toronto, Canada. pp. 26 + 7 appendices.

OECD. 1982. Organization for Economic Cooperation and Development. Eutrophication of Waters: Monitoring, Assessment and Control. OECD, Paris. pp. 154.

Rast, W. and Lee, G. F 1978. *Summary analysis of the North American (US portion) OECD eutrophication project: Nutrient loading-lake response relationship and trophic state indices.* Ecological Research Series, No. EPA-600/3-78-008, US Environmental Protection Agency, Environmental Research Laboratory, Corvallis, Oregon, USA. p. 454.

Rast, W. and Thornton, J. A. 1996. Trends in eutrophication research and control. Hydrologic Processes, 10, 295-313.

Ryding, S.-O. 1982. Lake Trehorningen restoration project: changes in water quality after sediment dredging. Hydrobiology, 92, 549-558.

Ryding S.-O. and Rast, W. 1989. The control of eutrophication of lakes and reservoirs. Man and the Biosphere Series. (Ed.) J. N. R. Jeffers. Volume I. The Parthenon Publishing Group. Park Ridge, New Jersey.

Salas, H.J. and Limón, G. 1985. Report of the 3rd meeting for the regional development project of simplified methodologies for the valuation of the eutrophication in temperate tropical lakes. Guadalajara, Jalisco, México. World Health Organization. Pan-American Health Organization. Environmental Health Program. No report number.

Sandoval, F. de P. 1994. *Pasado y futuro del Lago de Chapala.* In: *Gobierno del Estado de Jalisco,* Secretaría General, Unidad Editorial. Guadalajara, Jalisco. Mexico.

Sas, H., Ahlgren, I., Bernhardt, H., Bostrom, B., Clasen, J. , Forsberg, C., Imboden, D., Kamp-Nielson, L., Mur, L., de Oude, N., Reynolds, C., Schreurs, H., Seip, K., Sommer U. and Vermij S. 1989. Lake restoration by reduction of nutrient loading: Expectations, Experiences, Extrapolation. Academia-Verlag, Richarz.

Schindler, D.W. and Fee, E.J. 1974. Experimental Lakes Area: whole lake experiments in eutrophication. J. Fish. Res. Board Can., 31, 937-953.

Schnoor, J. L. 1996. Fate and transport of pollutants in water, air, and soil. In: Jerald L. Schnoor and Alexander Zehnder (Eds). Environmental Modeling.. John Wiley & Sons, Inc., New York. pp. 185-219.

SEMARNAP (Secretaría de Medio Ambiente, Recursos Naturales y Pesca) 1996. Diario Oficial de la Federación. Official norm establishing the maximum permissible values of contaminants in waste water discharged into national water reservoirs. NOM-001-ECOL-1996. Mexico City.

Simons, T. J. 1984. Effect of outflow diversion on circulation and water quality of Lake Chapala. Centro de Estudios Limnológicos. Secretaría de Recursos Hidráulicos. Guadalajara, Jalisco. México. Pan American Health Organization. Project MEX-CWS-010.

Smith, V.H., Tilman G.D. and Nekola J.C., 1999. Eutrophication: impacts of excess nutrient inputs on freshwater, marine and terrestrial ecosystems. Environmental Pollution, 100, 179-196.

Solana-Espinosa, G., Israde, I., de Anda, J., and Zarate-del Valle, P. F. 1998. Geochemistry data at Lake Chapala, Jalisco, Western México (ECOS M95 and SIMORELOS 970306003 (projects in progress): 1ra Reunion Nal. de Cs. de la Tierra, Soc. Geologica Mex., p. 124.

Thomann, R.V., Mueller, J.A. 1987. Principles of Surface Water Quality Modeling and Control. Ed. Harper and Row, New York, pp. 644.

USEPA (United States Environmental Protection Agency), 1999. Nutrient criteria technical guidance manual: Lakes and reservoirs. EPA 822-D-99-001. April, 1999. Washington, D.C.

Wetzel, R.G. 1975. Limnology. W.B. Saunders Co., Philadelphia.

Vollenweider, R.A. 1975. Input-output models with special reference to phosphorus loading concept in limnology. Scweiz. Zeit. Hydrol. 37:53-84

Cartographic and Statistical Sources

INEGI. (National Institute of Statistics, Geography, and Data Management), 1974. Digital Elevation Model. Geographic grads: 104-101W, 20-21N.

INEGI. (National Institute of Statistics, Geography, and Data Management), 1995. Espaciomapa Guadalajara. Hoja F13-12. Esc. 1:250,000. Image date: March-April, 1993.

CHAPTER

4

Detergents and Eutrophication of a Temperate Lake: Relevance for Indian Aquatic Environments

N.K. Kaushik
Department of Environmental Biology,
University of Guelph, Guelph, Ontario, Canada, N1G 2W1

ABSTRACT

A rapid increase in population and growth of industry and agriculature in the Lake Erie basin during 1920 to 1960 lead to dramatic increase in inputs of organic wastes and nutrients, especially in the western basin of the lake. By 1970 about half of the phosphorus loading was derived from detergents. Nutrient loading to Lake Erie was high and the western basin, which received about 74% of the total phosphorus load to the lake, was regarded as highly eutrophic. The western basin had phytoplankton biomass and taxa that were characteristic of eutrophic waters, with higher concentrations of chlorophyll a, higher primary production, and assimilation number. All eutrophic areas in Lake Erie with submerged rocky substrates promoted a profuse growth of filamentous algae, especially *Cladophora.* This affected water intakes; drinking water supplies had the taste of rotting vegetation. Stench arising from decomposing *Cladophora* affected tourism and fishing. Beaches were deserted and thus eutrophication caused a massive economic loss. Microbial decomposition of large quantities of dead algae and other organisms lead to hypoxic and anoxic conditions in the hypolmmion during lake stratification. Whereas only 3% of hypolommion in the central basis was anoxic during the summer of 1930, the value had increased to 93.7% by 1973. Under anoxic conditions phosphorus adsorbed on to sediment is released back into water and after the fall overturn and mixing of waters it promotes algal blooms. Anoxic conditions in hypolinnetic water eliminated benthic organisms as well and some highly valued commerical and sport fish.

In the mid 1960's, the media was inundated with features on pollution in Lake Erie and there was a tremendous public outcry both in Canada and the USA. By 1970, The International Joint Commission, initially asked to investigate the extent of pollution and to suggest remedial measures to stop eutrophication, had recommended a reduction in phosphates found in detergents as well those found in municipal sewage effluents. Following these recommendations some 30 years ago, phosphorus abatement programs in both Canada and the USA have had a significant impact on the oligotrophiciation of Lake Erie.

At present, most of the laundry detergent sold in North America and many European countries are phosphate free, despite regulations that allow the use of small quantities of phosphates. In India, the quantity of detergent-based phosphorus entering aquatic systems is relatively small when compared to the amount entering from other sources. Hence, except for water bodies in which detergents are a major source of phosphorus, regulations on phosphates in detergents will not help the overall oligotrophication of eutrophic Indian environments. However, the input of phosphorus from detergents consititutes a substantial nutrient pollution, and the extent of this pollution is likely to increase steadily. This would promote the eutrophication of unpolluted downstream reaches and even of coastal waters. Controlling the phosphorus content of detergents in India is not likely to be expensive because two or three main detergent manufacturers control a major share of the detergents market. In view of these facts, it could be concluded that the quantity of phosphates present in detergents in India should be regulated.

INTRODUCTION

Eutrophication is the most widespread water problem in many countries. In the US, it accounts for almost one-half of the impaired lake area and 60% of impaired river reaches (Smith et al., 1999). During the last four decades or so, a vast body of literature pertaining to eutrophication of aquatic environments—its causes, impacts and possible methods of remediation—has been generated. The term eutrophication refers to the addition of nutrients to aquatic environments, resulting in enhanced productivity. Natural eutrophication, or eutrophication resulting from the addition of nutrients that a water body receives from natural or unaltered systems, is a very slow process. It may take centuries for a body of water to change from an oligotrophic to a eutrophic state. However, a variety of human activities that affect a water body, e.g. deforestation, use of manure and fertilizers in agriculture, feed-lot operations, discharge of human and animal waste, and household wastewater, can drastically increase the inputs of nutrients, leading to a much faster rate of eutrophication. Nutrient enrichment resulting from anthropogenic sources is referred to as cultural eutrophication. These added nutrients promote the growth of algae and other aquatic plants and are divided into two groups: macronutrients and micronutrients. Macronutrients (e.g. carbon, nitrogen, phosphorus, hydrogen, oxygen, sulphur, magnesium, calcium) are required by algae and aquatic plants in relatively larger amounts as compared to micronutrients (iron, zinc, copper, manganese, etc.), which are required in much smaller quantities (Ryding & Rast, 1989). Insufficiency of any nutrient can become a limiting factor for algal growth. Compared to carbon and nitrogen, phosphorus is generally in short supply in most unpolluted freshwater systems, and a limiting factor for the growth of algae and other aquatic vegetation. In contrast, in coastal waters, it is generally nitrogen that is the limiting factor (see pages 65-80, National Research Council, 2000 for a detailed discussion).

Depending on the use of a waterbody, eutrophication can either be termed useful or it can be regarded as a pollution problem when its effects are undesirable. For example, eutrophication of an oligotrophic waterbody can enhance fish production but under extreme situations—resulting from cultural eutrophication—added nutrients can lead to a variety of harmful effects. Briefly, among other effects, increased availability of nutrients in freshwaters leads to an increase in algal production and algal biomass. This growth not only restricts the use of a water resource and affects water treatment procedures, but also makes a waterbody aesthetically unacceptable. In addition, decomposition of large quantities of organic matter resulting from dead algal and other plant biomass leads to low levels of dissolved oxygen in the water or even to anoxic conditions in some areas of impacted waterbodies. Obviously this results in the mortality of fish and other organisms.

Eutrophication of aquatic environments is not a recent phenomenon. Foulness of the River Thames in England was alluded to as early as 1620. As a result of continuous discharge of human and animal waste, the River Thames became so polluted that 1858 was known as 'the Year of the Great Stink' (Allan, 1995). In the first half of the last century, the extent and magnitude of cultural eutrophication of aquatic environments had vastly increased all across the world. The problem was so intense that the US National Academy of Sciences (1969) organized an international symposium to discuss the causes, consequences and corrective measures for eutrophication. This accelerated pace of cultural eutrophication resulted from factors such as population increase, intensification of agriculture through the use of fertilizers, industrial development, etc. Another factor that contributed significantly to the eutrophication of some of the waterbodies was the rapid increase in the use of phosphate-based detergents during the 1950s and 1960s. In North America as also many European countries, phosphate content in detergents is now regulated, which, in conjunction with controls on phosphate levels in effluents from sewage treatment plants, has led to oligotrophication of affected environments. While in recent times conditions have improved in North America and Europe in the last decade or so, the use of phosphate-based detergents has rapidly increased in many tropical countries. This has exacerbated the already rampant eutrophication problems. These countries are now experiencing many of the same environmental or cultural issues experienced in North America many years ago, despite the availability of information to manage the problem of eutrophication. This chapter discusses detergent composition, their contribution to the eutrophication of Lake Erie, the subsequent oligotrophication of Lake Erie through phosphate regulations of detergents and effluents from sewage treatment plants, and, finally, the implication of the use of phoshate-based detergents in India in particular and tropical countries in general.

DETERGENT COMPOSITION

Until the end of World War II, bar soaps, produced from animal fat and lye (sodium hydroxide) were the most common cleaning agents for laundries. However, because of shortages of raw materials for soaps during the war and also in order to improve the performance of the washing machines, new formulations of clothes' washing detergents, were developed. Synthetic detergents for household laundry with sodium tripolyphosphate (STPP) were first introduced in late 1940, (Litke, 1999). The main components of laundry detergents are: (i) surfactants; (ii) builders; and (iii) other ingredients (see Feisthauer et al.,

2002 for more details). Surfactants are the most important components of detergents and constitute from about 14 to 50% of formulations. They decrease the surface tension of water, and thus enhance loosening and removal of soil and dirt from laundry and also prevent redeposition of the soil on the cleaned clothing. The type and concentration of a surfactant used in a detergent is guided, among other factors, by washing temperature, type of builder used, and the type of soil to be removed. The quantity of surfactant in a detergent formulation is considerably more important in conferring detergency than is the type of builder present. The most commonly used surfactant in detergents is LAS or linear alkyl benzene sulfonate. In recent years, the use of polyoxyethylene alcohols or alcohol exthoxylates (AE) as surfactants has also increased (Feisthauer et al., 2002).

The function of builders is to enhance detergency by reducing water hardness, primarily by binding to Mg and Ca ions in hardwaters which would otherwise bind to surfactants and form so-called soap scum. Builders also provide buffering, maintain high pH and facilitate surfactant activity. The use of sodium tripolyphosphate (STPP) as a builder in detergents dates back to the late 1940s. Some formulations of detergents included as much as 60% by weight of STPP. However, since about 1980, other builders including nitriloacetic acid (NTA), Zeolite A, and polycarboxylate have been introduced (see Dwyer et al., 1990; Litke, 1999; Köhler 2001 for more details). Other ingredients in detergent formulations include whitening agents, enzymes, perfumes, etc., all of which are present in small quantities.

In this chapter, we are mainly concerned with the use of STPP as builders in detergent formulations because phosphates are a contributing factor for eutrophication of aquatic environments. Introduction of sodium tripolyphosphate in detergent formulations resulted in a much superior performance of detergents over soap-based products. Consequently, the demand for synthetic detergents grew very rapidly both in Europe and in North America (Kohler, 2001; Litke, 1999). Whereas, in 1940, phosphorus consumption for the manufacture of synthetic detergents in the USA was negligible; its use grew rapidly to about 220,000 metric tons by 1967. A similar increase in phosphate-based detergents occurred in Canada, including the Lake Erie basin. It has been estimated (Sly, 1976) that in the year 1966-67, phosphorus loading from detergents for Lake Erie was about 11,980 metric tonnes, constituting between 50 and 70% of the phosphorus loading from municipal and industrial wastes. Added nutrients from non-point sources as well as municipal wastes, including phosphates from detergents, became a contributing factor in eutrophication of the western basin of Lake Erie. As briefly discussed below, it was regulations on phosphorus loadings to the lake that initiated a reversal of the process of eutrophication, or oligotrophication.

LAKE ERIE

As pointed out by Charlton (in Chambers et al., 2001), Lake Erie is one of the most widely recognized examples of restoration of an aquatic ecosystem that had been damaged by eutrophication and exemplifies the recovery of a "dead lake". Lake Erie also exemplifies that even a very large system can get impacted by a constant input of nutrients. In addition, there is an extensive published literature on Lake Erie, not only since the time that the impact of eutrophication on this lake became apparent but from even earlier periods, and for the last three decades or so since the implementation of various measures for the recovery of the lake. A bibliography published in 1974 by the United States Army Corps of Engineers (Leach, 1999) listed a total of 3,627 papers dealing with biological, physico-chemical, engineering and socio-economic aspects of Lake Erie. Many new studies—including several books—have been added in the last thirty years (see, for example Burns, 1985; Weller, 1990; Munawar et al., 1999, McGucken, 2000). Since the southern part of Lake Erie lies in the USA and the northern part in Canada, recovery of Lake Erie also provides an excellent example of the importance of international cooperation in the management of environmental problems. Only a very brief review on Lake Erie is provided here and for a more detailed version, the reader is referred to Kaushik et al., (2001).

The Great Lakes of North America, (Lakes Superior, Michigan, Huron, Erie and Ontario) together cover an area of 244,000 km^2 and include 23,000 km^3 of freshwater, which is about 18% of the world supply, and comprise the largest system of freshwater on earth (Government of Canada and US EPA, 1995). Measuring 388 km in length from east to west, and 92 km in breadth from north to south, Lake Erie is twelfth in size among the world's largest lakes. The lake is divided into Western, Central and Eastern Basins. Lake Erie is the shallowest of the Great lakes and its depth is about 11 m, 25 m and 64 m in the western, central and eastern basins, respectively. The first documented sighting of Lake Erie by a known European was in 1669 and of Niagara Falls in 1678 (Sly, 1976). However, only after 1800 did the Europeans began to settle in the lake basin and the population rapidly increased after 1850. By 1890, the population in the Lake Erie basin was 2.7 million which rose to 5.2 million by 1920, 8.5 million by 1950, and 10.5 and 13.0 million by 1960 and 1980, respectively. The Lake Erie basin showed the greatest population growth among all the Great Lakes of North America. With the

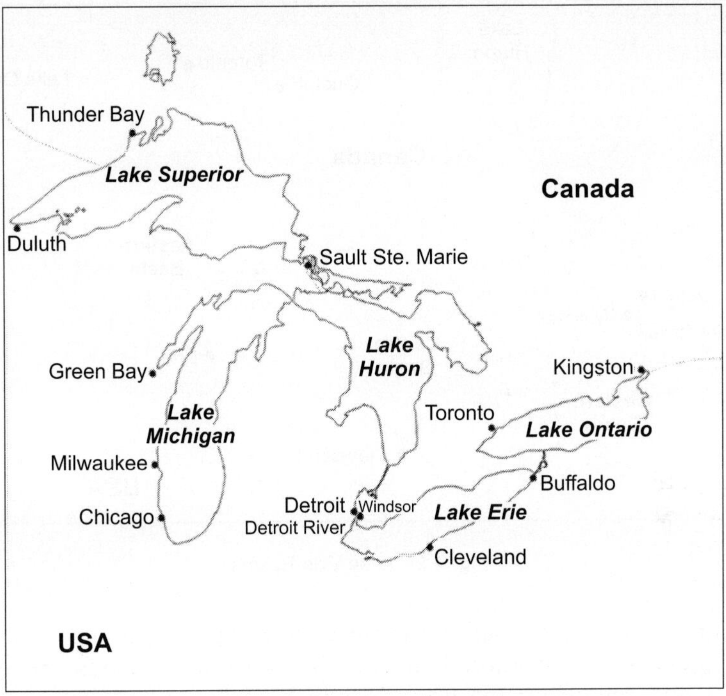

Fig. 4.1 The Great Lakes

development of industry, especially in the western basin, growth of population and extensive agricultural activities, the Lake Erie basin changed drastically. This rapid change, especially during the four decades from 1920 to 1960, lead to dramatic increases in inputs of organic wastes, sediments and nutrients, and transformed the western basin of this large water body into a highly eutrophic state. This, in turn, lead to a variety of significant environmental problems.

EUTROPHICATION OF LAKE ERIE

Beeton (1969) has collated the available data for the period 1910 to 1965 for the concentrations of dissolved solids and various ions in Lake Erie waters. Total dissolved solids increased by 56 ppm, reaching a value of about 200 ppm by 1965. Corresponding increases in the values for calcium, chloride, sodium plus potassium, and sulfites were 8, 16, 5 and 12 ppm. Most of these increases started from about 1930, paralleling the increase in population and various anthropogenic activities.

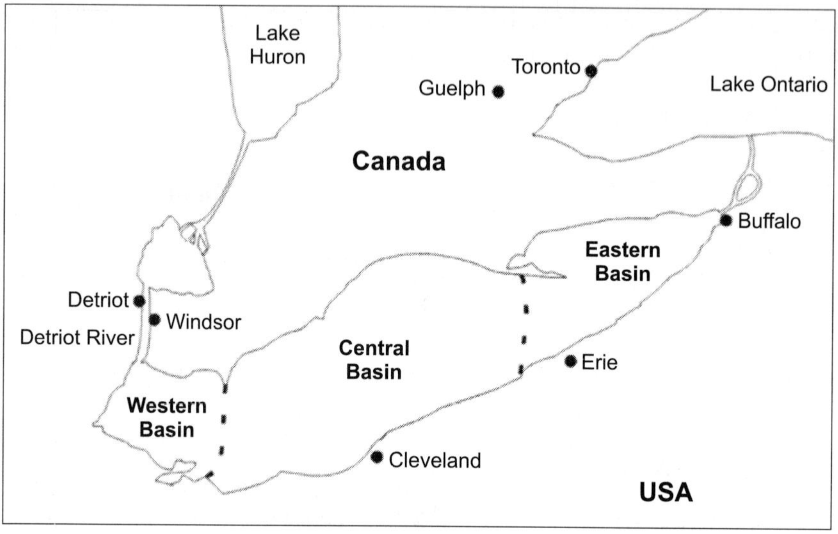

Fig. 4.2 Lake Erie Basins

Earlier data on the loadings of nitrogen and phosphorus to Lake Erie have not been considered useful for various reasons including uncertainty regarding the units in which earlier data were recorded (Beeton, 1969; Sly, 1976; Burns, 1985). Based on a few studies on the open lakewater of the western basin, ammonia increased fivefold and total nitrogen increased threefold from 1930 to 1958 (Beeton, 1969). Total nitrogen loadings to Lake Erie in 1966-67 was 160,969 metric tonne (International Joint Commission, 1975), which increased to 292,000 metric tons by 1976. The use of nitrogen fertilizers was regarded as the main cause (Burns, 1985) for this increase. Sly (1976) also estimated external phosphorus loadings to Lake Erie based on estimates of: (i) non-municipal loading (agricultural and industrial); (ii) municipal (human waste); and (iii) municipal (detergent) for the year 1966-1967. Loading from detergents was estimated on the basis of per capita use of detergents by the Canadian and US population in the Erie basin. Input for agriculture and other sources was derived from a report by the International Joint Commission. Sly (1976) concluded that phosphorus loadings for Lake Erie for 1966-67 from human waste was of the order of 6030 metric tonne, that from detergents was 11980 metric tonne, and from agriculture and other sources was 9300 metric tonne.

On a similar basis, Sly (1976) calculated loadings for years 1900 to 1970. He concluded that there was a gradual increase in phosphorus loading since 1900, corresponding to the population increase. However, beginning by about 1940 (when detergents came into use) phosphates from detergents rapidly changed the rate of increase and by 1970, about half the phosphorus loading was estimated to be derived from detergents. Sly (1976) has cited other studies that analyzed the sediment cores collected from Lake Erie. Results from these studies show that concentrations of organic carbon and total nitrogen continued to increase from about the mid-1700s to the 1950s. Total phosphorus content showed a similar trend up until the 1940s and then exhibited a sharp increase. By using published data pertaining to flows of major streams and rivers into and out of Lake Erie, Burns (1976) developed a water budget for the lake for 1970. Based on continuous flow data for various rivers and phosphorus concentration data, Burns determined a phosphorus budget for 1970 (Table 4.1).

Table 4.1 Phosphorus loading (metric tonne of P) for Lake Erie in the year 1970

Western Basin: Detroit River input (including flow from Lake Huron)	18,075
Western basin rivers, discharge and precipitation	5,723
Central basin rivers, discharges and precipitation	5,769
Eastern basin rivers, discharges and precipitation	2,494
Total input	32,061
Total output	3,848
Phosphorus net loading	28,218
Retention within the lake	88.0%

Source: Burns (1976)

The western basin received 56% of the total phosphorus loading for 1970 from the Detroit River, which, in addition to agricultural load from its watershed and sewage from the cities of Windsor and Detroit, also brings the outflow from Lake Huron. Thus, the western basin of the lake receives the bulk (about 74%) of the total phosphorus load and pollutants to Lake Erie.

Burns (1985) also collated reliable published data for the concentration of total phosphorus in the west basin. Values for various years in $\mu g/l$ were: 1942- 14.7; 1958- 33; 1959- 36; 1968- 50 and 1970- 50.9. These results clearly show more than a threefold increase in total phosphorus concentrations in the west basin from 1942 to 1970. In another study, Gächter et al. (1974) measured phosphorus concentrations at 25 stations distributed on the entire lake every month from April to December 1970. Total phosphorus concentrations showed a west-east gradient with the highest concentration in the western basin, where values were normally higher than 80 μg PO_4 /l. In the southern half of the western basin, the values ranged between 100 and 200 μg PO_4 /l, and the highest value was 450 μg PO_4 /l.

IMPACT OF EUTROPHICATION OF LAKE ERIE

Phytoplankton and Cladophora

Increased loading of nutrients to Lake Erie promoted algal growth. Although some of the earlier studies from the 1940s and 1950s (see Vallentyne, 1974; Burns, 1985) indicated that phytoplankton density in Lake Erie waters increased over a period of time, most of these studies suffered from one or more drawbacks including: limited duration of study, use of techniques that were not comparable, failure to sample nanoplankton, and data on biomass were lacking and were, therefore, not considered to be sufficiently comprehensive to accurately assess the issue of nutrient pollution (Vollenweider et al., 1974; Munawar & Munawar, 1976). A comprehensive lake-wide study that covered the western, central and eastern basins at offshore and inshore locations was carried out from April to December 1970. Samples were collected on 10 cruises and used to determine phytoplankton species composition, chlorophyll a, rate of primary production, and total biomass of algae (see Glooshenko et al., 1974a, 1974b; Vollenweider et al., 1974; Munawar & Munawar 1976). In brief, Munawar & Munawar (1976), who had also cited and listed earlier studies from the 1940s and 1950s, concluded that in the western basin of Lake Erie, floristic change paralleled eutrophication trends. Algal species like *Asterionella formosa, Tabellaria fenestrata* and *Melosira ambigua* that dominated before 1950 were insignificant in 1970, while eutrophic indicators like *Melosira binderana* and *Stephanodiscus* were frequent, and other eutrophic species such as *Aphanizomenon flos-aquae* and *Anabaena spiroides* were frequent during the summer in the western basin. As expected, chlorophyll a concentrations (Glooshenko et al., 1974b) were higher in the shallow western basin than in the central basin, especially during the summer months (Table 4.2). Higher values were also recorded on the south shore near large cities.

The results of the 1970 lake-wide study showed that between the period April and December, primary production in the western basin was 310 g C/m^3, as compared to 210 g C/m^3 in the central and 160g C/m^3 in the eastern basin. Average annual primary production in the western basin was higher (40 mg C/m^3 per hour) as compared to primary production (5 mg C/m^3 per hour) at the boundary of the central and eastern basins (Glooshenko et al. 1974a). Assimilation number is the primary production (i.e. the amount of carbon fixed) per unit weight of chlorophyll a (uncorrected) and is expressed as mg C/mg chlorophyll a/hour. Highest assimilation numbers were recorded (Glooshenko et al., 1974a) from the western basin, with mean values for the period April to December ranging from 2.5 to 3.0 mg C/mg chlorophyll a/hour.

Table 4.2 Average chlorophyll a (μg/l) in Lake Erie Basins, 1970

Cruise	*Eastern Basin*	*Central Basin*	*Western Basin*
April 7-11	4.1	5.3	5.9
May 6-10	3.5	2.9	4.4
June 2-6	5.4	2.7	5.9
July 3-7	1.5	2.5	11.9
July 28-Aug. 1	4.4	3.7	9.8
Aug. 25-29	2.7	4.1	19.3
Sept. 23-27	4.7	9.2	10.2
Oct. 21-25	1.4	3.0	7.8
Nov. 25-30	2.7	5.4	7.6
Dec. 14-18	2.6	5.7	3.3

Source: Glooshenko et al., 1974b

Corresponding values for the remaining lake were between 1.5 and 2.0 units, except near Cleveland, where it was 2.5. Glooshenko et al. (1974a) have cited references that show higher assimilation numbers for waters rich in nutrients. In Lake Erie, the gradient of these numbers correlated with the nutrient distribution. Munawar & Munawar (1976) reported on the phytoplankton biomass of Lake Erie based on a lake-wide study from April to December 1970. As expected, phytoplankton biomass values in the western basin during these months were higher and ranged between 0.8 and 13.2 g/m^3 compared to the values between 0.6 and 6.0, and between 1.0 to 4.2 g/m^3 for the central and eastern basins, respectively.

In summary, the western basin of Lake Erie, in comparison to the other two basins, had phytoplankton taxa that were characteristic of eutrophic waters, had higher concentrations of chlorophyll a, as also higher primary production, assimilation number, and phytoplankton biomass. As mentioned by Munawar & Munawar (1976), Vollenweider (1968) has suggested that aquatic environments with phytoplankton biomass greater than 10 g/m^3 should be regarded as highly eutrophic. Around 1970, nutrient loading to Lake Erie was high and the western basin was widely regarded as being highly eutrophic. Based on Vollenweider's classification, lakes with 3 to 5 mg/m^3 phytoplankton biomass are mesotrophic, and thus the eastern basin belonged in this category. The central basin fell between these two groups.

The preceding discussion was predominantly related to phytoplankton in open waters. Depending upon the nature of inshore areas, their depth, dilution factor, proximity to the source of nutrient loading , etc., the greatest impact can be on inshore regions. Munawar & Munawar (1976) have pointed out that on the basis of algal biomass, which ranged between 2 and 6 g/m^3 in offshore locations, central and eastern basins were not considered to be eutrophic. However, for the inshore locations of the central basin maximum value for algal biomass was 12 g/m^3 and for the eastern inshore locations 10 g/m^3. On this basis, the inshore locations of the central and eastern basins could be regarded as highly eutrophic. These areas, with submerged rocky substrates and rich in nutrients, also promote the growth of filamentous algae. One such alga, *Cladophora*, forms long strands of filaments. In normal aquatic ecosystems, *Cladophora* plays an important role by providing primary production as well as a shelter to aquatic organisms like amphipods, crayfish, molluscs, etc. These organisms may serve as food for the small fish that feed near the shore. However, under extreme eutrophic conditions, the rocky shorelines of Lake Erie were covered with thick growth of *Cladophora*. This created havoc for lakeshore residents and beach users. This aspect has been discussed by a number of authors, for example, by Ashworth (1986) in his book the "Late, Great Lakes" stated that as *Cladophora* grew profusely in Lake Erie in the 1950s and 1960s, "hundreds of square miles of it formed great mats upto two feet thick on the lake's surface; pieces of mats thousands of acres in size broke free and washed up on the beaches. Water intakes were fouled, drinking water supplies took on the distinctive disgusting taste of rotting vegetation." Similarly, Burns (1985) quotes from other authors: "I wish I could inundate you with pictures; pictures of deserted bathing beaches covered with layers of rotting *Cladophora* instead of bathers; pictures of bulldozers pushing *Cladophora* in mountains, pictures of front-end loaders scooping *Cladophora* and loading trucks for disposal elsewhere or burial on the beach..." It is obvious that the excessive growth of *Cladophora* due to eutrophication caused massive economic problems. Stench arising from dead and decomposing *Cladophora* was unbearable, beaches were deserted, and the tourism and fishing industries were profoundly affected. As a result of coverage of the problem in the popular press, *Cladophora* became a household term and, in the long run, this publicity helped the cause of eutrophication control.

Hypoxic and Anoxic Conditions

Increased algal and other plant biomass as well as that of other organisms in eutrophic waters leads to an increase in the demand for oxygen for respiration. Oxygen consumption also increases because of microbial activity involved in the decomposition of large quantities of dead organic matter resulting from dead algae and other plants. This increased oxygen consumption leads to hypoxia, i.e. when dissolved oxygen levels in water are less or equal to 2 mg/l. In extreme situations, all available oxygen may be consumed and the water becomes anoxic. In temperate lakes, such situations can arise in the hypolimnion during lake stratification because, at that time, there is no mixing of waters from the epilimnion to hypolimnion and so, diffusion of oxygen from surface layers to deeper zones is not possible. Anoxic and hypoxic conditions are not suitable for fish and other aquatic life in hypolimnetic water and result in mass mortality.

Based on an earlier work, Dobson & Gilbertson (1972), showed the gradual decline of dissolved oxygen in bottom waters of Lake Erie. They concluded that dissolved oxygen content in bottom waters of the central basin of Lake Erie in the beginning of June in 1929, 1949 and 1969 was about 10 mg/l. Based on inferred trends for depletion of dissolved oxygen for 1929 and 1949, it was estimated that by the end of September of these three years, 4 mg/l of oxygen was still available in 1929 and 2 mg/l in 1949. The trend for 1969, however, indicated complete anoxic conditions by the end of August, a time that is much earlier than the fall overturn (generally around mid-late September) when, because of decreasing temperatures, stratification breaks down and waters from the epilimnion and the hypolimnion mix.

Dobson & Gilbertson (1972) also calculated that during the summer, the mean depletion rate for dissolved oxygen for water of the hypolimnion of the central basin of Lake Erie in 1929 was 1.6 mg/l/month; by 1961, it was more than 3.0 and by 1967, almost 4.0 mg/l/month. Since the duration of summer stratification is about 110 days and dissolved oxygen in the hypolimnion at the onset of stratification is about 10 mg/l, a depletion rate above 3 mg/l/month was regarded as critical because, at this rate, hypolimnion becomes anoxic well before the onset of the fall overturn.

A detailed investigation of the hypolimnion of the central basin of Lake Erie was carried out under the name "Project Hypo" and the results were reported by Burns & Ross (1972) and by the International Joint Commission (1975). It was estimated (Table 4.3) that whereas in 1930 only 300 km^2, comprising 3.0% of hypolimnion, was anoxic during summer, by 1973 the anoxic area was 11,270 km^2 or about 93.7% of the hypolimnion (IJC, 1975).

It is well known that under anoxic conditions, phosphorus adsorbed onto the sediment is released back into the water. Thus, anoxia helps in the regeneration of phosphorus. Based on the duration of anoxia and area under anoxic conditions as well as other data available from the Canada Centre of Inland Waters, Burlington (Ontario), Burns & Ross (1972) estimated the extent of external and internal (from within the lake) loading of phosphorus in the central basin for the two months of anoxic conditions for the summer of 1970. The values were as stated below:

Table 4.3 Estimated anoxic area of hypolimnion in the central basin of Lake Erie for various years

Year	Anoxic Area (km^2)	*Percentage of Central Basin*	
		Hypolimnion	*Total*
1930	300	3.0	1.9
1959	3,600	33.0	22.3
1960	1,660	15.0	10.3
1961	3,640	33.0	22.5
1964	5,870	53.0	36.3
1967	7,500	68.0	46.4
1970	6,600	60.0	40.4
1972	7,970	72.5	49.3
1973	11,270	93.7	69.8
1974	10,250	87.0	63.4

Source: International Joint Commission (1975).

Table 4.4 Values of external and internal loading of phosphorus in central basin in 1970

External P loading (2 months):	1720 metric tonne
Internal P loading (2 months, area under oxic conditions):	437 metric tonne
Internal P loading (2 months, area under anoxic conditions):	1914 metric tonne
Total loading:	4071 metric tonne

These values indicate that during summer stratification, the central basin was not acting as a settling basin but rather, as a basin for regeneration of phosphorus. After the turnover and mixing of waters, this regenerated phosphorus promotes algal blooms. Thus, in order to eliminate internal loading of phosphorus resulting from anoxic conditions, the hypolimnion must be kept oxygenated. This can be achieved only if production of organic matter, i.e. algal biomass, is reduced by controlling the external inputs of phosphorus. On this basis, recommendations were made to the International Joint Commission (IJC) to reduce phosphorus inputs to the lake.

We have already seen that the eutrophication of Lake Erie, especially the western basin, caused many economic and aesthetic problems because of excessive production of phytoplankton and *Cladophora*. This excessive organic matter production was also responsible for hypoxia and anoxia in vast areas of the hypolimnion of the western and central basins. Anoxic conditions in bottom waters had direct biological effects. Important benthic macroinvertebrates that served as food for fish (e.g. burrowing mayflies, *Hexagenia limbata* and *H. rigida*) were eliminated (see references in Ludsin et al., 2001). Anoxic conditions in the hypolimnion also made this area unsuitable for cold-water fish of the central basin.

This resulted in a decline of some highly valued commercial and sport fishes (e.g. lake whitefish, *Coregonus clupeaformis,* and burbot, *Lota lota*) in Lake Erie by the 1960s (see references in Ludsin et al., 2001).

PUBLIC OUTCRY AND PHOSPHORUS REGULATIONS

In the 1960s, pollution of Lake Erie attracted public attention in both Canada and the USA, and there was a tremendous public outcry (see McGucken, 2000 for details). As early as 1963, the New York Times had reported the impact of phosphates and nitrates that entered Lake Erie from Detroit City and from the neighbouring communities. In 1964, the League of Women voters in the US published a report entitled "Pollution, the Other Side of Progress". In 1965, Newsweek published an article entitled "Great Lakes: The Dead Sea". Covering a conference on Lake Erie, the Time magazine in 1965 used the title "Dying Lake". Citizens from Ohio started a campaign to "Save Lake Erie Now" and also sent a signed petition to the Governor of the State. Similarly, in Canada, various groups contacted official government departments, newspaper articles were written and Society to Overcome Pollution (STOP) and another group "Pollution Probe" was established. In short, around the mid 1960s, in both countries, many newspapers, national magazines and television programs were inundated with articles, editorials, news and educational features on pollution in Lake Erie. Such terms as "giant cesspool" and "Dead Sea" were commonly used (see McGucken, 2000). The public outcry, both in Canada and the USA led to actions undertaken by various government organizations in both the countries.

The governments of Canada and the USA had signed a Boundary Waters Treaty in 1909. Under this treaty, the International Joint Commission (IJC) was created in 1911 to resolve problems pertaining to boundary waters. On October 7, 1964, the Governments of the United States and Canada requested the IJC to inquire and report, among other aspects, on whether or not the waters of Lake Erie were being polluted, the extent of pollution, its causes, and, if required, to suggest remedial measures (for more details see Vallentyne, 1974 and McGucken, 2000). The IJC submitted two interim reports (December 1965 and June 1968) so that the two governments could initiate steps to control the extent of pollution. A summary of the final report was released on October 8, 1969, followed by the release of a detailed report in 1970. In all, 19 specific recommendations dealing with various aspects of pollution in the lower Great Lakes were made, but to stop the growing trends of cultural eutrophication, the following recommendations were made to control phosphorus inputs (Vallentyne, 1974): (a) phosphates in detergents be immediately reduced to minimum practical levels with total replacement by environmentally less harmful compounds by 1972; (b) the remaining phosphates in municipal sewage effluents be reduced by not less than 80% prior to specific dates from 1972 to 1978 for different parts of the basins, with continued reduction thereafter to the maximum extent economically feasible; (c) programs be developed for the reduction of phosphorus inputs to water from agricultural sources; and (d) any new significant changes in the additions of phosphorus to water in the basins be regulated. Although both N and P were regarded as the most important nutrients responsible for eutrophication in the lower Great Lakes, it was considered easier to control phosphorus input because the bulk of the inputs (Table 4.4) of phosphorus in Lake Erie was from point sources (sewage) compared to nitrogen inputs that were mainly from non-point sources.

Table 4.5 Inputs of total phosphorus to Lake Erie in 1967 in Canada and the US (thousand metric tonnes).

Source	*US*	*Canada*	*Total*
Lake Huron			2
Municipal	16.2	1.1	17.3
Industrial	0.9	1.0	1.9
Lake drainage	4.4	1.8	6.2
Total			27.4

Source: Modified from McGucken, 2000

In addition, elimination of phosphates from detergents would remove about 50% of phosphates from the municipal waste and it was estimated that this would save $22 million as an annual cost of chemicals used in waste treatment to remove phosphates (Vallentyne, 1974).

As expected, the US Soap and Detergent Association and presidents of the Canadian branches of Proctor and Gamble, Lever Brothers, and of Colgate-Palmolive vehemently opposed controls on phosphates in detergents (for details see Vallentyne, 1974 and McGucken, 2000). But, after having considered various reports, the Government of Canada, nevertheless, announced a proposal to limit the amount of phosphate in detergents. By August 1, 1970, a regulation was in effect under the Canada Water Act which set a maximum concentration rate of 8.7% phosphorus (or 20% phosphate as phosphorus pentaoxide, P_2O_5). This was the first regulatory action taken in North America to control phosphate content in detergents. A new regulation came into effect on January 1, 1973, in which the maximum permitted phosphorus content was only 2.2% (or 5% P_2O_5) (see Vallentyne, 1974 and McGucken, 2000 for details).

In order to control eutrophication and maintain water quality in the lower Great Lakes, the Canada US Great Lakes Water Quality Agreement (GLWQA) was signed on April 15, 1972. Among other aspects, objectives of GLWQA included control of phosphorus loadings and to restore year-round aerobic conditions in the hypolimnion of the Central Basin. Based on modeling exercises, it had been predicted that in order to have a reasonable primary production and yet maintain aerobic conditions in the hypolimnion of the Central Basin, the phosphorus load from all sources (point and non-point sources, input from Lake Huron, and atmospheric inputs) would have to be reduced from the existing load of about 28,000 metric tonnes to about 11,000 metric tonnes. This entailed, in addition to regulation of phosphates in detergents, controls on phosphorus content from sewage treatment plants in the Lake Erie basin. Accordingly, sewage treatment plants with flows in excess of 38,000 m^3/day (3.8 million liters per day) were required to limit the total phosphorus content in the effluent to an average of 1.0 mg/l on an annual basis. To achieve this, facilities at the existing sewage treatment plants were upgraded and other treatment plants were constructed where none previously existed. In Ontario alone, nearly $1.5 billion was spent on the development of this infrastructure while it is estimated that the US spent $7.8 billion. Other estimates put this figure to $15 billion US (see Dolan, 1993; McGucken, 2000; Chambers et al., 2001).

OLIGOTROPHICATION OF LAKE ERIE

Decrease in Phosphorus Concentration

Decreases in phosphorus load to Lake Erie achieved through regulations on the phosphate content of detergents and through a mandatory phosphorus limit of 1.0 mg/l in effluents from sewage treatment plants had phenomenal effects on Lake Erie. As Charlton (in Chambers et al., 2001) puts it, results provide a dramatic example of the manners in which concerted efforts can reverse environmental change (for details see Munawar et al., 1999; McGucken, 2000). Regulations on phosphates in detergents was in place by January 1973 and by 1981, average phosphorus concentrations in effluent at each sewage treatment facility was less than 1.0 mg/l (Dolan, 1993; McGucken, 2000). Thus, municipal phosphorus loading to Lake Erie from the USA and Canada was reduced from 15,500 metric tonnes in 1972 to less than 2,500 metric tonnes in 1990 (Dolan, 1993). As already mentioned, according to the Great Lake Water Quality Agreement, the target for annual phosphorus load from all the sources was set at 11,000 metric tonnes. During the period 1981-1991, this target was met for all the years except for 1982, 1986 and 1990 (McGucken, 2000).

Obviously, reduced phosphorus loadings to the lake should result in lower phosphorus concentration in the lakewater. However, because of the vast size of this lake and inherent variability in phosphorus concentration, limited sampling and short duration of studies, it was difficult to properly assess the changes in phosphorus concentrations, especially during the initial recovery period. This was further complicated by the fact that in 1988, zebra mussels invaded parts of the Great Lakes. These mussels are prolific filter feeders and remove large quantities of plankton and chlorophyll a from the water column and, thus, have a strong effect on water. It is, therefore, essential to discriminate between the changes brought about as a result of a reduction in phosphorus loading and those resulting because of the zebra mussel. Data for the period 1971 to 1992 (McGucken, 2000) indicate that right until 1976, values for phosphorus concentrations for the western basin were above 40 μg/l but starting in 1977, values were close to 20 μg/l for most years. Similar trends were recorded for the central and eastern basins where concentrations in the early 1970s were around 15 μg/l but only around 10 μg/l for most years after 1976. Charlton et al. (1999) analyzed the integrated water samples taken from the surface to a depth of 10 or 20 meters during May to the end of August 1997 from 60 stations, covering all the basins of Lake Erie. Mean values for eastern and central basins were 9.8 and 9.6 μg/l, respectively, and those from the west central and west basins were 12.1 and 22.6 μg/l, respectively. These data show decreases in phosphorus concentrations compared to values reported in the 1960s and early 1970s. To delineate possible changes because of zebra mussels, Charlton (2001) concluded that phosphorus concentrations were about 20% lower in the east and central basins in 1988 than in 1968, i.e. prior to the introduction of zebra mussels. A more detailed comparison has been made by Charlton et al. (1999) who compared data on the total phosphorus available from the National Water Research Institute, Burlington, Canada, for the years 1968-72, 1984-88 (pre-mussel period) and 1994-96 (post-mussel period). In the west basin, the mean phosphorus concentration for the period 1968-72 was 41.2 μg/l and for the period 1984-88 it was 35.5 μg/l (Table 4.5). Since zebra mussels were not present in significant numbers during the period 1984-88, the decrease in phosphorus concentrations at the end of the period must have resulted from controls on phosphorus loadings. By 1994-96, phosphorus concentration in the west basin had further decreased to 17.7 μg/l. Similar changes were witnessed in the other basins also (Table 4.6).

It is interesting to note here that as phosphorus concentration decreased in all three basins of Lake Erie, the concentration of nitrates increased (Charlton et al., 1999; Charlton, 2001). In the central and eastern basins, the increase in nitrate concentrations from 1968 to 1988 was more

Table 4.6 Changes in total phosphorus concentrations (µg/l) and in chlorophyll a concentrations (µg/l) in Lake Erie between the periods 1968-72, 1986-88 and 1994-96

		Total Phosphorus			*Chlorophyll a*		
Basin	*Period*	*Mean*[x]	*Change*	*P(t-test)*	*Mean*[x]	*Change*	*P(t-test)*
East	1968-72	14.7			4.9		
East	1984-88	11.9	-2.8	0.003	2.8	-2.1	0.00
East	1994-96	8.0	-3.9	<0.001	2.1	-0.7	0.311
Central	1968-72	14.1			5.0		
Central	1984-88	10.9	-3.2	<0.001	2.3	-2.7	<0.001
Central	1994-96	8.4	-2.5	<0.001	1.8	-0.5	<0.001
West	1968-72	41.2			16.1		
West	1984-88	35.5	-5.7	0.003	8.3	-7.5	<0.001
West	1994-96	17.7	-17.8	<0.001	5.6	-2.7	<0.001

[x]Mean of June, July and August Values
Source: Charlton et al., (1999)

than 300%. This was believed to have resulted from the decrease in algal biomass and, hence, the demand for nitrogen.

Recovery of Phytoplankton and Decrease in Primary Production

These aspects have been touched upon in brief and, for more details, the papers cited here could be referred to. Results of a very extensive study on phytoplankton collected during spring, summer and autumn periods of 1983 to 1987 from offshore locations in all the three basins of the Lake Erie have been reported by Makarewicz (1993a). Phytoplankton were identified to the species level and the data were compared with those available from earlier years. In general, the occurrence of common species in 1970 and 1982-87 samples was similar. However, biomass of many nuisance and eutrophic species, e.g. *Stephenodiscus tenuis* and *S. binderanus* (diatoms), *Aphanizomenon flos-aquae* (blue green) and *Rhodomonas minuta* (Cryptophyta) had decreased by 70% to 98% in 1983 to 1987 period as compared to their biomass estimates for 1970 i.e. prior to regulation of phosphorus levels in detergents and municipal wastewater. This change was accompanied with an abundance of mesotrophic species such as *Asterionella formosa* (diatom) and *Rhizosolenia criensis* (diatom) during 1983 to 1987 and their densities were comparable to those during 1930s and 1940s, i.e. prior to the onset of eutrophication in Lake Erie. Additionally, compared to phytoplankton biomass in 1970, there was an overall reduction in phytoplankton biomass in all the three basins. Phytoplankton biomass (mean ± S.E.) in the western, central and eastern basins averaged 1.88 ± 0.12; 1.04 ± 0.075; and 0.63 ± 0.071 g/m^3, respectively, showing a 52% to 89% reduction compared to algal biomass in 1970 (Makarewicz, 1993a). Similar observations were recorded by Munawar & Munawar (1999) based on phytoplankton samples collected from all the three basins during July 1992 and the results compared with those from 1970 and 1978 studies. In addition to the overall trend in increase in number of species in 1978, 1992 when compared to those in 1970, there was also an increase in chlorophyta species. In 1992, chlorophyta biomass in the three basins accounted for 32% to 61% of the total phytoplankton biomass, whereas corresponding values for 1970, i.e. at about the peak of eutrophic conditions, ranged between 4% and 15%. Similar to the observations made by Makarewicz (1993a), a substantial decrease in phytoplankton biomass was also observed by Munawar & Munawar (1999). For example, mean July 1970 phytoplankton biomass for the western basin—the most eutrophic area of the lake—was about 5 g/m^3, whereas in July 1992 it was slightly less than 2 g/m^3. Corresponding changes in chlorophyll a (Table 4.6) have also been reported (Charlton et al., 1999).

It is obvious from the foregoing discussion that as a result of regulation on phosphate content in detergents and in effluents from sewage treatment plants, phosphate input to Lake Erie was considerably reduced, resulting in substantial decreases in phosphate concentration in water. This, in turn, affected species composition of phytoplankton and also reduced phytoplankton biomass in Lake Erie. Prior to the regulation of phosphates, the western basin of Lake Erie was regarded as eutrophic. However, as suggested by Makarewicz (1993a), based on the classification used by Vollenweider in 1968 (ultra-oligotrophic: maximum phytoplankton biomass <1 g/m^3; mesotrophic: 3-5 g/m^3; and highly eutrophic: >10 g/m^3), the western and central basins can now be considered as mesotrophic and the eastern basin as between meso- and oligotrophic.

Hypolimnetic Dissolved Oxygen

As already mentioned, eutrophication of Lake Erie had a serious impact on the hypolimnetic dissolved oxygen levels and during the summer stratification, a large area of hypolimnion in the central basin would become hypoxic or anoxic, which, furthermore, would affect the survival of benthic organisms and fish. A number of authors have reported hypolimnetic dissolved oxygen levels subsequent to the implementation of phosphorus abatement programs for Lake Erie (Bertram, 1993; Charlton, 1993, 1999; McGucken, 2000). Bertram (1993) plotted the dissolved oxygen depletion rates for the central basin of Lake Erie for most of the years from 1970 to 1991. For the years 1988 and 1989, the rates were less than 3.0 mg/l/month, lower than any figure reported during the previous 20 years. For the remaining years, the rates were higher than this value. Bertram also reported values for hypolimnetic dissolved oxygen for late August (prior to the fall overturn) for the years 1987 to 1991. For late August 1989, the mean hypolimnetic dissolved oxygen level was 4 mg/l, whereas for the remaining years, the corresponding values were about 1 mg/l or less. Thus, even after the phosphorus abatement programme for Lake Erie had been in place, hypolimnetic oxygen levels upto the year 1991 in the central basin had not recovered to levels that are essential for the survival of benthic organisms and fish.

Depletion of hypolimnetic oxygen has been discussed in detail by Charlton et al. (1993). Oxygen depletion rates depend upon a number of features such as the thickness of the hypolimnion, temperature, production of organic matter, internal phosphorus cycling, sediment resuspension, etc. A lag time between reductions in phosphorus loadings and changes in the dissolved oxygen depletion rate was expected (Charlton et al., 1993). Since 1989, it has also been suggested that dissolved oxygen depletion rates should not exceed 1.89 mg/l/month if the anoxic area is to be kept at 10% of the hypolimnion (see McGucken, 2000). From 1993 to 1997, dissolved oxygen depletion rates in the central basin were less than 2.10 mg/l/month and dissolved oxygen concentration at the end of August or in mid September ranged between 2 to 6 mg/l. However, in late August 1998, hypolimnetic oxygen concentrations were less than 1 mg/l for most of the central basin (Charlton et al., 1999). Thus, although hypolimnetic oxygen levels indicate a trend towards recovery, the improvement is not consistent.

Recovery of Fauna

A detailed comparison of zooplankton communities and biomass before and during eutrophication of Lake Erie as well as during oligotrophication of the lake has been reported by Makarewicz (1993b) and Johannsson et al. (1999). Mackarewicz (1993b) analyzed zooplankton samples from all three basins collected during 33 cruises covering spring, summer, and autumn periods of 1983 to 1987—a period that covers the oligotrophication of Lake Erie. These data were compared with available data from various studies covering the period 1939 to 1970, i.e. a period spanning before and during eutrophication of the Lake Erie. Mackarewicz (1993b) concluded that the lack of dominance of rotifer species that are regarded as indicators of eutrophic conditions from 1983 to 1987 suggested that Lake Erie as a unit was no longer eutrophic. Abundance of Cladophora during 1983 to 1987, a period of oligotrophication, was comparable to the abundance of this group during

1930s and 1940s, i.e a period prior to eutrophication of the lake. Also, zooplankton abundance and biomass decreased in 1983 to 1987 (except in 1985), this decrease correlated with a corresponding decrease in phytoplankton biomass during oligotrophication of the lake (Makarewicz, 1993b). In another extensive review of the zooplankton of Lake Erie, it was concluded that between 1928-30 and 1970, zooplankton abundance increased dramatically as a result of eutrophication. With reduced phosphorus loadings during 1970s and 1980s resulting from implementation of phosphorus abatement programme, zooplankton biomass declined (Johannasson et al., 1999).

Prior to eutrophication of Lake Erie and resultant anoxia in a vast area of the hypolimnion, nymphs of the mayfly *Hexagenia* spp. were very abundant and their densities generally averaged between 300 to 500/m^2 but in some locations exceeded 950/m^2. In 1940s and early 1950s, such densities on emergence often formed nuisance swarms. However, in 1953, these benthic mayfly nymphs died over a large area of western basin of Lake Erie (see Krieger et al., 1996 for details). This sudden loss was attributed to extensive hypolimnetic anoxia as a result of eutrophication leading to increased primary production. No swarms of mayflies were reported between 1953 and 1973. As we have seen, pollution abatement programmes and regulations on phosphorus loadings in Lake Erie in 1970s has resulted in a decrease in phyplankton production as well as some improvement in the dissolved oxygen levels in hypolimnetic waters during summer stratification. Krieger et al., (1996) have reported that *Hexagenia* have recolonized several nearshore locations and are extending to offshore areas of the lake. This is further substantiated by the number of large swarms of adult *Hexagenia* that have been observed between 1992 and 1996.

Recovery of Lake Erie from the impact of eutrophication has not been confined to phytoplankton and invertebrates only. Oligotrophication, achieved through phosphorus abatement programmes, has also been effective in the restoration of fish communities. An excellent paper on the subject (Ludsin et al., 2001) has examined, among other aspects, the relationship between system productivity and richness of fish species. Ludsin et al., (2001) have listed a number of references to support a recent contention that there exists a unimodal relationship between productivity and species richness, specifically for plants, invertebrates and certain species of vertebrates. Briefly, species richness is higher at low levels of productivity, it peaks at intermediate levels of productivity, and declines as productivity becomes too high. Ludsin et al., (2001) compared data for fish species composition and richness for more productive west basin and less productive central basin of Lake Erie on the basis of fish collections made by the Ohio Department of Natural Resources from fixed points for the period 1969 to 1996. This timespan covers the peak of the eutrophic condition of Lake Erie as well as the period of phosphorus regulation leading to oligotrophication. Their results suggested that phosphorus regulation influenced fish community structure and that system productivity (in terms of annual phosphorus loading) and total species richness were related within each basin. In the productive west basin, phosphorus abatement programmes resulted in declines in the abundance of six species that were tolerant to eutrophy (i.e. tolerant of hypoxia), whereas the abundance of three species that were intolerant of eutrophy increased through a period of time. In contrast, in the less productive central basin, only one tolerant species declined but four species intolerant of eutrophic condition recovered because of oligotrophication. Ludsin et al. (2001) concluded that reduced phosphorus loading to Lake Erie has led to fish community rehabilitation.

The foregoing review on the impact of phosphorus abatement programmes introduced about 30 years ago and included phosphorus reduction in effluents of sewage treatment plants and regulation of phosphate content in detergents—a regulation that was vehemently opposed by detergent industry—has had a very significant impact on Lake Erie. It has resulted in oligotrophication of the lake by decreasing phosphorus concentration in lakewater, leading to decreases in primary production and changes in algal species. This has, in turn, led to improvement in hypolimnetic oxygen levels as well as changes in zooplankton, survival of benthic organisms, and finally rehabilitation of fish communities. Whereas, as we have seen, some 30 years ago, publications on Lake Erie employed such phrases as "dead sea", "giant cesspool", etc., the current literature uses such words as "rehabilitation", "recovery", "lake in transition", "oligotrophication", etc. Mechanisms may have been different but oligotrophication of other temperate lakes has also been achieved, e.g. Lake Washington, USA; Lake Constance, Austria-Germany-Switzerland; Lake Ontario, Canada-USA.; Lake Maggiore, Italy; and Lake Veluwe, the Netherlands (Ludsin et al. 2001).

REGULATION OF PHOSPHATES IN LAUNDRY DETERGENTS IN NORTH AMERICA AND EUROPE

As already mentioned, in order to rehabilitate the waters of Lake Erie, phosphorus abatement programmes were initiated in Canada by mandating that by August 1970, phosphorus content in detergents were not to exceed 8.7% (or 20% phosphate as pentaoxide, P_2O_5). This regulation was modified in 1973 by limiting the phosphorus content to only 2.2% (or 5% P_2O_5). In the USA, five cities in the state of Illinois were the first to regulate phosphorus content (8.7%) in detergents in February 1971. Since then, phosphorus content in detergents have been regualted in 27 states and the District of Columbia (Litke, 1999). The last state to institute regulation was New Hampshire in 1995. In most of these states, the allowable limit for phosphorus is 0.5% by weight. However, as the number of states in the USA that regulated phosphates in detergents increased, manufacturers of detergents were forced to maintain inventories for different detergent products and this did not prove to be cost effective. As a result, by 1994, detergent producers decided to phase out the use of phosphates in domestic laundry detergents. By 1998, the trend in composition of laundry detergents sold in the USA was to use alternative builders or surfactants that were not affected by water hardness (Litke, 1999). The major manufacturers of laundry detergents in Canada are Proctor and Gamble, and Lever Brothers. They control 95% of the market in Canada. Since these manufacturers have eliminated phosphorus from laundry detergents sold in North America, 95% of the 150,000 tonne of laundry detergents used in Canada in 1996 were phosphate free even though regulations allow an amount upto 2.2% phosphates (Chambers et al., 2001 p. 23)

Many other countries in Europe have regulated the use of STPP in laundry detergents. Their use was banned in Switzerland and Austria and limited to small quantities in other European countries either by law or through voluntary agreements with the industry. In most European countries, and in some EU Accession countries, minimum levels of phosphates are used in detergents as a result of voluntary agreements (Köhler, 2001). Because of aggressive advertising for certain phosphate-free detergent brands and the perception of consumers that phosphates damage the environment, the use of STPP has now been abandoned by some manufactures. Consequently, laundry detergents that contain STPP are no longer sold in Germany, Italy, Switzerland, Austria and Norway in Europe as well as in Japan. It may be mentioned here that in Sweden, STPP use in laundry detergents is regulated. However, since Sweden now has a comprehensive system to remove phosphorus from waste waters, unrestricted use of STPP in detergents has again been allowed.

EUTROPHICATION AND DETERGENTS IN INDIA

Although the following discussion is primarily based on data available for aquatic environments in India, the main conclusions are applicable to tropical aquatic environments in general. Before we consider the role of detergent phosphates in India or tropical environments in general, it is essential to review the limonlogical aspects and conditions of tropical waters. As pointed out by Thornton (1987), more than a total of 100 countries from a larger part of African continent; large portions of Asia, Australia and South America are located in tropical and subtropical environments. The main climatic difference between temperate and tropical areas is of higher mean annual temperature in tropical areas, approximately 25°C in the tropics compared to about 10-15°C in temperate zones. Also, the annual temperature range in tropics is more limited. Higher temperatures affect the rate of various biological and limnological processes. In addition, most tropical countries fall in the category of "developing countries" with limited financial resources, and are also affected by population pressures. All these factors also have a bearing on eutrophication and its management.

Limnological Aspects in Tropical Regions

After an extensive survey of published literature on eutrophication and its management in tropical and subtropical regions, Thornton (1987) has discussed the modifying effects of climate, seasonality and hydrology on the eutrophication process in the tropics. He also summarized trophic state indices for temperate and tropical lakes (Table 4.7). In general, he concluded that the process of eurtrophication in tropical regions is mediated by many of the same factors. A total phosphorus concentration of 50-60 μg l^{-1} has been suggested as a boundary between mesotrophic and eutrophic tropical lakes.

Population Pressure

Population is a major factor in cultural eutrophication in tropical areas. In discussing the "causes, historical development, effect and future challenges of a common environmental problems; eutrophication", de Jonge et al. (2002) have briefly discussed the increase in world

Table 4.7 Comparison of trophic state indices for temperate and tropical lakes

Indicator	*Upper limit of Mesotrophy*	
	Temperate Lakes	*Tropical Lakes*
Mean primary productivity $(gC/m^{-3}d^{-1})^{-3}$	1.0	2-3
Chlorophyll a (mg m^{-3})	10-15	10-15
Total P (μg L^{-1})	30	50-60
Total N (mg m^{-3})	0.5-1.0	0.2-1.0
Photosynthetic limitation by	P	N
Dominant algal family	Baccilariophyceae	Cyanophyceac

Source: Thornton (1987)

populations. Population growth rate during the period 1950 to 2000 AD was over 350 fold, as compared to that in the period between 0 and 1500 AD. Only in 1960, the world human population was about three billion and by the year 2000, it had grown to nearly six billion. It is expected to touch the 8 billion mark within the next 30 years.

In the same vein, while discussing river conservation in the Indian subcontinent (Gopal, 2000) has alluded to the human impacts. With continuous human habitation for more than 5000 years, and with nearly 1.35 billion people, the Indian subcontinent accounts for 28% of the world's human population. In the last five decades the population in India has increased from 361 million to 1025 million; that in Pakistan from 40 million to 155 million; and in Bangladesh from 45.5 million to 132 million. Obviously, the population density in these countries has also tripled in the last five decades with about 900 persons km^{-2}, Bangladesh has the highest density, followed by India and Pakistan with about 300 and 175 persons km^{-2}, respectively. In places like Delhi, population density is greater than 8000 persons km^{-2}.

Conditions of Aquatic Environments

Large rivers and other flowing waters are major freshwater systems in the Indian subcontinent. Except for some lakes in the hilly areas, most waterbodies in India can be regarded as meso-or eutrophic. Even the larger rivers have become increasingly polluted from both point and non-point inputs (Gopal & Sab, 1993). Based on designated best use and related water quality criteria, five classes (A, B, C, D, and E; A being the best quality) have been recognized in India (Gopal, 2000). Water pollution has continued to increase with the result that rivers that were polluted over a few short stretches are now polluted upto their headwaters. Based on information from the Ministry of Environment and Forest for 1994, Gopal (2000) has listed 13 rivers that have grossly polluted stretches where water is fit only for irrigation or waste disposal. Thus, several sections of most rivers are now classified as 'E'. In River Yamuna upstream of Delhi, PO_4-P levels in 1958-59 were 0.1 mg l^{-1}; in 1989 the value had increased to 0.04 to 0.64 mg l^{-1}. Corresponding values for river stretch downstream from Delhi were 0-0.2 mg l^{-1} in 1958-59 but 0.08-2.35 mg l^{-1} in 1989. This was primarily because of input of sewage from Delhi city. The rivers in all other countries of the subcontinent are similarly polluted (Gopal, 2000). Many lakes and other water bodies are surrounded by urban sprawls and receive wastewater generated by cities. Similar situations pertain to many small lakes and ponds which are also eutrophic (Gopal, 1994; Kaushik et al., 2001).

Production of Laundry Detergents in India

Unlike in North America and Europe,where detergent production is controlled by a few multinational and large companies, manufacture of soaps and detergents in India involves not only large companies and multinationals but also small-scale indigenous producers. This is further complicated because of ambiguity as to what constitutes a synthetic detergent. As a result, it becomes difficult to obtain reliable data on total production, sale and composition of detergents. In 1982, the consumption of soaps in India was estimated at 940,000 metric tonne and that of detergents at 169,000 metric tonne or about 16% of soaps. However, in the last two decades, detergent use has considerably increased and according to the Centre for Monitoring the Indian Economy (CMIE), production of domestic "detergents and scourers" in 1996-97 was of the order of 0.99 million tonne. (Southey et al., 2001). In contrast, Indian Soap and Toiletries Association estimated the total production of detergents at 2.2 million tonne for 1997. The difference between the figures provided by CMIE and those from the industry is attributed to the fact that the industry includes larger estimates of localized products (Southey et al., 2001). A magazine published by VOICE (Voluntary Organization in Interest of Consumer Education) for consumer awareness (Consumer Voice 2001), has reported on the assessment of 14 leading brands of detergents for their performance on various aspects of laundry washing. The total production of detergents was estimated by VOICE at 2.7 million metric tonne.

Southey et al. (2001) have also commented on the per capita consumption of detergents in India, which in 1996 was 2.8 kg per year. This was considered very low compared to 5-6 kg per year in other developing countries like Pakistan, the Phillippines and Nigeria and much lower than the values for European (12-20 kg) or the US (28 kg) consumers. The detergent industry has projected that by 2005, per capita consumption in India would have risen to 4 kg per year.

Southey et al. (2001) have also cited figures obtained from CMIE for the share of detergent market by volume for the year 1997. Hindustan Lever controlled 30% and Nirma 35% of the market and the remaining 35% was shared by small-scale producers. There seems to be a little change in the market share of detergents reported by VOICE (2001). Hindastan Lever controlled 28%, Nirma 25% of the total 2.7 million tonnes of detergents produced. Other brands controlled 47% of the market of the total 2.7 million tonnes of detergents produced.

Phosphates in Detergents

Because of a large number of brands of detergents that are available in India, including those manufactured by small producers, it is difficult to gather information on the phosphate content of detergents. Southey et al., (2001) have quoted results from other studies on the STPP contents of six major brands of detergents and for ten brands of detergent bars that are used in India. For the four products that had higher levels of STPP, values ranged between 22% and 32% STPP by weight but for products manufactured by smaller producers who may use other builders like soda ash, values for STPP were only 5% or less. STPP content in eight leading detergent powders and four cakes has been reported by Trivedi et al. (2001). Four brands of detergent powders had STPP content ranging between 7 and 13%. For the cakes, STPP content ranged between 11 and 15%. VOICE, the non-government organization that had assessed performance of 14 leading brands of detergents in India, has also reported the STPP content of these brands in a magazine (Down to Earth, June 2001). Values for the top four brands in STPP content were 30.6%, 22.4%, 21% and 13.8% STPP by weight. These four brands included products from companies that control about 53% of the detergent market in India.

Based on data provided by the Centre for Monitoring Indian Economy, Southey et al. (2001) have assessed the growth rate of soap and detergent industry and also the output of STPP used as builder from the year 1993/94 to 1997/98. Average annual growth in synthetic detergent and scourer output during the period was nearly 10%. With an average growth in output at 12.5% for those years, the output of detergent STPP grew faster in most years than the output of soaps and detergents, which indicates an increase in the STPP content in products. This was attributed to the increased use of washing machines, which require high quality detergents. During the period 1993 to 1998, washing machine production in India increased by a remarkable 318%. Similar observations have been made for China, where ownership of washing machines in urban areas increased from 48.3% in 1985 to 78.4% in 1990 and rose to 88.9% in 1995.

With so many producers and brands of detergents that are available in India, it is difficult to assess the total quantity of phosphates annually used in detergents. However, it is obvious that there use has been increasing in the last decade or so. Southey et al. (2001) have quoted a figure of 33,400 tonne per year and, based on official data, they themselves arrived at a similar quantity for the year 1998-99. As mentioned earlier, VOICE Magazine (2001) has reported the total detergent production in India to be 2.7 million metric tonne. According to Southey et al. (2001) the detergent industry believes that 80% of all detergents in India, presumably from more localized, small producers, are phosphate free. Based on the above figure, one could conclude that only 20% of 2.7 million metric tonne of detergents, or about 540,000 metric tonne, have phosphates. As noted earlier, many leading detergent brands brands contain 13-30% phosphates by weight. Because of an increase in the use of laundry machines, it is also likely that the use of premium detergents with higher phosphate content has also increased. Even if we assume a figure of 10% STPP on an average, the quantity of phosphates annually used in detergents in India works out to be about 54,000 metric tonne. It is believed (Southey et al., 2001) that 80% of phosphate used in detergents finds its way into lakes and rivers.

Anthropogenic Sources of Phosphorus Loadings to Aquatic Environments in India

After a detailed consideration of various factors and on the basis of data gathered from a variety of sources, Southey et al., (2001) estimated the phosphorus loadings to aquatic environment in India. According to their survey, the bulk quantity of phosphorus comes from the use of detergents, human waste, fertilizers, the use of manure for agriculture and from livestock waste such as cattle, buffalo, sheep and goat. They also took into account the impact of dry season, wastewater treatment facilities and of rural and urban populations on the contribution of phosphorus to aquatic environment. According to their calculations, livestock waste with an annual input of 320,000 metric tonne of phosphorus or 64% of the total inputs, is the main source of phosphorus to aquatic environments. At 110,000 metric tonne or nearly 22% of the total, human waste is the second largest source of phophorus. These sources were followed by fertilizers: 61,000 tonne, or 12% metric of the total inputs. Detergents were responsible for only 12,000-18,000 metric tonne of the total inputs. Based on these estimates, Southey et al. (2001) discounted an earlier figure (Driver, 2000, cited in Southey et al., 2001) that detergents are responsible for only 0.23% of the total anthropogenic sources of phosphorus in aquatic environments in India. In fact, they believed that for metro cities, detergents may be responsible for as much as 10% and during the dry season, even 15% of the total anthropogenic sources of phosphorus to aquatic environments.

In North America, because of regulations on phosphates in detergents as well as on the phosphorus content in effluents from sewage treatment plants, phosphorus input resulting from human waste and from detergents are very low. In the USA, wastewater discharge for the 1980s contributed only 5.2% and fertilizers and manure together were responsible for 72% of total anthoropogenic inputs of phosphorus. The remaining 22% of inputs originated from non-point sources (Litke, 1999). Similarly, in Canada, only 8.5% of phosphorus inputs to aquatic environments is from household sewage (including detergents) and most of the remaining from agricultural activities (Chambers et al., 2001, p. 56).

In Europe also, human and animal wastes are the most important source of phosphorus in rivers and lakes. Köhler (2001) has cited papers which state that in EU countries, livestock waste contributes to 34%, human waste 24%, agricultural fertilizers 16%, and detergents 10%. He has listed a few examples for some of the European rivers where the bulk of phosphorus came from point sources. In general, we can conclude that in many of these European countries, detergents contribute a small fraction (upto 10% or so) of the anthropogenic inputs of phosphorus to aquatic environments. The reasons for the low input from detergents may vary in different countries. In India, because of the large animal and human population and the lack of sewage treatment facilities, the contribution from these sources comprises the main input. In North America, because of detergent regulation or the marketing for phosphate-free detergents coupled with the relatively low population and the availability of sewage treatment facilities, human waste is not an important source and agriculture sources rather become the dominant cotributors.

Controlling Phosphorus Pollution from Detergents in India

From the foregoing discussion, it is evident that for the eutrophication of Lake Erie in the early 1960s, detergents phosphates were identified as one of the major factors. Other important sources of nutrients were human waste entering from point sources and runoff from various agricultural operations entering from non-point or diffuse sources. Similar situations existed in other parts of North America and in European countries. Since it was easier to control the input from point sources and from detergents, regulations were put in place to eliminate or very significantly reduce phosphates from these sources. This resulted in the recovery or oligotrophication of Lake Erie. Similar steps led to the recovery of many other aquatic environments. It may be added here that the banning of detergent phosphates alone, without the regulations on effluents from sewage treatment plants, would not have been very effective. In India, as we have noted, the contribution of phosphorus from detergents is much smaller when compared to the inputs of human and animal wastes and from agricultural activities. Eliminating detergent phosphates, therefore, will not have a marked impact on the eutrophic conditions of most of the aquatic environments. Under such conditions, why bother with regulations on detergent phosphates? This has been the main point of contention by the industry against controls on detergents. Southey et al., (2001) have discussed various available options and the implications of these options under the existing Indian scenario. The options that have been used under existing conditions in Europe have been discussed briefly by Köhler (2001).

Wastewater Treatment

The installation of wastewater treatment plants with facilities for phosphorus removal have the greatest potential to control phosphorus inputs to aquatic environments. The existing technology can remove 95% or more of the phosphorus in wastewater (Köhler, 2001). This has been a major factor in the oligotrophication of Lake Erie and other lakes in the USA and Europe. Similarly, because of the installation of wastewater treatment facilities, the number of heavily polluted rivers in Western Europe has decreased from 25% in 1975-80 to about 5% in 1992-8 (Köhler, 2001). Such facilities not only control phosphorus resulting from human and animal waste, but also control phosphate emanating from the use of detergents. The availablity of phosphorus removal methods has led to the so-called "European Challenge" (see Southey et al., 2001), a well-articulated challenge to ban the use of phosphates in detergents. Briefly, it argues that the control of detergent phosphates alone may not result in the oligotrophication of impacted environments. Therefore, sewage treatment of phosphorus removal is essential. Once we have these facilities, the banning of phosphates in detergents becomes redundant.

As pointed out by Southey et al. (2001) the argument that phosphorus removal during wastewater treatment makes controls on detergents redundant seems irrelevant under the current conditions in India and other tropical countries. At present, not only is there a lack of sewage treatment plants but even sewage collection facilities do not exist. Until 1986, nearly 70% of the large cities (more than a population of 100,000) and small towns (above 50,000 people) had no sewage treatment facilities and only 2% of small towns had adequate sewage collection and treatment facilities (Gopal, 1994). It is true that since then there has been some improvement in the collection and provisions of wastewater treatment facilities. Providing sewage treatment facilities is also too expensive and beyond the capacity of all but a few cities (Southey et al., 2001). During the Lake Erie crisis, Ontario spent about $1.5 billion to build these facilities, while the adjoining American states to the Great Lake area spent about $7.5 billion. In order to improve the waste quality in the US, about $200 billion was spent on the construction of wastewater treatment plants and about $154 billion was spent on the operation and maintenance of waste treatement plants during the period 1972 to 1991

(Litke, 1999). In Canada, by 1996, 73% of Canadians were served by municipal sewer systems, 25% used septic beds for sewage treatment, and the remaining 2% were, in all likelihood, served by lagoons. In 1996, at least 94% of sewage collected by sewers received primary treatment or better. In the provinces from Alberta to Ontario, the majority of the sewage collected received secondary treatment or better. As a result of these treatment, phosphorus loadings from municipal wastewater treatment plants decreased by 37% during the period of 1983-1996. It is obvious that for some time to come, such facilities for the treatment of wastewater are not likely to be available. The argument that the tertiary treatment of wastewater makes controls on detergent phosphates redundant seems to be irrelevant to India for the foreseeable future (Southey et al., 2001). We may arrive at the same conclusions for many other tropical countries.

The Use of Sewage for Agriculture and Aquaculture

The use of sewage for agriculture and aquaculture have been discussed here very briefly. For a detailed review on the use of sewage for agriculture, the readers is referred to Southey et al. (2001) and a brief review on sewage and agriculture is provided by Jana and Jana (2002). Citing from a 1990 UNDP report on "wastewater irrigation in developing countries", Southey et al., (2001) point out that because of growing water shortages, particularly in dry regions, sewage farming has increased in recent years. This is especially useful for countries like Australia and Israel and areas like California that can maintain high levels of controls to avoid health risks associated with the use of treated wastewater. In 1982, Isreal used 42% of its treated sewage water in agriculture and the plan was to use upto 80% by the year 2001. India was a pioneer in the use of sewage farming which was introduced in Delhi in 1877 and soon thereafter in Kolkata (Southey et al. 2001).

In India, there are 142 class-1 cities with an average population of one million, with each city generating 10 mld (million liters per day) of domestic sewage. It is estimated that about 90 tonne of nitrogen, 32 tonne of phosphorus and 55 tonne of potassium valued at 61 million rupees could be recovered from this sewage daily (Edwards, 2000, cited in Jana & Jana, 2000). The Ganga Action Plan, Phase I (GAP I) involved 88 schemes for diverting wastewater to agriculture, involving some 873 mld, which was 35% of all the sewage generated by all urban areas in the Ganga basin. The remainder was to be dealt with under Phase II. Although it is believed that the initiatives under GAP I have been largely completed, in 1999, the diversions of sewage for farming were operating at about 50% of the planned capacity. Southey et al. (2001) believe that it is too likely that, faced with power shortages and other financial constraints, the treatment of wastewater will be inadequate. In addition to health problems associated with the use of inadequately treated sewage, there is also the likely problem with the uptake of heavy metals in the produce from farmers using sewage water. Southey et al. (2001) concluded that despite the efforts as mentioned in the National River Conservation Plan, India is unlikely to have sewage diversions large enough to make phosphorus controls in detergents redundant. They also calculated the extra cost that producers may have to incur if detergent phosphorus is removed from the sewage water through regulations of phosphate in detergents. Their own calculations, and the calculations based on the data from a 1997 study the Central Pollution Control Board, indicated that such costs would be modest.

Like sewage farming, the use of sewage for fish culture has also been practiced in India for decades, especially in West Bengal. Kolkata city, with a population of more than a 11 million people, discharges on an average, 110 mld of sewage, some of which is used by 130 sewage-fed fish farms covering an area of about 4000 h. It also provides a livelihood for a huge multitude of people. Many countries in Asia and Europe annually produce 3 to 4.5 tonne of fish per hectare based on sewage-fed fish farms. In addition to the production, wastewater aquaculture is highly effective in removing nutrients, including phosphorus, from water. Like sewage farming, wastewater-based aquaculture cannot be adopted on a very large scale.

Other Options

Other options like the taxation or the marketing of detergents under 'Ecomark' to promote environment-friendly and phosphate-free detergents, if workable, can only reduce the contribution of detergent phosphorus to aquatic environments but may not completely eliminate it. Plans for the marketing of detergents with 'Ecomark' were developed by the Bureau of Indian Standards nearly a decade ago but the manufacturer did not adopt this plan. A tax on detergent phosphates has been introduced in France in January 2000. It is not likely to have a major impact on the manufacturing of detergents with phosphates (Köhler, 2001).

REGULATIONS ON PHOSPHATES IN DETERGENTS

We have seen that although the use of phostate-based detergents has increased in India during the last few years, the quantity of phosphorus entering aquatic systems as a result of detergent use is small in comparison to the total amount entering from other sources. Obviously, expect for those small water bodies like village ponds and lakes where detergents phosphates could be a major source of phosphorus, controlling input of phosphorus from detergents will not help in the oligotrophication of eutrophic rivers and lakes in India. However, the input of phosphorus from detergents is, in fact, a substantial nutrient pollution and the extent of this pollution is likely to steadily increase. This pollution may not have an immediate impact on the already badly polluted and eutrophic reaches of rivers and streams, these nutrients would promote the eutrophication of downstream reaches and even of coastal waters. The coastal waters in many parts of the world are now going through a phase that the lakes and rivers went through in the 1960s (National Research Council, 2000).

As already stated, the phosphate content in detergents in Canada, the US, and many countries in Europe is regulated or banned. In Canada and parts of the US, this regulation has existed for more than three decades, although since then, these places have developed extensive and sophisticated wastewater treatment facilities. Controlling the phosphorus content of detergents in India is not likely to be expensive because a major share of the detergent market in India is controlled by two or three main detergent manufacturers (Southey et al., 2001). In view of these facts, one could conclude that phosphates present in detergents in India should be regulated and not more than 2.2% phosphorus should be allowed to begin with. In fact, many brands already have less than this phosphorus content. By the end of the next five years or so, no detergent product should have more than 0.5% phosphorus by weight. It appears strange that in countries like Canada and the US where phosphorus utilization in detergents is permissible and facilities for its removal from wastewater exist, manufacturers produce phosphate-free detergents. In contrast, in India, where sewage treatment facilities are far from adequate, detergent manufactures use as much as 32% STPP by weight.

ACKNOWLEDGMENTS

I am deeply thankful to Dr P. Sibley for critically reviewing the chapter and for many suggestions. I also greatly appreciate the financial support from CIDA, Canada and the Shastri Indo-Canadian Institute for an earlier study pertaining to detergents use in India.

REFERENCES

Allan, J.D. 1995. Stream Ecology: Structure and Function of Running Waters. Chapman & Hall, London.

Ashworth, W.A. 1986. The Late Great Lakes: An Environmental History. Collins Publishers, Toronto.

Beeton, A.M. 1969. Changes in the environment and biota of the Great Lakes. In: Eutrophication: Causes, Consequences And Correctives. National Academy of Sciences, Washington DC: 157-187.

Bertram, P.E. 1993. Total phosphorus and dissolved oxygen trends in the central basin of Lake Erie, 1970-1991. Journal of Great Lakes Research 19: 224-236

Burns, N.M. 1976. Nutrient budget for Lake Erie, 1970. Journal of the Fisheries Research Board of Canada 33: 520-536.

Burns, N.M. 1985. Erie: The Lake that Survived. Rowman and Allanheld Publishers, New Jersey.

Burns, N.M., & Ross, C. 1972. Project Hypo: an Intensive Study of the Lake Erie Central Basin Hypolimnion and Related Surface Water Phenomena. Canada Centre for Inland Waters, Paper No.6, United States Environmental Protection Agency, Technical Report, TS - 05 - 71 - 208-24.

Chambers P.A., Guy, M., Roberts, E.S., Charlton, M.N., Kent, R., Gagnon, C., Grove, G., & Foster, N., 2001. Nutrients and their Impact on the Canadian Environment. Agriculture and Agri-Food Canada, Environment Canada, Fisheries and Oceans Canada, Health Canada and Natural Ressources Canada.

Charlton, M.N. 2001. Lake Erie, Ontario: Recovery of a "Dead" Lake. In: Chambers P. A., M. Guy, E. S. Roberts, M. N. Charlton, R. Kent, C. Gagnon, G. Grove & N. Foster, Nutrients and their Impact on the Canadian Environment. Agriculture and Agrifood Canada, Environment Canada, Fisheries and Oceans Canada, Health Canada and Natural Ressources Canada: 65-66.

Charlton, M.N., Milne, J.E., Booth, W.G. & Chiocchio, F. 1993. Lake Erie offshore in 1990: Restoration and resilience in the central basin. Journal of Great Lakes Research 19: 291-309.

Charlton, M.N., Le Sage, R., & Milne, J.E., 1999. Lake Erie in Transition: The 1990s. In: Munawar, M., Edsall, T., & Munawar, I. F. (eds) State of Lake Erie: Past, Present and Future. Bakhuys Publishers, Leiden, The Netherlands: 97-124.

de Jonge, V.N., Elliott, M., & Orive, E., 2002. Causes, historical development, effects and future challenges of a common environmental problem: eutrophication, Hydrobiologia 475/476: 1-19.

Dobson, H.H. & Gilbertson, M., 1972. Oxygen Depletion in the Hypolimnion of the Central Basin of Lake Erie, 1929 to 1970. In: N. M. Burns, and C. Ross, Project Hypo: An Intensive Study of the Lake Erie Central Basin Hypolimnion and Related Surface Water Phenomena. Canada Centre for Inland Waters, Paper No.6, United States Environmental Protection Agency, Technical Report, TS - 05 - 71 - 208 - 24

Dolan, D.M. 1993. Point source loadings of phosphorus to Lake Erie: 1986-1990. Journal of Great Lakes Research 19: 212-223.

Thornton, J.A. 1987. Aspects of eutrophication management in tropical/subtropical regions. Journal of the Limnological Society of Southern Africa 13: 25-43.

Trivedi, R.K., Tripathi, R. and Jain, V. 2001. Environment risk assessment and synthetic detergents. In: Southey C., N. Kaushik and R. Trivedi (eds), Detergents and the Environment: The Canadian Experience and the Indian Scenario. Tata McGraw-Hill Publishing Company Limited, New Delhi: 183-199.

US National Academy of Sciences, 1969. Eutrophication: Causes, consequences, correctives. Proceedings of a symposium. Washington DC.

Vallentyne, J.R. 1974. The Algal Bowl: Lakes and Man. Miscellaneous Special Publication 22, Department of the Environment, Fisheries and Marine Service, Ottawa.

Vollenweider, R.A. 1968. Scientific fundamentals of the eutrophication of lakes and flowing waters, with particular references to nitrogen and phosphorus as factors in eutrophication. Organization for Economic Cooperation and Development, Paris.

Vollenweider, R.A., Munawar, M. & Stadelmann, P. 1974. A comparative review of phytoplankton and primary production in the laurentian Great lakes. Journal of the Fisheries Research Board of Canada 31: 739-762.

Weller, P. 1990. Fresh Water Seas: Saving the Great Lakes. Between the Lines, Toronto.

CHAPTER

5

Natural Eutrophication of a Shallow Tropical Freshwater Lake: Impact and Possible Restoration Strategy

V.P. Venugopalan[1], Rajesh Kumar[1], R. Rajamohan[1], R. Sekar[2], and K.V.K. Nair[3]

[1]Water and Steam Chemistry Laboratory, BARC Facilities, Kalpakkam, Tamil Nadu 603 102, India

[2]Max Plank Institute for Marine Microbiology, Celsiusstrasse 1, 28359-Bremen, Germany

[3]National Institute of Ocean Technology, Narayanapuram, Chennai, India

ABSTRACT

Eutrophication of aquatic systems is often caused by human activities. Here we shall discuss eutrophication and its effects in a shallow freshwater lake caused by purely natural factors involving aquatic birds. Excessive phosphate input into the Kokilamedu Lake via bird droppings led to eutrophication and associated bottom hypoxia and fish kill. Nutrient chemistry of the lake, characterized by low N : P ratio, facilitated the development of a dense chlorophycean-dominated phytoplankton standing stock. To analyze the influence of eutrophic conditions on periphyton, the development of periphyton in the Kokilamedu Lake was compared with that in another nearby freshwater reservoir. Parameters such as periphyton thickness, algal density, biomass, chlorophyll a, viable bacterial count, species composition and succession pattern of microalgae were compared. The results showed that periphyton thickness, algal density, species diversity, biomass, chlorophyll *a* and bacterial density were significantly different in the two lakes. Species composition as well as succession pattern of microalgae also varied between the two stations. In the Kokilamedu Lake, diatoms dominated the periphyton community throughout the study period; Chlorophyceae and Cyanobacteria were present in lesser numbers. In contrast, in the reservoir, the initial stage (upto three days) was dominated by Chlorophyceae, followed by domination by Bacillariophyceae and finally by Cyanobacteria. Nutrient characteristics of the bulk water influence periphyton development and species composition in the two environments. This chapter also discusses possible strategies for restoration of the lake. It is emphasized that management strategies must concentrate on stoppage of external nutrient loading via bird droppings and internal loading via sediment leaching.

INTRODUCTION

Lakes are important to mankind because of the services they provide: aesthetic pleasure, recreation, fish, transportation and water for irrigation, drinking and dilution of pollutants (Postel and Carpenter, 1997). Eutrophication, a word used increasingly while describing lakes and coastal marine ecosystems, is defined as the enrichment of a water body with nutrients with consequent deterioration of its quality, rendering it unfit for human purposes (Turk et al., 1974). Eutrophication, a major problem facing several of the world's freshwater lakes, endangers the very survival of many species of fishes and other aquatic organisms (Verschuren et al., 2002). Eutrophication has the potential to degrade the ecological functions of large aquatic systems that could result in long-term damages and also upset the ecological services provided by such water bodies. Unless active steps are continuously taken, the role of such water bodies as functioning ecosystems may be severely compromised. Anthropogenic influences on aquatic ecosystems play a major role in the degradation of ecosystem functions, largely through uncontrolled input of nutrients, especially nitrogen and phosphorus, into the system. Relative and absolute changes in the quantity of major plant nutrients have the ability to affect both species diversity and community composition in aquatic systems. The process of deterioration occurs naturally in all aquatic systems, whose nutrient inflow exceeds its outflow. Such natural eutrophication is an extremely slow process from the human point of view, often taking thousands of years. In contrast, anthropogenically caused eutrophication, aptly known as cultural eutrophication, is a relatively fast process. Some of the important sources of nutrients are increasing use of phosphorus detergents, agricultural runoff containing chemical fertilizers, and felling of forests that causes increased topsoil erosion and discharge of untreated or partially treated wastes into inland and coastal water bodies. Eutrophication, while causing an overall increase in population and biomass levels, results in a decline in the phytoplankton species diversity and change in species dominance. Reduced water transparency and excessive algal and macrophyte growth can interfere with uses and aesthetic quality of water and cause taste and odour problems in drinking water. Hypolimnetic anoxia caused by algal decomposition can result in fish kills and undesirable shifts in species composition of the biological community. Tropical lakes are particularly susceptible to cultural eutrophication, owing to increasing demand placed on scarce freshwater resources for domestic, industrial and waste disposal purposes. Moreover, in tropical areas, vector-borne diseases such as malaria may be increased by eutrophication because the water body may become a breeding ground for insect vectors.

There are plenty of reports available on the various aspects of cultural eutrophication of lakes and other inland water bodies as well as coastal marine ecosystems. However, reports on natural eutrophication are rare, probably due to the fact that the process is relatively very slow and takes place over generations of human life. In this chapter, we shall examine the causes and effects of naturally mediated nutrient enrichment in a shallow freshwater lake in south India.

FEATURES OF THE LAKE

The Kokilamedu Lake is a shallow (Z_{max} *ca* 2.8 m) perennial lake, located about 1.5 km from the seacoast at Kalpakkam (12°22′N, 80°11′E), on the east coast of India. With a water spread of about 1.4 km^2, this lake is located within a well-protected nuclear power complex and therefore, is not subjected to any kind of human activity (fishing, entertainment activity, etc.). A large-scale fish kill observed in this lake in May 1995 prompted a detailed study of the water body to understand the reasons for such biological catastrophe. The study confirmed the eutrophic status of the lake (Venugopalan et al., 1998).

EXPERIMENTAL STRATEGY

Analysis of water and sediment samples formed the most important strategy as in any eutrophication study (Parsons et al., 1984; APHA, 1989). Major parameters necessitating monitoring were temperature, dissolved oxygen, pH and plant nutrients (especially nitrite, nitrate, phosphate and silicate). Apart from this, analysis of chlorophyll and microscopic observation of microalgae gave a good indication of the phytoplankton standing stock and composition. Sediment core analysis was employed to study the vertical profile of organic carbon and sulphur, which are indicative of organic loading and anaerobic conditions. Another strategy employed was periphyton monitoring, using experimental coupons. Simultaneous exposure and sequential removal of experimental coupons is a simple and inexpensive method to study periphyton development in aquatic systems (Rao et al., 1997; Sekar et al., 1998; Sekar et al., 2002). The analytical methods employed are discussed by Bakke and Olsson (1986), Brown (1976), Desikachary (1959), Hustedt (1930), Philipose (1967) and Prescott (1978).

For the purpose of comparison, a nearby freshwater reservoir was also monitored. The Madras Atomic Power Station (MAPS) Open Reservoir (designated as Station 2) is a nearby (*ca* 3 km away from the Kokilamedu Lake) freshwater reservoir, which receives water from the Palar River and serves as source water for the cooling system of an electrical power plant. It has a water spread of 1.7 hectares and a storage capacity of 28,400 m^3, with depth of 2.3 m at overflow level.

LAKE EUTROPHICATION

Eutrophication of aquatic systems results from excessive nutrient inputs and is generally deleterious to the environment. Prominent symptoms of eutrophication are an increase in the concentration of plant macronutrients (total P > 20 µg/l, total N > 500 µg/l), turbidity (Secchi disc depth < 2 m) and chlorophyll (chlorophyll a > 10 µg/l) (Harrison, 1992). Eutrophication is a syndrome that develops when resilience mechanisms naturally available to the aquatic habitat are broken down (Harper, 1992). Extremely high concentrations of phosphate in the Kokilamedu Lake can be seen from the nutrient data presented in Table 5.1. These data may be compared with those shown in Table 5.2, collected from the nearby MAPS Open Reservoir (Station 2). Concentration of chlorophyll a in the lakewater was also high (33.6-69.5 µg/l). Phosphate levels in the open reservoir were relatively less, but nitrate levels were high. Fig. 5.1 shows the vertical profile of organic carbon and sulphur levels in the lake sediments. The profile indicated a recent increase in the organic carbon levels; sulphur levels did not show any discernible increase, though periodic fluctuations were visible.

Shallow eutrophic lakes display complex nutrient dynamics (Havens et al., 2001). Generally, inorganic phosphorus is limiting to phytoplankton in freshwater lakes (Schindler, 1977; Hecky and Kilham,

Table 5.1 Data on nutrient and water quality parameters in the Kokilamedu Lake (Data were collected during a cruise undertaken in June 1996).

Parameter	*Surface*		*1 m Depth*	
	Minimum	*Maximum*	*Minimum*	*Maximum*
Temperature (°C)	31.1	32.0	29.5	30.1
Nitrite (μg N/l)	1.2	5.1	2.1	3.8
Nitrate (μg N/l)	3.7	6.8	3.6	7.4
Phosphate (μg P/l)	53.9	76.3	50.5	78.4
Silicate (μg Si/l)	571	1496	103	1516
Dissolved O_2 (mg/l)	4.27	6.00	3.16	3.90
Chlorophyll a (μg /l)	33.6	69.5	32.8	70.4

Table 5.2 Nutrient and chlorophyll a levels in the MAPS Open Reservoir (Station 2). Ranges and mean are given for a six-month period (Jan.-Jun.) in 1996.

Nutrient/Pigment	*Minimum*	*Maximum*	*Mean*
Nitrite (μg N/l)	1.5	8.7	4.8±3.2
Nitrate (μg N/l)	17.5	79.3	44.1±25.8
Phosphate (μg P/l)	8.1	20.6	16.0±4.8
Silicate (μg Si/l)	3482	4171	3891±297
Dissolved O_2 (mg/l)	6.2	9.2	7.93±0.8
Chlorophyll (μg N/l)	9.07	18.5	13.3±3.8

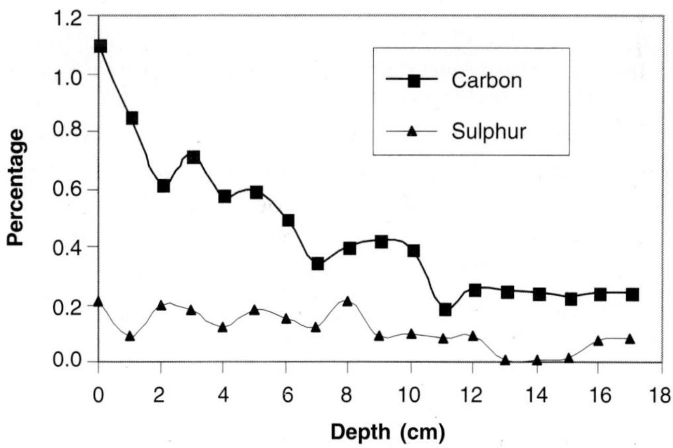

Fig. 5.1 Data on organic carbon and sulphur levels in sediment cores collected from the Kokilamedu Lake

1988), while excessive input of P may lead to eutrophication (Paerl, 1988; Smith et al., 1999). In Kokilamedu Lake, eutrophication has occurred due to natural factors and not as a result of human activities. Overhanging branches of trees (especially, *Ficus* sp. and *Casuarina equisetifolia*) growing on the banks of the lake are home to hundreds of aquatic birds, the most numerous among which are the little cormorants *Phalacrocorax niger* (Venugopalan et al., 1998). Cormorants are gregarious birds nesting on trees. They are well-known guano producers and use their guano for nest making. Analysis of nutrients in the bird guano confirmed that it is rich in nitrogen (total N 13.8%, ammonia 19.3 mg N/g, nitrite 3.7 mg N/g, nitrate 3.7 mg N/g dry weight) and phosphorus (orthophosphate 3.0 mg P/g dry weight). In the present study, the birds fed in the nearby Sadras and Edaiyur backwater systems and returned to their lakeside nests. Because of the large input of phosphate via bird droppings, the lakewater is rich in phosphate with a N : P (nitrate to o-phosphate) ratio much lower than the Redfield ratio. Lakewater phytoplankton at the time of study was observed to consist mostly of Chlorophyceae, with *Chlorella* sp. being the most dominant (Venugopalan et al., 1998). The results showed low levels of phytoplankton biodiversity (Shannon's species diversity index 0.15), typical of highly eutrophic lakes. Chlorophycean dominance in the lakewater was probably facilitated by the low lake N : P ratio. Earlier, Ryther and Dunstan (1971) had shown that Moriches Bay, receiving duck farm effluents and, consequently, possessing a low N : P ratio, had a phytoplankton standing stock consisting almost entirely of the chlorophycean species *Nannochloris atomus* and *Stichococcus* sp.

When nutrient levels in the Kokilamedu Lake were compared with those in the MAPS open reservoir (station 2), important differences were observed (Table 5.2). The reservoir water was rich in nitrate because of agricultural runoff into the Palar river. The chlorophyll a levels in the reservoir water were much lower than those in the Kokilamedu Lake. Phosphate levels were also lower than those in the lakewater. Analysis of the reservoir plankton showed diatoms to be the dominant community. Higher amounts of phosphate in a natural, rainfed lake such as the Kokilamedu Lake, with no sewage input or agricultural runoff, is an indication of organic pollution (Ryther and Dunstan, 1971). Eutrophication of the lakewater due to nutrient input via bird droppings is an interesting case of natural eutrophication (as against the usual cultural eutrophication). Earlier, we had reported a case of large-scale fish kill in this lake in the summer of 1995 (Venugopalan et al., 1998). The fish kill occurred due to a combination of bottom hypoxia (caused by natural eutrophication) combined with high surface temperatures. Verschuren et al. (2002) reported that eutrophication-induced loss of deep-water oxygen in Lake Victoria (East Africa) resulted in the collapse of indigenous fish stocks by eliminating suitable habitat for certain deep-water cichlids.

There are limited reports on nutrient enrichment of aquatic systems caused by bird faeces (Manny et al., 1994). Pettigrew et al. (1998) tested the hypothesis that inorganic nutrients released from waterfowl faeces would stimulate primary production and thereby, invertebrate grazer population in the Canadian prairie wetlands. They measured nutrient content of geese faeces and found it to be 5.7 mg N/g and 14.0 mg P/g dry wt. Mesocosm studies showed that addition of faeces produced an immediate increase in inorganic N and P content and that periphyton chlorophyll increased in response to faeces addition. An increase in the density of microcrustacean grazers was also observed in response to faeces addition. However, they concluded that nutrients leaching from waterfowl faeces might not have a significant impact under natural conditions.

Organic carbon and sulphur in sediment cores have been used by earlier researchers to reconstruct the eutrophication history of aquatic systems (Turner and Rabalais, 1994; Cooper and Brush, 1991; Brush 1984). Cooper and Brush (1991) used sediment cores to reconstruct the nutrient history of Chesapeake Lake and to understand anoxia and eutrophication in the overlying water column. Our studies showed (Fig. 5.1) that in recent years, there has been an increase in the deposition of organic carbon in the lake sediments. Sediment sulphur levels of > 1% are indicative of anaerobic sedimentary conditions (Cooper and Brush, 1991). However, the present results do not indicate the presence of anaerobic conditions at the sediment water interface.

PERIPHYTON STUDIES

Periphyton development in an aquatic environment depends on the characteristics of the ambient water (Rao et al., 1997). Nutrient and water quality parameters of the two freshwater systems during the periphyton studies are shown in Table 5.3. Major differences in the physical and biological characteristics of periphyton growth in the two environments are given in Fig. 5.2. Periphyton thickness and biomass were higher in the Kokilamedu Lake. The difference in the periphyton thickness at the two stations was significant ($F = 4.614$, $P = 0.0024$). Biomass increased with exposure time at both stations. However, after 15 days of growth, the biomass at station 1 was significantly greater than that at station 2 ($t = 3.665$, $P = 0.0105$). Algal density on the coupons was initially higher in the eutrophic lake but towards the end of the experiment, algal densities at stations 1 and 2 were comparable. The algal colonization pattern showed differences between the two stations. At station 1, the

Table 5.3 Water quality parameters of the Kokilamedu Lake and MAPS Open Reservoir during periphyton studies (data are maximum and minimum during the 15-day experiment duration).

Parameters	*Kokilamedu Lake*	*Open Reservoir*
Temperature (°C)	27.0-31.0	30.1-31.3
pH	8.7-9.15	8.6-8.9
Dissolved O_2 (mg/l)	5.8-9.7	6.2-7.4
Nitrite (μg N/l)	7.2-11.9	1.5-2.1
Nitrate (μg N/l)	0.48-7.1	15-37
Phosphate (μg P/l)	24.2-107	15-21
Silicate (mg Si/l)	1.54-9.5	3.48-8.5
Chlorophyll a (μg/l)	6.4-12.8	3.48-8.5

colonization declined after peaking on the 7th day, while at station 2, the algal density showed a continuous increase. Periphyton chlorophyll *a* at station 2 increased significantly after 7 days of growth, when compared to that at station 1, which reached a plateau after 7 days. Heterotrophic bacterial count associated with the periphyton was consistently higher at station 1; after 7days, it increased substantially. The number of algal species in the periphyton was temporarily higher at station 2, but it became comparable at both stations by the end of 15 days.

Chlorophyceae, Bacillariophyceae and Cyanobacteria were the major constituents of the periphyton. At station 1, the total number of species observed during the study period ranged from 18 to 23, whereas at station 2 it ranged from 18 to 32 (Fig. 5.2). The succession patterns of microalgae in the periphyton at station 1 and 2 are given in Fig. 5.3. At station 1, diatoms mostly dominated the periphyton throughout the study period, while Chlorophyceae and Cyanobacteria together contributed to 50% or less of the substratum coverage. The succession pattern at station 2 was characterized by a three-phased succession (Fig 5.3). The initial phase (1-3 d) dominated by Chlorophyceae, which was followed by Bacillariophyceae (4-10 d). After 10 d, diatom contribution reduced and Cyanobacteria became the dominant group in the periphyton.

Water chemistry, especially nutrient concentrations, influence colonization and species composition of periphytic communities (Pan and Lowe, 1994). The present study clearly showed the differences in periphyton development in the two freshwater environments. Periphyton thickness and biomass in the Kokilamedu Lake (station 1) were higher when compared to those of the periphyton developed in the open water reservoir (station 2). The higher levels may be attributed to the increased nutrient concentration of the eutrophic lake. This is supported by the

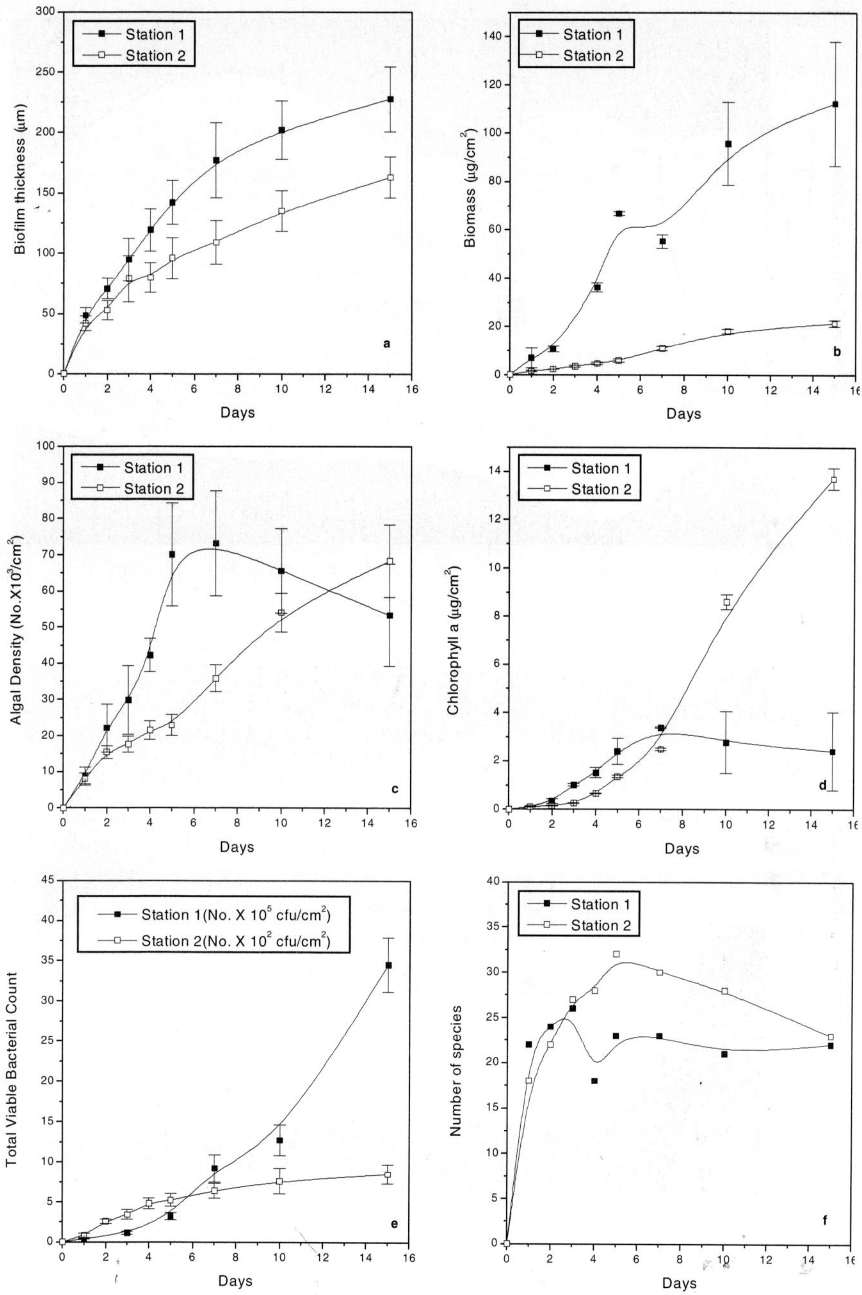

Fig. 5.2 Physical and biological characteristics of periphyton growth in the Kokilamedu Lake (station 1) and MAPS Open Reservoir (Station 2)

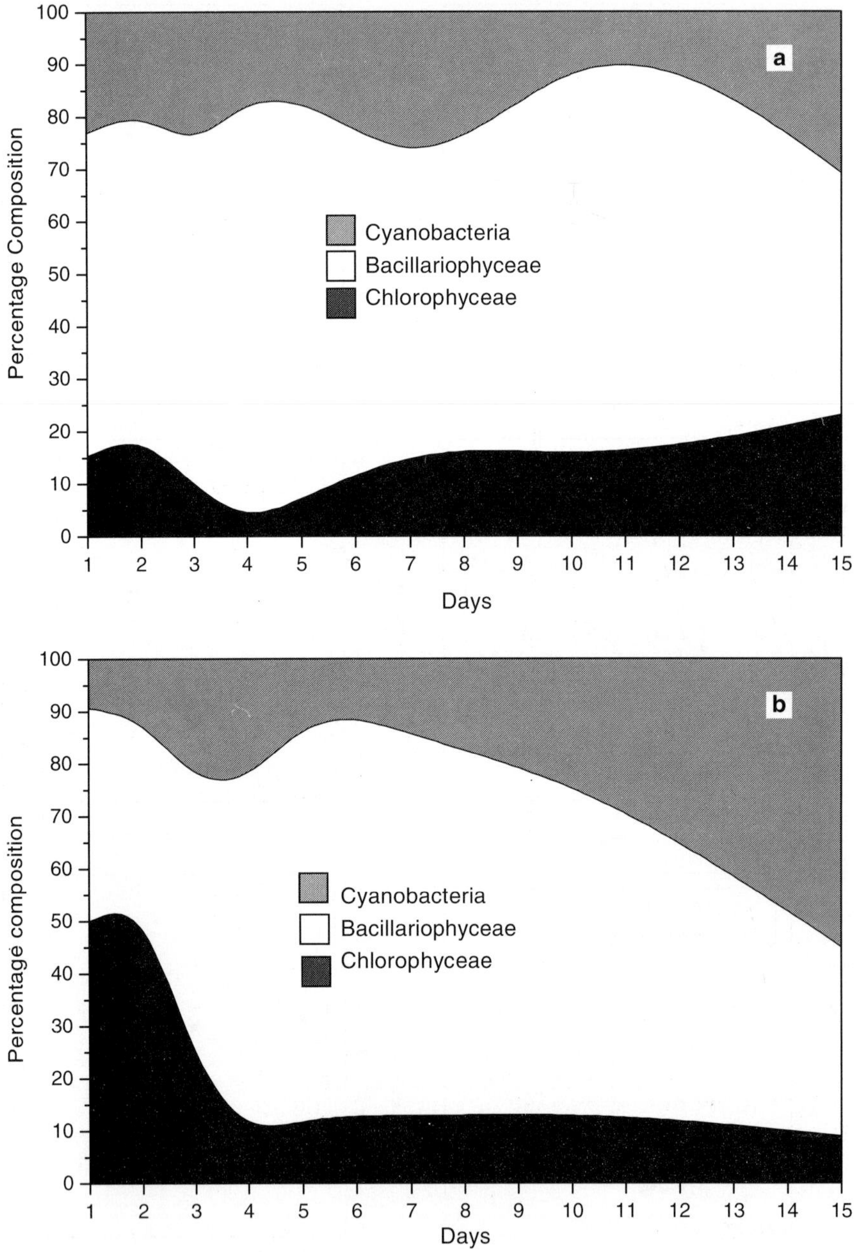

Fig. 5.3 Succession patterns of microalgae in the periphyton at station 1 (a) and 2 (b).

report by Guasch et al. (1995) who observed increased algal biomass and primary production in periphyton after nutrient enrichment, whereas non-enriched periphyton showed lower values. In the present study, the heterotrophic bacterial population associated with the periphyton in the eutrophic lake showed a rapid increase with exposure time. There was a concomitant decrease in the algal population, as indicated by a decrease in algal density and chlorophyll *a*. Reil & Arndt (2000) observed a reduction in autotrophic to heterotrophic microbial ratio with increasing eutrophication, while studying the effect of eutrophication on benthic microbial communities.

Nutrient concentration may be an important factor for the differences in species composition and periphyton succession observed in the two lakes. The eutrophic Kokilamedu Lake showed thicker periphyton growth and significant differences with respect to species composition and succession of microalgae and cyanobacteria in the periphyton, as compared to the periphyton developed in the open water reservoir. Thick periphyton growing on macrophyte leaves would reduce the light availability and thereby affect photosynthesis. There is some evidence that succession in periphyton (biofilm) communities is related to the changes in the nutrient chemistry of the periphyton matrix (Sekar et al., 2002). Therefore, it would be worthwhile to study the qualitative and quantitative aspects of nutrient dynamics within the periphyton matrix in eutrophic systems. Earlier, Mattila and Räisänen (1998) had showed that periphyton growth was positively correlated with concentrations of total nitrogen, total P and chlorophyll a. In their opinion, periphyton studies are a useful addition to conventional monitoring programmes of eutrophic lakes.

LAKE RESTORATION

Phosphorus enrichment, as observed in the case of the Kokilamedu Lake, can degrade the plant community in a lake, altering the competitive balance between aquatic flora, with consequences for the entire ecosystem (Mainstone and Parr, 2002). Lake restoration helps in restructuring and recreating the ecosystem to its former healthy state (Ramachandra et al., 2000). The main objective of a lake restoration programme is to convert a lake, characterized by high nutrient input, high primary production, murky water with anoxia and relatively low value of ecosystem services, into one with low nutrient inputs, low to moderate levels of plant nutrients, relatively clear water and high economic value of ecosystem services (Janssen, 2001). It is imperative that system behaviour—especially individual characteristics of the aquatic ecosystem—is properly understood before lake restoration is resorted to, in order to avoid costly mistakes (Sagehashi et al., 2001; Istvánovics et al., 2002; Gulati and van Donk, 2002). Often, restoration strategy requires site-specific research for effective management (Havens et al., 2001; Carpenter and Lathrop, 1999). Effective management programme, in turn, should focus on interception of nutrients. Under normal circumstances, ecosystem management would require the explicit treatment of interactions between humans and the ecosystem (Janssen, 2001). However, in the present case, this argument is not valid as the eutrophication is caused purely by natural factors and human interference is totally absent.

Tropical lakes are known to be more sensitive than temperate lakes to excessive nutrient supply and show higher proportionate changes in water quality and biotic communities in response to eutrophication (Lewis, 2000). Therefore, management of eutrophic tropical lakes needs to be approached in a different manner from that of temperate lakes. Tropical lakes are easily prone to oxygen depletion; this calls for stringent regulation of organic and nutrient loading. Lake management strategies should address allochthonous and autochthonous nutrient inputs into the lake (Muhammetoğlu et al., 2002). Unfortunately, control of allochthonous inputs via diffuse sources may not have any significant effect on lake revival. Delayed recovery observed during watershed management is often attributed to enhanced internal load during the transition to a new steady state between reduced external input and in-lake input (Istvánovics et al., 2002).

Aquatic ecosystems can process and assimilate P into long-term storage in bottom sediments, depending on a complex interplay of physical, chemical and biological processes (Havens et al., 2001). Nevertheless, a major route of loss of P from the water column is through planktonic uptake and subsequent settling (Stabel, 1985). While planktonic uptake of P is directly influenced by light availability (Nalewajko and Lee, 1983), sedimentary loss of phytoplankton is determined by its species composition, as different groups of organisms have different sinking properties (Havens and Schelske, 2001). Subsequent phosphate release from sediments rate would vary, depending on factors such as redox conditions at sediment-water interface, sediment resuspension rates, etc. (Sondergaard et al., 1992; Sas, 1989). Internal loading due to sediment release can become a very important source of P in lakes even after external inputs are controlled by proper management strategies.

An important reason as to why shallow, well-mixed lakes are less amenable to watershed-based management strategies is their predominant internal recycling processes. In other words, bottom sediments in eutrophic lakes serve as nutrient reservoirs, from where phosphorus can be released back into the water column. In the case of Lake Finjasjön, Sweden, it was shown that phosphorus-leaking black sediments was the cause of failure for lake recovery (Annadotter et al., 1999). Scharf (1999) also reported that restoration of the highly eutrophic Lingese Reservoir in Germany was less successful due to the failure to reduce P-recycling from sediments. It has been reported that shallow lakes with a long history of eutrophication require a substantial reduction in nutrient load before phytoplankton biomass starts to decrease (Istvánovics and Somlyódy, 2001). In fact, the relationship between P status of sediments and dynamics of a lake is still not fully resolved (Sas, 1989). Under such circumstances, removal of sediment could be considered as a viable option. Sediment dredging has been shown to be an effective method that can curtail the internal input of phosphorus into lakewater. In addition, sediment dredging also increases lake volume. Using a mathematical model, Sagehashi et al. (2001) tested various restoration methods available for shallow lakes. It was observed that sediment dredging was the most desirable restoration method from the management point of view. Restoration work in an urban hypereutrophic Lake Baton Rouge (Louisiana) consisted of dredging up of the lake sediment. An immediate improvement in water quality was observed (Ruley and Rusch, 2002). Recently, ecotechnological methods of lake restoration such as lake-internal P precipitation using aluminium or ferrous iron dosing are also increasingly being used for excess nutrient removal (Deppe and Benndort, in press).

Among other methods available for restoration of eutrophic lakes, biomanipulation promises prevention of harmful algal bloom due to eutrophication (Sagehashi et al., 2000). The principle is to keep zooplankton predator biomass high in order to prevent the excessive growth of algae (Shapiro et al., 1975). However, this is a palliative treatment, as it does not address the fundamental issue of excessive nutrient input into the lake. Dissolved silicate has been suggested as a remedy for eutrophication, for triggering diatom growth that would remove excess nutrients from the water column (Wassmann et al., 1996). Aquatic plant community is the foundation for a healthy and diverse ecosystem, providing food, shelter and breeding habitats for a range of animal species (Mainstone and Parr, 2002). As submerged macrophytes can moderate the structure of food webs, lake managers have adopted the option of increasing macrophyte abundance in order to restore eutrophic waters with a view of improving water quality, increasing water transparency and reducing phytoplankton biomass (Lau and Lane, 2002). Scharf (1999) reported that phosphorus management, along with appropriate fishery management (food web management), was very important for successful restoration of lakes.

In conclusion, restoration of Kokilamedu Lake should focus on significant reduction of nutrient entry into the lake. Since external P input is via aquatic birds, it is necessary that birds be prevented from nesting on the trees located on the banks of the lake. Phosphorus concentration in this small lake was observed to increase substantially following rainfall, due to washout of bird faeces from the tree branches into the lake below (data not shown). Management strategy, therefore, must include prudential cutting of trees. Once the external nutrient sources have been taken care of, the problem of internal loading has to be addressed too, in order to ensure that sustained nutrient addition into the lake is prevented. Going by the experience elsewhere, sediment dredging seems to be a good option in this regard. It is also very important that continuous long-term ecological monitoring (including periphyton monitoring) of the lake should go hand in hand with restoration methods.

REFERENCES

Annadotter, H., Cronberg, G., Aargen, R., Lundstedt, B., Nilsson, P. & Ströbeck, S. 1999. Multiple techniques for lake restoration. Hydrobiologia 359/396: 77-85.

APHA. 1989. Standard methods for the examination of water and wastewater, American Public Health Association, 17th edition, Washington DC.

Bakke R. & Olsson, P.Q. 1986. Biofilm thickness measurement by light microscopy. J. Microbiol. Methods, 5, 1-6.

Brown, H.D. 1976. A comparison of the attached algal communities of natural and artificial substrates. J. Phycol., 12, 301-306.

Brush, G.S. 1984. Stratigraphic evidence of eutrophication in an estuary. Wat. Resources Res., 20: 531-541.

Carpenter, S.R. & Lathrop, R.C. 1999. Lake restoration: Capabilities and needs. Hydrobiologia, 395/396: 19-28.

Cooper, S.R. & Brush, G.S. 1994. Long-term history of Chesapeake Bay anoxia. Science, 254: 992-993.

Deppe, T. and Benndorf, J. 2002. Phosphorus reduction in a shallow hypereutrophic reservoir by in-lake dosage of ferrous iron. Wat. Res., 36: 4525-4534.

Desikachary T.V. 1959. Cyanophyta. Indian Council of Agricultural Research, New Delhi.

Guasch H. & Sabater, S. 1995. Seasonal variation in photosynthesis-irradiance responses by biofilms in Mediterranean streams. J. Phycol., 31, 727-735.

Gulati, R.D. & van Donk, E. 2002. Lakes in the Netherlands, their origin, eutrophication and restoration: State-of-the-art review. Hydrobilogia, 478: 73-106.

Harper, D.M. 1992. Eutrophication of Freshwaters. Chapman & Hall, London, England.

Harrison, R.M. 1992. Pollution—Causes, effects and control. Second edition, Royal Society of Chemistry, p. 393.

Havens, K.E. & Schelske, C.L. 2001. The importance of considering biological processes when setting total maximum daily loads (TMDL) for phosphorus in shallow lakes and reservoirs. Environ. Pollut., 113, 1-9.

Havens, K.E., T. Fukushima, T. Iwakuma, R.T. James, N. Takamura, T. Hanazato & Yamamoto, T. 2001. Nutrient dynamics and the eutrophication of shallow lakes Kasumigaura (Japan), Donghu (PR China) and Okeechobee (USA). Envrion. Pollut., 111, 263-272.

Hecky, R.E. & Kilham, P. 1988. Nutrient limitation of phytoplankton in freshwater and marine environments: A review of recent evidence on the effects of enrichment. Limnol. Oceanogr. 33, 796-822.

Hustedt F. 1930. Bacillariophyta (Diatomeae). Die Süsswasser-Flora Mittleuropas (Ed. A. Pascher), pp. 1-466. Gustav Fischer, Jena.

Istvánovics, V. & Somlyódy, L. 2001. Factors influencing lake recovery from eutrophication—the case of basin 1 of Lake Balaton. Wat. Res. 35: 729-735.

Istvánovics, V., Somlyódy, L. & Clement, A. 2002. Cyanobacteria-mediated internal eutrophication in shallow Lake Balaton after load reduction. Wat. Res., 36: 3314-3322.

Janssen M. 2001. An exploratory integrated model to assess management of lake eutrophication. Ecol. Model., 140: 111-124.

Jeffrey S.W. & Humphrey, G.R. 1975. New spectrophotometric functions for determining chlorophylls a, b, c1, c2 in higher plants, algae and natural phytoplankton. Biochem. Physiol. Pflanzen, 167, 191-194.

Lau, S.S. S & Lane, S.N. 2002. Nutrient and grazing factors in relation to phytoplankton level in a eutrophic shallow lake: The effect of low macrophyte abundance. Wat. Res., 36: 3593-3601.

Lewis, W.M. Jr. 2000. Basis for the protection and management of tropical lakes. Lakes & Reservoirs: Research and Management. 5: 35.

Mainstone, C.P. & Parr, W. 2002. Phosphorus in rivers—Ecology and management. The Sci Total Env., 282-283: 25-47.

Manny, B.A., Johnson, W.C. & Wetzel, R. G. 1994. Nutrient additions by waterfowl to lakes and reservoirs: Predicting their effects on productivity and water quality. Hydrobiologia 279/280, 121-132.

Mattila, J. & Räisänen, R. 1998. Periphyton growth as an indicator of eutrophication: An experimental approach. Hydrobiologia 377: 15-23.

Muhammetoğlu, A., Muhammetooğlu, H., & Soyupak, S. 2002. Evaluation of efficiencies of diffuse allochthonous and autochthonous nutrient input control in restoration of a highly eutrophic lake. Wat. Sci.Technol., 45: 195-203.

Nalewajko, C. & Lee, K. 1983. Light stimulation of phosphorus uptake in marine phytoplankton. Mar. Biol., 74: 9-15.

Paerl, H.W. 1988. Nuisance phytoplankton blooms in coastal, estuarine and inland waters. Limnol. Oceanogr., 33: 823-847.

Pan, Y.D. & Lowe, R.L. 1994. Independent and interactive effects of nutrients and grazers on benthic algal community structure. Hydrobiologia, 291, 201-209.

Parsons, T.R., Yoshiaki, M. & Lalli, C.M. 1984. A manual of chemical and biological methods for seawater analysis, Pergamon Press, Oxford.

Pettigrew, C.T., Hann, B.J. & Goldsborough, L.G. 1998. Waterfowl feces as a source of nutrients to a prairie wetland: Response of microinvertebrates to experimental additions. Hydrobiologia, 362: 55-66.

Philipose, M.T. 1967. Chlorococcales. Indian Council of Agricultural Research, New Delhi.

Postel, S. & Carpenter, S. R. 1997. Freshwater ecosystem services. In: G. Daily (ed.), Ecosystem Services. Island Press, Washington, D.C.

Prescott G.W. 1978. How to Know the Freshwater Algae? Wm C Brown Publishers, Dubuque, Iowa, USA.

Ramachandra, T.V., Murthy, C. R. & Ahalya, N., (Eds). 2000. Proceedings of the International Symposium on restoration of lakes and wetlands. 27-29 November 2000, Indian Institute of Science, Bangalore, India. Online at http://ces.iisc.ernet.in/energy/water/ proceed/proceedings_gra.htm.

Rao T.S., Rani, P.G., Venugopalan, V.P. & Nair, K.V.K. 1997. Biofilm formation in a freshwater environment under photic and aphotic conditions. Biofouling, 11, 265–282.

Reil, M. & Arndt, L. 2000. Microbial metabolism in sediments of coastal inlets of the southern baltic sea—response to gradients of eutrophication. WVU-Workshop "Baltic Coastal Ecosystems—Structure, Function and Coastal Zone Management", Rostock, Germany, November 20 to 23, 2000.

Ruley, J.E. & Rusch, K.A. 2002. An assessment of long-term post-restoration water quality trends in shallow, subtropical, urban hypereutrophic lake.

Ryther, J.H. & Dunstan, W.M. 1971. Nitrogen, phosphorus and eutrophication in the coastal marine environment. Science 171: 1008-1013.

Sagehashi, M., Sakoda, A. & Suzuki, M. 2001. A mathematical model of a shallow and eutrophic lake (The Keszthely Basin, Lake Balaton) and simulation of restorative manipulations. Wat. Res., 35: 1675-1686.

Sagehashi, M., Sakoda, A. & Suzuki, M. 2000. A predictive model of long-term stability after biomanipulation of shallow lakes. Wat. Res., 34: 4014-4028.

Sas, H. 1989. Lake restoration by reduction of nutrient loading: Expectation, experiences, extrapolations. Academia Verlag Richarz, Germany.

Scharf, W. 1999. Restoration of the highly eutrophic Lingese Reservoir. Hydrobiologia, 416: 85-96.

Schindler, D.W. 1977. Evolution of phosphorus limitation in lakes. Science 195, 260-262.

Sekar, R., Nandakumar, K., Venugopalan, V. P., Nair, K.V.K. & Rao, V.N.R. 1998. Spatial variation in microalgal colonization on hard surfaces in a lentic freshwater environment. Biofouling, 13, 177-195.

Sekar, R., Nair, K.V.K., Rao, V.N.R., & Venugopalan, V.P. 2002. Nutrient dynamics and successional changes in a lentic freshwater biofilm. Freshwat. Biol. 47: 1893-1907.

Shapiro J., Lamarra, V. & Lynch, M. 1975. Biomanipulation—An ecosystem approach to lake restoration. In: Water Quality Management Through Biological Control (Eds) P.L. Brezonic & J.L. Fox, pp. 85-96. University Florida, Gainesville.

Sharma, M.O., Bhosle, N.B., & Wagh, A.B. 1990. Methods for removal and estimation of microfouling biomass. Indian J. Mar. Sci., 19, 174-176.

Smith, V.H., Tilman, G. D. & Nekola, J. C. 1999. Eutrophication: Impacts of excess nutrient inputs on freshwater, marine and terrestrial ecosystems. Environ. Pollut. 100, 179-196.

Sondergaard, M., Kristensen, P. & Jeppesen, E. 1992. Phosphorus release from resuspended sediments in the shallow and wind-exposed Lake Arreso, Denmark. Hydrobiologia 228, 91-99.

Stabel, H.H. 1985. Mechanisms controlling the sedimentation of various elements in prealpine lakes. In: Stumm, W. (Ed.), Chemical Processes in Lakes.

Turk, A., Turk, J., Witter, J.T. and Witter, R. 1974. Environmental science. WB Sanders Co., Philadelphia. P. 563.

Turner, R.E. & Rabalais, N.N. 1994. Coastal eutrophication near the Mississippi River delta. Nature 368: 619-621.

Venugopalan, V.P., Nandakumar, K., Rajamohan, R., Sekar, R., & Nair, K.V.K. 1998. Natural eutrophication and fish kill in a shallow freshwater lake. Curr. Sci., 74: 915-917.

Verschuren, D., Johnson, T.C., Kling, H.J., Edgington, D.N., Leavitt, P.R., Brown, E.T., Talbot, M. R. & Hecky, R. E. 2002, History and timing of human impact on Lake Victoria, East Africa. Proc. R. Soc. Lond. B Biol. Sci. 269: 289-294.

Wassmann, P., Egge, J.K., Reigstad, M. & Aksnes, D.L. 1996. Influence of dissolved silicate on vertical flux of particulate biogenic matter. Mar. Pollut. Bull. 33: 1-6.

PART 2

Biotic Environment

CHAPTER

6

The Impact of Eutrophication on Lake Chivero, Zimbabwe: A Tropical African Reservoir

B.E. Marshall

Department of Biological Sciences, University of Zimbabwe,
P.O. Box MP 167, Mount Pleasant, Harare, Zimbabwe

ABSTRACT

Lake Chivero is a eutrophic manmade lake situated near Harare, Zimbabwe, which receives sewage effluent discharged into its tributaries. The first evidence of eutrophication appeared in 1960 and the water quality deteriorated steadily until 1970, when a temporary improvement was brought about by the use of sewage effluent for irrigation. From 1980 onwards, water quality began to decline again because of the growth of Harare's population and a corresponding increase in sewage effluent. This decline was masked for a period by a major infestation of water hyacinth, which evidently removed nutrients from the water. This plant has been a problem on the lake because of the several major outbreaks in the fifty years, since its creation although biological control has now been effective in reducing its impact. Permanent blooms of blue-green algae have led to very high levels of primary productivity, now limited by reduced light penetration, while increased levels of organic matter in the hypolimnion have led to a strong oxycline, which would probably be absent if the lake was oligotrophic, owing to its weakly-developed thermocline. This has allowed ammonia—which has been implicated in fish kills—to accumulate in the hypolimnion. Zooplankton abundance is considerably greater than it is in oligotrophic Zimbabwean waters. The effect of eutrophication on the benthos is less clear since the impact of water level fluctuations is marked, but there does appear to have been an increase in molluscs and a change in the species composition of the oligochaetes over the last thirty years. Fish production is high and dominated by cichlid fishes that are able to utilize blue-green algae, but fish kills caused by deoxygenated water do occur from time to time. A major fish kill in March 1996 affected only one species, *Oreochromis macrochir*, and may have been caused by a combination of high ammonia levels, low dissolved oxygen at night and, possibly, algal toxins. The early control measures, which involved the use of sewage effluent for irrigation, have failed to cope with the increased quantity of sewage effluent and the lake continues to be eutrophic. An economic collapse in Zimbabwe has left the autorities without resources to invest in the treatment of sewage or control of nutrients and there is little prospect of the situation improving in the near future.

INTRODUCTION

Lake Chivero (formerly known as Lake McIlwaine) is a manmade lake located about 37 km southwest of Harare, the capital city of Zimbabwe (Fig. 6.1), was created in 1952 when a dam was built on the Manyame River to provide water for the city. At full supply level, it has an area of 26.3 km^2, a volume of 250 × 10^6 m^3 and a mean depth of 9.4 m. Its catchment area extends over 2,227 km^2 and most of the city of Harare, and its dormitory town of Chitungwiza (established in the 1970s) lies within the catchment. The remainder consists of mixed farming land with maize, tobacco and wheat as the principal crops, and a small area occupied by subsistence farmers.

Sewage effluent from Harare and Chitungwiza, which is discharged from three large sewage works into rivers supplying the lake (Fig. 6.1), and it began to show signs that it was becoming eutrophic soon after it was created. The first symptom of eutrophication, the appearace of algal blooms in winter, appeared in 1960 and by 1963, these algal blooms had become permanent and were causing difficulties in water purification (Munro, 1966; McKendrick, 1982). By 1962, there was some evidence that algal toxins were responsible for gastrointestinal problems amongst children in the city (Ziberg, 1996).

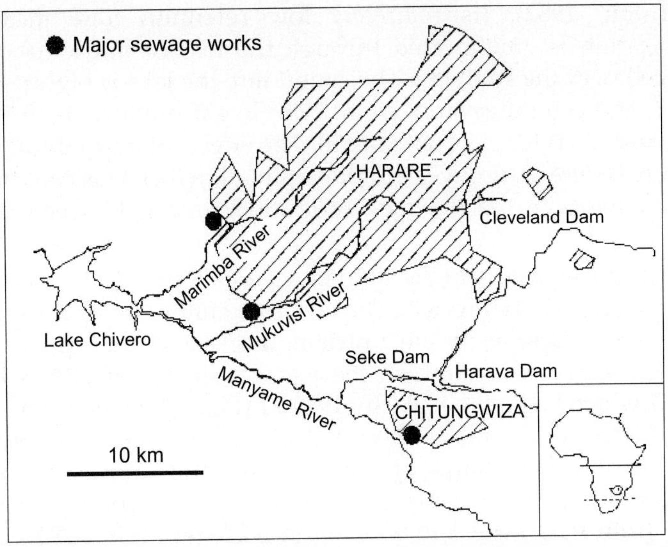

Fig. 6.1 The location of Lake Chivero in relation to the city of Harare and its dormitory town, Chitungwiza, and their principal sewage works

The eutrophication of the lake began when Harare was still a relatively small city with a population of about 300,000 people and discharged about 20,000 m^3 of sewage effluent per day (dry weather flow). Today, the population of Harare and Chitungwiza is around two million people and the dry weather flow of sewage effluent has risen to around 200,000 m^3 per day. Efforts to reduce the nutrient supply to the lake began in 1970 and were initially successful (Thornton & Nduku, 1982) but they failed when the quantity of sewage effluent being discharged increased rapidly after 1980 (Marshall, 1994). The situation is now essentially out of control since local governments, who are responsible for sewage treatment and water supply, lack the resources to expand and modernize the sewage treatment plants. Owing to the collapse to the Zimbabwean economy and currency, from 2000 onwards, there is little prospect of any improvement in the situation in the immediate future.

CHEMICAL HISTORY OF THE LAKE

The chemical status of the lake is influenced both by its hydrology and by the sewage effluent discharged into its tributaries. The mean annual runoff into the lake is 304.7×10^3 m^3 or about 1.22 times its volume, which means that it has a theoretical water retention time of 0.8 years (Ballinger

& Thornton, 1982). Its relatively low retention time means that eutrophication is ameliorated through the loss of nutrients in water discharged over the spillway. The runoff into the lake is highly variable, however, and can range from 20×10^6 m^3 in a drought year to 797×10^6 m^3 in a year of above average rainfall. The effects of eutrophication tend to be more severe during years of low rainfall, which was clearly shown by events that occurred during a series of dry years between 1991 and 1996.

The chemical history of the lake between 1958 and 2000 can be divided into four main periods (Fig. 6.2). The first lasted until 1970 and was marked by a general increase in the nutrient levels and the appearance of permanent algal blooms. In 1958, the lake was slightly alkaline with a pH of 7.5-8.0, which had risen to 9.0 by 1970-71 (Fig. 6.2a) while conductivity increased from 100 $\mu S\ cm^{-1}$ to about 170 $\mu S\ cm^{-1}$ over the same period (Fig. 6.2b). The concentrations of phosphorus and nitrogen increased significantly with the concentration of soluble reactive phosphorus rising fivefold from 0.04 mg l^{-1} in 1962-63 to 0.22 mg l^{-1} in 1970-71, while ammonia rose fourfold from 0.1 mg l^{-1} in 1962-63 to 0.4 mg l^{-1} in 1968-69 (Fig. 6.2c and 6.2d).

The second phase began with the diversion of sewage effluent for the irrigation of pastureland in 1970 and was marked by a rapid decrease in nutrient concentrations. By 1980, the pH had fallen to 7.5, conductivity to 126 $\mu S\ cm^{-1}$, and soluble reactive phosphorus to 0.04 mg l^{-1} (it had fallen to as low as 0.01 mg l^{-1} in 1978-79). Although ammonia concentrations fell rapidly after 1970, the level of nitrates rose until 1975, possibly because ammonia was oxidized, but both forms of nitrogen decreased thereafter until 1980, when the mean ammonia and nitrate concentrations were 0.04 and 0.06 mg l^{-1}, respectively. There was a visible improvement in the water quality during this period. The severity of algal blooms decreased to such an extent that the concentration of chlorophyll *a* declined to a level comparable to that in the oligotrophic Cleveland Dam (Table 6.1). This success led people to believe that the problem of eutrophication had been solved and Thornton (1981) was able to write of an "ecological disaster" being averted.

The next phase (1985-1991) was characterized by an extensive outbreak of water hyacinth *Eichhornia crassipes*, which at its peak, covered about 35% of the lake's surface (Chikwenhere and Phiri, 1999). Although the conductivity rose during this period, the water quality was generally good with the pH and the concentrations of phosphorus and nitrogen falling to their lowest levels since the mid-1960s. Although no data are available, it seems likely that the water hyacinth was taking up nutrients and masking the failure of the nutrient diversion programme.

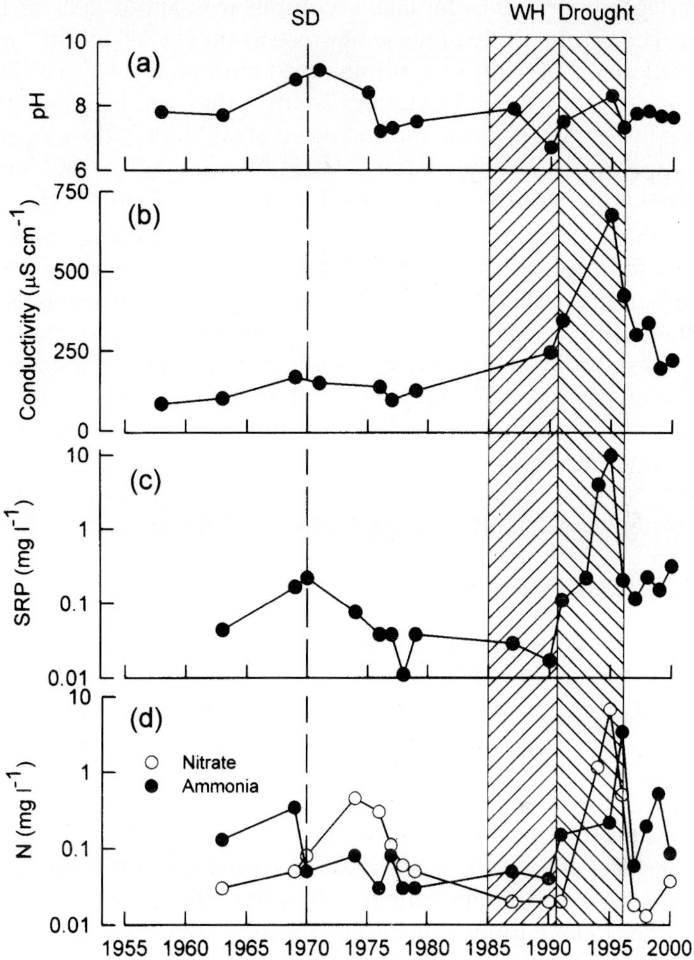

Fig. 6.2 The chemical history of Lake Chivero, 1958 to 2000, showing changes in: (a) pH, (b) conductivity, (c) soluble reactive phosphorus and (d) nitrogen, as ammonia and nitrates. The vertical line mariked SD indicates the diversion of sewage effluent that began in 1970, while the hatched blocks indicate a major water hyancinth infestation (WH) and prolonged drought (Drought). The data are mean values taken from various sources given in Marshall (1994, 1997), and the City of Harare (personal communication)

This failure became most apparent after the water hyacinth had been brought under control in 1991, and coincided with a series of five drought years when the inflow to the lake was greatly reduced. Indeed, in the driest months of these years, sewage effluent made up the sole inflow to the lake, as there was no natural flow in any of the tributary streams. During these drought years, the water level fell more or less continuously until by December 1995, the lake's volume was about 20% of its full capacity. The conductivity of the water rose to almost 700 μS cm^{-1} and the concentrations of phosphorus, nitrates and ammonia rose steadily until they were about 100 times greater in 1995 than they had been in 1991. The water quality was very poor and intensive algal blooms developed after the disappearance of the water hyacinth mats, with the levels of chlorophyll *a* exceeding those recoded in the 1960s (Table 6.1).

This phase came to an end with heavy rains during the 1995-96 monsoons that caused the lake to fill completely by February 1996 and then resulted in a very large and extensive fish kill (discussed later). This was followed by an immediate reduction in nitrogen and phosphorus and a corresponding fall in conductivity that was maintained by good rainy seasons in the following years. The levels of phosphorus and nitrogen were, nevertheless, higher than their pre-1970 levels and algal blooms remained a problem. It is clear that the diversion of sewage effluent is now having little effect on nutrient concentrations and that future droughts are likely to bring further problems.

THE TEMPERATURE AND OXYGEN CYCLE

Although Lake Chivero is situated in the tropics, the climate of the area is moderated by its altitude (the lake is 1,363 m above sea level). Mean air temperatures (mean minimum-mean maximum) range from 7-21°C in July to 15-29°C in October, while the water temperatures range from 16-25°C in the corresponding months (Torrance, 1972). Most Zimbabwean water bodies stratify during the summer months but the thermal stratification in Lake Chivero is weakly developed when compared to other reservoirs (Begg, 1970; Marshall and Falconer, 1973; Mitchell and Marshall, 1974; Donnelly 1980). Despite its barely discernible thermocline, Lake Chivero develops a strong oxycline and an anoxic hypolimnion (Fig. 6.3). It has been suggested that the oxycline might not have developed had the lake not been eutrophic (Beadle, 1981).

Table 6.1 Changes in the concentration of chlorophyll *a* (mg m^{-3}) in Lake Chivero, compared with Cleveland and Seke dams. Data from Thornton & Nduku (1982) and Ahmed (1992)

Reservoir	*Date*	*Chlorophyll a*
Lake Chivero	1968-69	50-150
	1979-80	2-29
	1992	48-221
Seke Dam	1992	88-134
Cleveland Dam	1992	2-14

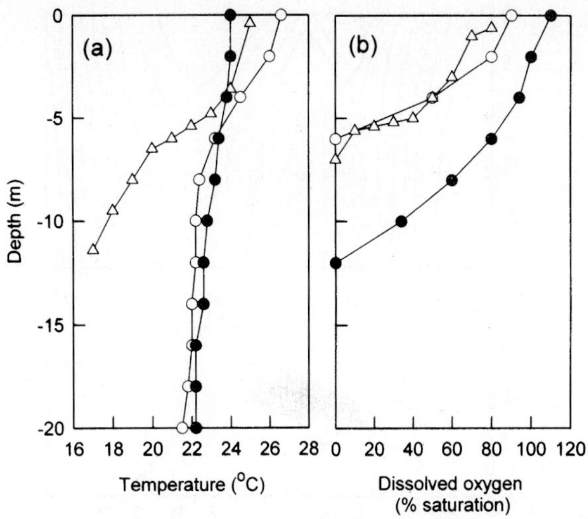

Fig. 6.3 (a) Temperature and (b) oxygen profiles in three Zimbabwean reservoirs in summer. Lake Chivero is eutrophic, while the Mazowe Dam is mesotrophic and the Mtsheleli Dam is oligotrophic.● = Lake Chivero, November 1968 (Marshall and Falconer, 1973), ○ = Mazoe Dam, February 1971 (Mitchell and Marshall, 1974), △ = Mtsheleli Dam, November 1974 (Donnelly 1980)

The lake is generally monomictic with stratification beginning in September-October and continuing until an overturn occurs in February-March (Fig. 6.4). Ammonia builds up in the hypolimnion during this time and the release of anoxic, ammonia-rich water has been implicated in fish kills. Polymicticism has been recorded, notably in 1977, when the hypolimnion was re-oxygenated on two occasions in January and February (Thornton and Nduku, 1982). On both occasions, nutrient upwelling and localized algal blooms confirmed that mixing had taken place. The most likely cause was a sudden inflow of river water following heavy rains, although river water may flow into the lake as a discrete body, displacing the hypolimnion upwards (Marshall and Falconer, 1973).

THE BIOTA

Primary Production and Plankton

Despite the problems caused by the algal blooms in the lake, the ecology of the phytoplankton is poorly understood. Falconer (1973) made the only detailed investigation of phytoplankton in the lake and found that it consists almost entirely of blue-green algae, principally *Microcystis*

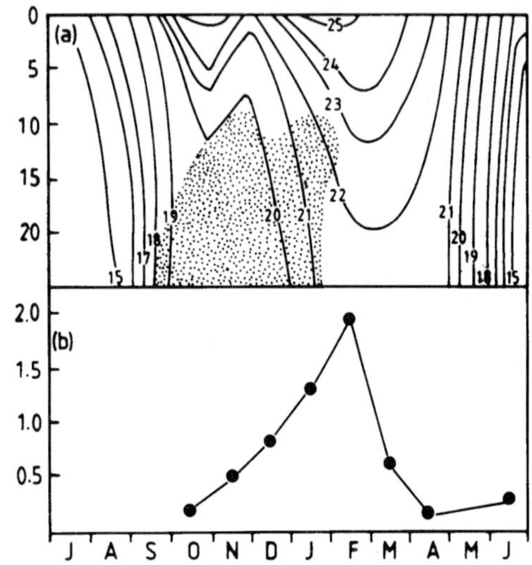

Fig. 6.4 The pattern of thermal stratification (isotherms in °C) in Lake Chivero, July 1968-June 1969. The shaded area denotes the extent of the anaerobic hypolimnion while the points show the buildup of ammonia in the hypolimnion. Redrawn from Marshall and Falconer (1973)

aeruginosa and a species of *Anabaena* (wrongly identified by Falconer as *A. flos-aquae*; G. Cronberg, personal communication). *Microcystis* was generally more abundant during the summer months, while *Anabaena* was dominant during the winter after the lake had become isothermal (Fig. 6.5). Unpleasant tastes and odours were prevalent at this time of year and water purification was made more difficult. Cases of gastroenteritis increased during these months and provided circumstantial evidence to support Zilberg's (1966) contention that algal toxins were responsible (Marshall, 1991). The problem of algal toxins is, indeed, a matter of concern because recent work has shown that the toxic species *Microcystis toxica* is present (G. Cronberg, personal communication) and microcystin and other toxins have been discovered both in the lake and in tapwater in the city (Johansson and Olsson, 1998). These findings have potentially serious implications for public health but have not so far been given much attention.

Primary productivity is extremely high as a result of the dense algal population in the lake. Productivity in the first two metres was around 1500 mg O_2 m^{-3} h^{-1} or more than three times as high as it was in the mesotrophic Mwenje Dam, where it was about 400 mg O_2 m^{-3} h^{-1} (Fig. 6.6). This was a striking contrast with the oligotrophic Lake Kariba where production was only about 60 mg O_2 m^{-3} h^{-1}. But primary productivity in Lake Chivero decreased rapidly with depth and it is likely that the

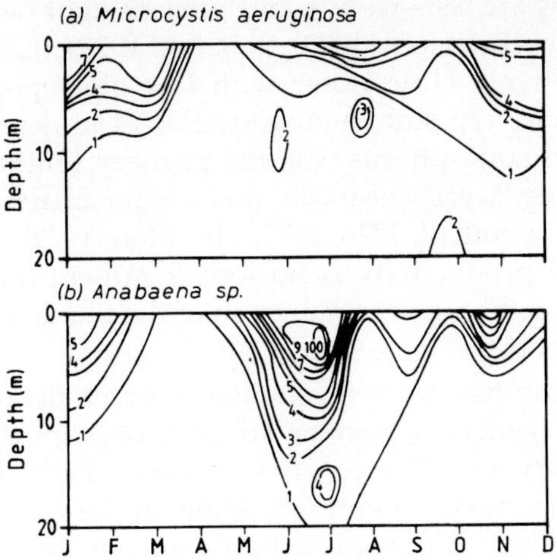

Fig. 6.5 The abundance of (a) *Microcystic aeruginosa* and (b) *Anabaena* sp. in Lake Chivero in relation to depth, 1969. Redrawn from Falconer (1973).

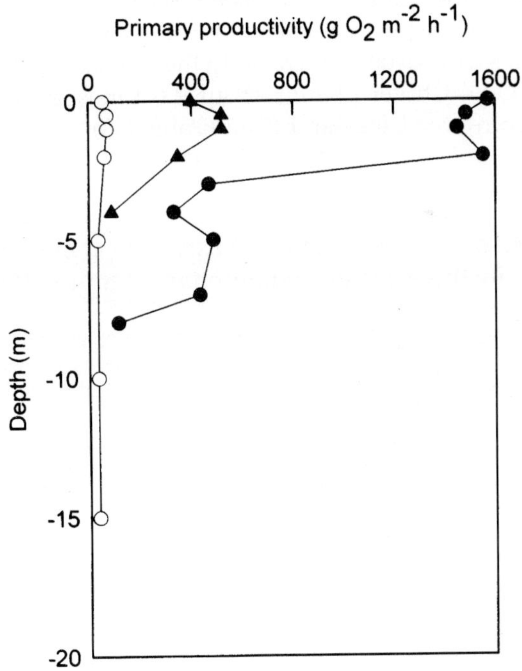

Fig. 6.6 Primary productivity in Lake Chivero, October 1971 (●), compared with the mesotrophic Mwenje dam, December 1971 (▲) and the oligotrophic Lake Kariba, February 1983 (○). From Mitchell and Marshall (1974) and Machena (1983).

dense algal blooms are self-shading and prevent light pentrating much below the surface (Falconer, 1973; Robarts, 1979). Lake Kariba, by contrast, has relatively clear water with low algal populations and primary productivity can still be measured at 14 m depth. As in most Zimbabwean lakes, phosphorus was the primary limiting nutrient in Lake Chivero during its early years but this was no longer the case by the 1970s (Robarts and Southall, 1976, 1977; Thornton, 1979). It seems likely now that primary productivity is no longer nutient-limited, but has become self limiting through the reduction of the photic zone caused by the dense algal blooms.

The zooplankton has not been studied in much detail. The species composition was generally similar to that of a number of other Zimbabwean reservoirs (Munro, 1966; Green, 1990; Elenbaas and Grundel, 1994). The impact of eutrophication on the zooplankton is not obvious, although there does seem to be a tendency towards a decrease in the importance of *Daphnia* spp. and an increase in *Ceriodaphnia* amongst the Cladocera (Table 6.2). Amongst the copepods, diaptomids seem to increase while cyclopiods decrease with increasing eutrophication, although these data need to be treated with caution and more work on this aspect still needs to be done. One clear impact of eutrophication seems to be an increase in the biomass of the zooplankton, with that in Lake Chivero being about 2.6 times greater than in the relatively oligotrophic Cleveland Dam (Table 6.3).

Macrophytes

In the 1960s, Lake Chivero supported extensive communities of rooted macrophytes, in three major communities (Munro, 1966). The most

Table 6.2 The relative importance (%) of certain zooplankton species in some Zimbabwean reservoirs. The trophic status of these reservoirs ranges from the most oligotrophic in the left-hand column to the most eutrophic in the right-hand column. The values for Marondera and Nyanga are the means of 8 and 10 small dams respectively. From Green (1990) and Elenbaas and Grundel (1994).

	Nyanga	*Marondera*	*Cleveland*	*Chivero*
Cladocera	54.1	53.2	18.8	30.0
Diaphanosoma	1.4	1.4	3.9	2.5
Daphnia	47.5	47.6	13.2	6.3
Ceriodaphnia	1.0	1.1	0.4	11.5
Bosmina	3.0	3.0	1.3	9.7
Others	1.2	1.2		
Copepoda	46.6	46.8	81.2	70
Diatomids	17.6	17.7	56.2	48.0
Cyclopoids	29.0	29.1	25.0	22.0

Table 6.3 The mean density (no. $m^{-3} \times 10^3$) and proportion (%) of the principal zooplankton groups in Cleveland Dam (oligotrophic) and Lake Chivero (eutrophic), 1990-91. From Elenbaas & Grundel (1994).

	Cleveland		*Chivero*	
Cladocerans	3.0	(9.3%)	12.7	(15.2%)
Copepods	13.4	(41.3%)	33.6	(40.6%)
Rotifiers	15.9	(49.4%)	37.2	(44.6%)
Total	32.4		83.6	

extensive (40% of the shoreline) was the *Nymphaea* community, followed by the *Polygonum* and *Typha* communities (25% and 20%, respectively) with the remaining 15% of the shoreline consisting of bare granite or ironstone rocks. The *Polygonum* community was then more or less restricted to the upper reaches of the lake, but it gradually spread over the rest of the shoreline to become the dominant community (90% of the shoreline) by 1973 (Jarvis, Mitchell and Thornton, 1982). The reasons for this spread are unclear but the impact of water hyacinth control measures may have contributed to the decline of the other communities. On the other hand, the water lilies (*Nymphaea*) and other semi-emergent species, as well as submerged ones like *Lagarosiphon*, which were commonly associated with them, could have died out as the algal blooms developed and reduced light penetration. There is some controversy about the loss of these communities since Junor (1969) suggested that the macrophagous cichlid fish *Tilapia rendalli* might have been responsible for it. But these communities occur along with this fish elsewhere in Zimbabwe and their disappearance may have been a combination of factors and the plants were perhaps unable to resist the impact of grazing by the fish under conditions of low light penetration.

Alien plants have been a matter of concern in the lake from the very beginning. Water hyacinth was first recorded in the Mukuvisi River in the 1930s (Brain, 1937) and a major outbreak occurred in the lake soon after its construction. This was brought under control with 2,4-D between 1953 and 1958, which may have been the earliest use of this herbicide against water hyacinth anywhere in the world; the government botanist at the time had worked on its development during the Second World war. This outbreak was kept under control for a numbers of years by a small team of men who patrolled the lake and destroyed any plants that appeared.

The next outbreak occurred in 1970 and followed a serious drought in 1967-68 when the lake level fell and exposed large numbers of water hyacinth seeds that were lying dormant at the bottom of the lake. This situation was aggravated by the fact that the control team had been disbanded, thereby teaching the authorities a lesson that was later to be forgotten. After some controversy over the use of 2,4-D, this outbreak was also brought under control by the end of 1971 through the use of this herbicide. The control team was set up again and successfully prevented reinfestation after the 1972-73 drought by destroying the germinating seedlings that appeared on the exposed mud flats as the lake level fell.

So successful was this control system that the authorities once again became complacent and disbanded the control team, with the result that another outbreak of water hyacinth began in 1985. This was the most severe in the lake's history with upto 35% of its surface being covered by the weed (Chikwenhere and Phiri, 1999). Expensive and ultimately futile attempts to remove it manually continued for some years until in 1991, when a large quantity of the weed (no reliable estimate was made but it may have amounted to many thousands of tonnes) was swept over the spillway of the dam. This damaged various structures extensively and threatened road and rail bridges downstream, and raised fears for the safety of the 36-m high dam itself. Consequently, it was decided to spray the weed with 2,4-D and it was brought under control in early 1992. At the same time, biological control with the weevil *Neochetina eichhorniae* and some other biocontrol agents had begun. These have been very successful and reduced water hyacinth coverage from about 35% of the lake's surface is 1991 to 3.5% in 2000 (Chikwenhere, 2001). An important factor in the success of the biological control programme was the fact that the authorities were unable to continue financing chemical control, thus allowing the weevils to proliferate.

Other invasive macrophytes have occasionally caused problems on the lake but none of them as serious as water hyacinth. A small outbreak of water lettuce *Pistia stratiotes* in the tributary streams was eradicated through biological control (Chikwenhere, 1994). The water fern *Azolla filiculoides* occurs around the shoreline but it is unable to colonize in the open water because its mats are broken up by wave action. The parrot's feather *Myriophyllum aquaticum* has been present in the lake since the 1960s but it has never become a problem. *Hydrocotyle ranunculoides*, known locally as 'spaghetti weed', is currently the most serious problem plant on the lake. It forms fringing mats along the shoreline that hinder access to the water. It has also contributed to a sharp reduction in the number of subsistence anglers who formerly caught large quantities of small fish from the water's edge. This has serious implications locally because most of these people are poor and unemployed, and rely on fish to supplement their diet.

Benthos

Benthic fauna can be good indicators of pollution but in Lake Chivero, the effects of eutrophication are not immediately obvious, perhaps because the benthos is affected by other factors. The first of these was the loss of the *Nymphaea* community during the 1960s, which brought about a considerable reduction in the diversity of benthic invertebrates (Table 6.4). If, as already mentioned, the loss of these plant communities could be attributed to the blooms, then the reduction in invertebrate diversity is one consequence of the lake's eutrophication.

The second important influence on the benthos is the effect of water level fluctuations on the fauna of the littoral zone. When the water level falls, the littoral becomes increasingly unsuitable for benthic insects and their numbers decline. If the water level drops by more than about 2 m in a year, large numbers of snails and mussels will be stranded on the exposed bed of the lake (Marshall, 1975). At this time of the year, oligochaetes are the dominant animals in the benthos. The lake can fill very quickly after the rains begin and the rise in water level causes nutrients to be released from the exposed mud and the vegetation growing on it. Chironomid larvae rapidly colonize the newly flooded land and can be found in very large numbers in shallow waters (Fig. 6.7).

Some changes in the benthic fauna that may be attributable to eutrophication have taken place over a period of time. Among the oligochaetes, there was a decline in the abundance of *Branchiura sowerbyi*

Table 6.4 The mean abundance (no m^{-2}) of benthic invertebrates in relation to the major communities of aquatic macrophytes in Lake Chivero. From Munro (1966) and Marshall (1971). The symbol + denotes densities less than 1 m^{-2}.

	1962-63		*1968-70*	
Vegetation type	*Nymphaea*	*Typha*	*Polygonum*	*No vegetation*
Branchiura sowerbyi	2480	1350	1853	769
Limnodrilus hoffmeisteri	+	+	+	141
Hirudinea			1	2
Ephemeropteran nymphs		1		1
Odonate nymphs	8	4		1
Trichoptera larvae	31	8	39	+
Chaoborus larvae and pupae	11	2	88	+
Chironomid larvae and pupae	1072	638	163	586
Tanypod larvae and pupae	222	263	136	156
Culicine larvae and pupae	3		10	10
Corbicula africana	200	228	536	
Total	4053	2529	2832	1687

(a trend already apparent in the 1960s: Table 6.4), while the more tolerant species *Limnodrilus hoffmeisteri* increased in numbers (Table 6.5). Another striking feature is a dramatic increase in the snail population that occurred between the 1960s and the 1990s; the prosobranch *Melanoides tuberculata* was particularly abundant, with densities exceeding 1,000 m^{-2} (Table 6.5). The reasons for the increase in snails is not immediately obvious but the population in Lake Chivero was much higher than in some other southern African lakes, which may reflect an increase in food available to them, as well as an increase in the calcium concentration of the water, which rose over 20 years from about 25 mg l^{-1} to 80 mg l^{-1} (Marshall, 1994). Both of these aspects reflect the eutrophic state of the lake.

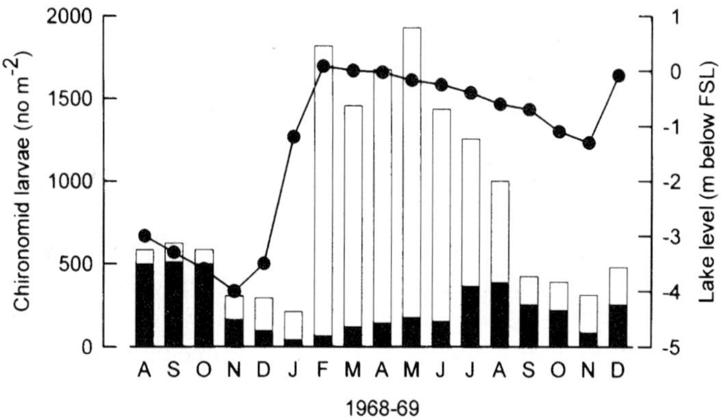

Fig. 6.7 The relationship between lake level and the abundance of chironomid larve and pupae in the littoral zone of Lake Chivero, 1968-69. Open histogram = Chironominae, solid histogram = Tanypodinae. From Marshall (1978).

Table 6.5 The mean density (no m^{-2}) of benthic invertebrates in Lake Chivero in 1968-69 and 1993. Data are for the low water period in both years. From Marshall (1978, 1995).

		1968-69	*1993*
Oligochaeta	*Branchiura sowerbyi*	816	
	Limnodrilus hoffmeisteri		493
Chironomidae	Chironominae	167	77
	Tanypodinae	41	82
	Unidentified		22
Mollusca	*Bellamyia capillata*		40
	Melanoides tuberculata		1036
	Other snails		6
	Corbicula africana		18
Total		10271	1774

Fish

Twenty-nine species of fish, eight of which are alien species introduced for various reasons, are known to occur in the lake (Table 6.6). The fish population has changed considerably during the lake's history but not all of these changes have occurred as a result of its eutrophication. The formation of the lake caused a decline in certain fluvial species, among them the cyprinids (*Barbus*, *Labeobarbus* and *Opasaridium*) that were unable to adapt to the standing water environment and they are present in small numbers and usually restricted to the mouths of the tributary streams. The rock and sand catlets (*Chiloglanis neumanni* and *Leptoglanis rotundiceps*) which are abundant in flowing water and are present throughout the catchment, were unable to survive in the lake at all and have never been recorded from it.

A second group of fishes includes the potamodrometic species (i.e. those that must run up flooding rivers to breed) such as *Brycinus imberi, Hydrocynus vittatus*, and *Labeo altivelis*. These species were abundant when the lake first filled but their numbers fluctuated, seemingly in relation to the degree of flooding in the tributary streams. Other factors that may have affected these species include the construction of weirs across the tributary streams and their increasing levels of pollution, both of which may affect their breeding. *Brycinus, Hydrocynus* and *L. altivelis* all decreased during the droughts of the early 1990s and it was feared that they had become extinct in the lake. A succession of good rainy seasons from 1995 to 2000 evidently improved the water quality in the tributary streams and a modest recovery in the populations in these fishes has occurred.

The most successful fishes are the cichlids, which are well adapted to living in standing waters but the most important of them are alien species. Two of them, *Oreochromis macrochir* (introduced in 1956) and *O. niloticus* (introduced around 1990) feed on, and can digest, the blue-green algae that constitute the algal blooms in the lake (Munro, 1967; Moriarty, 1973; Minshull, 1978). Until the 1990s, *O. macrochir* was the most important species in the lake, making up about 70% of the commercial catch (Marshall, 1978), but *O. niloticus* is now more important although the two species combined probably account for more than 80% of the total commercial catch from the lake (Department of National Parks and Wildlife Management, personal communication). The third important alien cichlid is the basslike species *Serranochromis robustus* that has become the main predaceous fish species in the lake, although it is not numerous as the two microphagous species.

Table 6.6 The fish species of Lake Chivero with an indication of their current status in the lake (Marshall, 1982) with the addition of some species recorded since then). The symbol * denotes an introduced species.

Family	*Species*	*Common name*	*Present status*
Anguillidae	*Anguilla bengalensis*	African mottled eel	Rare, possibly extinct
Mormyridae	*Cyphomyrus discorhynchus*	Zambezi parrotfish	Uncommon, localized
	Marcusenius macrolepidotus	Bulldog	Numerous, widespread
	Mormyrus longirostris	Bottlenose	Rare, localized
Characidae	*Brycinus imberi*	Imberi	Formerly abundant, now almost extinct
	Micralestes acutidens	Silver robber	Scarce, localized
	Hydrocynus vittatus	Tigerfish	Very scarce but showing some recovery
Cyprinidae	*Cyprinus carpio**	Common carp	Formely scarce, now becoming abundant
	*Carrasius auratus**	Goldfish	One specimen known
	Barbus lineomaculatus	Line-spotted barb	Scarce, localized
	Barbus paludinosus	Straight-finned barb	Scarce, localized
	Barbus radiatus	Beira barb	Rare
	Barbus trimaculatus	Three-spotted barb	Scarce, localized
	Labeobarbus marequensis	Yellowfish	Rare
	Opsaridium zambezenze	Barred minnow	Rare, possibly extinct
	Labeo altivelis	Manyame labeo	Once abundant declined in the 1980s
	Labeo cylindricus	Red-eyed labeo	Common but localized
Schilbeidae	*Schibe intermedius*	Butter catfish	Two specimens known
Clariidae	*Clarias gariepinus*	African catfish	Widespread but relatively scarce
Centrarchidae	*Micropterus salmoides**	Largemouth bass	Fairly numerous
Cichlidae	*Pharyngochromis acuticeps*	Dwarf bream	Numerous
	Pseudocrenilabrus philander	Pygmy bream	Numerous
	Oreochromis mossambicus	Mozambique tilapia	Rare, possibly extinct
	*Oreochromis macrochir**	Green-headed tilapia	Abundant
	*Oreochromis nilotocus**	Nile tilapia	Abundant
	*Tilapia rendalli**	Red-breasted tilapia	Numerous
	Tilapia sparrmanii	Banded tilapia	Abundant
	*Serranochromis robustus**	Yellow-bellied bream	Numerous
	*Sargochromis codringtonii**	Olive bream	Rare

A final species that may be benefiting from the lake's eutrophic status is the common carp, *Cyprinus carpio*. This exotic species has been present in Zimbabwe since the 1940s (Toots, 1970) but generally in very low numbers; a few had been caught in Lake Chivero in the 1960s after heavy rains had caused a fish pond to burst but they seemed to have disappeared by the early 1970s. However, carp began to breed in smaller reservoirs after the good rainy season of 1995-96 and their numbers began to increase in Lake Chivero shortly afterwards. They appear to have bred very successfully ever since and are becoming increasingly numerous in the lake. The reasons for the relatively sudden increase in the common carp are not clear but it may be related to changes in the lake's trophic structure that have created suitable conditions for it.

Fish production from the lake is extremely important to the local community and the lake support highly productive fisheries, some of which are licensed while others are unlicensed and illegal. This makes it extremely difficult to obtain reliable fishery statistics and the total catch is unknown. Marshall (1978) estimated that the total catch from all sources (the licensed commercial fisherman, as well as anglers and illegal fishermen), was around 300 t yr^{-1} or about 120 kg $h.^{-1}$ yr^{-1}. Since then, the number of licensed fisherman has been increased while the intensity of illegal fishing is probably also much greater, owing to an increase in the number of improverished people living close to the lake. It is probably not unrealistic to suggest that the productivity of the lake may now have reached 250 kg $h.^{-1}yr^{-1}$.

One final aspect of fish biology in Lake Chivero that should be mentioned are the fish kills that have taken place from time to time, usually after deoxygenated water from the hypolimnion is brought to the surface, often after high winds or heavy rains (Marshall, 1981). While these attract a great deal of attention from the public, they apparently have little effect on the fish populations in the lake and there is no evidence to suggest that the population of any species in the lake has been adversely affected by fish kills.

One possible exception to this generalization is the fish kill that occurred in March-April 1996 when vast numbers of fish, possibly exceeding 100 tonnes in weight, were killed within the space of few days (Moyo, 1996, 1997). The public reaction to this fish kill was feverish with assertions that poison—possibly through some kind of industrial accident—had been discharged into the lake and that the human population of the city was also about to be poisoned, being amongst the milder reactions. In reality, this fish kill was highly unusual in that only one species, *Oreochromis macrochir*, and only large (> 12 cm in length) individuals of that species, were affected. Had some kind of chemical poison been discharged into the lake, then a wider range of species and size classes would have been found dead.

The general explanation for this fish kill is clear. By the end of 1995, a series of droughts had affected Zimbabwe and the level of Lake Chivero was very low (about 20% of capacity) and the only water flowing into it was sewage effluent, frequently of very poor quality. Heavy rains began in December 1995 and the lake filled completely by February 1996, drowning large quantities of plant material that had grown on the exposed bed of the lake and a very dense algal bloom developed. It seems likely that the concentration of ammonia, and perhaps other toxic substances such as hydrogen sulphide, in the hypolimnion built up to a very high level. The lake became isothermal in March and it is likely that ammonia levels in the surface waters were very high, a situation aggravated by the fact that oxygen concentrations fell during the night (Fig. 6.8). It is not clear why the combination of these factors should have affected only *O. macrochir* of one particular size; the possibility that algal toxins were involved has been suggested by Moyo, (1996) who noted extensive liver damage in the dead fish.

The consequences of such a large fish kill were not obvious and fish catches were not significantly reduced, a testimony to the high reproductive capacity of cichlid fishes and the corresponding high level

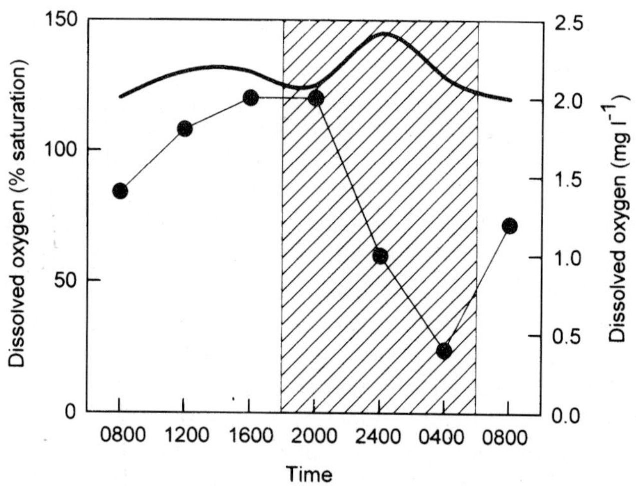

Fig. 6.8 The concentration of dissolved oxygen in surface waters of Lake Chivero over a 24-hour period in October 1970, expressed as per cent saturation (solid line) compared with March 1996, expressed as mg l^{-1} (points). The cross-hatching indicates the approximate hours of darkness (1800-0600). From Mitchell and Marshall (1974) and Moyo (1996).

of productivity of the lake. One possible consequence, that was not apparent at the time, was the replacement of *O. macrochir* by *O. niloticus* as the dominant species in the lake. Even this conclusion should be tempered by caution since *O. niloticus* has displaced indigenous tilapias in waters throughout Zimbabwe, regardless of their trophic status.

CONTROL MEASURES: SUCCESS AND FAILURE

Attempts to reduce the supply of nutrients began after experiments had shown that irrigating land with sewage effluent led to significant reductions in nitrogen and phosphorus (Table 6.7). In 1970, the Harare City Council purchased several farms between their sewage works and the lake and began large-scale irrigation of pastures with the intention of reducing the nutrient loading to the lake. These measures were successful at first and by 1980, the lake had retuned to a mesotrophic condition with a considerable improvement in its water quality. The algal blooms decreased, the levels of chlorophyll *a* declined, and submerged macrophytes such as *Lagarosiphon ilicifolius* increased. It was believed that the problem of its eutrophication had been solved (Thronton, 1982). But this was not the case; the water quality began to deteriorate in the mid-1980s (Fig. 6.2) and the present situation of the lake is probably as bad as it has ever been.

What went wrong? Quite simply, the authorities were unprepared for the increase in the population of greater Harare, which grew explosively and doubled between 1980 and 2000. The investment needed to deal with the increased quantity of sewage effluent was inadequate and became progressively less effective as economic decline in Zimbabwe accelerated from the mid-1990s onwards. Although the capacity of the sewage works has increased considerably, they are incapable of dealing with the volume of effluent and much of it is now very poorly treated. Similarly, the sewage farms were incapable of dealing with the increased volume of effluent and their effectiveness

Table 6.7 Reductions in phosphorus and nitrogen concentrations (mg l^{-1}) in sewage effluent before and after its use as irrigation water. Trials were carried out on sewage farms outside the cities of Bulawayo and Harare, Zimbabwe. Data from Thompson (1968) and McKendrick (1982).

	Bulawayo		*Harare*	
	Before	*After*	*Before*	*After*
SRP	20.6	0.9	9.0	0.9
NO_3–N	0.1	2.5	14.0	2.1
NH_3–N	52.9	0.1	12.0	0.5
Chloride			180.0	45.4

declined. Parts of these farms were converted to housing but no effort was made to overcome this loss through the development of new and better methods of nutrient removal, such as artiticial reed-beds or other forms of usable biomass. Furthermore, as the Zimbabwean economy began to collapse from 1997 onwards, inflation increased and the availability of foreign currency needed to import pumps and other equipment decreased. The authorities now find themselves in an impossible situation and are quite unable to control the discharge of sewage effluent into the lake.

There is little doubt the situation will become worse in the near future, since no Zimbabwean government or local government agency has the resources or capacity to deal with it effectively. The authorities now find themselves in the difficult position of having to spend more and more of their limited resources on water purification, i.e. they are being forced to deal with the symptoms of eutrophication rather than the cause.

REFERENCES

Ahmed, S. 1992. Bacterial abundance in relation to chlorophyll content in water bodies around Harare. BSc (Honours) dissertation, Department of Biological Sciences, University of Zimbabwe.

Ballinger, B.R. & Thornton, J.A. 1982. The hydrology of the Lake McIlwaine catchment. In: Thorton, J.A. (ed.). Lake Mcllwaine: The Eutrophication and Recovery of a Tropical African Lake. The Hague, Dr W Junk Publishers, pp. 34-41.

Beadle, L.C. 1981. The Inland Waters of Tropical Africa. 2nd ed. Longman, London.

Begg, G.W. 1970. Limnological observations on Lake Kariba during 1967 with emphasis on some special features. Limnol. Oceanogr. 15: 776-788.

Brain, C.K. 1937. Weeds of Southern Rhodesia. Part II : Water weeds. Rhodesia agric. J. 34: 612-615.

Chikwenhere, G.P. 1994. Biological control of water lettuce in various impoundments of Zimbabwe. J. Aquat. Plant Mgmt. 32: 27-29.

Chikwenhere, G.P. 2001. Current strategies for the management of water hyacinth on the Manyame River system in Zimbabwe. In: M.H., Julien, M.P. Hill, T.D. Center and J. Ding (eds). Proc 2nd IOBC Global Water Hyacinth Working Group. Australia Centre for International Agriculture Research, Canberra, pp. 105-108.

Chikwenhere, G.P. and Phiri, G. 1999. History of water hyacinth and its control efforts on Lake Chivero in Zimbabwe. In: M.P. Hill, M.H. Julien and T.D. Center, (eds). Proc 1st IOBC Global Water Hyacinth Working Group. Plant Protection Institute, Pretoria, pp. 91-100.

Donnelly, B.G. 1980. The energetics of a population of *Barbus mattozi* Guimaraes, 1884 (Pisces: Cyprinidae). PhD thesis, University of Zimbabwe.

Elenbaas, P.F.M. & Grundel, C. 1994. Zooplankton composition and abundance in two impoundments in Zimbabwe. Hydrobiologia 272: 265-275.

Falconer, A.C. 1973. The phytoplankton ecology of Lake McIlwaine, Rhodesia. M.Phil thesis, University of Landon.

Green, J. 1990. Zooplankton associations in Zimbabwe. J. Zool., Lond. 222: 259-283.

Jarvis, M.J.F., Mitchell, D.S. & Thornton, J.A. 1982. Aquatic macrophytes and *Eichhornia crassipes*. In: J.A. Thornton, (ed.). Lake McIlwaine: The Eutrophication And Recovery of A Tropical African Lake. The hague, Dr W Junk Publishers, pp. 137-144.

Johansson, S. & Olsson, M. 1998. Investigations of toxic algal blooms in a drinking water reservoir of Harare, Zimbabwe. MSc Report, Department of Ecology, Lund University, Sweden.

Junor, F.J.R. 1969. *Tilapia melanopleura* Dum. in artificial lakes and dams in Rhodesia with special reference to its undesirable effects. Rhod. J. Agric. Res. 7: 61-69.

Machena, C. 1983. Primary productivity in the Sanyati basin of Lake Kariba, Zimbabwe, and development of a regression model for predicting fish yields. M.Sc thesis, University of Zimbabwe.

McKendrick J. 1982. Water supply and sewage treatment in relation to water quality. In: J.A. Thorton, (ed.). Lake McIlwaine; The Eutrophication and Recovery of a Tropical African Lake. The Hague, Dr W Junk Publishers, pp. 202-217.

Marshall, B.E. 1975. Observations on the freshwater mussels (Lamelliabranchia: Unionacea) of Lake McIlwaine, Rhodesia. Arnoldia (Rhod.) 7 (16): 1-16.

Marshall, B.E. 1981. Fish and eutrophication in Lake McIlwaine, Zimb. Sci. News 15: 100-102.

Marshall, B.E. 1978. Aspects of the ecology of benthic fauna in Lake McIlwaine, Rhodesia. Freshwat. Biol. 8: 241-249.

Marshall, B.E. 1991. Toxic cyanobacteria in Lake Chivero: A possible health hazard? Trans. Zimb. Sci. Ass. 65: 16-19.

Marshall, B.E. 1994. Ecology and mangement of the Manyame lakes. In: T. Matiza & S.A. Crafter (eds). Wetlands Ecology and Priorities for Conservation in Zimbabwe. Gland, Switzerland, IUCN, pp. 55-67.

Marshall, B.E. 1997. Lake Chivero after forty years: The impact of eutrophication. In: N.A.G. Moyo, (ed.). Lake Chivero: A Polluted Lake. Harare, University of Zimbabwe Pubcations, pp. 1-12.

Marshall, B.E. & Falconer, A.C. 1973. Physico-chemical aspects of Lake McIlwaine (Rhodesia) a eutrophic tropical impoundment. Hydrobiologia 42: 45-62.

Minshull, J.L. 1978. A preliminary investigation of the ecology of juvenile *Sarotherodon macrochir* (Boulenger). MSc thesis, University of Rhodesia, Salisbury.

Mitchell, D.S. & Marshall, B.E. 1974. Hydrobiological observations on three Rhodesian reservoirs. Freshwat. Biol. 4: 61-72.

Moriaty, D.J.W. 1973. The physiology of digestion of blue-green algae in the cichlid fish *Tilapia nilotica*. J. Zool., Lond. 171: 25-39.

Moyo, N.A.G. 1996. The Poisoned lake. Zimbabwe Review 2(4) : 3-5.

Moyo, N.A.G. (ed.). 1997. Lake Chivero. A polluted lake. Harare, University of Zimbabwe Publications.

Munro, J.L. 1966. A limnological survey of Lake McIlwaine, Rhodesia. Hydrobiologia 28: 281-308.

Munro, J.L. 1967. The food of a community of East African freshwater fishes. J. Zool., Lond. 151 : 389-415.

Robarts, R.D. 1979. Underwater light penetration, chlorophyll a and primary production in a tropical African lake (Lake McIlwaine, Rhodesia). Arch.Hydroboil. 86: 423-444.

Robarts, R.D. & Southall, G.C. 1976. Algal bioassays of two tropical Rhodesian reservoirs. Acta Hydrochim. Hydrobio. 3: 369-377.

Robarts, R.D. & Southall, G.C. 1977, Nutrient limitation of phytoplankton growth in seven tropical manmade lakes, with special reference to Lake McIlwaine, Rhodesia. Arch. Hydrobiol. 79: 1-35.

Thompson, J.G. 1968. Report on the use of municipal sewage effluents for irrigation on Aisleby and Good Hope, Bulawayo. Paper presented to the Symposium on the re-use of Sewage Effluent. Department of Conservation and Extension, Salisbury, Rhodesia.

Thorton, J.A. 1979. Some aspects of the distribution of reactive phosphorus in Lake McIlwaine, Rhodesia: Phosphorus loading and abiotic responses. J. Limnol. Soc. stn. Afr. 5; 65-72.

Thorton, J.A. 1981. Lake McIlwaine: An Ecological disaster averted. Zimb. Sci. News 15:87-88.

Thornton, J.A. (ed.), 1982. Lake McIlwaine: The Eutrophication and Recovery of a Tropical African Lake. The Hague, Dr W Junk Publishers.

Thornton, J.A. & Nduku, W.K. 1982. Water chemistry and nutrient budgets. In: Thornton, J.A. (ed.). Lake McIlwaine: The Eutrophication and Recovery of a Tropical African Lake. The Hague, Dr W Junk Publishers. Pp. 43-59.

Toots, H. 1970. Exotic fishes in Rhodesia. Rhod. Agric J. 67: 83-88.

Torrance, J.D. 1972. Malawi, Rhodesia and Zambia. In: Griffiths, J.F. (ed.). Climates of Africa. World Survey of Climatology. Volume 10, pp. 409-460.

Ziberg, B. 1966. Gastro-enteritis in Salisbury European children. Cent. Afr. J. Med. 12: 164-168.

CHAPTER

7

Invasive Aquatic Weeds and Eutrophication: The Case of Water Hyacinth in Lake Victoria

Adrian E. Williams[1] and Robert E. Hecky[2]

[1]APEM Ltd., Enterprise House, Manchester Science Park, Lloyd Street North, Manchester, M15 6SE, UK

[2]Biology Department, University of Waterloo, 200 University Avenue West, Waterloo, Ontario, N2L 3G1, Canada

ABSTRACT

Invasive aquatic weeds have spread throughout the world's waterways as a result of anthropogenic activities. The three most problematic species are water hyacinth, floating water fern and water lettuce. With the potential to double in biomass in a matter of days and the ability to migrate easily due to their free-floating vegetative form, they can successfully colonize new habitats, form dense mats along shorelines and displace native vegetation. In doing so, they effect ecological processes within lakes and decrease the biodiversity. Moreover, for human populations living around such lakes, they reduce the access and quality of available drinking and irrigation water, prevent fishermen from making a living, clog water intakes at hydro-electric dams, increase vector-borne diseases and generally lead to an increase in human suffering. Cultural eutrophication of lakes is also a major problem around the world and can amplify the problem of nuisance floating plants. Weed growth can expand rapidly with an increase in loading of nutrients such as nitrogen and phosphorus. In Africa, there are many examples of eutrophication leading to elevated weed biomass and the subsequent problems that this entails. In Lake Victoria, East Africa, water hyacinth was a major problem in the 1990s that has now largely subsided. However, there are signs that eutrophication, as a result of anthropogenic changes in land use, is still increasing. At a minimum, this will probably continue to maintain the presence of water hyacinth within the lake but it also has the potential to bring about a resurgence to its nuisance status. If this occurs then the human suffering experienced could be above and beyond that seen during the 1990s. Methods to manage water hyacinth include physical, chemical and biological forms of control. However, for the long-term sustainability of Lake Victoria, including the control of water hyacinth, only improved land, air and watershed management can provide the solution to reducing eutrophication and the potential of water hyacinth outbreaks. The measures needed to effect the changes necessary for nutrient control require a massive financial injection. The countries bordering Lake Victoria have a direct responsibility towards it and should contribute in every way to its restoration, as the result of non-action will be regionally devastating. However, there are also implications for the global economy; therefore, solutions also need to involve a wider arena.

INVASIVE AQUATIC WEEDS

Freshwater macrophytes are a diverse conglomeration of macroscopic vascular plants including some relatively "large plants" living either in, on or at the periphery of freshwaters. These aquatic plants include ferns, mosses and flowering plants as well as large trees (Sculthorpe 1967) that require saturated conditions or access to standing freshwater to thrive. Understandably, whilst morphological characteristics enable this wide ranging group to be delineated into a plethora of subdivisions, the entire array of aquatic macrophytes can be loosely divided into four main groups: (i) those that grow either totally submerged and are often, although not always, rooted; (ii) those that float freely on the surface and typically remain unrooted; (iii) those that have an immersed and rooted basal structure but with upper immersed parts with leaves either emergent; or (iv) floating on the water's surface. Macrophytes are diverse in form and structure not merely between species but also within, as morphological differences can occur within a species, depending upon the environmental conditions in which it finds itself (Smith, 1982). Having adapted to nearly all aquatic habitats, they are a common sight within freshwater systems. However, this diversity and adaptability has also enabled aquatic macrophytes to establish in areas outside of their natural range. With increases in anthropogenic migration and commerce, plants and animals, aquatic and terrestrial, have been globally distributed, both intentionally and accidentally. Horticulturists have intentionally introduced many aquatic plants. Water hyacinth [*Eichhornia crassipes* (Martias) Solms] (Fig. 7.1), a hugely problematic aquatic weed, was taken from its native range in the Amazon River basin and spread around the world during the nineteenth century. It was seen as a botanical speciality and, because of its beautiful and delicate inflorescence, used to decorate ornamental ponds (Penfound & Earle, 1948; Lindsey & Hirt, 1999). On the other hand, some plants have been taken by settlers due to the benefits they provided in the homeland. Furthermore, the inevitable influence of global travel has resulted in plants and their spores becoming accidental travellers within many diverse forms of transport such as suitcases or tanker ballast waters (Hayes, 2002). As such, non-indigenous plants can and indeed often have colonized new aquatic habitats.

Various terms have been used to describe these aquatic macrophytes such as alien and invasive. Synonyms for alien species include exotic, introduced, non-indigenous and non-native species. Alien species include those introduced and present in areas beyond their known historical range. This includes introductions from other continents, bioregions and also those not native to the local geographic region

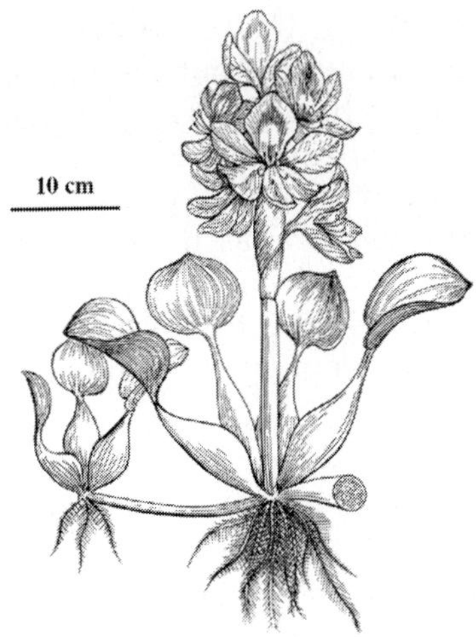

Fig. 7.1 Water hyacinth [*Eichhornia crassipes* (Martias) Solms]. (Based on Greathead & de Groot, 1993).

(Melvin, 1999). Clearly, many alien species are not problematic and, indeed, coexist peacefully within native ecosystems, providing habitat and sustenance to native wildlife and human populations. Invasive species, however, are those alien species that invade habitats and displace other species through rapid growth and spread. Invasive species are easily propagated asexually by root or stem fragments and/or rapidly mature; are typically prolific seed producers and have high seed germination rates (Auld et al., 1987). Alien species that are inclined to invasiveness also have the added advantage of being relatively free from natural controls such as disease, parasites or even herbivores because they are outside of their natural biogeographic ranges (Melvin 1999). Therefore, alien species can expend more energy on growth and reproduction; thus aiding their invasive nature (Bazzaz et al., 1987).

Invasive species have been referred to as a form of biological pollution because they can upset the equilibria between native species that have formed slowly over millennia within natural ecosystems. However, how do invasive species interfere with waterbodies and wetlands? Although every species behaves differently in the environment as well as in its interaction with other species, there are general patterns of influence that occur commonly. When an invasive plant enters a new habitat, it will outcompete and thus displace native vegetation, thereby becoming established. Some of the ways such displacements affect the native biota include a reduction in biodiversity, a decrease in habitat and food and an overall change to ecological processes (de Groot, 1993). Furthermore, invasive species may even hybridize with native plants, further changing the ecosystem (Melvin, 1999).

As invasive species can rapidly out-compete natives and dominate an ecosystem, a reduction in biodiversity follows. Indeed, it is estimated that alien, invasive species are implicated in 42% of the species listed as endangered or threatened under the United States Endangered Species Act (Melvin, 1999). Furthermore, as native flora and fauna have co-evolved over long periods of time, replacement of native vegetative systems with the ones dominated by invasive species typically alters these relationships with accompanying reductions in available habitat and food.

Ecosystem processes may change as a result of invasive plants. Native ecosystems have developed under and adapted to particular abiotic factors and ecosystem processes such as rates of nutrient cycling, rainfall patterns and fire regimes. The presence of alien species can alter these processes, which has a knock on effect to the ecosystem as a whole (Flack & Benton, 1998). For instance, where water hyacinth has been introduced, it alters the natural hydrologic cycle by transpiring greater quantities of water, upto three times more, than native vegetation in the same habitat (de Groot, 1993). Water tables and some surfacewater habitats are thus reduced and native species are impacted.

Invasive water weeds also have more direct impacts on anthropogenic activities. Due to their typically fast growth rate and the absence of predators, they can make waterways unnavigable, decrease the generating capacity of hydro-electric stations, reduce fishing catches and block irrigation canals. By clogging sewage and drainage systems, they can cause flooding, contaminate drinking water and create breeding grounds for harmful insects and bacteria (Twongo et al., 1995; de Groot, 1993). In West Africa, the inland waters of Ghana, Burkina Faso and Togo have been particularly hard hit by invasive macrophytes. The weeds slow down the boats of fishermen and make landings difficult; as a result more of their catch goes off. Fishermen also use more fuel, upto three times as much, to push their boats through the heavy weeds (Akinyemiju, 1987; FAO, 2000). These increased costs are then reflected in consumer prices.

Whilst invasive aquatic macrophytes such as submerged hydrilla (*Hydrilla verticillata* (L. fil.) Royle) and Eurasian watermilfoil (*Myriophyllum spicatum* Linaeus) are problematic around the world (Gangstad 1978), the weeds that have caused most nuisance, especially in the tropics, are the floating plants such as water hyacinth, floating water fern (*Salvinia molesta* Mitchell) (Fig. 7.2) and, to a lesser extent, water lettuce (*Pistia stratiotes* Linaeus) (Fig. 7.3). These three species are well known for their ability to colonize, establish and alter physical and biological functions within aquatic systems (de Groot, 1993, Howard & Harley, 1998).

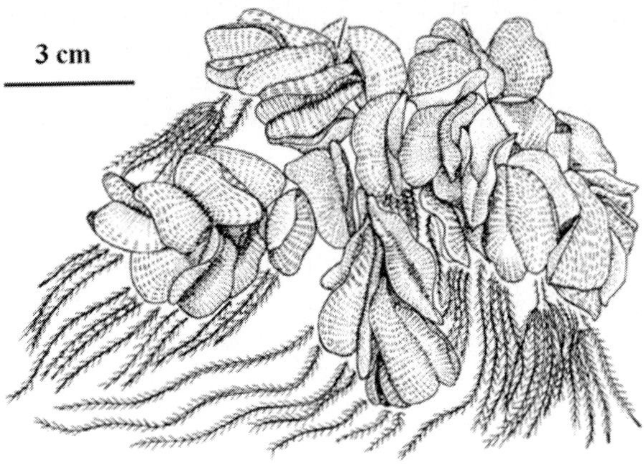

Fig. 7.2 Floating water fern (*Salvinia molesta* Mitchell). (Based on Greathead & de Groot, 1993).

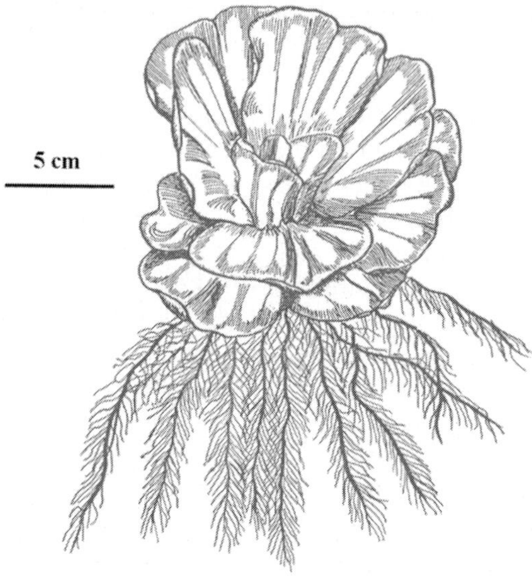

Fig. 7.3 Water lettuce (*Pistia stratiotes* Linaeus). (Based on Ivens, 1967).

Water hyacinth, a perennial, herbaceous member of the pickerelweed family (Pontederiaceae), is native to tropical America. It can grow upto a metre or so in height and can double its biomass in six days (Lindsey & Hirt, 1999). It has an attractive purple flower that has made it a favourite amongst horticulturists. As a result, it has been spread widely. It was recorded in Egypt in the 1870s, the USA in the 1880s, Australia and southern Asia by the 1890s, China and the Pacific by the early 1900s, South Africa in 1910, East Africa by the 1930s and West Africa by the 1970s. It is now established throughout tropical and warm temperate regions of the world (Gopal, 1987; Julien et al., 1999). Reproduction is largely by vegetative growth with daughter plants forming from thin stolons that break, allowing dispersal by wind and current. However, large quantities of seeds are also produced, which sink and remain viable in the benthic substrate for upto 20 years (Das, 1969).

Floating water fern is a free-floating, branched member of the Salviniaceae family. It has originated from southeastern Brazil and is probably the hybrid of two species, *S. biloba* and *S. auriculata* (Forno & Harley, 1979). Individual plants are upto 30 cm long with numerous leaves that are capable of rapid growth, doubling in biomass in just over two days (Harley & Mitchell, 1981) and forming dense mats. Common in aquaria, floating water fern has been spread to most tropical and subtropical areas of the world. It arrived in Asia in 1939 (Williams, 1956),

Southern Africa, namely the Zambezi River, in 1949 (de Groot, 1993), East Africa in 1957 when it was accidentally introduced into a dam near Kitale in Kenya (Ivens, 1967) and Australia in 1953. The plant is considered to be a sterile pentaploid and only reproduces vegetatively by parent plants breaking and forming numerous daughter plants, although other members of its family form sporocarps on their trailing underwater stems (Mitchell, 1978).

The origin of water lettuce is uncertain and while it may have originated in South America, it has been used in Africa as a medicine and fodder for cattle for centuries being recorded in Egypt in 77 AD. (Sculthorpe 1967). As such, although it is now considered a native of Northern Africa (Harley, 1993), it has spread over the rest of Africa and parts of Asia and in the 1970s, also found its way to Australia (Mitchell, 1978). Water lettuce reproduces both vegetatively and sexually. It is less aggressive and smaller than either water hyacinth or floating water fern and has, in places, been replaced by these more aggressive plants. However, it can still cause serious problems in tropical and subtropical regions such as blocking inlet grilles of hydro-electric stations as it has done on the Pangani River in Tanzania (Ivens, 1967).

EUTROPHICATION AND INVASIVE AQUATIC WEEDS

Nutrients are continuously lost from lakes. This loss occurs through outflows, burial within sediments or to the atmosphere (Moss et al., 1997). Therefore, allochthonous inputs of nutrients into a lake are extremely important to the production within the specific lake. Eutrophication can be defined as the gradual increase in the concentration of phosphorus, nitrogen and other plant nutrients in ageing aquatic systems, such as a lake, which results in the system becoming more productive. Eutrophication was originally identified as an inevitable and natural phenomenon and the process usually takes thousands of years to progress, eventually reaching an end point as the lake is converted to a wetland (Lindeman, 1942; Deevey, 1942). However, later investigations found that lakes may enter phases where their waters become more oligotrophic or more eutrophic due to biogeochemical changes in their catchments (Haworth, 1969, Macan, 1970). For example, climate change can influence chemical weathering thus, alter nutrient loading (Pennington, 1981; Whiteside, 1983). As such, lakes naturally high in nutrients do exist dependent on the yield of nutrients from their catchments and the morphology of their basins (McGowan et al., 1999; Moss et al., 1994). Overall, however, the most pervasive cause of nutrient enrichment within waterbodies is usually considered to be anthropogenic in nature. Whilst technically this is termed cultural or anthropogenic eutrophication, it is more readily just called eutrophication and refers to increasing nutrient concentrations in receiving waterbodies and the consequent rise in organic productivity.

In nearly all the cases analyzed, the expansion of human colonization has brought eutrophication and other ecological shifts within lakes. The expansion of the Roman Empire resulted in algal species changing in some lakes, indicating that eutrophication had occurred (Jones et al., 1978). More recently, the settlement and growth of cities like Seattle has resulted in similar changes within the algal community of associated lakes (Stockner & Benson, 1967). Sawyer (1947) and Hasler (1947) were among the first scientists to see that industrialization and urbanization had serious effects on lake ecosystems. Sawyer (1947) investigated a chain of lakes in Wisconsin, USA and found that sewage effluent from the town of Madison was the cause of algal blooms and resulting odours in Lakes Monona and Waubesa. Nutrients can, however, enter a system from several sources. Point sources, as the name implies, enter at a single point and include outputs from such things as sewage treatment works that can carry large volumes of industrial as well as domestic waste. When a contaminant comes from a widespread area rather than from a specific point, the source of the contaminant is known as diffuse or non-point source (Nebel & Wright, 1998).

Palaeolimnological records show that eutrophication has occurred globally and diachronically (Fritz, 1989; Zullig, 1989; Anderson, 1995). As such, these records identify that eutrophication is not just a recent phenomenon occurring only within the last hundred years or so. However, the increase in eutrophication due to industrial, urban and agricultural policies and phosphorus and nitrogen usage has wrought changes in lakes over the last few decades that have caused great concern to scientists, governments and the public alike. These concerns are based around problems of water supply, recreation, aesthetic value, public health and loss of biodiversity. However, what are the impacts of eutrophication on aquatic macrophytes? Aquatic plant biomass often increases with a corresponding increase in the nutrient loading from within a catchment (Mitchell, 1974). As rooted emergent and floating leaved macrophytes typically gain a substantial amount of their nutrient requirements from sediments via their roots, it is the unrooted, submerged and free-floating plants that generally respond most dramatically to eutrophication. Free-floating plants depend almost entirely on the water beneath them for their nutrients and can respond in a similar manner to phytoplankton (Denny, 1980). However, phytoplankton, being surrounded by water and having a smaller size and mass, have shorter diffusion pathways than macrophytes and are thus usually quicker to respond to increased nutrient loads. As such, phytoplankton can often out-compete macrophytes for increased nutrients. In some circumstances, though, macrophytes prevent phytoplankton blooms from developing, thus enabling macrophytes to survive and dominate in relatively high nutrient conditions. Ultimately, as eutrophication increases, submerged macrophytes disappear because of light limitation through shading and a phytoplankton-dominated state takes over (Timms & Moss, 1984; Scheffer, 1993). Floating plants, however, are not affected by decreasing light penetration within the water body and can persist and indeed benefit from increasing nutrient supplies. In fact, if phytoplankton growth becomes light limited because of self shading or shading from floating plants, then floating plants can take advantage of the increased nutrient availability and expand their coverage even further. Overall, eutrophication can increase floating plant biomass of both native as well as invasive macrophytes but as invasives are superior competitors with fewer parasites or grazers, they consequently flourish to a greater extent than the natives. This can lead to severe problems within lakes.

Within Africa, there are several incidences of eutrophication leading to an increased biomass of invasive water weeds that subsequently have resulted in detrimental impacts upon lakes. When new lakes are formed, waters that once ran through now deposit their nutrient-rich sediment in the recently formed lake. In addition, increased water retention behind the dam reduces the washout of plankton populations. Newly waterlogged soils release nutrients and the terrestrial vegetation, now inundated, rots and also releases nutrients. Oxygen levels can also fall as a result of decomposition and this can further increase the release of soil-bound nutrients from mineral particles. After a while, deltas form at the lake periphery and sediment is deposited there whilst submerged soils no longer leach nutrients and terrestrial vegetation has all been broken down. Once the lake has reached its capacity, nutrient loss via outflow streams will remove a considerable amount of the nutrient load. But in the initial flooding phase, there can be a considerable loading of nitrogen and phosphorus compounds within the lake (Moss, 1991).

The construction of the Kariba Dam on the Zambezi River along the border between Zimbabwe and Zambia was completed in 1959 and provides a prime example of how eutrophication, due to reservoir creation, can favour the expansion of invasive macrophytes. Construction of this dam was a massive undertaking and formed Lake Kariba, a 300 km long waterbody with a maximum surface are of 5,580 km^2 (Fig. 7.4), much of it over flooded land. Shortly after the dam's completion, during the flooding and nutrient loading phase, the invasive weed *S. molesta* appeared. At the time the lake's maximum storage capacity was reached in 1962; weed infestation was at its peak, covering

22% of the lake's surface, i.e. in excess of 1,000 km^2 (Denny, 1985). In the mid 1960s and as a result of flushing of plants and nutrients, that could only occur once the lake had filled, the infestation declined to about 12%. Then followed a collapse attributed to several factors such as a grasshopper (*Paulinia acuminata*) that was introduced as a biological control in 1970 as well as the large-scale removal of *Limnothriss miodon*, a clupeid introduced from Lake Tanganyika, whereby as much as 50 tonnes of P was harvested from the lake annually (Marshall, 1993). Whatever the cause of the collapse in 1980, *S. molesta* only covered 1% of the lake surface and was considered to no longer be problematic. Since then, *P. stratiotes* has appeared but not to any great extent. In more recent years, however, water hyacinth has begun to cause problems. It first appeared in the lake in 1984 but reached the eastern shores of Gache Gache Bay by 1994 where plants covered over 4 km^2 of the bay. This has had serious socio-economic impacts for the local communities there (Chikwenhere et al., 1999).

Lake Chivero, formerly known as Lake Mcllwaine, in Zimbabwe, has had similar problems. It is a small manmade lake with a surface area of 26 km^2 (Fig. 7.4). It was dammed in 1952, in all likelihood undergoing a similar initial phase of eutrophication as other dammed waterbodies. However, in addition, the Manyame River system washed soil and agricultural nutrients into the lake and enrichment was further increased as a result of industrial and sewage effluent entering from the city of Harare (Chikwenhere & Phiri, 1999). Water hyacinth was first identified in the lake in 1953, one year after the dam's completion. However, the first major outbreak was not until 1955. This was controlled with herbicides but the weed returned with outbreaks in 1971 and 1986 with over 20% of the lake surface being covered with the weed. Then followed a period where the macrophyte community was dominated by *P. stratiotes* but this was later replaced again with water hyacinth and by 1990, it covered over a third of the lake surface. Undoubtedly, high nutrient levels in the lake were a main causative factor for the water hyacinth invasion (Marshall, 1993; Moyo, 1997). By 1997, the weed had diminished to less than 5% cover but whether this was as a result of control methods or that the lake was flooded in 1996, potentially flushing weed and nutrients out of the system, has not been determined.

Several lakes in Ghana have undergone weed invasions following impoundment. The Akosombo Dam on the River Volta in Ghana formed Lake Volta. Completed in 1966, the lake is around 400 km long and 25 km wide, covering a surface area of about 8,500 km^2 (Fig. 7.4). The entire lake became anoxic for a while after completion as a result of decomposing softwood forests. The resulting de-oxygenated hypolimnion probably led to the release of phosphorus from iron and manganese mineral complexes in the sediment, thus, bringing them into the water column (Mortimer, 1941). The lake has infamously been troubled by the invasive *P. stratiotes* (Okali & Hall, 1974). Other manmade lakes in Ghana have had similar problems such as the Weija and Kpong reservoirs (Odei, 1973; de Graft-Johnson, 1993) (Fig. 7.4). The identification of water hyacinth in Ghana in 1984 has caused further concerns and throughout Ghana, the implications of eutrophication on invasive weeds are well understood (de Graft-Johnson, 1993).

In Tanzania, the dam reservoir Nyumba ya Mungu (Fig. 7.4) went through a similar initial eutrophic phase upon completion and filling of the reservoir (Taylor, 1993). However, an initial phase of eutrophication does not always occur. In Lake Nasser, Egypt and Sudan (Fig. 7.4), the basin, comprising largely of rock and sand, showed no initial signs of eutrophication. Indeed, it appears that Lake Nasser is still free of any major infestations of invasive weeds (Fayad, 1999).

As previously mentioned, nutrients within a lake either tend to be washed out, locked within the sediment or lost to the atmosphere. As such, following a period of disturbance, manmade lakes often begin to become more oligotrophic as the internal loading subsides. However, this is only true if additional anthropogenic eutrophication does not occur. In Lake Kainji, Nigeria (Fig. 7.4), water retention is short and the loss of nutrients by flushing rapid (Moss, 1991). As such, nutrient build up and problems with invasive weed were not initially apparent. The situation changed however and in 1994 large floating mats of water hyacinth became prevalent in the lake (Farri & Boroffice, 1999). Agricultural and urban runoff containing high nutrient loads was

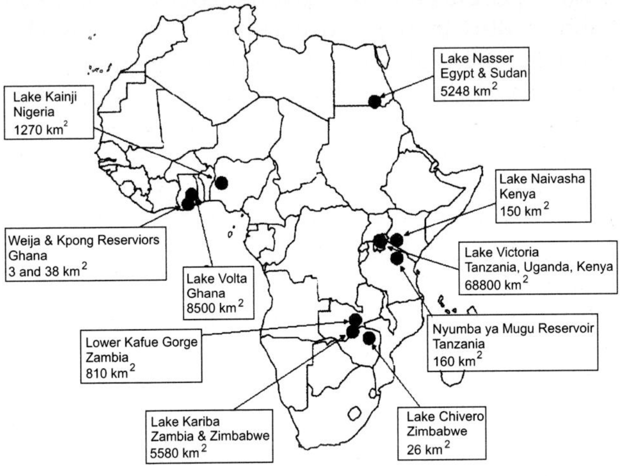

Fig. 7.4 Location and size of African water bodies mentioned in the text

blamed for these later infestations of invasive weeds. The same process can be seen within naturally occurring lakes.

Lake Naivasha in Kenya is a natural freshwater lake with a surface area of 150 km^2 (Fig. 7.4). It has faced severe *S. molesta* infestations since 1962. At times, upto 25% of the lake surface has been covered by the weed (Ivens, 1971). In the early 1990s, water hyacinth began to proliferate and has now largely replaced *S. molesta* (Ivens, 1971; Njuguna & Thitai, 1993). Eutrophication resulting from agricultural and urban runoff within the lake catchment has aided the invasion of these aquatic weeds (Harper, 1992). Likewise, in Zambia, water hyacinth on the Lower Kafue River has been directly attributed to anthropogenic nutrient loading. Water hyacinth first appeared in Zambia in the 1960s. In the early 1970s, it was seen on the Lower Kafue River and by 1998, the government had declared it a national disaster (Kampeshi & Shantima, 1999). The main infestation was near the large commercial farms in the Mazabuka area some 100 km from the Kafue Gorge Dam and as such was considered to have been encouraged by farm runoff rather than being influenced by the dam.

LAKE VICTORIA

Located at the equator at an altitude of 1,134 m, Lake Victoria is the second largest lake in the world with a surface area of almost 69,000 km^2 and 3,450 km of shoreline. It has a residence time of 21 years and a flushing time of almost 150 years. Lake Victoria is shared by Kenya, Tanzania and Uganda. Living within its catchment area of over 250,000 km^2, which includes parts of Rwanda and Burundi (Fig. 7.5) are 28 million people. The lake system is a vastly important socio-economic resource to these people and their governments. On a local level, it is an important source of drinking and irrigation water as well as food. In addition, the lake offers an effective medium for transportation. On a national level, it provides export revenue from fisheries (both live and processed exports), electricity from hydro-power generation and income from tourism for both governments and local populations alike. Furthermore, the lake is a convenient disposal site for human, agricultural and industrial waste (Balirwa, 1998). The area surrounding Lake Victoria is rich in agriculture, animal husbandry and industrial growth. It is also the most densely populated area in the region (Bootsma & Hecky, 1993; Hecky, 1993). Annual population growth is in the region of 3% and yet the population is among the poorest in the world with a per capita income of under US$ 300 per annum (UNECA, 1995).

Lake Victoria is not an isolated waterbody as its catchment area clearly crosses several national boundaries. Indeed, the lake contributes to the hydrological regime of the Nile basin and the ecohydrology of the

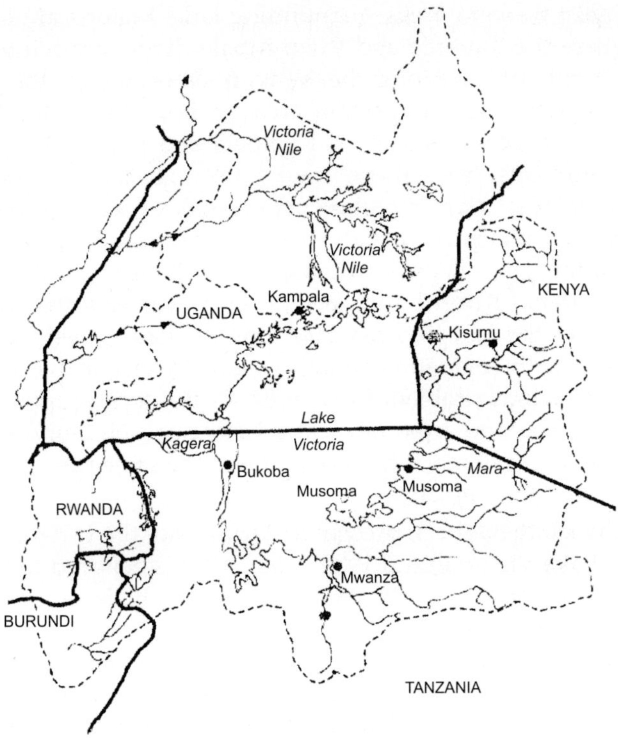

Fig. 7.5 Lake Victoria and its surrounding catchment. (Taken from Scheren et al., 2000).

entire region (Bugenyi & Balirwa, 1998). Moreover, the vegetated littoral habitats are closely linked to the ecology of the lake (Beadle, 1981). Indeed, many freshwater lakes are dynamically dependent upon their riparian zone. In contrast to smaller lakes and even large reservoirs, which typically have long dendritic forms with relatively simplistic shorelines, Lake Victoria is an old flooded land surface (Johnson et al., 1996) with a complex shoreline of many local embayments and protected channels. These bays and channels can be more directly affected by human activities than the broad open expanses of other waterbodies and also provide different ecological conditions for plant growth (Kling et al., 2001). The wetland ecotone, being the transition zone between land and water and thus two very different ecological systems, is complex and any damage to it is highly undesirable. Without the wetland acting as a buffer, the influence of one of the bounding ecosystems on the other can be both large and potentially detrimental (Holland, 1988). The impacts can go in either direction i.e., either increased runoff and eutrophication from the land to water or increased flooding of land adjacent to water.

The largest wetland areas surrounding Lake Victoria are in western Uganda where the Katonga and Ruizi-Kibale Rivers drain into the lake (Beadle & Lind, 1960). Along the Kenyan shoreline are the Yala and Nyando wetlands. Other important areas include the wetlands around major urban conurbations such as Kampala and Jinja that help reduce inputs of nutrients from those towns through natural purification processes (Balirwa 1998). The plant community of all these wetlands is dominated by emergent species such as papyrus (*Cyperus papyrus* Linaeus), hippo grass [*Vossia cuspidada* (Roxb.) Griff.], bulrush [*Typha domingensis* (Pers.) Kunth] and reeds (*Phragmites mauritianus* Kunth). However, over the last two decades, one invasive weed has become prevalent. Of the three floating aquatic weeds outlined above, only water hyacinth is a serious problem in Uganda. Water fern is apparently not present in Uganda and although water lettuce can be found in most of Uganda's aquatic systems, it has not attained the same pest status as has water hyacinth (Twongo, 1993).

Water hyacinth has been present in Africa since the 1870s but was not reported in Lake Victoria until 1989, although it is believed to have been present since at least the early 1980s (Twongo & Balirwa, 1995). It was present upstream of Lake Victoria on European plantations within the Kagera catchment since at least the 1940s and reached nuisance levels here, choking riverine wetland lakes, by the 1980s. Migration via the Kagera River was the most likely point of entry for water hyacinth into Lake Victoria. Indeed, the Kagera River even now remains an active immigration point for water hyacinth. The problems associated with water hyacinth became apparent in Lake Victoria in the early 1990s. As demonstrated in the Democratic Republic of Congo where two plants were seen to produce 1,200 daughter plants over four months (Ivens, 1967), water hyacinth has massive growth potential. By 1995, 80% of the Ugandan coastline was fringed with the plant (Matagi, 2002). The weed usually formed mats that extended out from the shore for at least 15 m but could, in places, completely cover small embayments. In Murchison Bay, which receives the urban runoff and sewage of the city of Kampala, the fringe crept out beyond 50 m and formed dense floating mats, some of which were over 300 ha. in size.

Water hyacinth, with its characteristic thick fringes and floating mats, has had serious impacts on many aspects of life around Lake Victoria (Twongo, 1996). The weed has impeded navigation, even in motorized boats, and transportation costs have risen due to the necessity of taking alternative indirect routes or as a result of motoring through the mats which, if physically possible, increases fuel consumption. The weed has affected fishermen by fouling their nets and traps, reducing catches and preventing access to landing sites. Under the mats, reduced light and oxygen has decreased the plankton primary production and native fish have died. Fish are an important source of protein to lakeside communities and without it, health suffers. Moreover, the mats are breeding grounds for carriers of human and animal disease such as malaria, schistosomiasis, encephalitis and river blindness. Irrigation schemes have been put under pressure and irrigation channels blocked. The weed has reduced the supply of clean potable water and caused difficulties in water extraction. Unsurprisingly, its presence has increased disputes between local communities and on an overall basis, it has been responsible for the translocation of some communities away from Lake Victoria when local watering points and lake access became unavailable. Within the wetland fringe, native plants have been eliminated through successional shifts, and generally, a loss of biodiversity is apparent (Willoughby et al., 1996). Apart from the reduction in national revenue from the reduced export of fish produce, tourism has declined and intermittent closure of the hydro-electric plant in Jinga, due to weed build up, has caused disruption within Uganda's capital Kampala (Twongo & Balirwa, 1995; Twongo et al., 1995; Lindsey & Hirt, 1999).

However, despite the prolific and aggressively invasive nature of water hyacinth and its dominance in Lake Victoria during the 1990s, by the end of 1998, 80% of the weed had gone (LVEMP 2001) and by 2000, although small pockets could be seen, the weed was no longer the nuisance it had been. Various control methods (see below) had been utilized and its demise was largely attributed to the use of a biological control. However, recent work strongly suggests that the climatic influence of *El Nino* also played a major role in the weeds subsidence (Williams et al., 2005). Furthermore, despite its disappearance, it is likely that the weed will return and indeed recent evidence indicates that this may already be happening with water hyacinth clogging the mouth of Murchison Bay at Port Bell and preventing fishermen from launching their boats (IMPECCA, 2001; TNV, 2002). It might be considered that at worst, any water hyacinth resurgence event would not exceed the extent to which it reached in 1998. However, Lake Victoria is a changing system and as such it is important to consider the potential impact of water hyacinth resurgence within the lake under these changing conditions. Furthermore, it is essential to define the best possible methods for its control so that they may be applied in a timely manner so as to prevent another resurgence.

The growth rate of water hyacinth is strongly dependent upon the concentration of dissolved nitrogen (N) and phosphorus (P) within which it grows (Debusk & Dierberg, 1989; Moorhead et al., 1988; Reddy et al., 1989; 1990). Reddy & Tucker (1983) found that rates of N and P uptake were similar to one another and correlated directly with growth. Desougi (1984) found that low levels of total P (0 - 0.62 mg l^{-1}) significantly inhibited growth relative to higher concentrations and plant productivity did not increase with concentrations of total N above 21 mg l^{-1} and total P above 62 mg l^{-1}. Sato & Kondo (1981), on the other hand, reported that water hyacinth achieved its maximum growth rate at nutrient concentrations of 28 mg l^{-1} total N and 7.7 mg l^{-1} total P.

Gophen et al. (1995) found average concentrations of total N and total P within Lake Victoria (Kenya) to be 0.64 and 0.07 mg l^{-1}, respectively. Likewise, Guildford and Hecky (2000) reported a mean total P concentration for offshore Lake Victoria (Uganda) of 0.08 mg l^{-1}. Although measurements taken from Napoleon Bay in 2001 (A.E. Williams, pers. obs.) found the concentrations, then to be a higher and in the region of 0.75 and 0.18 mg l^{-1} for total N and total P, respectively. In Murchison Bay and other inlets fed with nutrient-rich waters from land runoff and sewage outfalls from large municipalities, the concentrations can be an order of magnitude higher than these (Fisheries Resources Research Institute Uganda, unpublished data). However, despite this range (0.64 - 7.5 mg l^{-1} total N and 0.07 - 18.0 mg l^{-1} total P) the levels are still below those required for maximum weed growth. This is particularly so for N.

The C : N : P ratio from plant tissues provides more information concerning the relative scarcity of certain nutrients. In July 1995, the average C : N : P ratio (atoms) for water hyacinth plants from the northeastern area of Lake Victoria was 995 : 31 : 1 (n = 23, SEM = 208 : 6 . 4 : 0.21) (R.E. Hecky unpublished data) whilst in July 2001, it was 327 : 19 : 1 (n = 8, SEM = 116 : 6.6 : 0.35) (A. E. Williams unpublished data). These ratios are not identical and Duarte (1990) suggested that the variability in N and P content could be due to plants growing in different nutrient concentrations found within their local environments. This fact is not helped by the mobility of the plants in that where a plant is found does not necessarily indicate where it grew. Moreover, the analyzed plants were taken during different growing seasons. However, there is a similar pattern between the two groups of plants in comparison with the Redfield ratio of 106 : 16 : 1 (Redfield, 1934), which is often considered optimal for phytoplankton growth, i.e. water hyacinth requires a great deal of carbon relative to the other two macronutrients. As a large macrophyte, the requirement for more carbon relative to phytoplankton is clearly understandable and this nutrient is available to water hyacinth in an inexhaustible supply from the atmosphere. Atkinson & Smith (1983) found that the mean C : N : P ratio for 92 benthic marine macroalgae and seagrasses was 700 : 35 : 1, whilst using data from 26 other species described in the literature, the ratio was 430 : 17 : 1. These high carbon ratios were accredited to the structural complexity of macroalgae relative to phytoplankton. Therefore, overall the higher carbon requirements of water hyacinth in 1995 prior to population collapse, in comparison to 2001 when populations were low, may have been due to the size of plants. In a crowded environment, the need to grow above your neighbours and, therefore, your competitors is paramount to obtain sufficient solar radiation. In fact, water hyacinth is known to adapt morphologically to reduced light conditions by extending their petioles (Methy et al., 1990; Methy & Roy, 1993; Williams et al., 2005). In 1995, when dense mats were present within Lake Victoria and plants were in strong competition with one another, the plants were probably, on the whole, taller. Furthermore, Atkinson & Smith (1983) found that plants grown in low nutrient conditions had higher C : N and C : P ratios than those grown in nutrient-rich waters. As such, in 1995, the higher biomass of weed probably reduced the total amount of available nutrients and thus raised the C : N and C : P ratios. Moreover, growth in dense mats might also explain the higher N : P ratios seen in 1995 and suggests that in dense stands, P may ultimately limit growth. In 2001, when relatively few plants were present, the N : P ratio was much closer to the Redfield ratio and, as such, it may indicate that neither N nor P were limiting water hyacinth growth to the same extent as they seemed to be in 1995.

Other plant nutrients are also important. Water hyacinth grown in a calcium-deficient medium did not reproduce and a threshold of 5 mg l^{-1} was determined. Likewise, Fe deficiency (< 0.3 mg l^{-1}) inhibited growth and resulted in chlorosis of the leaves (Gopal 1987). In 1995, Fe levels in Lake Victoria were less than 0.05 mg l^{-1}, well below the published threshold. However, in the near shore environment, iron may be mobilized from insoluble iron oxides as a result of the anoxic conditions existing beneath floating weed mats (Hill et al., 1999), although equally during prolonged anoxia, iron is lost to the sediment as it binds with sulphide. In June 2001, concentrations of calcium within the waters of Thruston and Napoleon Bay, including both open water and near shore sites, was on average 4.7 mg l^{-1} (n = 16, SEM = 0.12 mg l^{-1}) and Fe 0.32 mg l^{-1} (n = 16, SEM = 0.04 mg l^{-1}), whilst plant tissues (leaf and root combined) contained 10.6 and 6.1 mg g^{-1} of calcium and iron (n = 8, SEM = 1.8 and 2.9 mg g^{-1}, respectively). Likewise, the sediment contained 5.0 and 9.3 mg g^{-1} of calcium and iron (n = 13, SEM = 1.3 and 2.2 mg g^{-1}), respectively (A.E. Williams unpublished data).

Overall, water hyacinth plants (at the low population densities observed in 2001) did not seem unduly stressed and in competition for nutrients. Although it is apparent that water hyacinth can proliferate in a wide range of nutrient regimes, it is also clear that water hyacinth thrives in high levels of nutrients. In the 1990s, water hyacinth was particularly abundant in Speke Gulf, Tanzania (Fig. 7.5). This was most likely as a result of discharge wastes from Mwanza Town (APS, 1991). Likewise, in Murchison Bay, Uganda, where nutrient levels are typically 10 times higher than that found in open water as a result of inputs from the city of Kampala, weed proliferation was and still is greater than in many other areas of the lake. As such, any increase in nutrients within Lake Victoria is highly undesirable from many ecological perspectives, not least of which is the potential massive proliferation of water hyacinth under such conditions.

Lake Victoria is becoming more eutrophicated, a condition that may have serious consequences for the return of the weed. Evidence for this eutrophication is well documented. During the 1950s and 60s, the distribution of oxygen within Lake Victoria followed an annual cycle of thermal stratification. From a thermally stratified state developing in September, the lake's isotherms began to destabilize and usually by January, wind and internal waves could mix the lake, making it almost isothermal between July and August. Oxygen concentrations were only low during the stratified period and then too, only below a depth of 55 m. Complete anoxia was rare. Throughout the year, surface oxygen levels were, as a result of atmospheric diffusion, high, typically being in the region of 95-100% and even during periods of mixing, surfacewaters were still at 90% oxygen saturation (Talling, 1957; Fish, 1959; Newell, 1960; Kitaka, 1971). However, by the 1980s, things had changed. Oxygen concentrations in the top 20 m of the lake had risen in comparison to the 1960s. Indeed, the surfacewaters were often supersaturated. However, at a depth below 40 m, oxygen levels had fallen and during the thermally stratified period from October to March, between 30 to 50% of the bottom of the lake was anoxic (Hecky, 1993; Hecky et al., 1994) (Fig. 7.6).

As a result of anoxic conditions in the hypolimnion and in general near the lake bottom, denitrification has probably increased (Hecky et al., 1996). In turn, this results in low N : P ratios beneath the thermocline and throughout the water column once the system turns over. Furthermore, due to the de-oxygenated hypolimnion, phosphorus is more likely to be released from iron and manganese complexes in the sediment, thus bringing more P into the water column and further decreasing the N : P ratio (Mortimer, 1941) as seen from the nutrient concentrations described above (Gophen et al., 1995). However, most (> 90%) P in the sediments is organically bound (R. E. Hecky unpublished data) and release from sediments due to anoxia may be a minor process in the P cycle. Closer to shore, the situation is slightly different because light is available for nitrogen fixation (Mugidde et al., in press) throughout the shallower waters and oxygen is higher, thus reducing denitrification and phosphorus release and increasing the N : P ratio (Bootsma & Hecky,

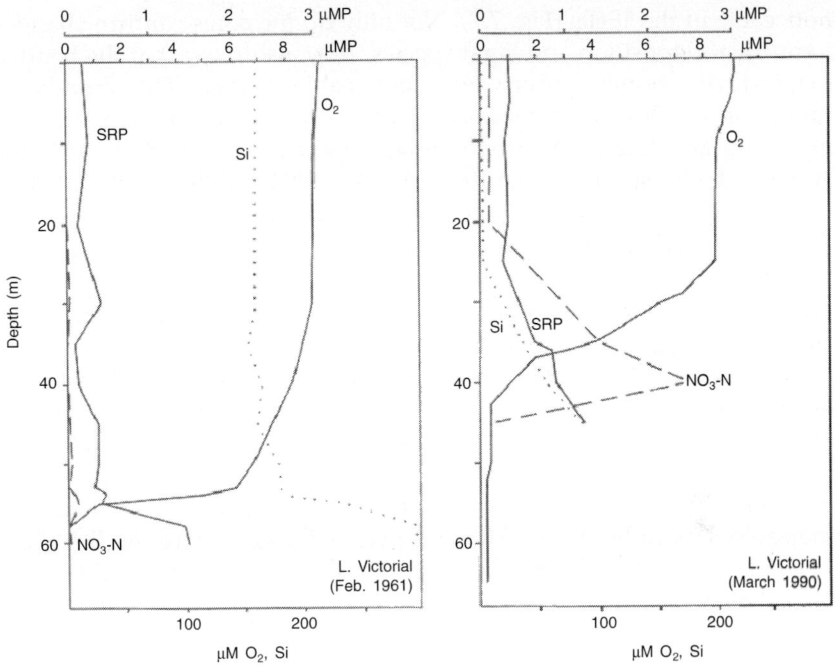

Fig. 7.6 Depth profiles of O_2, soluble reactive phosphorus (SRP), NO_2-N and Si in February 1961 and March 1990 during the stratified season. (Based on Hecky 1993).

1993; Hecky, 1993). Nonetheless, a decreasing N : P ratio will encourage the development of nitrogen-fixing cyanophyta and, indeed, the algae community has changed. Talling (1965) recorded a clear seasonal succession within the phytoplankton, which consisted largely of diatoms (*Aulacoseira* and *Cyclostephanus*), cyanophyta (*Anabaena*) as well as chlorophyta. Now the system is dominated by more nitrogen-fixing cyanophyta such as *Cylindrospermopsis*, *Planktolyngbya* and *Anabaena* as well as low-siliceous diatoms such as *Nitzschia* (Ochumba & Kibaara, 1989; Lehman & Branstrator, 1993; Kling et al., 2001). The changes in the algal community have resulted in higher rates of photosynthesis and nitrogen fixation. Indeed, a twofold increase in pelagic productivity and a four or fivefold biomass increase has been contemporaneous with the shift to nitrogen-fixing cyanophyta. This increase has led to surfacewaters being supersaturated with oxygen (Hecky, 1993; Mugidde, 1993; Lung'ayia et al., 2001).

The changes over the last few decades have been given further credence with the aid of palaeolimnological evidence gleaned from sediment cores from the bottom of Lake Victoria. These identify that the changes now seen in the lake began in the 1940s, becoming very noticeable in the 1960s (Fig. 7.7). Not only do the cores confirm changes in nutrient deposition and algal species, they also show that the benthic macro-invertebrates underwent a noticeable change. The abundance ratios of *Procladius brevipetiolatus* and *Chironomus imicola*, anoxia-intolerant and tolerant Chironomidae respectively, shifted towards the anoxia-tolerant species in the 1960s (Hecky, 1993; Vershuren et al., 2002).

For successful management of the system and invasive weeds, it is imperative to understand why these biotic and abiotic changes occurred. Initially, it was considered that fish introductions, in particular the voracious predator Nile perch (*Lates niloticus*) introduced in the mid 1950s, had impacted the food chain and disrupted the ecological stability of the system from a top-down perspective. This, in turn, affected nutrient regeneration and recycling (Ogutu-Ohwayo, 1990; Carpenter et al. 1986; Williams & Moss 2003). However, whilst fish introductions did no doubt have an impact (Kaufman, 1992) on the fish community, it is now understood that changes within the lake began to manifest themselves by at least the 1940s, i.e. over a decade before the fish were introduced (Verschuren et al., 2002). Therefore, some other factor was responsible. Natural climatic changes have been observed through Lake Victoria's history in the sense that water temperatures have risen in the

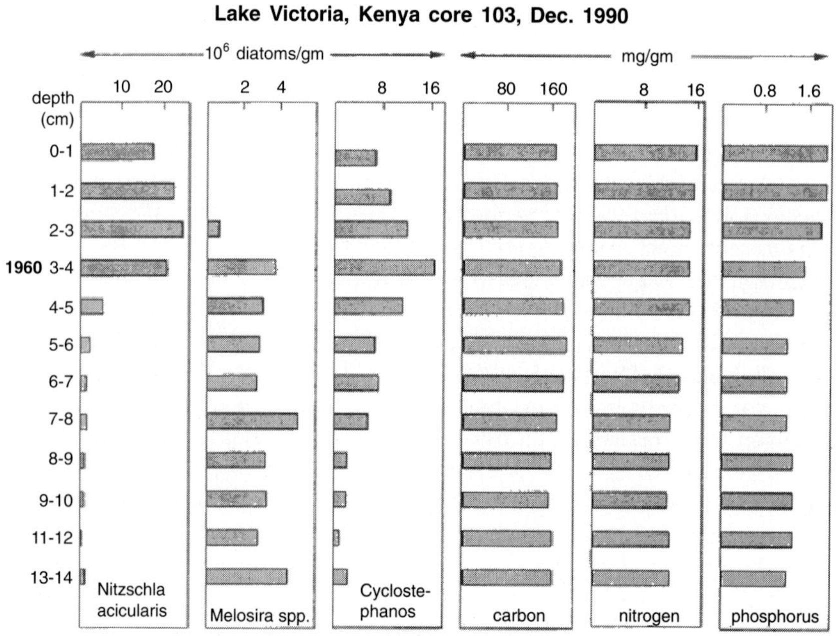

Fig. 7.7 Diatom stratigraphy and sedimentary C, N and P content of the recent deep water sediments (55 m) in the northeastern sector of Lake Victoria. Date at 13 cm is 1800 AD (Taken from Hecky, 1993).

later half of the nineteenth century to the present (Hecky, 1993; Hecky et al., 1994) and changes in water level are also apparent—a 2.5 m rise between 1961-1964 and another significant rise of 1.5 m between 1977-1980 (Kite, 1981). Such climatic alterations could impact upon the processes within the lake. However, the rising lake level would only have increased the surface area by about 3% (Bootsma & Hecky, 1993). This in itself would not have increased N and P deposition of the like seen when dams are built because a lake as large as Victoria is well buffered against small oscillations in lake level. Furthermore, whilst these climatic events have probably occurred on a number of occasions within the last 10,000 years, palaeolimnological evidence identifies that algal species have remained fairly uniform for at least the last 12,400 years, only beginning to change within the last 80 years (Stager & Johnson, 2000; Hecky, 1993). As such, the changes in the biological structuring of Lake Victoria and increases in N and P sedimentation, whilst almost certainly interlinked and involved in various and complicated feedback and cascading mechanisms (Williams & Moss, 2003), have been primarily driven by causative agents other than natural climatic variations or fish introductions.

Phosphorus deposition has increased in sediments and total P concentrations have risen in lakewaters. Indeed, there is no doubt that P loading has increased by a factor of at least two and that planktonic primary production has increased to a similar extent with a shift to nitrogen-fixing cyanophyta. The lake is now light limited for phytoplankton growth due to self shading by the high algal turbidity and deep mixed layers (Mugidde, 1993; Mugidde et al., in press). It now appears that anthropogenic activities around the lake have led to the changes seen today and are the primary cause of eutrophication (Hecky, 1993; Bootsma & Hecky, 1993; Scheren et al., 2000). The lake has undergone a classic cultural eutrophication (Hecky, 1993) with consequences for the upper food web as well (Seehausen et al., 1997). The whole process of bottom-up versus top-down control in lakes is a complex one (Williams & Moss, 2003) and not within the remit of this chapter to explore. However, rather than a top-down fisheries-driven ecosystem, it seems that Lake Victoria is a bottom-up nutrient-driven system.

From the population of the Lake Victoria catchment being in the region of 4.5 million people between 1900 and 1932, the figure rose to 27.7 million by 1995 (United Nations, 1995). This increase was largely due to the completion in 1930 of the Ugandan railroad and the resulting immigration and better availability of medicine that it bought. It is, therefore, unsurprising that changes within Lake Victoria, evident within the fossil record, began in the 1930s. The increase in population around

Lake Victoria has increased point source pollution, resulting from both industrial and municipal waste. Industrial pollution originates mainly from large towns on the periphery of the lake and, whilst many Kenyan factories do have waste treatment plants, most factories in Uganda and Tanzania do not. Moreover, treatment facilities for municipal waste in Uganda are considered, to be either inadequate for the growing population or non-existent (UNEMA, 1995). However, the majority of waste from Ugandan factories does at least filter through a wetland buffer before entering the lake. On an overall basis, the urban waste in its entirety accounts for only a small proportion of the nutrient input into Lake Victoria contributing, as it does, for 6 and 9% of the total N and P entering the lake (Scheren et al., 2000).

Of more concern to Lake Victoria is its diffuse pollution. With 90% of the growing population relying directly on subsistence farming and animal husbandry, any demographic increase brings a corresponding increase in agriculture. In the early stages of agricultural development, land is often deforested and burnt. Atmospheric inputs of nutrients are, therefore, increased (Bootsma & Hecky, 1993). The average rainfall over the lake is approximately 100 km^3 yr^{-1} (Piper et al., 1986) and, based upon calculations by Scheren et al. (2000) this accounts for 72% of the N and 36% of the P entering the lake.

A further impact of agriculture is that with the removal of vegetation so soil erosion increases and, moreover, with Lake Victoria supplying water for irrigation, many agricultural plots are situated adjacent to the lake. Soil as well as organic matter from crop debris therefore enters the system (Lindenschmidt et al., 1998). Agriculture itself brings with it increased fertilizer use, which also enters the lake. In places, cattle densities are above those of the human population with manure also contributing to the nutrient input of the lake (Bootsma & Hecky, 1993). As the population increases and land becomes less available, agriculture encroaches into the wetlands. Riparian vegetation surrounding Lake Victoria prevents erosion and the transport of solutes into the lake (Lindenschmidt et al., 1998). The removal of this vegetation for agricultural and fuel purposes increases sediment and nutrient loading into the lake. Wetlands, posing as the natural purification system between land and water, are critical to the lake's functioning. Whilst soils themselves are efficient phosphorus retainers, nitrogen compounds are extremely soluble and plant uptake and denitrification within wetlands reduces nitrogen inputs dramatically. The reclamation of wetlands due to the growth of human populations within the Lake Victoria catchment over the last few decades has probably decreased the buffering capacity of the wetlands, further exacerbating eutrophication (Crul, 1995). It has been estimated that agricultural runoff accounts for the remaining 22% of

N and 55% of P entering Lake Victoria, the total of which is in the order of 117,000 t yr^{-1} of N and 14,000 t yr^{-1} of P (Scheren et al., 2000). However, these estimates do not allow for biological nitrogen fixation within the lake nor atmospheric dry deposition that will substantially increase these estimates (R. E. Hecky unpublished data).

On the whole, it is clear that whilst water hyacinth has largely disappeared within Lake Victoria, it is a known fact that water hyacinth growth and production is highly favoured by an environment rich in nutrients. The current and ongoing eutrophication of Lake Victoria is, therefore, of great concern on many levels, not least of which is the potential for an enormous explosion in water hyacinth biomass above and beyond that previously seen at its peak in the mid 1990s.

If the weed does return, there are several methods that can be used to control it. Some of these used in Lake Victoria during the 1980s and 90s are believed to have been quite effective.

THE CONTROL OF WATER HYACINTH

Once established, invasives may be difficult or impossible to control or eliminate and often attempts to do so adversely impact the remnant native vegetation (Njuguna & Thitai, 1993). If water hyacinth returns to Lake Victoria, procedures will be undertaken, as they were before, to remove it. But the elimination of water weeds is an expensive process, both in fiscal and temporal terms. Three basic methods exist to remove the alien invaders—physical, chemical and biological controls.

Physical control involves employing manual labour and mechanical harvesters to remove weeds. The use of physical barriers such as floating booms to coral weeds and prevent their spread as well as aiding ease of physical removal can also be implemented (Akinyemiju, 1987). In Lake Victoria, manual removal at landing stages was previously undertaken extensively by locals being provided, as they were, with equipment such as rakes and wheelbarrows via the Lake Victoria Environmental Management Project (LVEMP) and other emergency response programmes. Aquatic vegetation cutters or "Swamp devils" were used around Kisumu Port and Homa Bay Pier, Kenya as well as within the vicinity of the hydro-electric dam at Jinja. Booms were also used, particularly at the hydro-electric dam, to reduce weed expansion (LVEMP, 2000). These techniques clearly worked in the sense that the weed was removed. However, benefits of such a method are typically localized due to the sheer scale of the problem on Lake Victoria and short lived, as a cleared site can be inundated with new weed in a matter of days, either through growth of any remaining weed or immigration of plants from outside of the site. Furthermore, booms can become over burdened from the weight of the weed (Goodland, 1995) and machinery can breakdown. Unfortunately, with a lack of expertise, tools and spare parts, the expensive harvesters were, on occasion, left to rust on the shore. However, physical control has been used successfully throughout the world to manage invasive weeds (NAS, 1976) and in Nigeria, such methods are considered relatively effective (Juweto, 1993). Overall, however, physical control has to be intensive and regular and is thus, ultimately, expensive. The cost of clearing a hectare of weed is estimated to be between US $ 2,400-30,000 (Haller, 1996). Farri & Boroffice (1999) when, also considering physical control in Nigeria, considered it be too costly (US $5,000 ha^{-1}). Moreover, none of these estimates take account of the health risks associated with clearing operations as faced by the clearing crews. Therefore, for a water body the size of Lake Victoria, physical control is not a feasible long-term control option.

Chemical control is, worldwide, one of the most commonly used methods of macrophyte repression. Chemicals can be applied from the air, water or land and some degree of accuracy as to where the herbicide lands can be achieved. Chemicals such as diquat, glyphsophate, amitrole and the amine and acid formulations of 2,4-D are the most effective and commonly used chemicals against water hyacinth. Application needs to be undertaken by trained individuals using correct dosages and applied at appropriate stages of growth. When they are applied incorrectly, problems can be encountered not least of which is killing non-targeted plant species, be they aquatic macrophytes or agricultural crops (Njuguna & Thitai, 1993). Herbicides themselves typically only cost in the region of US $50 ha^{-1} of treated area. However, application costs by plane, boat or hand are in addition to this estimate (Howard & Harley, 1998). Therefore, herbicide use is only viable over small areas, as large-scale application can become prohibitively expensive. However, whatever the operational costs, if weeds grow faster than chemicals can be applied, then the control of invasives is clearly not accomplished (de Groot, 1993). Moreover, if the dead or dying weed is not removed, the rotting biomass can lead to localized de-oxygenation which can, in turn, lead to other detrimental impacts upon the water body (Mallya, 1999; LVEMP, 2000). Furthermore, whilst plants may appear to have been controlled the problem may not have been solved if, as in the case of water hyacinth, the seeds remain buried in the sediment and propagate some years later or, as is more typical, plants remaining in the marginal vegetation re-colonize the waterbody. Repeated applications are, therefore, required (Julien et al., 1999).

The main problem associated with herbicide use and indeed the reason they are not used as widely as they might be is the potential effect, real or perceived, of chemicals on humans either directly through ingestion or indirectly through reduced income. Lake Victoria is a major source of drinking water and, therefore, the problem of ingestion is clear. Also, the sites most affected by water hyacinth and so most in need of herbicide application, are typically adjacent to lakeshore communities and farmlands which increases the risk of destroying non-targeted plants and crops. Furthermore, fear concerning the impact on native biota, not just the loss of non-targeted plants but also the potential for bioaccumulation of contaminant dioxins through the food chain, is also an area of concern for governments considering their own populations health and the sale of fish to international exporters (Murty & Mulendema, 1999). The lake supports an extremely valuable export fishery that earns critical foreign exchange for the riparian countries. The European Union, which is a major importer of fish from Lake Victoria, made it very clear that widespread application of chemicals could threaten the marketability of the fish. Finally, if biological control is used (see below), the application of herbicides may impact biological agents. Whilst experiments have found that diquat, glyphosphate, and 2,4-D do not increase the mortality rates of *Neochetina eichhorniae*, they did move away from treated plants to untreated ones (Haag, 1986). In addition, Patnaik et al. (1987) found that *N. bruchi* mortality was increased by several of the commonly used herbicides.

On the whole, correct application of chemicals, can be very effective in the short-term but as with physical control, the use of herbicides is too expensive for repeated use on the scale required for Lake Victoria. Whatever the human and ecosystem effects of herbicide application are, other than limited trials in 1997 within Ugandan waters, no official large-scale application of herbicides has been carried out within Lake Victoria. Currently, an agreement among the three riparian states forecloses application within any one country without prior consultation with the other countries because of possible economic damage that could arise from loss of the valuable fish export industry.

Biological control involves the use of host specific agents that are the naturally occurring enemies of invasive weeds in their native bioregion. The United States Department of Agriculture initiated biological control in Argentina in 1961 and since then, weevils, moths and fungi have been tested for host-specificity (Center, 1994). Host-specific agents have now been successfully used to control several aquatic weeds such as water hyacinth, floating water fern and water lettuce in a variety of localities around the globe (Harley, 1990; de Groot, 1993; Julien et al., 1999). Indeed by 1998, a total of seven agents had been released in 33 countries (Julien & Griffiths, 1998). Biological control of water hyacinth has, therefore, become very popular and two weevils in particular, *N. eichhorniae* and

N. bruchi, have been used extensively and successfully in North America, Asia, Australia and Africa.

In the long term, once the weed is under control, a natural flux between the agent and weed should exist, thus keeping in balance the density of weed and numbers of biological agent present (Howard, 2000), although as with all dynamic systems, fluctuations in both populations will occur (Harley, 1993). Biological control can provide an environmentally sensitive, cost effective and permanent solution to invasive weed growth. However, the two drawbacks (assuming that host-specificity is guaranteed) are that biological control usually takes between three to five years for the insect populations to reach a level from which control can be achieved (Howard & Harley, 1998; Julien et al., 1999) and, moreover, it does not always work (Salmah et al., 1991; McFadyen, 2000; Julien et al., 2001).

In Lake Victoria, both species of *Neochetina* weevil were introduced on a large scale in December 1996 and an 80% decrease in water hyacinth abundance has been accredited to these introductions (Ochiel, 1999; LVEMP, 2000). However, the rapid decrease in water hyacinth biomass from December 1996 to 1998 (two years) is extraordinarily fast and not usual of typical weevil control. Whilst the introduction of weevils into Lake Victoria probably did weaken the water hyacinth mass over large areas, the driving force behind the sudden and concurrent loss of weed from Lake Victoria and other lakes in Uganda during 1997/1998, was most probably a result of *El Nino* events which occurred at that time (Williams et al. 2005). *El Nino* related events, in all likelihood, had multiple affects on water hyacinth. Since the late 1980s, water levels had fallen in Lake Victoria, however, the *El Nino* caused a dramatic increase (> 1 m) in the lake level. Rising water levels dislodged and mobilized shore-bound plant mats and carried the now free floating mats to open waters, where wave action helped to destroy them. In addition, extreme river discharges flushed water hyacinth from neighbouring wetlands and rivers, thereby reducing immigration potential in subsequent years. Moreover, light reaching the lake's surface was reduced by a thick and extensive cloud cover which limited plant productivity because of the plant's high light saturation for photosynthesis (Williams et al., 2005). All these effects on growth and mortality caused a rapid decline in hyacinth populations. In any case, during 2001, observations over a six-month period found weevil abundance to be very low, with the majority of plants hosting no weevils. Feeding scars were also minimal (A. E. Williams pers. obs.). There is a possibility that when the water hyacinth mats sank, they took with them many of the weevil larvae and eggs (IMPECCA, 2001). Therefore, remaining populations may not be in a natural state of flux, being too few and dispersed, and might be unable to combat any resurgence event. Reintroduction of the weevil from rearing centres will be necessary to insure adequate populations so as to effect control. However, even at low densities, weevils are now present in the lake and these may reduce growth rates from those experienced during the original colonization of Lake Victoria by water hyacinth. A thorough study of the population dynamics of water hyacinth in Lake Victoria is required for effective management of the weed and also to apportion the affects of weevil, light, wave action and nutrients on controlling the weed in Lake Victoria.

Of most use in the control of water hyacinth is an integrated method of control. Integrated control employs all three of the above control methods in an overall management strategy. Each of the three is tailored to suit a specific problem and the extent to which each is used varies with location. In general, however, mechanical and chemical controls are used initially to bring short-term benefits, whilst the biological approach, taking longer to implement and develop, follows after the first two control strategies (de Groot, 1993; Howard & Harley, 1998). In this regard, maintaining weevil-rearing centres established by the LVEMP will be an essential feature of active management to insure that weevil populations remain at an adequate level for control.

There is one other type of control, however, that is of key importance but yet is not mentioned as often as the above control methods—the control of nutrient inputs. The levels of available nitrogen and phosphorus have often been cited as the most important factors in limiting water hyacinth growth (Carignan & Neiff, 1994, Heard & Winterton, 2000; Musil & Breen, 1985, Reddy et al., 1989, 1990, 1991). Indeed, a Commonwealth Science Council survey of exotic floating weeds throughout Africa (Harley, 1993) recommended "that action be taken at a national and subregional level to reduce nutrient input (particularly from urban and industrial waste) into water bodies, and to develop land-use schemes for long-term utilization and sustainable development of watersheds". However, in 1993, at a symposium on water hyacinth control, only three papers from 13 country and regional reports mentioned the importance of managing nutrient inputs through improved watershed management. In 1998, during a global working group on biological and integrated control of water hyacinth, in a session dedicated to integrated management, only two out of the 13 papers presented identified the importance of nutrients in controlling water hyacinth. In 2000, at the second meeting of the global working group, just two from 22 papers contained the word "eutrophication" in the context of it encouraging water hyacinth, whilst the impact of nutrient inputs on weed growth was only mentioned in a further five. However, the concerns over nutrient inputs were often mentioned in the summaries and overviews of the symposiums and, indeed, within the global arena, the problems of eutrophication and nutrient pollution are well known by scientists, governments and the public alike. It is understood that nutrient-rich waters offer prime conditions for weed growth and, if unabated, eutrophication and lack of management will lead to future invasions.

THE MANAGEMENT OF LAKE VICTORIA

It is understandable and necessary that from within affected countries scientists and practitioners of weed control turn to mechanical, chemical and biological means of short-term control and management. Moreover, in the short term, all biological possibilities should be vigorously explored as appropriate insects are available and have been used elsewhere, often to good effect. Yet the only real long-term solution to the water hyacinth problem is not biological control but rather the control of eutrophication and nutrient inputs through sustainable management of watersheds and resources. The reliance of developing nations on biological control as a long-term solution rather than sustainable management and development of their treasured resources is of major concern.

The control and remediation of eutrophication is difficult due to the complexities of the systems involved. Many attempts have been made to reduce the impact as well as the causative agents of eutrophication through biological and chemical manipulations. Biological approaches of controlling eutrophication typically involve the manipulation of the biological system such as changing the dominance of certain fish species, particularly piscivorous fish through top-down biomanipulations (Shapiro et al., 1975; Carpenter et al., 1985). Within the context of Lake Victoria, however, chemical manipulations are of more pertinence. Such measures include the treatment of waters flowing from sewage works and factories and reducing diffuse nutrient loading by changing land-use practices. These approaches can lower nutrient inputs into a water body and thus reduce problems of eutrophication through bottom-up processes (McQueen, 1990). However, whilst the control of point sources is relatively easy, the control and treatment of diffuse sources is much more challenging and non-point sources dominate the current loading of N and P into the lake (Scheren et al., 2000). Within an African context and in particular Lake Victoria, land management is key and must be fully planned to minimize the impact of current practices and to reduce risks inherent in the expansion of agriculture, which at present is at 1% each year (Scheren et al., 2000). By reducing slash-and-burn and deforestation, atmospheric deposition should decrease as the land surface recovers vegetative cover. Wetlands provide a natural buffer and their preservation is essential. Reduced and slow release fertilizer use and crop/land rotation should be encouraged through education. Overall, there needs to be a change in anthropogenic land use and a reduction in polluting events, both chemical and biological. Big steps need to be made towards a greater level of sustainability. Although these steps are complex, expensive and time consuming, in the final synopsis it is the only sustainable solution for continued long-term alleviation of current and future problems within Lake Victoria. All the systems: macrophytes, nutrients, fisheries, plankton and microbial are interlinked and holistic management involving land, water and airsheds is required. Whilst the solutions are not easy, the consequences of not doing anything may be even worse.

Due to the scale of Lake Victoria, a massive financial commitment must be undertaken. Unfortunately, whilst governments work on four-or five-year cycles, desiring re-election gained primarily from creating and spending wealth, ecosystems work on much longer time scales. Moreover, the continuing growth of the population in Lake Victoria's catchment means that greater sustainability is unlikely to be accomplished in the near future (Marshall, 1993). The United Nations (1995) projects that the Lake Victoria catchment will contain 53 million people by 2020. It is quite likely, therefore, that further degradation of the Lake Victoria system will be encountered. However, whilst people and governments understand the requirements to invest in research on areas such as fisheries, there is less understanding as to why a shift in algal species, for instance, is of any cause for concern. The responsibility of education is, therefore, on ecologists from around the world. Due to the enormous cost in human and natural resources that will be incurred if the current situation is not amended and the potential economic knock on effect this will likely have on the global economy, the financial burden for implementation must be shared by all the countries immediately affected and the world's governments at large.

REFERENCES

Akinyemiju O.A. 1987. Invasion of Nigerian waters by water hyacinth. Journal of Aquatic Plant Management, 25, 24-26.

Anderson N. J. 1995. Naturally eutrophic lakes: Reality, myth or myopia? T.R.E.E., 10, 137-138.

APS 1991. Harmful hyacinth resists elimination. Science and Technology Features Service, All Africa Press Service, 31st January 1991, Nairobi, Kenya.

Atkinson M.J. & Smith S.V. 1983. C : N : P ratios of benthic marine plants. Limnology and Oceanography, 28, 568-574.

Auld B.A., Menz K.M. & Tisdell C.A. 1987. Weed Control Economics. Academic Press, London, pp. 177.

Balirwa J.S. 1998. Lake Victoria Wetlands and The Ecology of The Nile Tilapia Oreochromis Niloticus Linne. PhD thesis, Wageningen Agricultural University, pp. 247.

Bazzaz, F.A., Chiariello, N.R., Coley, P.D. & Pitelka, L.F. 1987. Allocating resources to reproduction and defense. New assessments of the costs and benefits of allocation pattern in plants are relating ecological roles to resource use. BioScience, 37, 58-66.

Beadle L.C. & Lind E.M. 1960. Research on the swamps of Uganda. Uganda Journal, 24, 84-98.

Beadle L.C. 1981. Inland Waters of Tropical Africa. Longman, London, pp. 475.

Bootsma H.A. & Hecky R.E. 1993. Conservation of the African great lakes: A limnological perspective. Conservation Biology, **7**, 644-656.

Bugenyi F.W.B. & Balirwa J.S. 1998. East African species introductions and wetland management: Socio-political dimensions. In: "Science in Africa Emerging Water Management Issues". American Association for the Achievement of Science Symposium Proceedings.

Carignan R. & Neiff J.J. 1994. Limitation of water hyacinth by nitrogen in subtropical lakes of the Paraná floodplain (Argentina). Limnology and Oceanography, 39, 439-443.

Carpenter S.R., Kitchell J.F., Hodgson J.R., Cochran P.A. Elser J.J., Elser M.M., Lodge D.M., Kretchner D., He X. & von Ende C.N. 1986. Regulation of lake primary productivity by food web structure. Ecology, 68, 1863-1876.

Carpenter S.R., Kitchell J.F. & Hodgson J.R. 1985. Cascading trophic interactions and lake productivity. BioSciences, 35, 634-639.

Center T.D. 1994. Biological control of weeds: Water hyacinth and water lettuce. In: "Pest Management in The Subtropics. Biological Control–A Florida Perspective". (Eds) D. Rosen, F.D. Bennet & J.L. Capinera Intercept Ltd, Andover, UK, pp. 737.

Chikwenhere G.P. & Phiri G. 1999. History of water hyacinth and its control efforts on Lake Chivero in Zimbabwe. In: "Proceedings of The First IOBC Global Working Group Meeting for The Biological and Integrated Control of Water Hyacinth". (Eds) M.P. Hill, M.H. Julien & T.D. Center Weeds Research Division, ARC, South Africa, pp. 182.

Chikwenhere G.P., Keswani C.L. & Liddel C. 1999. Control of water hyacinth and its environmental and economic impacts at Gache Gache in the Eastern reaches of Lake Kariba, Zimbabwe. In: "Proceedings of The First IOBC Global Working Group Meeting for The Biological and Integrated Control of Water Hyacinth". (Eds) M.P. Hill, M.H. Julien & T.D. Center Weeds Research Division, ARC, South Africa, pp. 182.

Crul R.C.M. 1995. Conservation and Management of the African Great Lakes Victoria, Tanganyika and Malawi. UNESCO/IHP-IV M-5.1. Comparative and comprehensive study of great lakes, pp. 129.

Das R.R. 1969. A study of reproduction in *Eichhornia crassipes* (Mart.) Solms. Tropical Ecology, 10, 195-198.

de Graft-Johnson K.A.A. 1993. Aquatic plants infestation in water bodies in Ghana. In: Control of Africa's floating water weeds. (Eds) A. Greathead & P.J. de Groot. Commonwealth Science Council, Series no. CSC (93) AGR-18, Proceedings 295, pp. 187.

de Groot, P.J. 1993. Introduction and summary. In: Control of Africa's Floating Water Weeds. (Eds) A. Greathead & P.J. de Groot Commonwealth Science Council, Series no. CSC (93) AGR-18, Proceedings 295, pp. 187.

Debusk T.A. & Dierberg F.E. 1989. Effects of nutrient availability on water hyacinth standing crop and detritus deposition. Hydrobiologia, 174, 151-159.

Deevey E.S. 1942. Studies on Connecticut lake sediments III. The biostratification of Linsley Pond. American Journal of Science, 240, 313-24.

Denny P. 1980. Solute movement in submerged angiosperms. Biological Reviews, 50, 65-92.

Denny P. 1985. The Ecology and Management of African Wetland Vegetation: A Botanical Account of African Swamps and Shallow Waterbodies. Geobotany 6, Dr W Junk Publishers, The Netherlands, pp. 344.

Desougi L.A. 1984. Mineral nutrient demands of the water hyacinth (*Eichhornia crassipes* (Mart.) Solms) in the White Nile. Hydrobiologia, 110, 99-108. (1984

Duarte C.M. 1990. Seagrass nutrient content. Marine Ecology Progress Series, 67, 201-207.

Farri T.A. & Boroffice R.A. 1999. An overview on the status and control of water hyacinth in Nigeria. In: Proceedings of The First IOBC Global Working Group Meeting for The Biological and Integrated Control of Water Hyacinth. (Eds) M.P. Hill, M.H. Julien & T.D. Center Weeds Research Division, ARC, South Africa, pp. 182.

Fayad Y.H. 1999. Water hyacinth infestations and control in Egypt. In: Proceedings of the first IOBC global working group meeting for the biological and integrated control of water hyacinth. (Eds) M.P. Hill, M.H. Julien & T.D. Center Weeds Research Division, ARC, South Africa, pp. 182.

Fish G.R. 1959. A seiche movement and its effect on the hydrology of Lake Victoria. Fish Publications, London, 10, 1-68.

Flack S.R. & Benton N.B. 1998. Invasive Species and Wetland Biodiversity. National Wetlands Newsletter, 20, 7-11.

Food and Agriculture Organisation 2000. Fighting water weeds in West Africa. News and highlights, no. 705.

Forno I.W. & Harley K.L.S. 1979. The occurrence of *Salvinia molesta* in Brazil. Aquatic Botany, 6, 185-187.

Fritz S.C. 1989. Lake development and limnological response to prehistoric land-use change in Diss, Norfolk, UK Journal of Ecology, 77, 182-202.

Gangstad, E.O. 1978. Weed control methods for river basin management. CRC Press, Florida, pp. 229.

Goodland R. 1995. Uganda: Owen Falls Hydroproject: An Environmental Reconnaissance of The Water Hyacinth Problem. Environment Department Report, World Bank, Washington, pp. 23.

Gopal B. 1987. Water hyacinth: Aquatic plant studies. Elsevier Science Publishers, Amsterdam, pp. 471.

Gophen M., P.B.O Ochumba & L.S. Kaufman 1995. Some aspects of perturbation in the structure and biodiversity of the ecosystem of Lake Victoria (East Africa). Aquatic Living Resources, 8, 27-41.

Greathead A. & de P.J. Groot 1993. Control of Africa's Floating Water Weeds. Commonwealth Science Council, Series no. CSC (93) AGR-18, Proceedings 295, pp.187.

Guildford S.J. & Hecky R.E. 2000. Total nitrogen, total phosphorus and nutrient limitation in lakes and oceans: Is there a common relationship. Limnology and Oceanography, 45, 1213-1223.

Haag K.H. 1986. Effects of herbicide application on mortality and dispersive behaviour of the water hyacinth weevils, *Neochetina Eichhorniae* and *Neochetina Bruchi*. Environmental Entomology, 15, 1192-1198.

Haller W.T. 1996. Evaluation of the Kelpin 800 aquatic weed harvester. Aquatic, 18, 10-15.

Harley K.L.S. & Mitchell D.S. 1981. The biology of Australian weeds 6. *Salvinia molesta*. Journal of Australian Institute of Agricultural Science, 47, 67-76.

Harley K.L.S. 1990. The role of biological control in management of water hyacinth, *Eichhornia crassipes*. Biocontrol News and Information, 11, 11-22.

Harley K.L.S. 1993. Biological control of floating aquatic weeds in Africa. In: Control of Africa's floating water weeds. (Eds) A. Greathead & P.J. de Groot Commonwealth Science Council, Series no. CSC (93) AGR-18, Proceedings 295, pp. 187.

Harley K.L.S. 1993. Commonwealth Science Council survey project on exotic floating African water-weeds: survey report. In: Control of Africa's floating water weeds. (Eds) A. Greathead & P.J. de Groot Commonwealth Science Council, Series no. CSC (93) AGR-18, Proceedings 295, pp. 187.

Harley K.L.S. 1993. Floating aquatic weeds: What are they? Where are they? How should we manage them? In: Control of Africa's Floating Water Weeds. (Eds) A. Greathead & P.J. de Groot Commonwealth Science Council, Series no. CSC (93) AGR-18, Proceedings 295, pp. 187.

Harper D.M. 1992. The ecological relationships of aquatic plants in Lake Naivasha, Kenya. Hydrobiologia, 232, 65-71.

Hasler A.D. 1947. Eutrophication of lakes by domestic drainage. Ecology, 28, 383-395.

Haworth E.Y. 1969. The diatoms of a sediment core from Blea Tarn, Langdale. Journal of Ecology, 57, 429-439.

Hayes K.R. 2002. Identifying hazards in complex ecological systems. Part 2: Infection modes and effects analysis for biological invasions. Biological Invasions, 4, 251-261.

Heard T.A. & Winterton S.L. 2000. Interactions between nutrient status and weevil herbivory in the biological control of water hyacinth. Journal of Applied Ecology, 37, 117-127.

Hecky R.E. 1993. Peter Kilham memorial lecture: The eutrophication of Lake Victoria. Verhandlungen Internationale Vereinigung fur Theoretische und Angewandte Limnologie, 25, 39-48.

Hecky R.E., Bugenyi F.W.B., Ochumba P., Talling J.F., Mugidde R., Gophen M. & Kaufman L. 1994. Deoxygenation of the deep water of Lake Victoria, East Africa. Limnology and Oceanography, 39, 1476-1481.

Hecky R.E., Bootsma H.A., Mugidde R. & Bugenyi F.W.B. 1996. Phosphorus pumps, nitrogen sinks, silicon drains: plumbing nutrients in the African Great Lakes. In: The Limnology, Climatology and Paleoclimatology of the East African Lakes. (Eds) T.C. Johnson & E. Odada Gordon and Breach, Toronto, pp. 205-224.

Hill G., Day R., Phiri G., Lwanda C., Njaya F., Chimatiro S. & Hill M.P. 1999. Water hyacinth biological control, in the Shire River, Malawi. In: Proceedings of the first IOBC global working group meeting for the biological and integrated control of water hyacinth. (Eds) M.P. Hill, M.H. Julien & T.D. Center Weeds Research Division, ARC, South Africa, pp. 182.

Holland M.M. 1988. Technical consultations on landscape boundaries. Report of a SCOPE/MAB workshop on ecotones. In: A New Look at Ecotones: Emerging International Projects on Landscape Boundaries. (Eds) F. Di Castri, A.J. Hansen & M.M. Holland Biology International, Special Issue 17, Paris, pp. 163.

Howard G.W. & Harley K.L.S. 1998. How do floating aquatic weeds affect wetland conservation and development? How can these effects be minimised? Wetlands Ecology and Management, 5, 215-225.

Howard G.W. 2000. Control options: Freshwater invasives. In: Best Management Practices for Preventing and Controlling Invasive Alien Species. (Eds). G. Preston, A.G. Brown, E. van Wyk Symposium Proceedings, Cape Town, South Africa, pp. 316.

IMPECCA 2001. Biological and integrated control of *Eichhornia crassipes*. Water hyacinth news, 4. The IMPECCA programme, CABI Bioscience, UK.

Ivens G.W. 1967. East African Weeds and Their Control. Oxford University Press, Nairobi, pp. 244.

Johnson T.C., Scholz C.A., Talbot M.R., Kelts K., Ngobe G., Beuning K. Ssemanda I. & McGill J.A. 1996. Late Pleistocene desiccation of Lake Victoria and rapid evolution of cichlid fishes. Science, 273, 1091-1093.

Jones R., Benson-Evan K., Chambers F.M., Seddon B.A. & Tai Y.C. 1978. Biological and chemical studies of sediments from Llangorse Lake, Wales. Verhandlungen Internationale Vereinigung fur Theoretische und Angewandte Limnologie, 20, 642-648.

Julien M.H. & Griffiths M.W. 1998. Biological control of weeds. A world catalogue of agents and their target weeds. Wallingford, UK, CAB International, pp. 240.

Julien M.H., Giffiths M.W. & Stanley J.N. 2001. Biological control of water hyacinth 2. The moths Niphograpta albiguttalis and Xubida infusellus: biologies host ranges and rearing, releasing and monitoring techniques for biological control of Eichhornia crassipes. ACIAR Monograph No. 79, pp. 90.

Julien M.H., Giffiths M.W. & Wright A.D. 1999. Biological control of water hyacinth. The weevils Neochetina bruchi and N. eichhorniae: biologies, host ranges and rearing, releasing and monitoring techniques for biological control of Eichhornia crassipes. ACIAR Monograph No. 60, pp. 87.

Juweto E. 1993. Mechanical control of water hyacinth: The Nigerian experience. In: Control of Africa's floating water weeds. (Eds) A. Greathead & P.J. de Groot. Commonwealth Science Council, Series no. CSC (93) AGR-18, Proceedings 295, pp. 187.

Kampeshi C. & Shantima M. 1999. Nutrient loading and water hyacinth infestation in the Lower Kafue River. In: Proceedings of The First IOBC Global Working Group Meeting for The Biological and Integrated Control of Water Hyacinth. (Eds) M.P. Hill, M.H. Julien & T.D. Center Weeds Research Division, ARC, South Africa, pp. 182.

Kaufman L. 1992. Catastrophic change in species-rich freshwater ecosystems. BioScience, 42, 846-858.

Kitaka G.E.B. 1971. An instance of cyclonic upwelling in the southern offshore waters of Lake Victoria. African Journal of Tropical Hydrobiology and Fisheries, 1, 85-92

Kite G.W. 1981. Recent changes in level of Lake Victoria. Hydrological Sciences Bulletin, 26, 233-243.

Kling H.J., Mugidde R. & Hecky R.E. 2001. Recent changes in the phytoplankton community of Lake Victoria in response to eutrophication. In: The Great Lakes of the World (GLOW): Food-web, Health and Integrity. (Eds) M. Munawar & R.E. Hecky Backhuys, Netherlands, pp. 47-65.

Lehman J.T. & Branstrator D.K. 1993. Effects of nutrients and grazing on the phytoplankton of Lake Victoria Verhandlungen Internationale Vereinigung Fur Theoretische Und Angewandte Limnologie, 25, 850-855.

Lindeman R.L. 1942. The trophic-dynamic aspect of ecology. Ecology, 23, 399-418.

Lindenschmidt K.E., Suhr M., Magumba M.K., Hecky R.E. & Bugenyi F.W.B. 1998. Loading of solute and suspended solids from rural catchment areas flowing into Lake Victoria in Uganda. Water Research, 32, 2776-2786.

Lindsey K. and Hirt H.M. 1999. Use water hyacinth—A practical handbook of uses for the water hyacinth from across the world. Marianum Press, Kisubi, Uganda, pp. 114.

Lung'ayia H., Sitoki L. & Kenyanya M. 2001. The nutrient enrichment of Lake Victoria (Kenyan waters). Hydrobiologia, 458, 75-82.

LVEMP 2000. Water Hyacinth and Other Invasive Weeds in Lake Victoria: A status Report. Lake Victoria Environmental Management Project, Fisheries Resources and Research Institute, Jinja, Uganda.

LVEMP 2001. Water hyacinth control component. First Lake Victoria Environmental Management Programme (LVEMP) Regional Scientific Conference. Kisumu, Kenya.

Macan, T.T. 1970. Biological studies of the English Lakes. Longman Group, London, pp. 260.

Mallya G.A. 1999. Water hyacinth in Tanzania. In: Proceedings of The First IOBC Global Working Group Meeting for The Biological and Integrated Control of Water Hyacinth. (Eds) M.P. Hill, M.H. Julien & T.D. Center Weeds Research Division, ARC, South Africa, pp. 182.

Marshall B.E. 1993. Floating water-weeds in Zimbabwe, with special reference to the problem of water hyacinth in Lake Chivero. In: Control of Africa's Floating water weeds. (Eds) A. Greathead & P.J. de Groot Commonwealth Science Council, Series no. CSC (93) AGR-18, Proceedings 295, pp. 187.

Matagi S.V. 2002. Some issues of environmental concern in Kampala, the capital city of Uganda. Environmental Monitoring and Assessment, 77, 121–138.

McFadyen R.E.C. 2000. Successes in Biological Control of Weeds. In: Proceedings of the X International Symposium on Biological Control of Weed. (Ed.) N.R. Spencer Montana State University, Bozeman, Montana, USA, pp. 149.

McGowan S., Britton G., Haworth E. & Moss B. 1999. Ancient Blue-Green Blooms. Limnology and Oceanography, 44, 2, 436-439.

McQueen D.J. 1990. Manipulating lake community structure: Where do we go from here? Freshwater Biology, 23, 613-620.

Melvin N.C. 1999. Noxious, Invasive and Alien Plant Species: A Challenge in Wetland Restoration and Enhancement. Wetland restoration information series 1, Wetland Science Institute, Washington.

Methy M. & Roy J. 1993. Morphogenetic changes induced by a low red : far-red ratio and their growth consequences in water hyacinth (*Eichhornia crassipes*). Journal of Experimental Botany, 44, 1275-1280.

Methy M., Alpert P. & Joy J. 1990. Effects of light quality and quantity on growth of the clonal plant *Eichhornia crassipes*. Oecologia, 84, 265-271.

Mitchell D.S. 1974. Aquatic Vegetation and Its Use and Control. UNESCO, Paris, pp. 135.

Mitchell D.S. 1978. Aquatic Weeds in Australian Inland Waters. Australian Government Publishing Service, 189 pp.

Moorhead K.K., Reddy F.R. & Graetz D.A. 1988. Water hyacinth productivity and detritus accumulation. Hydrobiologia, 157, 179-185.

Mortimer C.H. 1941. The exchange of dissolved substances between mud and water in lakes. Journal of Ecology, 29, 280-329.

Moss B., Beklioglu M., Carvalho L., Kilinc S., McGowan S. & Stephen D. 1997. Vertically challenged limnology: contrasts between deep and shallow lakes. Hydrobiologia, 342/343, 257-267.

Moss B., McGowan S & Carvalho L. 1994. Determination of phytoplankton crops by top-down and bottom-up mechanisms in a group of English Lakes, the West Midland Meres. Limnology and Oceanography, 39, 1020-1029.

Moss, B., 1991. Ecology of Fresh Waters: Man and Medium. Blackwell Scientific Publications, Oxford, pp. 417.

Moyo N.A.G. 1997. Causes of fish deaths in Lake Chivero. In: Lake Chivero: A Polluted Lake. Proceedings of a workshop held in Harare. (Ed.) Moyo N.A.G. University of Zimbabwe Publications, Harare, pp. 103.

Mugidde R. 1993. The increase in phytoplankton primary productivity and biomass in Lake Victoria (Uganda). Verhandlungen Internationale Vereinigung Fur Theoretische Und Angewandte Limnologie, 25, 846-849.

Mugidde R., Hendzel L. & Hecky R.E. (In press). Pelagic nitrogen fixation in Lake Victoria, Uganda. Journal of Great Lakes Research.

Murty A.S. & Mulendema B.C. 1999. Growth of the water hyacinth on Lake Kariba and planning for integrated control. In: Proceedings of The First IOBC Global Working Group Meeting for The Biological and Integrated Control of Water Hyacinth. (Eds) M.P. Hill, M.H. Julien & T.D. Center Weeds Research Division, ARC, South Africa, pp. 182.

Musil C.F. & Breen C.M. 1985. The development from kinetic coefficients of a predictive model for the growth of *Eichhornia crassipes* in the field. I. Generating kinetic coefficients for the model in greenhouse culture. Bothalia, 15, 689-703.

NAS 1976. Making aquatic weeds useful: Some perspectives for developing countries. National Academy of Sciences, Washington, pp. 174.

Nebel B.J. & Wright R.T. 1998. Environmental Science: The Way the World Works. Prentice Hall, New Jersey, pp. 630.

Newell B.S. 1960. The hydrology of Lake Victoria. Hydrobiologia, 15, 363-382.

Njuguna S.G. & Thitai G.N.W. 1993. The control of floating water-weed infestations in Kenya. In: Control of Africa's floating water weeds. (Eds). A. Greathead & P.J. de Groot Commonwealth Science Council, Series no. CSC (93) AGR-18, Proceedings 295, pp. 187.

Ochiel G.R.S., Mailu A.M., Gitonga W. & Njoka S.W. 1999. Biological control of water hyacinth on Lake Victoria, Kenya. In: Proceedings of The First IOBC Global Working Group Meeting for The Biological and Integrated Control of Water Hyacinth. (Eds) M.P. Hill, M.H. Julien & T.D. Center Weeds Research Division, ARC, South Africa, pp. 182.

Ochumba P.B.O & Kibarra D.I. 1989. Observations on blue-green algal blooms in the open waters of Lake Victoria, Kenya. African Journal of Ecology, 27, 23-34.

Odei M.A. 1973. Observations on some weeds of pharmacological importance of the Volta Lake. *Bulletin of IFAN*, 35, 57-66.

Ogutu-Ohwayo R. 1990. The decline of the native fish in lakes Victoria and Kyoga (East Africa) and the impact of introduced species, especially the Nile perch, *Lates niloticus*, and the Nile tilapia, *Oreochromis niloticus*. Environmental Biology of Fishes, 27, 81-96.

Okali D.U.U & Hall J.B. 1974. Die-back of *Pistia stratiotes* on Volta Lake Ghana. Nature, 248, 452-453.

Patnaik N.C., Mohapatra L.N., Ghode M.K. & Misra B.K. 1987. Effects of some weedicides on the water hyacinth weevil (*Neochetina bruchi* Hustach) and its weed host. Plant Protection Bulletin - India, 39, 23-26.

Penfound W.T. & Earle T.T. 1948. The biology of water hyacinth. Ecological Monographs, 18, 447-472.

Pennington W. 1981. Records of a lake's life in time: the sediments. Hydrobiologia, 79, 197-219.

Piper B.S., Plinston T. & Sutcliffe J.V. 1986. The water balance of Lake Victoria. Hydrological Sciences Journal, 31, 25-37.

Reddy K.K. & Tucker J.C. 1983. Productivity and nutrient uptake of water hyacinth, *Eichhornia crassipes*. I. Effect of nitrogen source. Economic Botany, 37, 237-247.

Reddy K.R., Agami M. & Tucker J.C. 1989. Influence of nitrogen supply rates on growth and nutrient storage by water hyacinth (*Eichhornia crassipes* (Mart.) Solms) plants. Aquatic Botany, 36, 33-43.

Reddy K.R., Agami M. & Tucker J.C. 1990. Influence of phosphorus on growth and nutrient storage by water hyacinth (*Eichhornia crassipes* (Mart.) Solms) plants. Aquatic Botany, 37, 355-365.

Reddy K.R., Agami M., D'Angelo E.M. & Tucker J.C. 1991. Influence of potassium supply on growth and nutrient storage by water hyacinth. Bioresource Technology, 37, 79-84.

Redfield A.C., 1934. On the proportions of organic derivatives in sea water and their relation to the composition of plankton. James Johnstone Memorial Volume. Liverpool University Press, Liverpool, pp. 192.

Salmah M. R., Mansor M. & Ahmad A. B. 1991. A preliminary study of the distribution of *Neochetina eichhorniae*; a possible biological agent for water hyacinth in Kerian District, Perak. In: Proceeding Asas dan Gunaan dalam Biologi. (Ed.) A. Bidin A. UKM.

Sato M. & Kondo S. 1981. Biomass production of water hyacinth and its ability to remove inorganic minerals from water. I: Effect of the concentration of culture solution on the rates of plant growth and nutrient uptake. Japanese Journal of Ecology, 31, 257-267.

Sawyer C.N. 1947. Fertilization of lakes by agricultural and industrial drainage. New England Water Works Association, 61, 109-127.

Scheffer M., Hosper S.H., Meijer M-L., Moss B. & Jeppesen, E. 1993. Alternative equilibria in shallow lakes. T.R.E.E., 8, 275-279.

Scheren P.A.G.M, Zanting H.A. & Lemmens A.M.C. 2000. Estimation of water pollution sources in Lake Victoria, East Africa: Application and elaboration of the rapid assessment methodology. Journal of Environmental Management, 58, 235-248.

Sculthorpe C.D. 1967. The Biology of Aquatic Vascular Plants. Arnold, London, pp. 610.

Shapiro J.H., Lamarra V. & Lynch M. 1975. Biomanipulation: An ecosystem approach to lake restoration. In: Water Quality Management Through

Biological Control. (Eds) P.L. Brezonik & J.L. Fox Symposium proceedings, University Florida, pp. 164.

Seehausen O., van Alphen J.J.M. & Witte F. 1997. Cichlid fish diversity threatened by eutrophication that curbs sexual selection. Science, 277, 1808-1811.

Smith H. 1982. Light quantity, photoperception and plant strategy. *Annual Review of* Plant Physiology, 33, 481-518

Stager J.C. & Johnson T.C. 2000. A 12400 ^{14}C yr offshore diatom record from east-central Lake Victoria, East Africa. Journal of Palaeolimnology, 23, 373-383.

Stockner J.G. & Benson W.W. 1967. The succession of diatom assemblages in recent sediments of Lake Washington. Limnology and Oceanography, 12, 513-532.

Talling J.F. 1957. Some observations on the stratification of Lake Victoria. Limnology and Oceanography, 2, 213-221.

Talling J.F. 1965. The photosynthetic activity of phytoplankton in East African lakes. *Internationale Revue der Gesamten Hydrobiologie*, 501, 1-32.

Taylor A.R.D. 1993 Floating water-weeds in East Africa, with a case study in Northern Lake Victoria. In: Control of Africa's Floating Water Weeds. (Eds) A. Greathead & P.J. de Groot Commonwealth Science Council, Series no. CSC (93) AGR-18, Proceedings 295, pp. 187.

Timms, R.M. & Moss B. 1984. Prevention of growth of potentially dense phytoplankton populations by zooplankton grazing, in the presence of zooplanktivorous fish, in a shallow wetland ecosystem. Limnology and Oceanography, 29, 472-486.

TNV 2002. Hyacinth Back With Vengeance. The New Vision, October 8th 2002, Uganda.

Twongo T. 1996. Growing impact of water hyacinth on near shore environments on Lakes Victoria and Kyoga (East Africa). In: The Limnology, Climatology and Paleoclimatology of the East African Lakes. (Eds) T.C. Johnson & E. Odada Gordon and Breach, Toronto, pp. 633-642.

Twongo T. & Balirwa J.S. 1995. The water hyacinth problem and the biological control option in the highland lake region of the upper Nile basin—Uganda's experience. The Nile 2002 Conference: Comprehensive Water Resources Development of The Nile basin - Taking Off. Arusha, Tanzania Arusha, Tanzania.

Twongo T. 1993. Status of the water hyacinth in Uganda. In: Control of Africa's floating water weeds. (Eds) A. Greathead & P.J. de Groot Commonwealth Science Council, Series no. CSC (93) AGR-18, Proceedings 295, pp. 187.

Twongo T., Bugenyi F.W.B. & Wanda F. 1995. The potential for further proliferation of water hyacinth in Lakes Victoria, Kyoga and Kwania and some urgent aspects for research. African Journal of Tropical Hydrobiology and Fisheries, 6, 1-10.

UNECA 1995. African Compendium of Environment Statistics. Addis Ababa, United Nations Economic Commission for Africa.

UNEMA (1995). The National Environment Action Plan for Uganda. Ugandan National Environment Management Agency, Ministry of Natural Resources, Kampala.

United Nations (1995). World Population Prospects: The 1994 Revision. ST/ESA/SER.A/145, Department for Economic and Social Information and Policy Analysis, Population Division, New York, pp. 886.

Verschuren D., Johnson T.C., Kling H.J., Edington D.N., Leavitt P.R., Brown E.T., Talbot M.R. & Hecky R.E. 2002. History and timing of human impact on Lake Victoria, East Africa. Proceedings of the Royal Society, London Series B, 269, 289-294.

Whiteside M.C. (1983). The mythical concept of eutrophication. Hydrobiologia, 103, 107-111.

Willoughby, N.G., Watson, I.G. & Twongo, T. 1996. Preliminary studies on the effects of water hyacinth on the diversity, abundance and ecology of the littoral fish fauna in Lake Victoria, Uganda. In: The Limnology, Climatology and Paleoclimatology of the East African Lakes. (Eds) T.C. Johnson & E. Odada Gordon and Breach, Toronto, pp. 643-655.

Williams A.E. & Moss B. 2003. Effects of different fish species and biomass on plankton interactions in a shallow lake. Hydrobiologia, 491, 312-328.

Williams A.E., Duthie H.C. & Hecky R.E. 2005. Water hyacinth (*Eichhornia crassipes* (Martias) Solms) in Lake Victoria—Why did it vanish so quickly and will it return? Aquatic Botany.

Williams R.H. 1956. *Salvinia auriculata* Aublet: The chemical eradication of a serious aquatic weed in Ceylon. Tropical Agriculture, 33, 145-157.

Zullig H. 1989. Role of carotenoids in lake sediments for reconstructing trophic history during the late Quaternary. Journal of Palaeolimnology, 2, 23-40.

CHAPTER

8

Phytoplankton Characteristics, Trophic Evolution, and Nutrient Dynamics in an Urban Eutrophic Lake: Kandy Lake in Sri Lanka

E.I.L. Silva

Institute of Fundamental Studies, Hantana Road, Kandy, Sri Lanka

ABSTRACT

One of the most crucial water pollution problems prevalent in lakes and reservoirs located in urban environments is cultural eutrophication, resulting in impairment of water for intended uses such as drinking, food processing, recreation, aesthetic purposes, etc. Cultural eutrophication is a result of nutrient enrichment (N and P) anthropogenic events rather than natural processes. A manmade water body, Kandy Lake, located in the heart of the second largest city adjoining the world famous Buddhist temple in Sri Lanka was affected by an outbreak of a cyanobacteria bloom (*Microcystis aeruginosa*) in May 1999 with the onset of the southwest monsoonal winds.

Although seasonal change in water temperature was within the tropical range, Kandy Lake showed a marked diel variation temperature/density stratification. Alkaline to acidic water column from surface to bottom of the deep basin was characteristic with a clinograde oxygen profile, which collapsed only on a few occasions under prevailing high wind during peak southwest monsoon. The calcium and bicarbonate dominant lakewater was about threefold more conductive than the inland surfacewater found at a similar elevation. Long-term enrichment resulting from both anthropogenic and natural processes led the water body to collapse due to an outbreak of cyanobacteria (*Microcystis aeruginosa)* bloom in May 1999. The most likely hypothesis for the out break of cyanobacteria bloom was the sudden drop in the water level, which facilitated rapid uptake of nutrients from bottom layers due to *M. aeruginosa*. Nitrogen, phosphorus and silica dynamics, coupled with annual rainfall pattern and monsoon-bound wind mixing determined the seasonal periodicity of phytoplankton species during pre- and post-blooming periods. *Pediastrum simplex*, a Chlorophyte commonly found in eutrophic tropical waters, which dominated the phytoplankton community under dry weather in Kandy Lake, became virtually absent following the outbreak of *M. aeruginosa* bloom, while *Aulacoseira granulata*, a chain forming centric diatom, known to be a climax species in the tropics, declined dramatically during post-blooming. The outbreak of cyanobacteria changed the functional phytoplankton groups from a rhythmically oscillating Chlorophyte and a Diatomophyceae to a multiple assemblage of Cyanophytes. The upper level of the annual mean chlorophyll-a, which showed an exponential increase over a period of seven years, had a negative correlation with the lower limit of the annual mean water transparency. Implementation of either top-down or bottom-up strategies or a combination of both for the restoration of Kandy Lake should be based on a clear understanding of food web interactions and external and internal nutrient loading of this unique water body.

INTRODUCTION

Urban water bodies are susceptible to high anthropogenic pressures and water quality attributes are tied up with hygienic and economic conditions of the riparian communities. Management of these water bodies are neglected to a greater extent in developing countries primarily due to poor awareness and financial constraints. Since these water bodies receive a variety of untreated effluents (human excreta, agricultural, industrial, domestic and clinical wastes), progressive eutrophication with intermittent blooming is inevitable (Silva, 2003a). The chemicals used in parks, gardens and lawn care carried by runoff could be toxic to aquatic life (Straškraba & Tundisi, 1999). Blooming of cyanobacteria with formation of surface scum is a severe socio-economic problem in many water bodies located in urban centers with no central sewerage treatment facilities. Curiously, this unique problem of lakes and reservoirs especially located in the humid tropics that are distinctly urban in character have received no attention in the limnological and watershed management literature.

Algal blooms disturb the oxygen balance and secrete toxins of low molecular weights, which are harmful to many aquatic animals, including mammals. They also cause gill clogging of some surface feeding fishes (Codd et al., 1989; Carmichael, 1992). Nevertheless, cyanobacterial blooms are an important source of food for phytoplanktivorous fishes feeding on water column suspension (Hofer & Schiemer, 1983; Phillips et al., 1985; Rabergh et al., 1991). In non-harvesting water bodies with poor grazing efficiency by zooplankton and extremely low predator pressure, concurrent ichthyo-eutrophication is very likely to prevail due to excretion by dense fish population. Fish populations dominated by prolific herbivores like tilapia in non-harvesting urban water bodies that primarily serve aesthetic purposes may suffer from hypereutrophication. Although progressive eutrophic process is well understood, the timing and magnitude of sudden outbreaks of cyanobacteria blooms still remains a bizarre problem. Schiemer et al., (2001) demonstrate a mesotrophic to eutrophic shift with decreasing water level in shallow irrigation reservoirs in Sri Lanka but the occurrence of cyanobacteria blooms in Sri Lankan inland water bodies was not uncommon (Silva & Wijeyaratne, 1999). A dense dark green bloom of *Spirulina* sp and *M. aeruginosa* coexists in the Beira Lake (Colombo Lake), located in the heart of the capital of Sri Lanka since the early eighties. In 1991, a thick bloom of *M. aeruginosa* covered the entire Kotmale reservoir, the uppermost hydropower reservoir in the Mahaweli River basin during the dry spell of 1991 before the reservoir reached its 10 years after commissioning. This bloom disappeared gradually with increasing water level during the northeast monsoonal rain. In contrast, a scum of *Anabaena aphanizomenoids* was scattered in the northern basin of the Parakrama Samudra, a large dry zone irrigation tank (Schiemer, 1983) during the high water level in 1993; it disappeared with the release of surfacewater to the command area. Therefore, it is evident that sudden emergence of algal blooms and subsequent ecosystem collapse in the tropics is not necessarily a time series phenomenon of cultural eutrophicaion. It may also appear unexpectedly in manmade water bodies when favourable hydraulic balance is coupled with certain limnological factors (e.g. high nutrients and alkalinity) and prevailing climatic conditions such as wind and irradiance.

The abundance, composition and temporal succession of planktonic algae vary among water bodies to a grater extent in the tropics. *Aulacoseira granulata,* a chain forming centric diatom is a climax taxon in most freshwaters in the tropical latitude, while *Microcystis aeruginosa,* a bloom-forming cyanobacteria with toxigenic strains (Carmichael, 1992) is the most common in galloping eutrophication. Monsoon-driven rainfall, coupled with water release for power generation and irrigation, are the primary driving forces of mesotrophic to eutrophic shift in tropical reservoirs (Schiemer et al., 2001) but abrupt changes in hydraulic balance will certainly result in outbreaks of blooms provided that the other environmental conditions are favourable (Silva, 2003a). However, information is scarce on hydrological balance, chemical environment, nutrient dynamics, temporal and spatial variations of phytoplankton assemblage, light climate, photosynthetic characteristics, heterotrophic potential, food web relationships, etc., of freshwater bodies subjected to anthropogenic pressures in the tropics. An attempt is made in this chapter to provide some information to fill this long-standing limnological lacunae on tropical urban water bodies. The paradigm and uniqueness of trophic evolution of Kandy Lake in Sri Lanka is discussed here, highlighting seasonal/diel and spatial patterns of chemical environment, phytoplankton composition and periodicity, nutrient dynamics and trophic characteristics in relation to its landscape ecology, hydrology and other environmental attributes.

KANDY LAKE

Kandy Lake is an ornamental water body, located in the hill capital, Kandy, the second largest city in Sri Lanka. Situated in the heart of Kandy, enhanced with a paramount scenic value, adjacent to the world famous Buddhist Temple *Dalada Maligawa* where the sacred tooth relic of Lord Buddha is placed, the lake has become one of the largest tourist attractions of the country. The lake was built by the last king of the Singhalese monarchy (1789-1815), Sri Vikrema Rajasinghe, between 1810 and 1812 using forced labour to enhance the panoramic beauty of the Royal Palace Complex and surrounding temples. Kandy, the symbolic holy city of the country, is renowned as one of the key heritage cities in the world because of its cultural legacy, archeological importance and aesthetic value.

The Lake has been in existence since its construction as a perennial water body and the water from the lake had been used to augment the city water supply during the mid seventies for a short period. Since this lake is located adjoining the most esteemed religious centre, fishing and bathing are prohibited and, at present, the lakewater is neither used for irrigation nor for other domestic purposes. People make pleasure trips by motorboats, which are being in use for several years. A chronic cyanobacteria species (*M. aeruginosa*), emerged as a bloom and formed into a thick scum in Kandy Lake with the onset of the southwest monsoon in May 1999. The outbreak of this bloom became a major socio-political issue because the Kandy Lake is a national asset.

Kandy Lake (7° 18´N; 80° 39´E), located 510 m above mean sea level, drains a small watershed of 2.87 km^2 which is confined to the southern fraction of Kandy city and southwest slope of the *Udawattekelle* forest reserve behind the *Maligawa* (temple). A small islet is located more or less at the centre of the Lake and the decorated wall which borders the western and northern boundaries of the lake also contributes to enhance its panoramic outlook. This northwestwardly elongated water body with 18 ha. of surface area is skirted by a circumference of 3.05 km and has a maximum fetch of 1.16 km (Fig. 8.1). The mean depth is 8.5 m, while the deep area goes up to 12.5 m and about 75% of the basin is less than 4 m with the rest being more than 10 m, dividing the Lake roughly into shallow and deep basins (Fig. 8.1). The volume of the Lake at full supply level (FSL) is 0.3839 MCM. The lake has no prominent littoral zone and the edges differ from place to place. On the western and northern sides, there is no shore as such and the water level seldom recedes from the embankment. Four major silt traps have been constructed across the inflows and several road surface silt traps are also located at various points. The bank around the lake has been planted with large trees belonging to different families of flowering, ornamental and shading tropical species.

The sluice gate, the only outlet of the water body located at the southwest corner of the Lake, is usually kept closed. Two small brooks feed the Lake, excess water spills over only during the rainy season and evaporation losses are high through the dry months. Kandy experiences the highest rainfall with the onset of northeast monsoon from October until January, whereas the first inter monsoon (Feb.-Mar.) is normally dry, the second dry spell occurs again during July-August (Fig. 8.2). However, the rainfall pattern in this area, which is orographic in nature, varies from year to year (Fig. 8.2). Erosion and subsequent transport of sediment has resulted in substantial siltation at the immediate outfalls of the inflows. About 1.5m thickness of sediment has built up in the central part of the deep basin. The wastewater discharged from small-scale hotels, hospitals, schools and temples drains into the lake via thirteen inlets. The present land use in the watershed is essentially urban housing and government and private sector buildings. About 70% of the drainage basin shows high-density residential land use and a few typical Kandyan home gardens also exist towards the southwest border of the watershed.

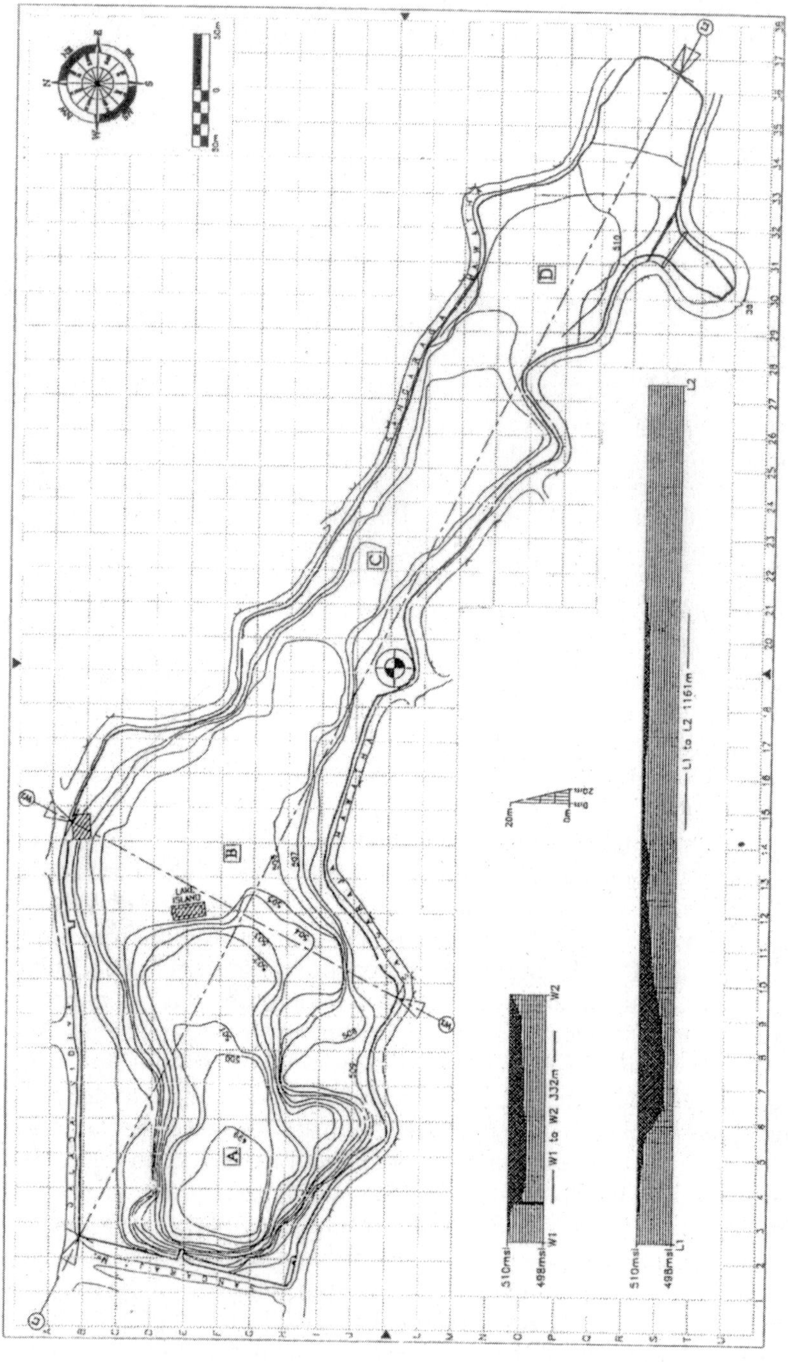

Fig. 8.1 Map of Kandy Lake illustrating depth contours, cross section of the inflow outflow direction and sampling sites

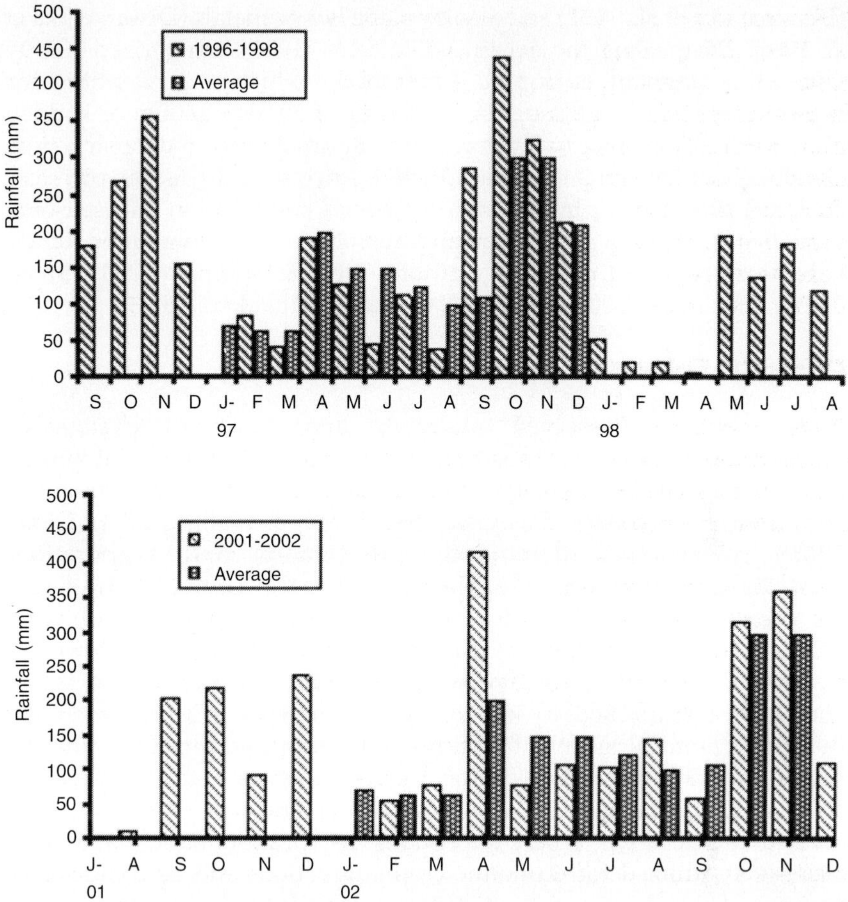

Fig. 8.2 Monthly rainfall during pre- and post-blooming periods in relation to average rainfall in Kandy Lake

The rest of the drainage basin of the Lake has a dense natural forest cover confined to the *Udawattekelle* sanctuary. Neither private houses nor other buildings in the watershed are connected to a central sewer system.

The Lake had not been subjected to a regular monitoring for its limnology or water quality since its construction up until late seventies. A preliminary limnological study carried out during 1979-1980 on some physico-chemical characteristics also described the meso- and macro-aquatic fauna (De Silva & De Silva, 1984). This study found no significant seasonal or temporal variation in physico-chemical features. Nevertheless, it was reported that Kandy Lake has been enriched with P and N compounds and polluted with faecal matter to a great extent

(Dissanayaka et al., 1982) and also by some heavy metals (Dissanayaka et al. 1986). Diagnosing the existing data Silva (1996), categorized Kandy Lake as a stagnant eutrophic water body which is susceptible for hypereutrophication although an outbreak of an algal bloom or sudden mass mortality of fish had never been reported since its construction. Limnological investigations conducted since mid 1996s on physico-chemical limnology, phytoplankton (species composition and seasonal periodicity), primary production and microbial consortium in the Kandy Lake were reported by many authors (Silva & Schiemer, 2001; Silva, 2003a; Silva et al., 2003; Sharaff, 2003; Samaradiwakara, 2003).

PHYSICO-CHEMICAL CHARACTERISTICS

Water contains dissolved materials and gases in changeable concentrations and composition in space and time. Catchment geochemistry has been shown to be the most important determinant of surfacewater chemistry (Douglas, 1968; Webb & Walling 1974). Gibbs (1970) coined rock dominance, precipitation and evaporation-crystallization processes to explain the chemical composition of epi-continental waters. Altitude can also influence the surfacewater chemistry due to variations in water balance (Vitousek, 1977). However, nothing is definitive; spatial and temporal variation in the water chemistry conditioned by the overall framework within which the hydrogeochemical system functions but, within a particular climatic region, rainfall, watershed characteristics, vegetation and soil become more significant (Gower, 1980). Tropical watersheds, especially those located in the areas where monsoons are predominant, have been intensively utilized for economic crop cultivations and agro-industries because of the characteristic distribution and intensity of rainfall. The quality of surfacewater in the downstream, therefore, has been affected mainly by irrigation-related agriculture practices and alterations in land use. In urbanized areas, unplanned human settlement is the largest component of the land use. Hence, nutrient enrichment and subsequent eutrophication, faecal contamination and trace metal pollution have been identified as the major issues associated with water quality problems of urban centres in the tropics.

Surface temperature in Kandy Lake was relatively low during the winter months (Oct.-Jan.) compared to dry months (Feb.-Apr.), but annual range was typical for tropical water bodies located in mid elevations (Table 8.1). Although there was no marked spatial variation, a narrow vertical gradient of temperature had been established in the deeper basin (Fig. 8.3). This microstratification does not represent a typical thermocline with notable temperature zones (epilimnion,

Table 8.1 Annual ranges of physicochemical parameters and nutrients of surfacewater in Kandy Lake from 1996-2002

Parameter	*1996*	*1997*	*1998*	*1999*	*2000*	*2001*	*2002*
Tem. ºC	25.4-30.2	25.2-30.5	26.6-30.6	25.6-31.6	25.2-30.8	25.1-28.8	25.2-28.2
EC μS	200-240	173-250	228-290	222-286	198-246	188-226	148-244
pH	7.64-8.11	7.77-8.73	7.73-8.37	7.83-8.87	7.02-8.87	6.84-8.87	7.10-8.49
Alk. mgl^{-1}	122-135	104-122	105-124	115-144	109-281	79-321	116-356
DO mgl^{-1}	4.32-9.86	4.04-9.76	4.92-12.7	3.94-13.8	3.42-9.87	2.42-9.71	1.72-12.3
TSS mgl^{-1}	9.9-24.7	3.0-20.9	3.0-18.4	34.8-22.4			
SO_4^{2-} mgl^{-1}	0.06-4.46	3.09-6.17	3.04-9.74	3.04-9.74			
Cl^- mgl^{-1}	19.0-39.2	15.2-39.8	18.1-25.3	18.1-25.3			
Na^+ mgl^{-1}	9.82-13.8	10.5-25.2	12.4-22.6	12.4-22.6			
Ca^{++} mgl^{-1}	12.8-18.9	7.46-18.8	1.3-20.8	1.3-20.8			
Mg^{++} mgl^{-1}	5.52-7.75	6.04-8.18	6.76-8.12	6.76-8.12			
K^+ mgl^{-1}	4.98-6.24	4.31-6.95	4.25-8.56	4.25-8.56			
dP μgl^{-1}	nd-22	nd-9	nd-11	nd-10	nd-10	nd-12	nd-13
tP μgl^{-1}	23-56	14-137	25-52	25-52	34-86	23-105	20-91
NO_3-N μgl^{-1}	37-668	29-816	33-376	33-376	11-448	01-148	6-612
NH_4-N μgl^{-1}	24-489	15-774	50-568	50-774	16-604	26-304	83-824
DSi mgl^{-1}				1.48-5.68	2.48-5.12	3.48-6.12	0.26-6.08

metalimnion and hypolimnion) as in a temperate region. Slight seasonal changes of midday surface temperature in Kandy Lake were linked with the seasonality of incident radiation, which is also distinct in the tropics to some extent. However, the diel changes in temperature in both surface and bottom water were significant in this small water body (Table 8.2). The surface and bottom water temperatures changed by 12.8°C and 11.5°C, respectively, within a day and the lowest temperature occurred after midnight under dry weather condition. This day and night range in temperature may be narrow during November-January. The diel cycle of temperature/density stratification can have large effects on vertical distribution of cyanobacteria with varying positive and negative buoyancy (Talling & Lemoalle, 1998). Temperature/density stratification resulting from solar heating during the daytime will be subjected to nocturnal turnover. Such convectional motions generate redistribution of phytoplankton through the mixed layer. In Kandy Lake, the mixed depth was confined to upper six metres and did not show any marked seasonal variation.

Surfacewater in the Lake was alkaline and it was further triggered by the photosynthetic reaction induced by high algal biomass. Although the surfacewater was alkaline, it had been buffered to a certain extent with the increasing depth in the deep basin resulting in acidic water at the bottom (Table 8.1 and Fig. 8.4). When surface pH was dropped with

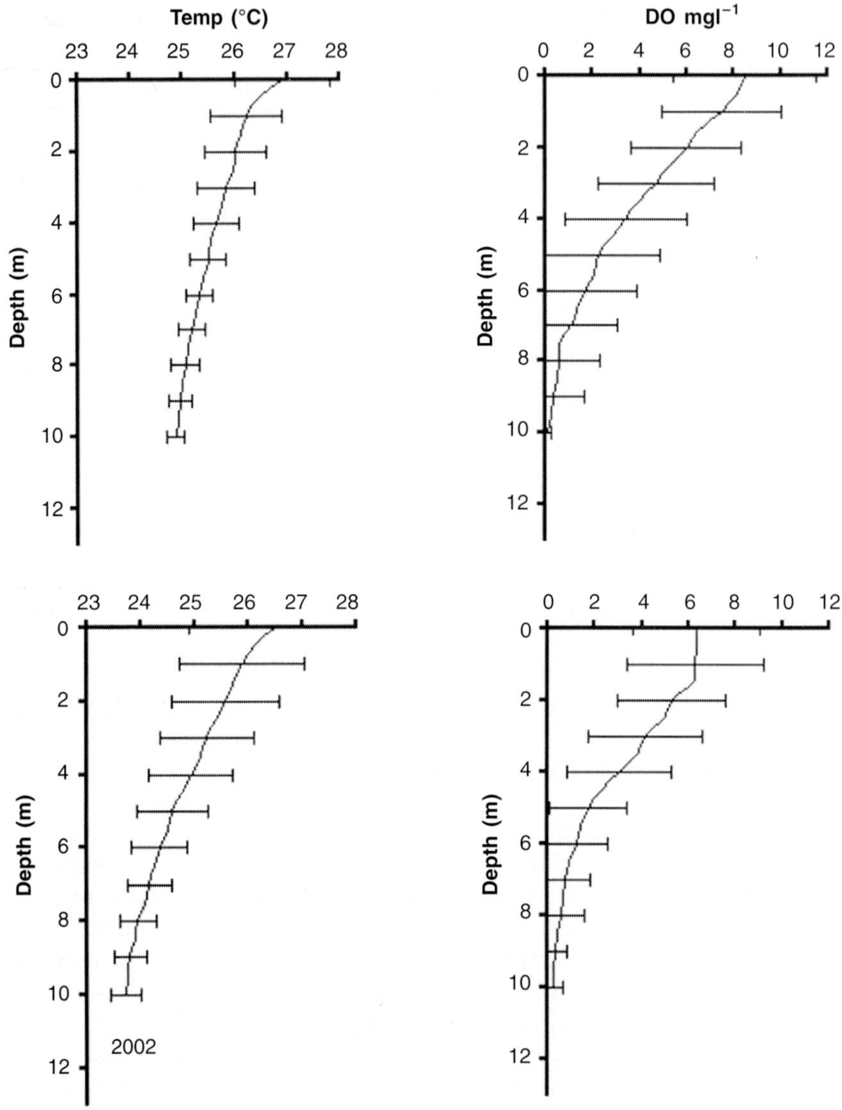

Fig. 8.3 Vertical profiles of annual mean temperature and oxygen concentration of the deep basin in 2001 and 2002 (horizontal line represent the standard deviation)

freshwater in-fluxes, the bottom water became more acidic. Surface water pH in Sri Lanka showed a marked monsoon-bound seasonality of pH in relation to the annual rainfall pattern and acidic pH prevails even in downstream irrigation tanks during the northeast monsoon (Silva & Davies, 1987; Silva & Gamlath, 2000). pH of surface water in Kandy Lake

Table 8.2. Diel variation of physico-chemical parameters and nutrients in Kandy Lake in July 1999

Time	*12:00*		*16:00*		*20:00*		*0:00*		*4:00*		*8:00*	
Parameter	*Surface*	*Bottom*	*Surface*	*Bottom*	*Surface*	*Bottom*	*Surface*	*Bottom*	*Surface*	*Bottom*	*Surface*	*Bottom*
Tem. °C	32.2	29.0	25.9	28.4	26.0	23.9	27.0	22.9	19.4	17.5	21.2	20.5
DO mgl^{-1}	7.95	0.48	9.00	0.55	8.80	0.44	9.07	0.49	6.52	0.57	7.22	0.40
pH	8.41	7.10	8.2	6.98	8.41	7.01	8.05	6.93	8.06	7.02	7.89	6.94
EC µS	223	234	216	244	212	238	220	230	220	236	222	242
dP μgl^{-1}	nd	nd	nd	1	5	2	3	3	2	7	nd	4
tP μgl^{-1}	45	118	53	110	48	48	130	68	181	45	68	81
NO_3-N μgl^{-1}	290	234	177	244	243	238	276	230	226	236	202	242
NH_4-N μgl^{-1}	33	429	75	1503	126	794	85	755	50	93	526	626

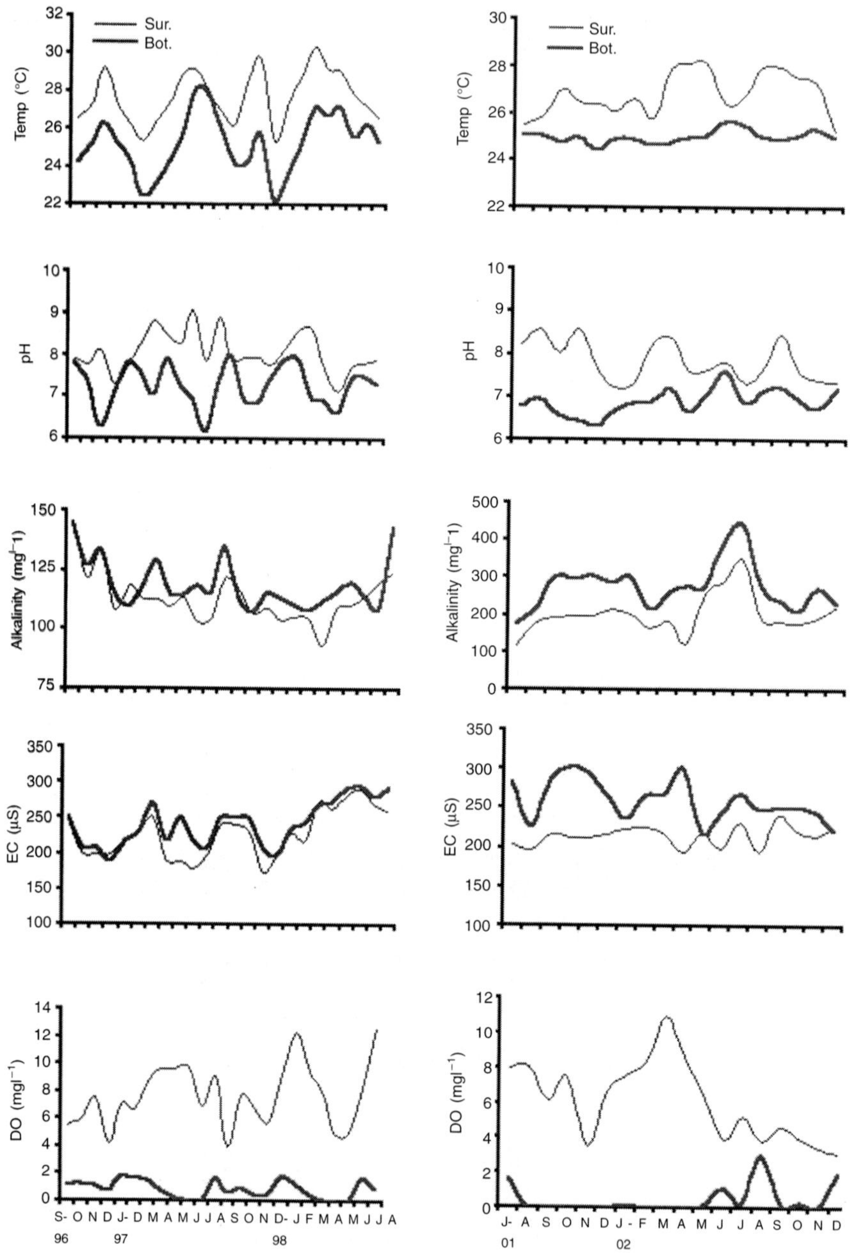

Fig. 8.4 Monthly variation of physico-chemical parameters of Kandy Lake during pre- and post-blooming periods

also decreased considerably after rainfall (Figs 8.2 and 8.4). Usually, when photosynthetic organisms consume free carbon dioxide, bicarbonate concentration will increase in the surfacewater while microbial decomposition in the aphotic zone may result in acidic pH (Wetzel, 1983). This phenomenon indicates that heterotrophic processes taking place in the bottom layers of the deep basin were quite different from the autotrophic processes in the surface water. On diel basis, pH decreased with decreasing photosynthetic rate (Table 8.2). Possible biological implications of CO_2 accumulate in the deep layers or elevated pH in surface layers induced by photosynthetic consumption of CO_2 still remain poorly documented from tropical lakes and reservoirs (Talling & Lemoalle, 1998). Seasonal and spatial patterns in bicarbonate alkalinity was more or less similar to that of pH, but the pattern in the deep basin has been reversed and bicarbonate concentration in the bottom water was significantly higher than the surface water (Fig. 8.4).

The electrical conductivity (EC), essentially a function of bicarbonate concentration at a given temperature in soft inland waters in the tropics, was relatively high in Kandy Lake as compared to the other water bodies located at similar elevations (Silva & Gamlath, 2000). The lowest value coincided with the highest rainfall and it increased progressively during dry weather, perhaps due to high evaporation losses (Fig. 8.4). Relatively high EC in bottom water of the deeper basin may also be attributed to a relatively high concentration of bicarbonate. However, there was no marked diel change in EC (Table 8.2). Relatively high EC of Kandy Lake compared to the water bodies located at similar altitude indicates the recipient nature of high amount of wastewater and eutrophic condition of the water body. In general, urban water bodies contain more cations and anions generating from household activities. Total suspended solids (TSS) in the lake was relatively high during the rainy season and there was no marked difference in TSS between dry and moderately rainy seasons (Samaradiwakara, 2003). The particles suspended in the water column in the inflow areas during the rainy season were mainly inorganic due to sediment loading. They settled down to the bottom within a few days but suspended particles in the euphotic depth were essentially planktonic algae during most of the year.

Dissolved oxygen concentration in the surface and bottom water during the midday varied within a wide range with intermittent fluctuations (Table 8.1) but did not show a particular seasonal pattern (Fig. 8.4). Although a horizontal gradient was not prominent, a clinograde vertical profile (Fig. 8.3) had been established in the deep basin, which is typical for productive water bodies (Ganf, 1974; Lewis, 1987). The occurrence of a thick layer with reduced oxygen below six metre depth in the deep basin during most of the year could be a combined product of high photosynthetic rate in the trophogenic zone, poor mixing in the water column and high consumption of oxygen by bacteria in the tropholitic zone. It is worthwhile mentioning that oxygen depletion of the deep basin collapsed only during a few occasions under prevailing high wind in the peak southwest monsoon. Since the bottom layer of the deep basin was either anoxic or contained less oxygen during most of the year, only *Chaoborus*, a Dipteran larvae inhabited the lake bottom (Silva et al., 2003). Apparently, when anaerobic conditions occur in the bottom of the deep basin, they affect the biogeochemical processes and nutrient dynamics of the entire lake ecosystem (Barica, 1980; Conley, 2000). The surface water became saturated with the dissolved oxygen when the sky was clear during the daytime and the oxygen depletion on the following morning was not significant in the diel cycle under such a situation (Table. 8.2). But, oxygen depletion in the surfacewater during sunrise was common in this water body with complete or near complete anoxia in the bottom layers. Although the depletion in oxygen in deep water occurred in a regular manner, it did not lead to a phenomenon of fish kill.

Of the major cations, calcium was the most dominant cation and magnesium concentration was higher than the sodium ions. Potassium retained the lowest concentration of cations. Calcium plays a central role in the pH-CO_2-HCO_3^- equilibrium and hence influences the supply of photosynthetically available carbon and the capacity of water to buffer pH fluctuations. Among the anions, bicarbonate was dominant in Kandy Lake as in other tropical freshwaters. However, with regard to composition and ratio of anions, the chloride ion was subdominant while sulphate was the least abundant in the lake water (Samaradiwakara, 2003). Apart from HCO_3^-, the other major anions do not appear to regulate phytoplankton populations in freshwater. Most of the surface waters in Sri Lanka are rich in sodium, and the concentration of calcium and magnesium are more or less similar but potassium has the lowest concentration (Dissanayaka & Weerasooriya, 1986). Calcareous formations like limestone are not well represented over much of the tropics, which partly accounts for the domination of Na^+ over Ca^{++} in tropical waters. Even though the bicarbonate concentration was the highest compared to chloride and sulphate, a substantial concentration of chloride was present in Sri Lankan freshwaters (Silva, 1999; Silva & Gamlath, 2000). Ionic composition and relative ionic ratios in lowland dry zone reservoirs in Sri Lanka are also more or less similar to that of the highland water bodies but concentrations are relatively higher (Silva, 1998, 2003b). Calcium and magnesium are higher in this Sri Lankan water body when it drains the watershed dominated by calcareous rocks. In general, sodium should be higher when the water body is receiving urban wastewater. Relatively higher concentration of calcium and magnesium ions compared to sodium ions in the Kandy Lake may be attributed to major alterations of land use in the watershed, especially due to its urbanized nature without waste treatment facilities.

PHYTOPLANKTONS

Phytoplankton is a key component of the aquatic (microbial) food web in the pelagic zone reflecting the trophic status of the environment in both temperate and tropical waters. In the humid tropics, they play an additionally important role in the pelagic food web as a direct and valuable source of food for seston-feeding fish, which are common in many lakes and reservoirs (Hofer & Schiemer, 1983; Payne, 1986). Although there is a large diversity of standing water bodies in the tropical latitude as a result of climatic and geomorphologic heterogeneity, detailed phytoplankton investigations are rare. Very few studies have addressed phytoplankton composition and their temporal changes under eutrophic condition (Thornton, 1982; Sugunan, 1980: Kanan & Job, 1980; Zafar, 1986; Osborne, 1991; Alvarez-Cobelas & Jacobsen, 1992). The accepted phenomenon is that temporal variability of phytoplankton in tropical lakes and reservoirs are driven by seasonal rainfall and mixing, but diel patterns in shallow water bodies and long-term stable types with sudden shifts are poorly understood.

Phytoplankton species belonging to eight classes were registered from Kandy Lake during pre- and post-blooming period (1996-2002) with a maximum of 14 taxa in both Chlorophytes and Cyanophytes/ Cyanobacteria (Table 8.3). Five taxa belonging to Diatomophyceae with a common centric form, *A. granulata* were found dominant or sub-dominant before the outbreak of *M. aeruginosa* bloom in 1999. A new species of centric diatom, *Urosolenia denticulata* (the freshwater counterpart of genus *Rhizosolenia*) described from Victoria and Minneriya reservoirs in Sri Lanka (Rott et al., in press) was also found in Kandy Lake as an extremely rare species. Except *A. granualata*, the other diatoms (*Cyclotella, Navicula* and *Synedra* spp) found were either rare or of minor importance to the population abundance. Dinoflagellates and desmids were represented only by a few and partly uncommon taxa but never plentiful in the Lake (Table 8.3). A Chrysophyte, genus *Mallamonas* and a unicellular Xanthophyte (*Isthmochloron lobulatum*) had only a minor contribution to species richness. Two species of Euglenophytes belonging to the genera *Euglena* and *Phacus* appeared in the Lake during dry months sometimes in moderate numbers. Desmids were associated with low Ca^{++} and a low N/P ratio (Reynolds, 1990), while the abundance of

Table 8.3 Phytoplankton species list of Kandy Lake from 1997 to 2002

	YEAR					
Cyanophyceae/Cyanobacteria	*1997*	*1998*	*1999*	*2000*	*2001*	*2002*
Anabaena sp.	e	e	e	e	-	-
Aphanizomenon sp.	-	-	e	c	c	c
Aphanocapsa delicatissima	e	e	e	-	-	-
Chroococcus limneticus	d	d	d	d	d	d
Coelosphaerium sp.	e	e	e			
Cylindrospermopsis philippinensis	-	-	-	c	b	b
Merismopedia elegans	d	d	d	d	d	d
Merismopedia punctata	b	b	b	b	b	b
Microcystis aeruginosa	c	c	a	a	b	c
Microcystis flos-aquae	-	-	-	c	c	c
Microcystis incerta	-	-	-	d	d	d
Microcystis wesenbergii	-	d	d	d	a	b
Oscillatoria raciborskii	-	-	-	-	b	b
Planktolyngbya circumcreta	d	d	d	d	d	d
Diatomophyceae						
Aulacoseira granulata	a	a	b	b	b	b
Cyclotella comta	e	e	e	e	d	d
Navicula sp.	e	e	e	e	e	e
Synedra sp.	e	e	e	e	e	e
Urosolenia denticulata	e	e	e	e	e	e
Chrysophyceae						
Mallomonas sp.	e	e	e	e	e	e
Xanthophyceae						
Isthmochloron lobulatum	-	-	-	e	d	d
Dinophyceae						
Peridiniopsis penardiforme	-	-	e	e	e	e
Peridinium gatunense	-	-	e	e	e	e
Chlorophyceae						
Ankistrodesmus bernardi	d	d	d	d	d	d
Botryococcus sp.	e	e	e	e	e	e
Coelastrum astroideum	d	d	d	d	d	d
Coelastrum reticulatum	d	d	d	d	d	d
Dictyosphaerium pulchellum	d	d	d	d	d	e
Golenkinia radiata	e	e	e	-	-	-

(Table 8.3 Contd.)

(Table 8.3 Contd.)

	YEAR					
	1997	*1998*	*1999*	*2000*	*2001*	*2002*
Kirchneriella dianae	e	e	e	e	e	e
Monoraphidium contortum	e	e	e	e	e	e
Oocystis parva	d	d	d	d	d	d
Pediastrum duplex	e	e	e	e	e	e
Pediastrum simplex	a	a	b	b	b	b
Pediastrum tetras	e	e	e	e	e	e
Scenedesmus tropicus	e	e	e	e	e	e
Tetraedron incus	e	e	e	e	e	e
Zygnemaphyceae						
Closterium acutum	d	d	d	d	d	d
Closterium apertum	e	e	e	e	e	e
Staurastrum brachioprominens	d	d	d	d	d	d
Staurastrum brevispina	d	d	d	d	d	d
Euglenophyceae						
Euglena acus	d	d	d	d	d	d
Phacus sp.	d	d	d	d	d	d

a: common throughout the year and became dominant or subdominant; b: common throughout the year in low numbers; c: occasionally became common; d: rare in occurrence; e: very rare in occurrence

cyanobacteria is correlated with the high concentrations of organic nitrogen.

From September 1996 to January 1997, *A. granulata,* dominated the species assemblage with *Pediaustrm simplex* as the subdominant species. *Merismopedia punctata* was the common Cyanophyte during this period but registered in relatively low numbers. *M. aeruginosa,* although present, recorded in small numbers with the other minor species (Fig. 8.5 & Table 8.2). *A. granulata* decreased considerably from February to May in 1997 while *P. simplex* became the dominant taxon. *A. granulata* increased again towards the wet season and progressively decreased during February-March 1998. The abundance of *P. simplex* was suppressed by a colony forming cyanophyte, *M. punctata* and registered as the subdominant species between May and July, 1998. *M. aeruginosa* succeeded as the subdominant species in August 1998 (Fig. 8.5). The abundance of *M. aeruginosa* and the other species decreased again during the rainy season (Nov.-Dec.) in 1998 perhaps due to mixing or changes in chemical environment but re-established again during February 1999. Phytoplankton minimum was common during the northeast monsoon

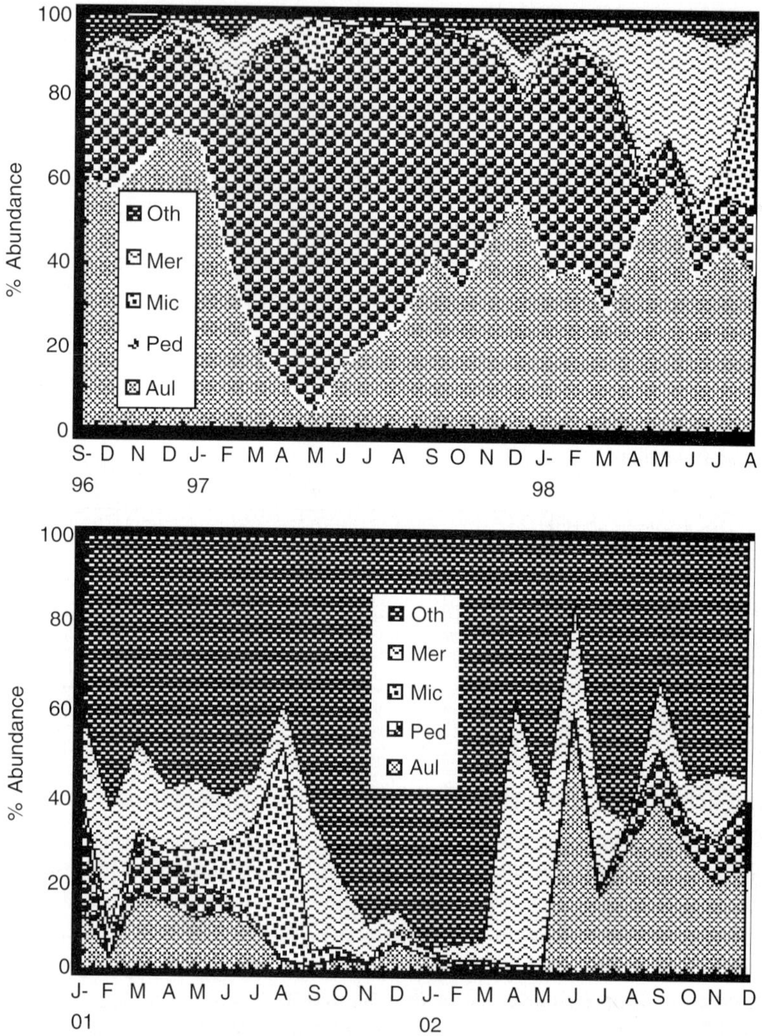

Fig. 8.5 Monthly variation of the relative abundance of phytoplankton species in Kandy Lake during pre- and post-blooming periods (Aul, *Aulacoseira granulata*; Ped, *Pediastrum simplex*; Mic, *Microcystis aeruginosa*; Mer, *Mersmodepia punctata*; Oth, Other species)

(Oct.-Jan.) in Sri Lankan standing water bodies (Silva & Davies, 1987; Rott et al., in press).

The water level of the Lake was dropped by about 1 m in March 1999 for the purpose of constructing a lower level sluice gate at the outflow. Towards the beginning of May, *M. aeruginosa* appeared as a bloom at the shallow inflow area with the onset of the southwest monsoonal wind.

The bloom concentrated mainly in the shallow inflow area and the scum drifted to the small edges and bays. The intensity of the bloom was maximum during early July and chlorophyll a concentrations increased upto 150 μg l^{-1} in the shallow area. While manually removing the dead scum of the bloom to a larger extent, the water level increased gradually and reached its maximum towards August with several scattered showers (Fig. 8.2). The surface scum disappeared gradually but *M. aeruginosa* was retained as the dominant species since then in moderate densities. In the following inter-monsoonal rain (Oct.-Dec.) in 1999, the two species of phytoplankton exhibiting a rhythmic oscillation (*A. granulata* and *P. simplex*) before the blooming re-appeared in the lake in moderate numbers but did not re-establish. *P. simplex* remained in small numbers and the abundance of other cyanophytes such as *M. wesenbergii*, *M. incerta*, *Chroococcus*, *Coelosphaerium* and *Cylindrospermopsis* species increased progressively in the phytoplankton assembles (Fig. 8.5 & Table 8.3). This resulted in a complete shift of phytoplankton species assemblage dominated by *A. granulata*, *P. simplex* and *M. punctata* during pre-blooming to species mainly dominated by cyanophytes (Fig. 8.5). Although *M. aeruginosa* appeared throughout the following years and became subdominant at irregular intervals, it did not bloom as in 1999.

With *M. aeruginosa,* three clearly distinct but cosmopolitan taxa belong to the same genus (*M. flos-aquae, M. incerta* and *M. wesenbergii*), were also found in Kandy Lake in small numbers during pre-blooming. Their abundance, specially *M. wesenbergii* increased progressively in 2000 and it became the dominant species during July-October. Consequently, blue-greens dominated the phytoplankton community in Kandy Lake instead of greens and diatoms following the outbreak of bloom (Fig. 8.5). *M. wesenbergii* became the most dominant species in the phytoplankton assemblage for several months in 2001 under dry weather. Several other species of the genus *Microcystis* (*M. lamelliformis, M. comperei, M. smithi,* etc.,) that reaches the highest morphological variability in tropical freshwaters (Komarek et al., 2002), were not reported from Sri Lanka (Rott & Lenzenweger, 1994, Rott et al., in press); nor were they recorded from the Kandy Lake.

Phytoplankton communities in tropical lakes and reservoirs represent summer communities of temperate lakes with a large number of tropical taxa including pantropical and regional endemic elements (Vyvermann, 1996). However, there is little information on the distribution, composition and succession of tropical phytoplankton communities and their diversity in relation to environmental gradients in tropical lakes and reservoirs (Lewis, 1978; Biswas, 1978; Henry et al., 1984; Ramberg, 1987; Mukankomeje et al., 1993; Branco & Senna, 1994,

1996). In small ponds and reservoirs in Asia, phytoplankton minimum can be characteristic during the washout and elevated turbidity in the wet months (Sugunan, 1980; Kanan & Job, 1980; Fatimah et al., 1984; Khondker & Parveen, 1993). The role of seasonal hydrology on phytoplankton abundance is evident in Sri Lanka, where plankton studies have a long history (Apstein, 1907; Holsinger, 1955; Rott, 1983; Rott et al., in press). Lewis (1996) suggested a progressive decline in phytoplankton diversity towards the tropics. On the contrary, extremely high diversity of phytoplankton has been shown for floodplain lakes in Papua New Guinea (Vyvermann, 1996). Phytoplankton communities in very large lakes are mainly dominated by non-motile species (Lewis 1978; Carney et al., 1987; Talling 1986). Species sequences have been reported during the early development of manmade lakes (van de Heide 1973; Biswas, 1978; Matsumara-Tundisi et al., 1991; Branco & Senna, 1996). However, dissimilarity in species composition and diversity of phytoplankton in adjoining reservoirs have been reported from Sri Lanka (Silva & Wijeyaratne, 1999). This indicates that although cascading reservoirs show similarity in basic limnological characteristics, their microhabitat and food web structures and spatial and temporal environmental gradients may vary.

The oscillation of two species of phytoplankton in Kandy Lake may be attributed to the rainfall pattern. The concentration of dissolved silica during the rainy season and its relative proportion to the total phosphorous may promote the growth of *A. granulata*. A greater seasonal change of phytoplankton was seen in Lake Naivasha, Kenya, influenced by variable re-suspension of sediment with nutrient exchange induced by the wind regime (Kalff & Watson, 1986; Kalff & Brumelis, 1993). Change in density and species composition also occur in tropical lakes and reservoirs with long-term enrichment (Henry et al., 1984; Ramberg, 1987; Osborne, 1991; Hecky, 1993), although few studies are well documented (Talling & Lemoalle, 1998). The progressive increase of *M. aeruginosa* in the Kandy Lake may be attributed to long-term enrichment. Further, internal loading in shallow water bodies is higher than that of in deeper ones. Apparently, the accumulation of organic sediment with progressive land use changes in the watershed over the last two hundred years is substantially high in the Kandy Lake, which has no low level sluice to wash out the bottom sediment. As mentioned earlier, the climatic conditions of the area where Kandy Lake is located include a pronounced dry weather during Feb.-Mar. and Jul.-Aug. (Fig. 8.2). This has a direct effect upon water mixing and nutrient regeneration, which in turn, affect the physio-chemical parameters and the phytoplankton community.

Clearly, the distribution of cyanobacteria varies from reservoir to reservoir, irrespective of their geographical location and hydrological regime. A completely different spectrum of phytoplankton could be seen even in a reservoir cascade (Hart, 1996). Attempts have also been made to correlate the occurrence of cyanobacteria to catchment characteristics and land use changes (Martin, 1993). The growth of cyanobacteria was promoted in nitrogen-limited aquatic environments when phosphorus is not limited. In Kandy Lake, nitrogen limitation may not be the single factor that enhanced the growth of cyanobacteria. Colonization of non-heterocystous blue-green algae is an indication of N-saturation and low nitrogen to phosphorous ratios which favour dominance by blue-green algae in lake phytoplankton (Smith, 1983). The relative abundance and the species composition of cyanobacteria may be determined by the dynamic changes of the relative proportion of nitrogen species. Indeed, dissolved silica and its relationship to other reactive micronutrients may be a key actor regulating the growth of the centric diatom *A. granulata*. Multivariate techniques have proved to be a viable tool to identify the most likely scenario for the abundance and distribution of phytoplankton in aquatic ecosystems in relation to ecological variables (Varis et al., 1989; van Tongeren et al., 1992; Alvarez-Cobelas et al., 1994; Romo & van Tongeren, 1995). To date, neither environment characteristics nor community interactions have been able to correctly explain the abundance and distribution of phytoplankton species in eutrophic water bodies. Also, the biology and ecology of phytoplankton assemblage in productive waters in the tropics still remain partly uncertain. Several factors seemed to influence the increased occurrence of cyanoprokaryotic microflora in Kandy Lake. *M. aeruginosa* was correlated positively to higher concentrations of nitrogen compounds and total alkalinity which occurred during the dry period. During this period, eutrophic conditions could have been increased by the lower dilution of nutrient fluxes along with the establishment of a marked temperature/density gradient.

In Kandy Lake, phytoplankton species composition showed abrupt changes over time but biomass was low during the peak rainy season (high water levels) and high during dry season (low water level). The phytoplankton growth and subsequent increase in algal biomass with decreasing water level led into bloom formation. The type of dominant cyanobacteria species and prevailing weather condition primarily determined the formation of surface scum from a bloom. Although the biomass of filamentous form of cyanoprokaryotic microflora, (e.g. *Cylindrospermopsis* and *Oscillatoria*) was extremely high (hypereutrophic level) at several instances during post-blooming, they did not form a surface scum. Altogether, five functional species groups of phytoplankton of both k- and r- strategists (Reynolds et al., 2002) found in Kandy Lake are listed below:

(a) Large chain forming centric diatom, mainly *A. granulata*, dominant during mixing periods before the blooming (k-strategist).
(b) A colony forming Chlorophyte is common in eutrophic waters, with somewhat higher frequency of some varieties in the tropics primarily *P. simplex*, which almost disappeared during the post-blooming (k-strategist) period.
(c) Colony forming Cyanobacteria, genus *Microcystis*, mainly scum forming *M. aeruginosa* and *M. wesenbergii* (r-strategists).
(d) Small colony forming and non-heterocystous Cyanobacteria which does not aggregate into surface scum and appeared in moderate numbers during pre- and post-booming, *M. punctata* (k-strategist).
(e) Filamentous heterocystous forms of Cyanophytes/Cyanobacteria dominant during low water periods with a highly flexible ecological niche (Rott et al., in press): *Cylindrospermopsis raciborskii*, *Oscillatoria* sp (r-strategists).

The species groups a and b dominated the phytoplankton assemblage during the rainy and dry seasons, respectively, before the outbreak of *M. aeruginosa* bloom. *M. aeruginosa*, species group c showed a progressive increase and caused a dramatic change in phytoplankton species composition in May 1999. Following the blooming, phytoplankton composition was characterized by species groups a, c and d but suppressed the group b. Species of group e established in the lake in 2002 but the increasing freshwater input and mixing resulted in the emergence of *A. granulata* during the rainy season.

NUTRIENTS

Few of the elements such as N, P and Si known to be essential macro- or micro-nutrients have been discussed under nutrients in this chapter in relation to regulating or limiting factors of phytoplankton growth and periodicity in the Kandy Lake. The major sources of external nutrient inputs into the Lake includes wastewater drains, runoff from lawns and gardens, faecal matter from roosting cormorants and other birds, and a large resident population of fruit bats and leaf falling from trees around the bank (Fig. 8.6). Nitrate input through seepage is probably the largest from an urban watershed which is not facilitated with a central sewer system. Although it is assumed that feeding of fish by pilgrims contributes a fair amount of organic matter into the water body, the fish population itself, dominated by tilapia, is more important as a

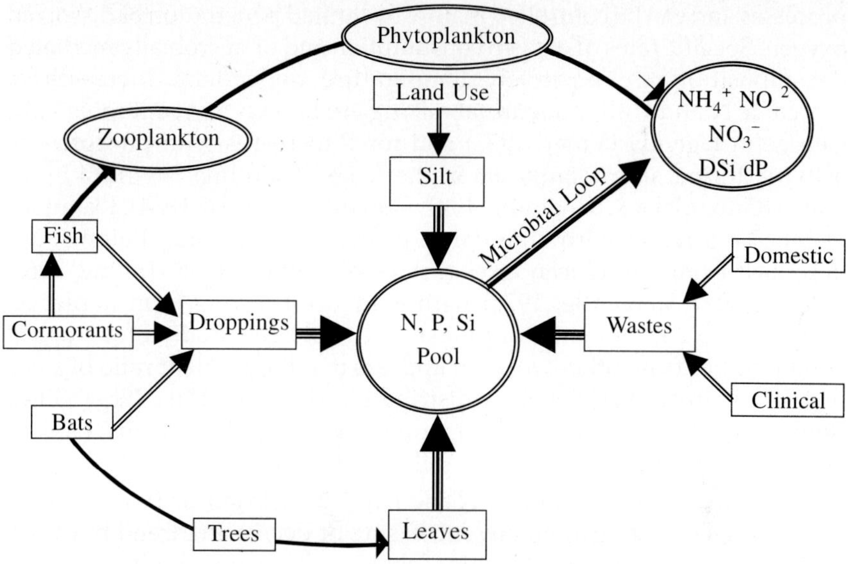

Fig. 8.6 Schematic diagram of nutrient relationships in Kandy Lake

contributor for the internal nutrient pool. Atmospheric sources of phosphorous and nitrogen can be generally discounted in small water bodies located in non-industrialized areas. Year-round sedimentation of dead organic biomass due to sinking was one of the major suppliers for the nutrient pool in a water body where the surface water flush-out only during rainy periods. Apparently, the nutrient budget was unique (Fig. 8.6). Shallow waters can be very productive with intense metabolism including mineralization within the water mass. Talling and Talling (1965) suggested prevalence of N-limitation rather than P-limitation in tropical inland waters. This is not to say that P-limitation is impossible in the tropics (Kalff, 1983; Bootsma & Hecky, 1993). Under eutrophic conditions, nutrients may not be a limiting factor of phytoplankton growth; however, depletion may occur due to intensive utilization and sporadic dilution.

The seasonal pattern either in tropics or temperate regions is mainly annual but in Sri Lanka, which experiences monsoonal rains, the rainfall pattern is predominant (Silva & Davies, 1987). In eutrophic lakes and reservoirs, high fluxes of carbon, nitrogen, and phosphorus combined with diel cycles of stratification (temperature/density) plus physiological activities of autotrophic algae and heterotrophic bacteria can lead such diel fluctuations to become predominant. Decomposition within the sediments in the bottom may be subordinate to these above processes and can be controlled mainly by limited penetration of dissolved oxygen. Specific rates of oxygen consumption and of microbially mediated decomposition are positively temperature dependent. Increases in inorganic N and P after nocturnal mixing are known from eutophic Lake George in Uganda (Viner, 1977) and for P increases with phosphatase activity during afternoon hours in the Lake Parakrama Samudra in Sri Lanka (Gunatilaka & Senaratna, 1981; Gunatilaka, 1983 & 1984). Examples of Silica depletion with abundance of main abstractors, diatoms, are described from the African lakes (Prawse & Talling, 1958; Talling, 1966; Adeniji, 1977; Lemoalle, 1978) with evidence for long-term depletion (Hecky, 1993). The kinetics of uptake have not been studied for tropical situations but from other works, it appears that the ambient ratio of silica or P : Si is influential for their distribution (Kilham et al., 1987). Silica depletion following the diatom growth is very clear in Sri Lankan reservoirs, including Kandy Lake (Silva unpublished data).

The long-term variations of N and P nutrients in surface waters of the Kandy Lake did not exhibit any increasing or decreasing trend but year-round ranges were extremely wide (Table 8.1) . The upper limits of NH_4-N and NO_3-N concentrations in both surface and bottom waters were extremely high with rapid fluctuations as compared to those irrigation and hydropower reservoirs in Sri Lanka. However, changes in NH_4-N and NO_3-N did not show a marked seasonal pattern during pre- and post-blooming periods (Fig. 8.7). The concentration of NH_4-N was always higher in the bottom waters as compared to the surface in this water body, indicating the role of microbial loop under anoxic conditions (Fig. 8.7). An extremely high concentration of NH_4-N above 1400 μgl^{-1} was registered before blooming in 1999. In contrast, the lower limit in the surfacewater was less than 50 $\mu g\ l^{-1}$. Dominance of NH_4-N as nitrogen species is an indication of that organic nitrogen was converted into ammonia through the process of ammonification. In the anoxic hypolimnion of eutrophic lakes, nitrate is reduced to NH_4-N. The concentration of NH_4-N can develop upto more than 1 mg l^{-1} in the anoxic hypolimnion of small eutrophic lakes as a result of heterotrophic degradation of organic matter and direct animal excretion (Reynolds, 1990). This transformation is carried out by microorganisms such as *Nitrocci* spp. Sharaff (2003) reported the occurrence of nitrification and denitrifying bacteria in high densities in Kandy Lake.

Since NH_4-N is the most energetically favourable source of combined nitrogen, plankton uptake NH_4-N as their nitrogen source in preference to NO_3-N when both source of elements are available (Brezonik, 1972; Zervenboonm & Mer, 1981). When the concentration of NH_4-N or other nitrogen species are high in a water body, it may result in colonization of

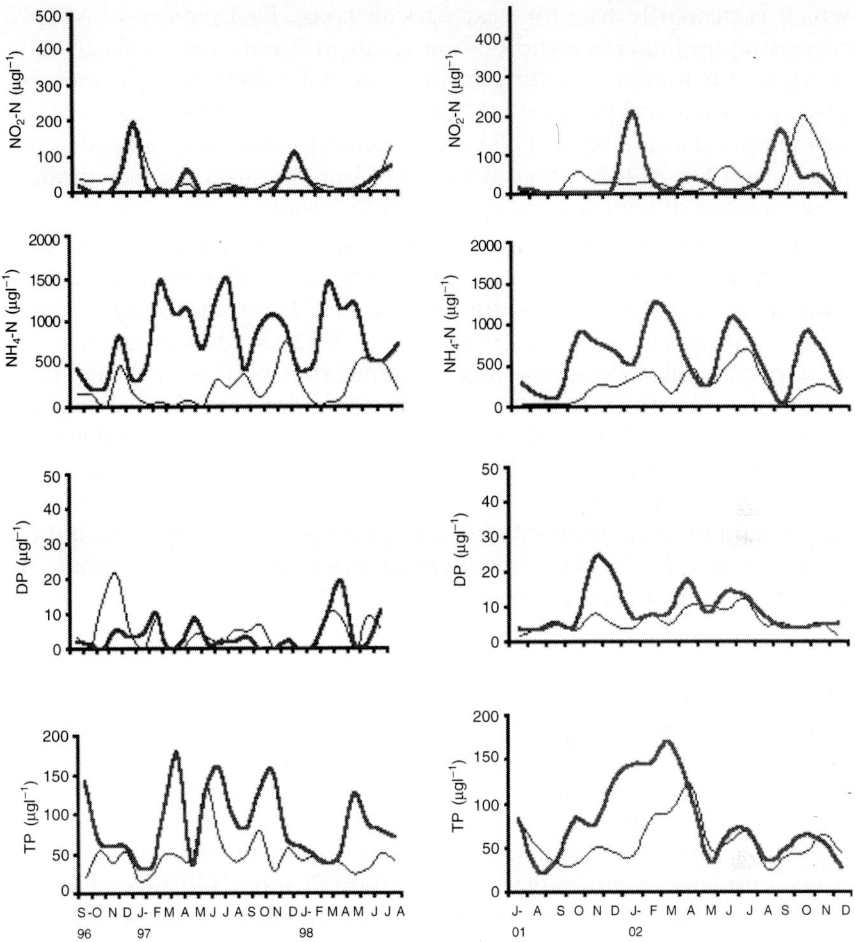

Fig. 8.7 Monthly variation of micronutrient (DP: dissolved phosphorus, TP: total phosphorous, NO_3-N: nitrate nitrogen, NO_2-N: nitrite nitrogen, NH_4-N: ammonia nitrogen in Kandy Lake during pre- and post-blooming periods

non-nitrogen fixing cyanobacteria such as genus *Microcystis*, *Merismopedia*, etc. But the accepted phenomenon is that the abundance of nitrogen fixers are high in tropical waters and the element ratio and biomass are often suggestive of nitrogen deficiency (Viner, 1977; Mukankomeje et al., 1993). However, the low concentrations always do not reflect N deficiency because the ions can be continually regenerated in aquatic ecosystems. Tropical water appears to have a lower ratio of dissolved inorganic nitrogen to soluble reactive rather phosphorus than waters in temperate regions (Viner, 1975; Wood et al., 1984; Talling, 1992) which is primarily true for East African lakes. Diel pattern of NH_4-N concentration in surface and bottom water in Kandy Lake indicates the dynamic fluctuation resulting from microbial activities, utilization by phytoplankton and physical and chemical processes (Table 8.2). Spatial and temporal availability of NH_4-N potentially occur on small scale than are typical for NO_3-N (Reynolds, 1990), but the surface concentration may increase following a collapse of algal bloom

The concentrations of NO_3-N in the bottom water were relatively low when compared to the surface in Kandy Lake. The highest surface concentration of NO_3-N was 816 μg l^{-1} in 1998 before the blooming and a maximum of 612 μg l^{-1} was registered in 2002 after blooming. Nitrate-N is present at a generally low concentration (<60 μg l^{-1}) and may accumulate locally when oxygen tensions are less than 1 mg l^{-1} (Reynolds, 1990). Extremely low levels or even undetectable levels of NO_3-N in the bottom water have been registered during pre- and post-blooming periods (Fig. 8.7). Relatively high concentrations of nitrogen species and their intermittent fluctuation indicate gross organic pollution of the water body. This may result either from allocthonous or autochthonous organic loading. Seepage of NO_3-N into the water body may be extremely high from the surrounding watershed. It has been observed that some hotels release their waste, even night soils to the Lake directly during the night through underground canals. Enrichment by sewage and urban runoff in the tropics is not uncommon (Osborne, 1991). Nitrate enrichment may come from erosion of parks, gardens and lawns, other altered land use pattern and faecal pollution. Faecal pollution in crowded urban areas is more important in southeast Asia. Nitrogen transformation and the nitrogen budget of tropical eutrophic water have been not properly investigated yet, although large subtropical water bodies have been subjected to detailed investigations (Ashton, 1985).

The concentrations of dissolved phosphorus (d-P) ranged between undetectable (ND) level and 22 μg l^{-1} in the surfacewater and the range was between ND and 19 μg l^{-1} in the bottom water. The concentrations of total phosphorus (t-P) were high in the bottom waters (36 - 171 μg l^{-1}) when compared to the surfacewater (14 - 131 μg l^{-1}) (Fig. 8.7). The upper limit of d-P in Kandy Lake was fairly higher than the reported values for Sri Lankan reservoirs (Gunatilaka, 1983, 1984; Silva & Gamlath, 2000; Schiemer et al., 2001; Silva & Schiemer, 2001; Silva et al., 2002), but the mean value was within the acceptable range for tropical eutrophic water bodies. An undetectable concentration of dissolved phosphorus indicates the rapid turnover within the water body. Eutrophic water bodies with higher algal biomass uptake dissolved phosphorus from the nutrient pool very rapidly (Wetzel, 1983). In many instances, uptake by algae uptake exceed the rate of external loading and thermal stratification restrict internal recycling (Reynolds, 1990). However, it is important to realize that a low concentration or undetectable levels of dissolved nutrients do not mean that the nutrients are limited for phytoplankton growth. Extremely high concentrations of t-P in the bottom waters of the lake indicate accumulation of organic matter in the bottom. This may result from both autochthonous and allochthonous processes mentioned earlier. When fish biomass is extremely high in non-harvesting water bodies, their contribution to the nutrient pool is colossal. Relatively low concentration of total phosphorous in the surface layer compared to the bottom of the water body may be attributed to rapid uptake by phytoplankton. A complex set of eco-physiological and biogeochemical processes should occur in the Lake to resulting dynamic fluctuation of phosphorus without a particular spatial or seasonal patterns.

Further, a slight increase in t-P and inorganic nitrogen in the surfacewater in the Lake was found following the rainfall, although there was no clear seasonal pattern in N and P. This suggests that wastewater generated in the watershed also contributes a substantial amount of nutrients to the water body. Furthermore, the droppings of aquatic birds such as roosting cormorants, pelicans and the visiting birds (e.g. egrets) and the bats residing on trees around the Lake will certainly contribute to the nutrient loading of the lake. In addition, the foliage parts (e.g. leaves) fallen from the trees around the lake also contribute to the lake as a nutrient source. The lake ecosystems is also adapted to recycling nutrients within the water body by enhancing the microbial activity. Particularly, browsing behaviour of cichlid fish (e.g. tilapia), the most dominant fish in the water body, stirs up the bottom sediment. Vertical gradient of nutrient in the deep basin of the lake may be attributed to the chemical gradient (e.g. pH, DO, temperature) and also to the distribution pattern of microbial population.

Information on dissolved silica (DSi) is not available for the pre-blooming period but the concentration and its seasonal changes are shown in Fig. 8.8. Many studies have shown that the diatom population responds to DSi concentration or its relative proportion to t-N and t-P (Werner, 1977; Paasche, 1980). DSi had a positive correlation with the population density of *A. granulata* (Figs. 8.5 & 8.8) during the post-blooming period. When diatom was least abundant, silica concentration was high and the concentration decreased largely during the period when the diatom population was re-established in the following dry period in 2002. In temperate lakes, DSi is subjected to marked seasonal variations that increase during high winter discharge and water

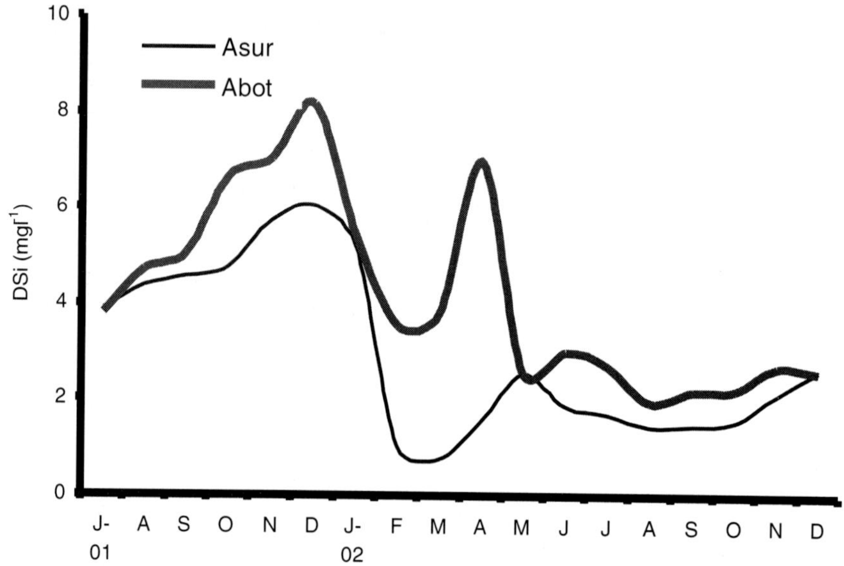

Fig. 8.8 Monthly variation of dissolved silica concentrations in Kandy Lake during post-blooming periods

circulation, sometimes decreasing to very low levels as a direct result of biological uptake (Reynolds, 1990). The findings from Kandy Lake at the very least support the view that the concentration and the amount of DSi regulate the growth of diatoms in eutrophic water bodies, as a consequence of species composition of phytoplankton.

TROPHIC EVOLUTION AND ECOSYSTEM COLLAPSE

Evidently, there were two clear phases in Kandy Lake during pre- and post-outbreak of *Microcystis* bloom. Of course, eutrophication is diverse in its apparent consequences but it may be unique with respect to the characteristics of the water body. Chlorophyll-a (chl-a) content, an index of phytoplankton biomass, was above 20 $\mu g\ l^{-1}$ since mid 1996 except in a few occasions under rainy conditions (Fig. 8.9). The chl-a content remained more or less consistent at relatively low levels from May to December 2002 when the water level was maintained at FSL. Water level raised up to the FSL level in May, 2002 after heavy rains occurred in April, which is normally unusual (Fig. 8.2). With respect to the chl-a content, Kandy Lake remained eutrophic since 1996, according to the classification forwarded by Dillon & Rigler (1974). Further, the annual mean values showed a progressive increase before and after blooming in 1999 with an exponential growth in the upper limit from 1997 to 2002

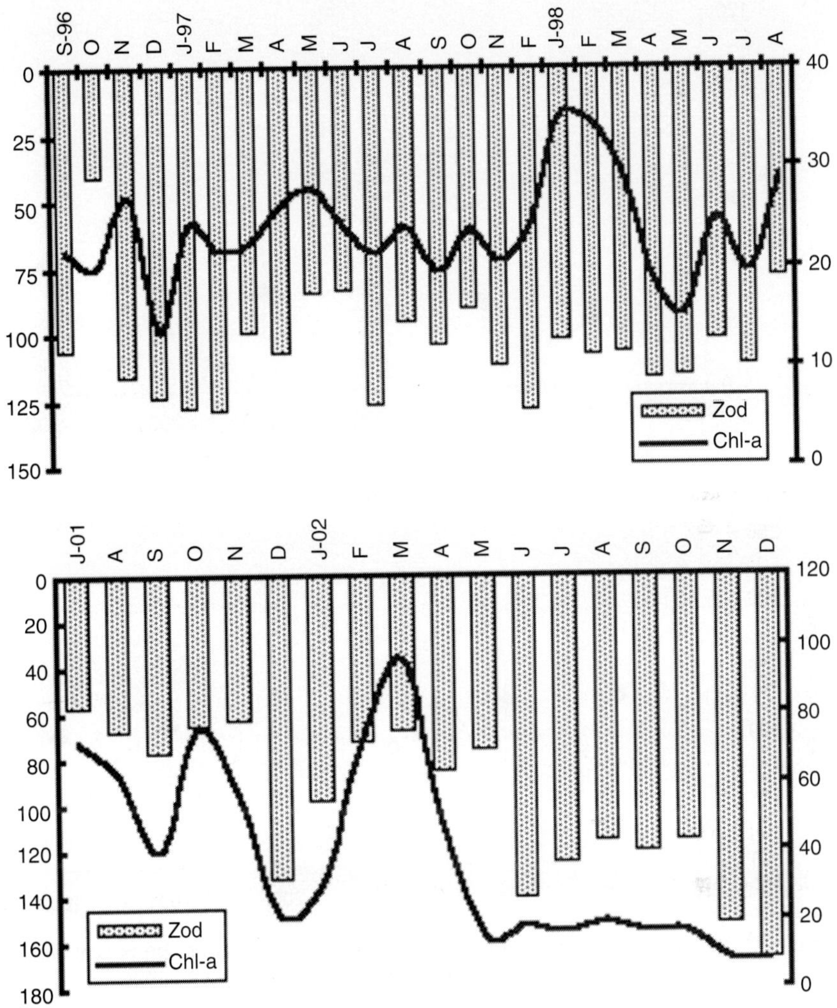

Fig. 8.9 Monthly variation of Secchi disc visibility and Chlorophyll-a content in Kandy Lake during pre- and post-blooming periods

(Fig. 8.10). The annual upper limit was above hypereutrophic limit after blooming and the lower limit coincided with the rainfall associated filling of the lake. In December 2002, the lower limit came up to the level of mesotrophic (< 10 μg l^{-1}) lakes.

Long-term changes in chl-a content and water transparency in terms of Secchi depth, an approximation of underwater light penetration are good predictors of trophic evolution. Secchi depth visibility of a water body is a function of both phytoplankton density and inorganic turbidity.

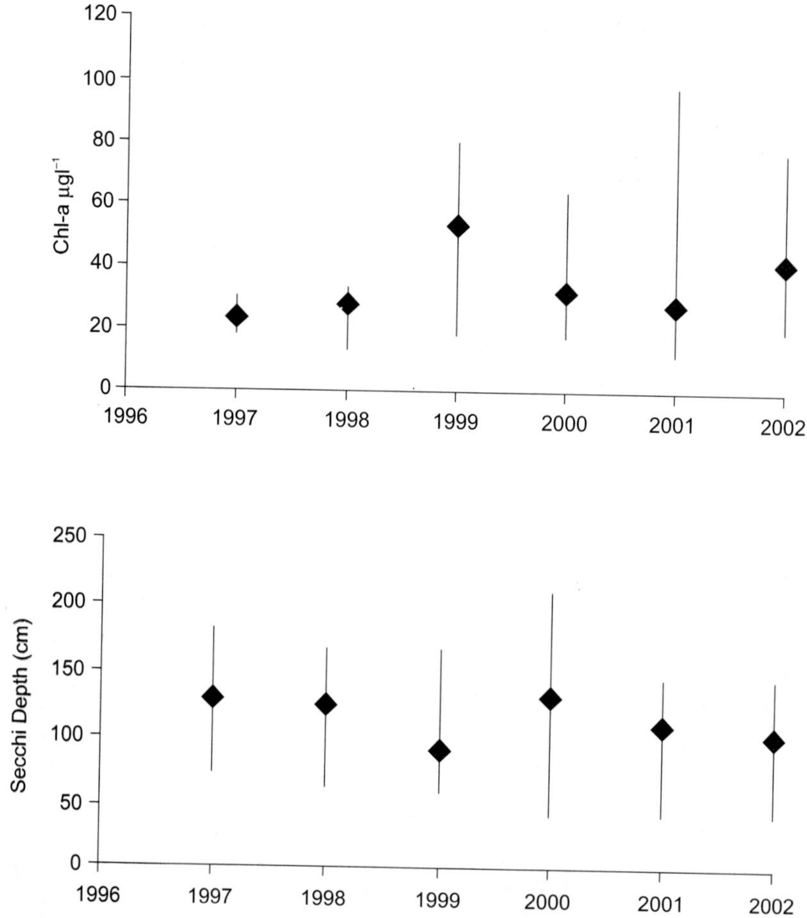

Fig. 8.10 Long term trends in mean chlorophyll-a content and average Secchi disc visibility in Kandy Lake. Vertical lines represent the annual range

An enormous amount of silt and suspended solids are transported to the lake with the storm water runoff decreasing the water transparency. Sediments will settle within a short period, improving the water clarity. In contrast, during the extreme dry spell (Feb.–Mar.), underwater light penetrations decreased rapidly due to scattering and absorption by phytoplankton. Low light penetrating conditions can be extended to several months if freshwater is not brought in. Long-term changes of annual mean values of Secchi depth were decreasing from 1997 to 2002, indicating a higher prevalence of phytoplankton densities (Fig. 8.10). As mentioned earlier, there was an oscillation of *A. granulata* and *P. simplex* with progressive increase of *M. aeruginosa* in the Kandy Lake during pre-blooming. The oscillation of a diatom and a green algae may be an indication of trophic equilibrium while the progressive increase of *M. aeruginosa* could be considered as indicative of trophic evolution. Kandy Lake is one of the oldest water bodies in Sri Lanka with continuous water for about 200 years but an outbreak of an algal bloom had not been reported until May 1999. A marked stratification with a notable micro-stratification (Talling & Lemoalle, 1998), continual oxygenation depletion of deeper layers with no seasonal changes (van der Heide, 1973; Matsumara-Tundisi et al., 1991) and often with an accumulation of NH_4-N near the bottom are evidence for permanent eutrophication leading to hypertrophic conditions.

It is not possible to diagnose the time series in the enrichment process since there are huge gaps in the available data on nutrients and trophic characteristics. Nevertheless, the Kandy Lake had been subjected to a progressive trophic evolution from an eutrophic state to a hypertrophic one from 1997 to 2002. With its large pool of nutrients, there was an ecological equilibrium perhaps due to a balance in removal and loading of essential nutrients such as N, P and DSi for the growth of phytoplankton. Apparently, several factors had caused the enrichment of the water body. Increase in human population with altered land use could have had direct or indirect contributions to the nutrient pool. In addition, the resident populations of bats, roosting cormorants and other birds and tree leaves make direct contributions. Some of the tourist hotels used to discharge untreated sewage directly to the Lake, though it is illegal. The peculiar feeding behaviour of cichlid fish also mobilize sediment-bound phosphorus from the shallow area, which can be considered a top-down effect from introduced changes in the fish population.

Outbreak of a bloom: Eutrophication, which is a global phenomenon, usually involves the replacement of diverse biota of low abundance by those of few species at high abundance (Talling & Lemoalle, 1998). The eventual impacts are widespread in the tropics, although documentation of the process is rarely available. In the tropics, as elsewhere, it has been triggered by gross organic pollution. The outbreak of a cyanobacteria bloom (*M. aeruginosa*) that occurred in Kandy Lake during the dry spell in 1999 is an excellent example to demonstrate the event of an ecosystem collapse in the process of trophic evolution from eutrophic to hypertrophic. The most likely limnological interpretation for the outbreak of *M. aeruginosa* is as follows: the lowering of the water level during the dry spell of 1999 may have resulted in more access for *M. aeruginosa* to utilize phosphorus mobilized in the bottom layers (Silva, 2003a). Figure 8.11, a schematic diagram of the vertical

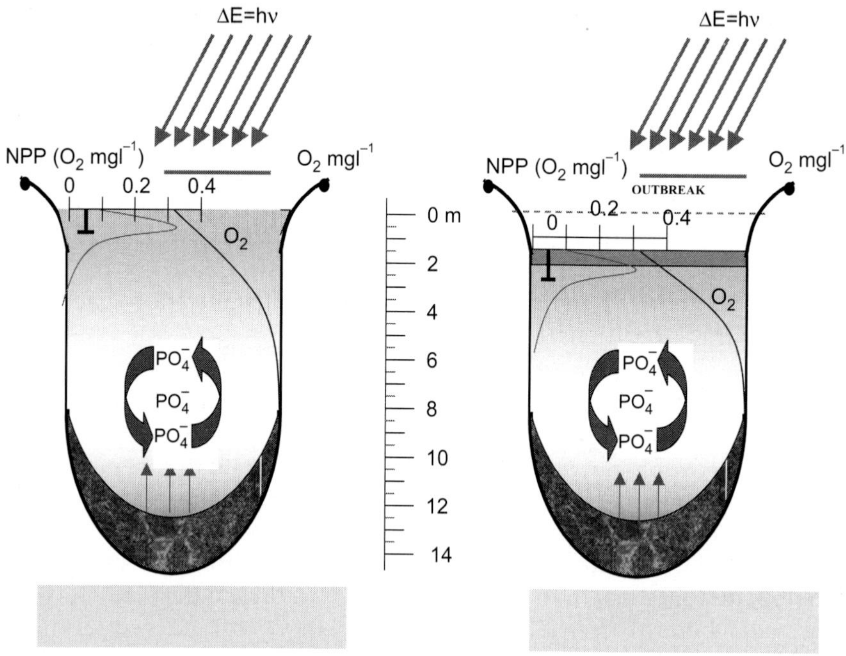

Fig. 8.11 A schematic diagram of limnological processes taking place in the Kandy Lake during high and low water levels

profile of the deep basin, shows the oxygen distribution, photosynthetic profile and Secchi disk visibility during high and low water level and proportionate sediment layer. It is known that gas-vacuolated cyanobacteria (e.g. *Microcystis*) have developed buoyancy regulation, enabling them to move rapidly through the water column.

The unique feature is that when water level was high, there was no turbulent mixing due to windless conditions, especially during November and early December. Moderate turbulence mixing occur from late December to February as a result of the northeast monsoon. Due to the runoff, the external nutrient loading becomes predominant and highly convective forces mediate sediment erosion, but the magnitude of the erosion may be comparatively less. The deeper light penetration with euphotic depths of more than 2 metres, ensures higher turnover rates of algal biomass, but this is subjected to both dilution and some loss through relatively high and balanced flushing rates (Duncan et al., 1993). The apparent predominance of phosphorus in man-induced eutrophication of temperate lakes is not necessarily matched in the tropics where natural N : P concentration ratio is often higher (Schiemer, 1983).

On the other hand, availability of chl-a in the entire water column is indicative of rotation of phytoplankton with the convective currents. The reserve potential of tropical phytoplankton to uptake additional nutrients is very high in an absolute sense and the response of the water body to hypereutrophication of tropical latitude may, therefore, be stronger than it is for temperate areas. The mix layer is more dynamic in a tropical setting than in a temperate area (Lewis, 1996). The diel cycle of temperature/density stratification can have large-scale effects on the vertical distribution of blue green-algae (Cyanobacteria). Day-to-day variation of wind speed was positively related through turbulence and re-suspension to the abundance of diatom *Aulacoseira italika* in a shallow Brazilian reservoir (de Lima et al. 1983). Phytoplankton is inherently susceptible to both the radiation/temperature and water balance complexes of environmental factors. Its own reaction upon the physical and specially the chemical environment can also be profound and often cyclic in time. The biotic interactions that include grazing can influence further temporal changes.

APPROACHES OF RESTORATION

Restoration and management of eutrophic water bodies located in urban centers are priorities in the socio-economic agenda of local governments, urban councils, city missions, etc. A variety of bottom-up and top-down strategies have been tested (e.g. diversion of effluent outfalls, dredging, aeration, chemical treatment, bio-manipulation, watershed management.) around the world and there are both success stories and failures (McQueen et al., 1989; Lammens, 1999; Annadotter et al., 1999). Failures in restoration and sustainable management of eutrophic urban water bodies, especially in developing countries, are primarily linked with poor knowledge on hydraulic balance, limnological processes and dynamics of a particular aquatic system and the disturbance in the watershed or within the water body triggered by human activities. Trophic evolution and subsequent hypereutrophication is a fundamentally chronological phenomenon. However, sudden outbreaks of algal blooms in eutrophic waters may be linked with other unknown factors. Silva and Schiemer (2001) emphasizes the human factor as the fourth dimension of reservoir limnology in the tropics. In the tropics, hydraulic balance is rigorously manipulated, resulting in sudden changes in limnological processes and dynamics (Schiemer et al., 2001). Shallow manmade water bodies in monsoon Asia show a distinct annual trophic shift from mesotrophic to eutrophic, resulting from rainfall-bound filling and progressive water release to meet the demand in the command area.

A variety of technological and biological methods have been established for both bottom-up and top-down control. Successful results have been achieved in many instances by implementing a combination of top-down and bottom-up controls. The present knowledge on Kandy Lake is sufficient to identify the appropriate strategies to be implemented in order to reverse the hypereutrophication or to control the bloom-forming cyanobacteria through either top-down or bottom-up strategies or a combination of both. The appropriate and cost-effective methods described briefly below have their intrinsic worth and limitations. The schematic model shown in Fig. 8.11 and the available limnological information clearly demonstrate that the biomass regulation or prevention of emergence of *Microcystis* bloom in Kandy Lake is a combination of both top-down and bottom-up approaches (Silva, 2003a). Multiple techniques may also be relevant in the case of restoration since urban lakes are very different from other manmade water bodies. Nevertheless, sound knowledge on watershed characteristics, hydrology, limnological processes and the human activities taking place in the watershed play a key role in restoration and management.

a. Wastewater Treatment and Diversion

This is a bottom-up control, which involves collection of wastewater and domestic sewage, and treating them in a central wastewater treatment plant that discharges the effluent into the Mahaweli River at an appropriate place through a convey outfall. Although it is a long term and sustainable approach, it could be implemented step wise and involves correct planning and a large sum of financial resources.

b. Hypolimnetic Aeration (Artificial Circulation)

Hypolimnetic aeration or artificial circulation, which is referred to as destratification, is another bottom-up control technique. This reduces the mobilization of phosphorus from bottom sediment under anoxic conditions and promotes chemical oxidation of organic and inorganic substances. In this case, the anoxic layer of the Kandy Lake (below 6 m) in the deep basin should be aerated during the dry season using aerators of suitable capacity. This is a well-established technology applied in many countries to restore small water bodies like Kandy Lake. The success of this method has been shown by various private sector organizations that have established this treatment technology. They also treat the specific water body with microorganisms in order to expedite the process of decomposition. Therefore, care should be taken if genetically modified microorganisms are used in this process. Regular monitoring of water quality is required during the aeration process to determine the efficiency of the method. There are several adverse effects of artificial circulation. If nutrients are limiting the productivity in the euphotic zone, then circulation may increase particulate phosphorus, which could be mobilized to usable forms and water transparency could become worse than before if the bottom sediment is stirred up and it may prevent the phytoplankton successes ion from blue greens to greens due the changes in carbon dioxide and pH in the euphotic zone.

c. Hypolimnetic Withdrawal

Aeration can also be coupled with siphoning out water from the deeper layers with less oxygen. In this case, it is necessary to raise the water level of the lake to FSL with the onset of the inter-monsoonal rains and siphon water from the deep layers (below 6 metres) where permanent oxygen depletion occurs. Siphoning helps both bottom-up and top-down controls since it removes the algal biomass while oxygenating the deeper layers. Already available local technologies can be adopted to achieve this task. Therefore, the cost involved in this method could be minimum compared to the other high-tech methods.

Release of hypolimnetic water that may contain high concentrations of phosphorus, ammonia, and hydrogen sulphide and trace metals and less oxygen may cause a water quality problem downstream. In the case of Kandy Lake, the outflow canal, which is about 4 km, contains no important fishery and is neither used for recreation nor water supply. Therefore, special precautions will be not necessary to maintain downstream water quality.

d. Dredging (Sediment Removal)

Properly conducted dredging (sediment removal) is one of the most effective bottom-up control techniques of eutrophic water bodies, commonly used in developed countries. There are four major reasons for removing sediment from a lake bottom: deepening, nutrient control, reduction of macrophytes, and removal of regimented toxic materials. The main constraint involved in dredging of urban lakes is disposal of the sediment. It should be noted that the Kandy Lake has about 1.5 m sediment thickness in the deeper basin.

There is little doubt that dredging can be successful as a lake restoration technique. If there is a thorough pre-implementation evaluation of the lake setting, proper equipment is selected, disposal areas are designed for end-of-treatment and dredging is conducted by conscientious and competent operators, then restoration can be achieved successfully to a considerable extent through dredging. However, in the case of Kandy Lake, it is necessary to remove the sediment and silt accumulated at the outfalls of major silt traps and sediment from the shallow basin on a regular basis. This can also be achieved by using locally available technologies at a relatively low cost.

e. Establishment of Wetlands

Artificial wetlands appear to be an attractive and valuable water quality management strategy and their use in upgrading wastewater and storm water discharge quality is increasing. Nutrient availability to cyanobacteria can also be reduced by establishing wetlands within small bays in the shallow basin of the Kandy Lake with floating macrophytes such as water hyacinth and other species native to Sri Lanka. These wetlands should be constructed using gabion walls allowing inter-permeability between the wetland and open water. In such a case, it is necessary to harvest the vegetative parts on a regular basis with proper maintenance.

f. Fish Removal

The population density of fish in this non-harvesting water body is extremely high but the fish diversity is limited to a few species. It seems that the Kandy Lake has become more or less a monoculture of introduced cichlids. They are omnivorous but ingest cyanobacteria very effectively. They also stir bottom sediment, mobilizing nutrients when they feed on bottom substrate. Therefore, it is necessary to remove this fish species periodically, maintaining their population density below the carrying capacity as a measure of biomass removal (including P and N) and introduce suitable carnivorous species to reshape the food chain.

CONCLUSIONS

Processes and dynamics of inland standing water bodies in the tropics are poorly studied compared to the present density of manmade water bodies throughout the tropics. The poor knowledge on Asian lakes and reservoirs is a result of the low density of natural lakes in the region and poor utilization of inland waters as a biological resource, which has been the case till quite recently, but the lack of expertise also seems to be a major factor. Besides, the available information has deviated a little from the conventional limnological studies and a few have highlighted the effects of human activities on limnological processes in Asian reservoirs.

Nevertheless, limnological studies on urban water bodies are becoming an increasingly important issue since they are being subjected not only for eutrophication and pollution but there are also evidences for catastrophic collapse of urban water bodies. Therefore, the knowledge on impact of manipulation of hydraulic balance on limnological processes and dynamics of tropical manmade water bodies need to be reorganized.

The limnological processes and dynamics of Kandy Lake are regulated by the local climate because of its small catchment, reduced flow-through and subsequent long residence time or low flushing. A prolonged stagnant nature of Kandy Lake, which receives substantial inflows mainly during October-December has led the water body to become more or less a closed lentic ecosystem. Mixing of the entire water body has been hindered by morphometric features and its very location in the landscape results in an establishment of permanent oxygen depletion in the deep basin. Dynamic diel changes in temperature/density stratification have led the nutrient regenerated from the bottom to become available in the surface layers. This also gave the access for phytoplankton to utilize the nutrient in the entire mixed depth. The small size of the lake has hindered horizontal environmental gradients but vertical gradients for some parameters were prominent from surface to bottom in the deep basin. The levels of some physico-chemical attributes were relatively high when compared to those standing water bodies located at a similar elevation, indicating gross organic pollution. A progressive trophic evolution towards hypereutrophic and subsequent outbreak of a *Microcystis* bloom demonstrates the rate and magnitude of enrichment of this non-harvesting urban water body.

The functional groups of phytoplankton composition and their relative abundance, concentration of micronutrients and their dynamic fluctuations nature, are inherent characteristics of an eutrophic water body enriched by both allochthonous and autochthonous processes. Although it looks more productive and healthy, the Kandy Lake is partially a dead water body polluted mainly by organic wastes. Eutrophication and subsequent hypereutrophication are necessarily centred upon long-term nutrient loading and enrichment associated with thermal stratification and wind-driven convective forces and internal loading to a greater extent. Therefore, nutrient loading from the watershed is also not the only cause of hypereutrophication in the tropical urban water bodies. The sudden drop of water level by man for various purpose and colonization of exotic cichlids are directly or indirectly linked with the impulsive emergence of algal blooms in eutrophic tropical manmade lakes when lake ecosystems and local climate are favourable for both growth and sustenance of bloom-forming phytoplankton. The role of bioturbation and grazing pressure on impulsive emergence of algal blooms in cichlid-dominant tropical reservoirs needs a more detailed study. More attention should be paid on eutrophic and hypertrophic waters in the tropics since more than half of the bloom-forming cyanobacteria strains are toxic and the organic compounds found in eutrophic waters form some carcinogenic compounds during the process of chlorinating.

ACKNOWLEDGMENTS

This chapter is a product of a long-term research carried out on various aspects of limnology of the Kandy Lake by the Ecology and Environmental Biology Project of the Institute of Fundamentals Studies. I am very much thankful to Ms Fathima Sharaff, Ms Charukshi Karunathilake, Mr Namal Athukorale, Ms Iranganie Thumpela, Mr Rohana Weerakone and Mr Wajira Kangara for various help given during the preparation of this chapter. Dr Shirani Nathanael was kind to go through the draft manuscript and made very valuable suggestions and constructive comments. I am also thankful to the Director IFS for continuous financial support and encouragement throughout.

REFERENCES

Adeniji, H.A. 1977. Preliminary study of the Silica content and diatom abundance in Kainji Lake, Nigeria. In: H.L. Golterman, (ed.), Interaction Between Sediments And Freshwater, The Hague: Junk & PUDOC: 331-334.

Alvarez-Cobelas, M. & Jacobsen, B. 1992. Hypertrophic phytoplankton: An overview. Freshwater Biological Association. Freshwater Forum 2: 184-199.

Alvarez-Cobelas, M., Velasco, J.L., Rubio, A. & Rojo, C. 1994. The time course of phytoplankton biomass and related limnological factors in shallow and deep lakes: A multivariate approach. Dev. Hydrobiol. 94: 139-151.

Annadotter, H., Cronberg, G., Aagren, R., Lundstedt, B., Nilsson, P. & Ströbeck, S. 1999. Multiple techniques for lake restoration. Hydrobiologia, 395/396: 77-85.

Apstein, C. 1917. Das plankton im Colombo-See auf Ceylon. Sammelausbeute von A. Borget, 1904-1905, Zool. Jahrb. (Abt. System) 25: 201-244.

Ashton, P.J. 1985. Nitrogen transformations and the nitrogen budget of a hypertrophic impoundment (Hartbeespoort Dam, South Africa). J. Limnol. Soc. Sth. Afr., 11, 32-42.

Barica, J. & Mur, L.R. (eds). 1980. Hypertrophic Ecosystems. Developments in Hydrobiology. 2 Dr. W. Junk Publishers. The Hague: pp. 363.

Biswas, S. 1978. Observations on phytoplankton and primary productivity in Volta. Lake. Ghana. Verh. Internat. Verein. Limnol., 20: 1672-1676.

Bootsma, H.A. & Hecky, R.E. 1993. The conservation of the African Great Lakes: A limnological perspective. Cons. Biol. 7: 1-13.

Branco, C.W.C. & Senna, P.A.C. 1994. Factors influencing the development of *Cylindrospermopsis raciborskii* and *Microcystis aeruginosa* in the Paranoa Reservoir, Brasilia, Brazil. Arch. Hydrobiol. (Suppl.). 75: 85-96.

Branco, C.W.C. & Senna, P.A.C. 1996. Phytoplankton composition, community structure and seasonal changes in a tropical reservoir (Paranoa Reservoir, Brazil). Arch. Hydrobiol. (Suppl.). 114: 69-84.

Brezonik, P.L. 1972. Nitrogen: sources and transformations in natural waters. In: H.E. Allen, & J.R. Kramer (eds), Nutrients in Natural Waters, New York: Wiley, 1-47.

Carney, H.J., Richerson, P.J. & Eloranta, P. 1987. Lake Titicaca (Peru/Bolivia) Phytoplankton: Species composition and structural composition with other tropical and temperate lakes. Arch. Hydrobiol., 110: 365-85.

Carmichael, W.W. 1992. Cyanobacteria secondary metabolites—The cyanotoxins. J. Appl.Bacteriol., 72: 445-459.

Codd, G. A., Brooks, W. P., Lawton, L. A. & Beattie, K. A. 1989. Cyanobacterial toxins in European waters: occurrence, properties, problems and requirements. In: D. M. J. Richardson. & J. Bridges (eds), Watershed 89. The Future for Water Quality in Europe. Vol. 2, Wheeler, Oxford: Pergamon Press, 211-220.

Conley, D.J. 2000. Biochemical nutrient cycles and nutrient management strategies. Hydrobiologa, 410: 87-96.

de Lima, W.C., Marins, M. de. A & Tundisi, J.G. 1983. Influence of wind on the standing stock of *Melosira indica* (Ehr.) Kutz. Revta. Bras. Bioliologie. 43: 317-320.

De Silva, P.K. & De Silva, K.H.G.M. 1984. An ecological study of the meso and macrofauna of the littoral region of Lake Kandy in Sri Lanka. Arch. für Hydrobiol. 102: 53-72.

Dissanayake, C.B., Senaratne, A., Weerasooriya, S.V.R. & De Silva, S.H.G. 1982. The Environmental pollution of Kandy lake, Sri Lanka : A case study from Sri Lanka. Environ. Internat., 17: 343-351.

Dissanayake, C.B & Weerasooriya, S.V.R. 1986. Environmental chemistry of Mahaweli River, Sri Lanka. Int. Nat. J. Env. Stud., 28: 207-233.

Dissanayake, C.B., Rohana Bandara, A.M. & Weerasooriya, S.V.R. 1986. Heavy metal abundance in the Kandy Lake: An environmental case study from Sri Lanka. Environ. Geol. Wat. Sci. 10(2): 81-88.

Dillon, P.J. & Rigler, F.H. 1974. The phosphorus-chlorophyll relationship in lakes. Limnol. Oceanogr., 19: 767-773.

Douglas, I. 1968. The effects of precipitation chemistry and catchment area lithology on the quality of river water in selected catchments in Eastern Australia. Earth Sci. Journal, 2: 126-144.

Duncan, A., Gunatilaka, A. & Schiemer, F. 1993. Limnological Aspects of Landscape Management in Sri Lanka. In: W. Erdelen, C. Peru, N. Ishwaran & C.M. Madduma Bandara (eds), Ecology and Landscape Management in Sri Lanka. Margraf Scientific Books, 381-395.

Fatimah, M.Y., Mohsin, A.K.M. & Kamal, A.S.M. 1984. Phytoplankton composition and productivity of a shallow tropical lake. Pertanika, 7: 101-113.

Ganf, G.G. 1974. Rates of oxygen uptake by the planktonic community of a shallow equatorial lake (Lake George, Uganda). Oecologia 15: 17-32.

Gibbs, R. J. 1970. Mechanisms controlling world water chemistry, Science 170: 1088-1090.

Gower, A.M. (ed.). 1980. Water quality in catchment ecosystems. John Wiley & Sons, Chichester: pp. 335.

Gunatilaka, A. 1983. Phosphorus and phosphatase dynamics in Parakrama Samudra based on diurnal observations. In: F. Schiemer, (ed.), Limnology of Parakrama Samudra-Sri Lanka: A Case Study of an ancient man-made lake in the tropics. Dev. Hydrobiol. 12: 35-47.

Gunatilaka, A. 1984. Observations of phosphorus dynamics and orthophosphate turnover in tropical lake Parakrama Samudra, Sri Lanka. Verh. Internat.Verein. Limnol., 21: 994-1000.

Gunatilaka, A. & Senaratne, C. 1981. Parakrama Samudra (Sri Lanka) Project, a study of tropical lake ecosystem II. Chemical environment with special reference to nutrients. Verh. Internat. Verein. Limnol. 21: 1000-1006.

Hart, R.C. 1996. Comparative ecology of plankton in cascading warm-water reservoirs: aspects of relevance to tropical limnology. In: F. Schiemer, & K.T. Boland (eds), Perspectives in Tropical Limnology, SPB Academic Publishing, Amsterdam: 113-130.

Hecky, R.E. 1993. The eutrophication of Lake Victoria. Verh. Internat. Verein. Limnol., 25: 39-48.

Henry, R., Tundisi, J.G. & Curi, P.R. 1984. Effects of phosphorus and nitrogen enrichment on the phytoplankton in a tropical reservoir (Lobo Reservoir, Brazil), Hydrobiologia, 118: 177-85.

Hofer, F. & Schiemer, R. 1983. Feeding ecology, assimilation efficiencies and energetics of two herbivorous fish: *Sarotherodon* (*Tilapia*) *mossambicus* (peters) and *Puntius filamentosus* (cuv. Et. Val.). In: F. Schiemer, (ed.), Limnology of Parakrama Samudr, Sri Lanka, A case study of an ancient manmade lake in the tropics. Dev. Hydrobiol. 12: 155-164.

Holsinger, E.C.T. 1955. The distribution and periodicity of the phytoplankton of three Ceylon lakes. Hydrobiologia, 7: 25-35.

Kalff, J. 1983. Phosphorus limitation in some tropical African lakes. Hydrobiologia, 100: 102-112.

Kalff, J. & Brumelis, D. 1993. Nutrient loading, wind speed and phytoplankton in a tropical African lake. Verh. Internat. Verein. Limnol., 25: 860.

Kalff, J. & Watson, S. 1986. Phytoplankton and its dynamics in two tropical lakes: a tropical and temperate zone comparison. Hydrobiologia, 138: 161-176.

Kanan, V. & Job, S.V. 1980. Diurnal, seasonal and vertical study of primary production in Sathiar Reservoir. Hydrobiologia, 70: 103-117.

Khondker, M. & Parveen, L. 1993. Daily rate of primary productivity in hypertrophic Dhamondi Lake. In: Tilzer, M.M. & Khondker, M. (eds),

Hypertrophic and Polluted Freshwater Ecosystems : Ecological bases for water resources management. Bangladesh Department of Botany, University of Dhaka, 181-191.

Kilham, P., Kilham, S.S. & Hecky, R.E. 1987. Hypothesized resource relationships among African planktonic diatoms. Limnol. Oceanogr., 31: 1169-81.

Komarek, J., Komarkova-Legnerova, J., Sant'Anna, C.L., Azevedo, M.T. & Senna, P.A.C. 2002. Two common *Microcystis* species (*Chroococcales, Cyanobacteria*) from tropical America including *M. panniformis* sp.nov. Cryptogamie, Algologie 23: 159-177.

Lammens, E.H.R.R. 1999. The central role of fish in lake restoration and management. Hydrobiologia. 395/396: 191-198.

Lemoalle, J. 1978. Relations silice-diatomes dans le lac Tchad. Cah. ORSTOM, sér. Hydrobiol., 12: 137-141.

Lewis, W. M. 1978. Dynamics and succession of the phytoplankton in a tropical lake: Lake Lanao, Philippines, J. . Ecol., 66: 849-880.

Lewis, W.M. 1987. Tropical Limnology. Ann. Rev. Ecol. Syst., 18: 159-184.

Lewis, W.M. 1996. Tropical Lakes: How latitude makes a difference. In: F. Schiemer & K. T. Boland, (eds), Perspective in Tropical Limnology, SPB Academic Publishing, Amsterdam, 43-64.

Martin, A.B. 1993. The contribution and relative significance of terrestrial leaf influx to the Carbon and trihalomethane precursor concentrations in drinking water lakes and reservoirs. Ph.D Dissertation, Kent State University, Kent. OH.

Matsumura-Tundisi, T., Tundisi, J.G., Saggio, A., Oliveira Neto, A.L. & Espindola, E. G. 1991. Limnology of Samuel Reservoir (Brazil, Rondonia) in the Filling Phase. Verh. Internat. .Verein. Limnol., 24: 1482-1488.

McQueen, D.J., Johannes, M.R.S., Post, J.R., Stewart, T.J. & Lean, D.R.S. 1989. Bottom-up and top-down impacts on freshwater pelagic community structure. Ecol. Monogr., 59(3): 289-309.

Mukankomeje, R., Plisner, P.D., Descy, J.P. & Massault, L. 1993. Lake Muzahi, Rwanda limnological features and phytoplankton production. Hydrobiologia, 257: 107-120.

Osborne, P.L. 1991. Seasonality in nutrients and phytoplankton production in two shallow lakes: Waigani lake, Papua New Guinea, and Barton Broad, Norfolk, England. Int. Rev. ges. Hydrobiol, 76: 105-12.

Paasche, E. 1980. Silicon. In: I. Morris, (ed.), The Physiological Ecology of Phytoplankton, Oxford: Blackwell, 259-84.

Prawse, G.A. & Talling, J.F. 1958. The seasonal growth and succession of plankton algae in the White Nile. Limnol. Oceanogr., 40: 718-29.

Payne,A.I. 1986. The ecology of tropical lakes and rivers. J. Wiley. Chichester. pp. 301.

Phillips, M. J., Roberts, R. J., Stewart, J. A. & Codd, G.A. 1985. The toxicity of the cyanobacterium *Microcystis aeruginosa* to rainbow trout *Salmo gairdneri* Richardson. J. Fish Diseases 8: 399-344.

Ramberg, L. 1987. Phytoplankton succession in the Sanyati Basin, Lake Kariba. Hydrobiologia, 153: 193-202.

Rabergh, C. M. I., Bylund, G. & Eriksson, J.E. 1991. Histopathological effects of microcystin-LR, a cyclic peptide toxin from the cyanobacterium (blue-green alga) *Microcystis aeruginosa*, on common carp. Aquat. Toxicol., 20: 131-146.

Reynolds, C.V. 1990. The Ecology of Freshwater Phytoplankton. Cambridge University Press. pp. 284.

Reynolds, C., Huszar, V., Kruk, C., Naselli-Flores, L. & Melo, S. 2002. Towards a functional classification of the freshwater phytoplankton. J. Plankton. Res., 24: 417-428.

Romo, S. & van Tongeren, O.R.F. 1995. Multivariate analysis of phytoplankton-related environmental factors in a shallow hypertrpohic lake. Hydrobiologia, 299: 93-101.

Rott, E. 1983. A contribution to the phytoplankton species composition of Parakrama Samudra, an ancient manmade lake in Sri Lanka, In: F. Schiemer, (ed.), Limnology of Parakrama Samudra- Sri Lanka, A case study of an ancient manmade lake in the tropics. Dev. Hydrobiol. 12: 209-226.

Rott, E. & Lenzenweger, R. 1994. Some rare and interesting plankton algae from Sri Lanka. Biologia, Bratislava. 49/4: 497-500.

Rott E., Kling, H. & McGregor, in press. A LM and SEM study of *Urosolenia* including the description of tropical and temperate new taxa. Diatom Research.

Rott, E., Silva, E.I.L., Enriquez, E. & Ingthamjitr, S. (in press). Phytoplankton community-structure (species composition, diversity, chlorophyll, key variables) in five SE Asian water bodies. Dev. Hydrobiol.

Samaradiwakara, S.R.M.S. 2003. Some aspects of limnology of Kandy lake. M.Phil. Thesis, University of Kelaniya.

Schiemer, F. (ed.). 1983. Limnology of Parakrama Samudra- Sri Lanka, A case study of an ancient man-made lake in the tropics. The Hague: Junk. pp. 236.

Schiemer, F., Amarasinghe, U.S., Frouzova, J., Sricharoendham, B. & Silva, E.I.L. 2001. Ecosystem structure and dynamics—A management basis for Asian reservoirs and lakes. S.S. De Silva (ed.), Reservoir and Culture-based Fisheries: Biology and Management. ACIAR Publication No 98, Canberra, Australia, 215-226.

Sharaff, F.F. 2003 Investigation on the state of pollution and the bacterial consortia of surface water in Kandy lake. M.Sc. thesis, University of Kelaniya, 119.

Silva, E.I.L. 1996. Water Quality of Sri Lanka. A review of twelve water bodies, IFS, Kandy, Sri Lanka: pp141.

Silva E.I.L. 1999. Status of surface water quality in Sri Lanka. SCOPE 82: 111-124.

Silva, E.I.L. 2003a. Emergence of a *Microcystis aeruginosa* bloom in an urban water body, Kandy Lake, Sri Lanka. Current Science 85: 723-725.

Silva, E.I.L. 2003b. Quality of Irrigation Water in Sri Lanka—Status and Trends. In: R.P.D. Silva, (ed.), Water Resources Research in Sri Lanka. PGIA, University of Sri Lanka: 35-44.

Silva E.I.L. & Davies, R.W. 1987. The seasonality of monsoonal primary productivity in Sri Lanka. Hydrobiologia, 150: 165-175.

Silva E.I.L & Wijeyaratne, M.J.S. 1999. The Occurrence of Cyanobacteria in the Reservoir of the Mahaweli River Basin in Sri Lanka. Sri Lanka J. Aquat. Sci. 4: 51-60.

Silva, E.I.L. & Gamlath, G.A.R.K. 2000. Catchment Characteristics and Water Quality of three FISHSTRAT Reservoirs in Sri Lanka (Victoria, Minneriya and Udawalawe). Sri Lanka J. Aquat. Sci. 5: 55-73.

Silva, E.I.L. & Schiemer, F. 2001. Human Factor: Fourth Dimension of Reservoir Limnology in the tropics. In: S.S. De Silva, (ed.), Reservoir and Culture based Fisheries: Biology and Management. ACIAR Publication No 98, Canberra, Australia: 111–125.

Silva, E.I.L., Amarasinghe, U.S., De Silva, S.S., Nissanka, C. & Schiemer, F. 2002. Some aspects of photosynthetic characteristics in a set of perennial irrigation reservoirs located in five river basins in Sri Lanka. Hydrobiologia. 485: 19-33.

Silva, E.I.L., Ekanayake, M. & Karunathilake, K.M.B.C. 2003. Seasonal abundance of two species of rotifers (*Brachionus calyciflorus* and *Keratella tropica*) in a tropical urban water body, Kandy lake in Sri Lanka Sri Lanka J. Aquat. Sci., (in press).

Smith, V.H. 1983. Low nitrogen to phosphorus ratios favour dominance by blue-green algae in lake phytoplankton. Science 221: 669-671.

Straškraba & Tundisi 1999. Guidelines of Lake Management Vol. 9. Reservoir Water Quality Management IELC: 229.

Sugunan, V.V. 1980. Seasonal fluctuation of plankton of Nagarjunasagar reservoir A.P., India. J. of Inland fish. Soc., India, 12(1): 79-91.

Talling, J.F. 1965. The photosynthetic activity of phytoplankton in East African lakes. Int. Revue. ges. Hydrobiol. 50: 1-32.

Talling, J.F. 1966. The annual cycle of stratification and phytoplankton growth in Lake Victoria (East Africa). Int. Revue. ges. Hydrobiol. 51: 545-621.

Talling, J.F. 1986. The seasonality of phytoplankton in African lakes. Hydrobiologia, 138: 139-160.

Talling, J.F. 1992. Environmental regulation in African shallow lakes and wetlands. Rev. Hydrobiol. trop. 25: 229-56.

Talling, J.F. & Talling, I.B. 1965. The chemical composition of African lake waters. Int. Rev. ges. Hydrobiol. 50: 421-63.

Talling, J.F., Wood, R.B., Prosser, M.V. & Baxter, R.M. 1973. The upper limit of photosynthetic productivity by phytoplankton: Evidence from Ethiopian soda lakes. Freshwat. Biol. 3: 53-76.

Talling, J.F. & Lemoalle, J. 1998. Ecological Dynamics of Tropical Inland Waters. Cambridge University Press: pp. 441.

Thornton, J.A. (ed.) 1990. Lake McIlawaine, The eutrophication and the recovery of a tropical African manmade lake. Mongr. Biologicae 49. The Hague: Junk.

van der Heide, J. 1973. Plankton development during the first years of inundation of the Van Blomestein (Brokopondo) Reservoir in Suriname, Verh. Internat. .Verein. Limnol., 24: 1171-1173.

van Tongeren, O.F.R., Van Liere, L., Gulatti, R.D., Postema, G. & Boesewinkel-DeBruyn, P.J. 1992. Multivariate analysis of the plankton communities in the Loosdrecht lakes: relationship with the chemical and physical environment. Developments in Hydrobiology. 74 : 105-117.

Varis, O., Sirvio, H. & Kettunen, J. 1989. Multivariate analysis of lake phytoplankton and environmental factors. Archiv für. Hydrobiologie 45: 1-12.

Viner, A.B. 1975. The supply of minerals to tropical rivers and lakes (Uganda). In: A.D. Hasler, (ed.), Coupling of Land and Water Systems, New York: Springer, 227-261.

Viner, A.B. 1977. Relationships of nitrogen and phosphorus to a tropical phytoplankton population. Hydrobiologia, 52: 185-196.

Vitousek, P.M. 1977. The regulation of element concentration in mountain streams in the northeastern United States. Ecological Monographs, 47: 65-87.

Vyvermann, W. 1996. The Indo-Malaysian North-Australian phycogeographical region revised. In: J. Kristiansen, (ed.), .Biogeography of Freshwater Algae, Dev. Hydrobiol., 118: 107-120.

Werner, D. 1977. Silica metabolism. In: D. Werner, (ed.), The Biology of Diatoms, Oxford: Blackwell: 110-149.

Webb, B.W. & Walling, D.E. 1974. Local variation in background water quality. Science of the Total Environment 3: 141-153.

Wetzel, R.G. 1983. Limnology, 2nd Edition, Saunders Coll. Philadelphia pp. 858.

Wood, R.B., Baxter, R.M. and Prosser, M.V. 1984. Seasonal and cooperative thermal stratification in some tropical crater lakes, Ethiopia, Freshwat. Biol., 14: 551-573.

Zafar, A.R. 1986. Seasonality of phytoplankton in some south Indian lakes. Hydrobiologia, 138: 177-187.

Zevenboom, W. & Mur, L.R. 1981. Simultaneous short-term uptake of nitrate and ammonium by *Oscillatoria agardhii* grown in nitrate- or light-limited continuous culture. J. gen. Microbiol., 126: 355-63.

CHAPTER

9

The Importance of Nutrient Input, Invertebrate Predation and Oxygen Deficit Governing the Temporal and Spatial Distribution of Plankton Community in Tropical Reservoirs

Ricardo Motta Pinto-Coelho[1], José Fernandes Bezerra-Neto[2] and Laura Rull del Aguilla[3]

[1]Departamento de Biologia Geral, Instituto de Ciências Biológicas, Universidade Federal de Minas Gerais, Belo Horizonte, Brazil

[2]Programa de Pós-graduação em Ecologia, Conservação e Manejo da Vida Silvestre, PG-ECMVS, UFMG, Belo Horizonte, Minas Gerais, Brazil

[3]Programa de Posgrado em Ecología y Oceanografia, Agência Espanhola de Cooperación Internacional, AECI, Alicante, España.

ABSTRACT

Tropical zooplankton communities have quite different structural properties when compared to temperate counterparts. The aim of this study is to assess the effects of nutrient influx, oxygen deficit and invertebrate predation on the structure of tropical zooplankton. Three reservoirs in of tropical South America were selected as study cases. The effects of long-term eutrophication on zooplankton community were investigated on a small urban reservoir, the Pampulha Reservoir. This system suffered a steady increase in its trophy along the last decade. A data set consisted of monthly samples of zooplankton and major limnological factors such as nutrients N and P taken between the years of 1993 and 1998 were used. The effect of a spatial gradient of trophy was investigated in one of the largest tropical reservoirs of Brazil, the Furnas Reservoir (1440 km^2). A data set consisted of twelve sampling points located in one of the reservoir axis (Sapucaí River) of 220 km that were visited every three months between 1999 and 2000. Finally, the effect of oxygen deficit and invertebrate predation (*Chaoborus brasiliensis*) on the structure of crustacean zooplankton were investigated in a small, urban but protected mesotrophic artificial lake. The lake was located inside a dense *Eucaliptus* forest. A data set of a complete annual cycle (1999-2000) was used in the analysis. Zooplankton suffered similar modifications in its structure, considering both the temporal and spatial scale. The influx of nutrient caused the disappearance of calanoids, and biomass increase of some cyclopoids (specially of *Thermocyclops*). Further, it was also demonstrated that oxygen and invertebrate predation are key factors in structuring the vertical distribution of zooplankton and the water column of a tropical lake.

A GLOBAL PROBLEM: THE EUTROPHICATION OF TROPICAL ECOSYSTEMS

Freshwater of good quality is an increasingly limiting resource in the world. This holds specially for the "tropical world" (Tundisi, 2003). Most lakes, rivers and wetlands of low latitude areas are suffering with the input of inorganic sediments and limiting nutrients such as nitrogen and phosphorus. Also, there are many studies demonstrating the fact that the inadequate management of fauna resources with overexploitation in some cases and/or introduction of exotic species, in other cases, are also causing a rapid depletion of water quality in these systems (Pinto-Coelho 1994; Wetzel & Gopal, 2001; Henry, 2003).

In several tropical countries, the rapid population increase also threatens the water resources by inappropriate disposal of domestic garbage. The human occupation of these areas is also related to the contamination of lakes and reservoirs by untreated wastewater. Additionally, polluted effluents from mining areas, oil refineries and other industries, with special emphasis on the food processing facilities, are also present in large areas of recently industrialized tropical countries. Primitive techniques of farm management based on forest clearance followed by ignition of the remaining biomass also cause the influx of inorganic sediments and nutrients into natural freshwater tropical ecosystems. The recent and intense human development of many tropical areas may be the cause of the rapid eutrophication of warm lakes and reservoirs around the world. As elsewhere, eutrophication in tropical areas is reflected by increased primary production (phytoplankton and macrophytes), decrease of water transparency and the appearance of a stable oxygen deficit in the hypolimnion (Henry et al., 1989; Thomaz & Bini, 2003). However, the eutrophication process in the tropics has many specific characteristics that have led to some important and still open questions: (1) Does eutrophication cause similar responses at the community level if a larger geographic area is considered? (2) Is there a group of organisms (i.e. from plankton community) that is always affected in a similar manner by eutrophication? (3) Is the rate of eutrophication higher in tropical systems for a given external load of nutrients? (4) Is the control of eutrophcation more difficult in tropical areas?

The magnitude of biological responses to the nutrient load and the impact of fish as well invertebrate predation on the structure of plankton community also needs to be better understood in the tropics. This can only be achieved by studies comparing the effects of eutrophication on larger scales of time and space. There is also demand for more generalist approaches such as the investigations concerned to quantify the mutual relationships between Limnetic and littoral communities or the effects of riparian vegetation on the structure of aquatic communities with special reference to the plankton.

IMPACT OF EUTROPHICATION ON TROPICAL ZOOPLANKTON

Zooplankton is known to have a high metabolic rate and thus this community must react fast to environmental signals. Zooplanktons have long been used as an indicator of lake trophy (Gannon & Stemberger, 1978). Nevertheless, the responses of zooplankton to the nutrient load are often contradictory (Ravera, 1996). Despite these contradictory responses, it is widely accepted that zooplankton is a key community in the development of new techniques aiming towards quality water recovery in freshwater ecosystems (Carpenter & Kitchell, 1992).

The objective of this contribution is to assess the responses of zooplankton community structure to external nutrient load, oxygen deficit and invertebrate predation at three different ecological scales: (a) long-term (1992-1998) changes to eutrophication (Pampulha Reservoir); (b) spatial (horizontal) responses to nutrient increase (Furnas Reservoir); and (c) this investigation will also deal with the effects of oxygen deficit and invertebrate predation on vertical structure of zooplankton (Nado Reservoir).

SELECTED STUDY CASES

Study Area 1 (Pampulha Reservoir)

Pampulha reservoir is located in the northern part of Belo Horizonte city (Fig. 9.1), the capital of Minas Gerais State. This reservoir has a volume of 11×10^6 m^3 and a surface are of 2.1 km^2. The reservoir was built in 1938 for water supply but in the early eighties, the water plant had to be closed due to the incidence of cyanobacteria blooms in the lake. Since then, the reservoir has exhibited the "classical" signs of eutrophication such as outbreaks of the macrophyte *Eichhornia crassipes* (1992-1995) and successive blooms of *Microcystis aeruginosa* (1980-1998). The steady increase of trophy observed during the last decade in Pampulha Reservoir and the long-term monitoring program covering not only the basic limnological factors such as physical and chemical parameters but also a detailed inventory of plankton and macrophyte communities in this lake make it possible to assess the long-term effects of eutrophication on the ecosystem.

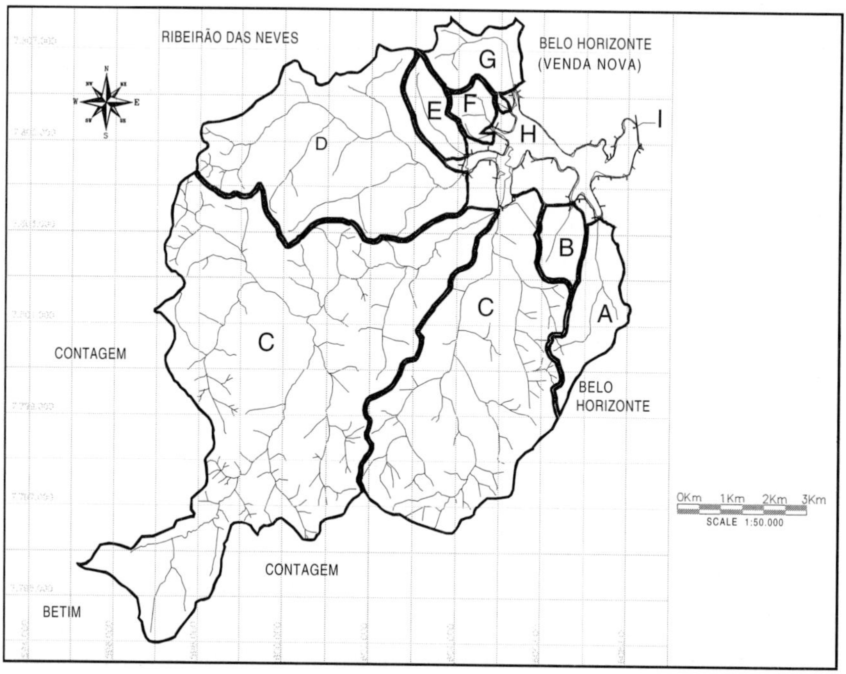

Fig. 9.1 Map of Pampulha Reservoir (H), its hydrographic basin and the major tributaries: (A) Mergulhão stream, (B) Tijuco stream, (C) Ressaca/Sarandi streams, (D) Água Funda stream, (E) Baraúna stream, (F) AABB stream, (G) Olhos D'Água stream and the outlet (I)

Study Area 2 (Furnas Reservoir)

Furnas Reservoir (20°40'S, 46°19'W), the largest reservoir of southeastern Brazil (Fig. 9.2), is located in the upper Rio Grande in the state of Minas Gerais. It has a surface of 1450 km^2. The lake can be divided in two axis of ca. 250 km each, covering the basins of Rio Grande and Rio Sapucaí. The maximum depth is 90 m and the mean depth is 13 m (Pinto-Coelho and Corgosinho, 1998).

Figure 9.2 shows a Landsat image (channels 3, 4 and 5) of the reservoir obtained in August 1999 after processing and classifiying the available data. This image was geo-referenced using a IBGE carthografic map (1 : 50.000). A number of 20 control points were used for this procedure. A final RMS error of just 28.3 mts was obtained. The sampling stations were plotted using their geographic coordinates obtained with a 12-channel Garmin GPS device available during the sampling excursions. The software used was IDRISI *for Windows 3.2* (Clark University).

Most regions of the Furnas Reservoir are still oligotrophic. However, a clear process of eutrophication is taking place on the distal region of the Sapucaí River subaxis as a result of the rapid economic development of its upper basin. Larges areas along the upper Sapucaí River are covered by coffee, sugarcane and orange plantations. Besides, some nearby cities like Alfenas, Varginha, Pouso Alegre and Machado are experiencing a rapid increase due to the recent establishment of several medium and small industries as well as some service providing enterprises. The eutrophic gradient of and its long residence time (around one year) of the Furnas Reservoir makes this environment suitable for describing the effects of eutrophication on the structure of zooplankton in a tropical lake.

Study Area 3 (Nado Reservoir)

The Nado Reservoir is another small urban water body located in the northern part of Belo Horizonte city (19°49'56"S, 43°57'34"W), Minas Gerais state. This water body is surrounded by an artificial forest of old and tall *Eucaliptus* trees (Fig. 9.3). It is a small mesoeutrophic reservoir with a total area of 15 ha and maximum depth of 7.6 m. Due to the surrounding dense vegetation, the reservoir exhibits a remarkable thermal stratification during most of the year. The epilimnion is well oxygenated and the hypolimnion has anoxic conditions during the period. There is a short period of circulation in the dry season (June-August) wherein, oxygen reaches the bottom of the lake (Bezerra-Neto & Pinto-Coelho, 2002). Furthermore, there is a large population of the

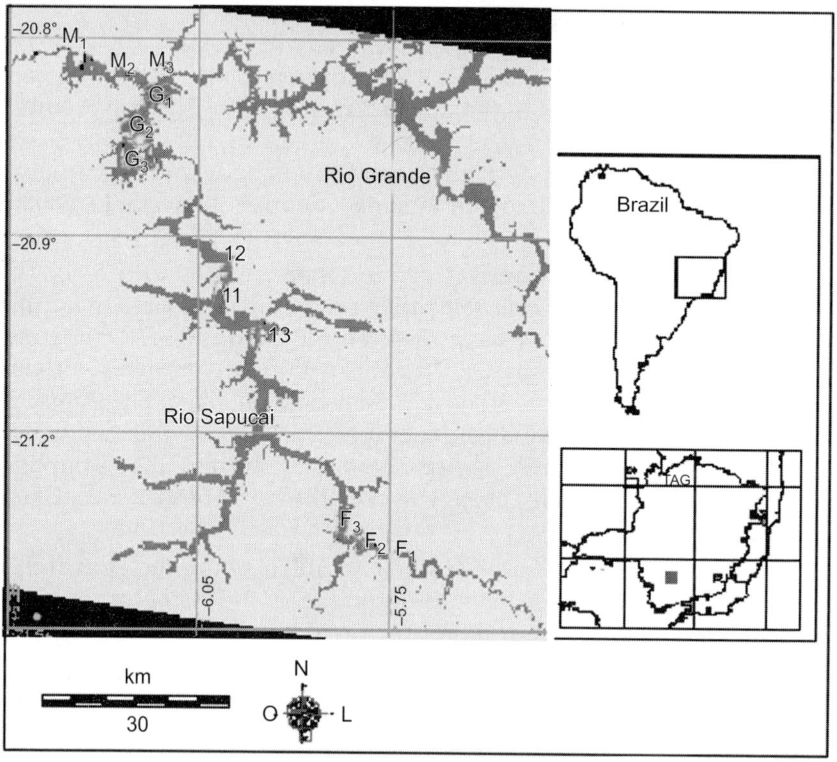

Fig. 9.2 Map of Furnas Reservoir. The four selected sampling areas are t: M-points (M1, M2 and M3); oligotrophic region ; G-points (G1, G2, G3): mesotrophic ; I- (I1, I2, I3); and F-points (F1, F2, and F3): eutrophic areas. At the right, two charts showing the position of Minas Gerais sate in relation to South America (top) and the position of Furnas Lakes in the state of Minas Gerais (bottom).

diptera *Chaoborus brasiliensis* in this lake that undergoes a clear seasonal cycle (Bezerra-Neto, 2001). Due to these conditions, this is a suitable reservoir for investigating the effects of oxygen and invertebrate predation on the structure of zooplankton.

SAMPLING STRATEGIES

Pampulha Reservoir

Water samples for chemical analysis and plankton organisms were collected at a central station (z = 6-8 m) in the Pampulla Reservoir, which has been regularly monitored for more than 15 years. Different depths covering the water column were considered: 0, 2, 4 and 6 m. The

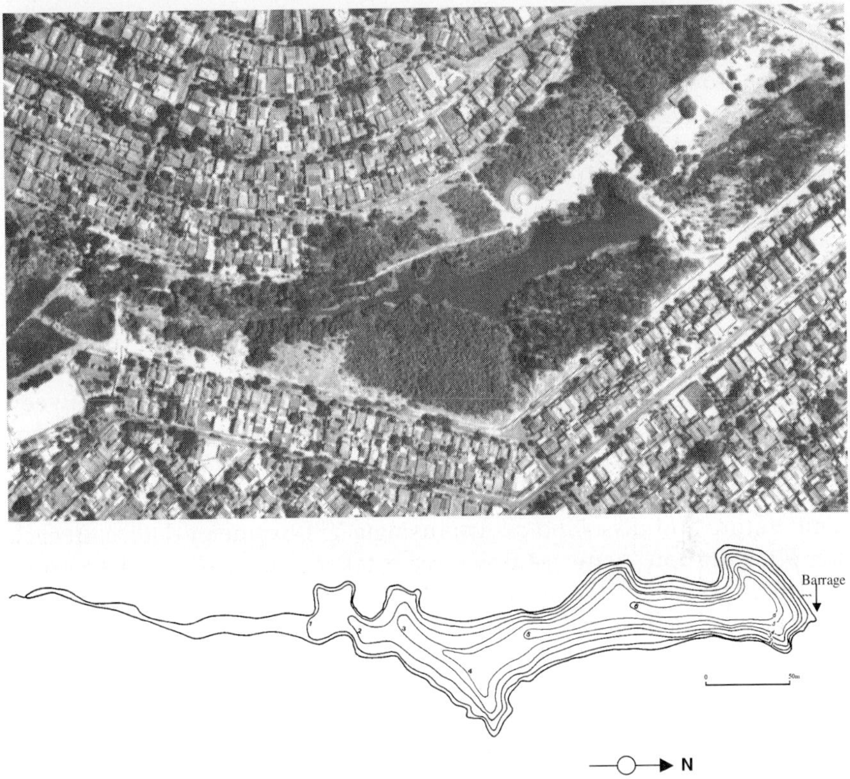

Fig. 9.3 Airborne photography of northern part of Belo Horizonte city taken in August 1989. In the central region is Lado Reservoir and the surrounding park (courtesy: PRODABEL S/A). Bottom: depth curves of Nado Reservoir obtained with a bathimetric sonde

sampling was done on a monthly basis between 1993 and 1995 and every two weeks from 1996 to 1998. Zooplankton was collected utilizing a Schindler-Patalas trap of 5.1 l of capacity, equipped with a collecting net of 90 μm mesh size. In this contribution, only zooplankton data from 2.0 m were used in analyzes. Water transparency and electric conductivity were measured in situ by means of a Secchi disk of 35 cm of diameter and a Yellow Springs device, respectively.

Water samples for chemical analysis were deep frozen (-25 °C) right after collection. The determination of inorganic nitrogen (ammonium) was done according to Mackereth et al. (1978). Total phosphorus and Chlorophyll-*a* were determined using the procedure described by Murphy & Riley (1962) and Lorenzen (1967), respectively. POC was estimated using the chemical digestion method of oxidation using $K_2Cr_2O_7$, considering a new procedure improved by Torres et al. (1998).

Zooplankton was fixed using a buffered formalin solution (pH = 7.0), saturated with sucrose. A minimum of 700 individuals were enumerated and measured in each sample using a M3C Leica dissecting stereomicroscope. The biomass estimation was done using the equations relating length to dry weight provided by Bottrell et al., (1976); Rosen (1981), Malley et al., (1989); Culver et al., (1985) and Masundire (1994). Early evaluations of the allometric coefficients for the dominant Pampulha zooplankters showed a significant and positive relationship between allometric and gravimetric determinations of biomass.

Furnas Reservoir

Twelve sampling stations were visited every three months between August 1999 and July 2000. These points were grouped in four different classes according to their geographic regions (Fig. 9.1).

A depth profile was taken for the determinations of water temperature and dissolved oxygen using a YSI oxymeter device. In each sampling station, water samples were taken with a 2-liter Van Dorn Hydrobios bottle at the surface and at the bottom. The water samples were immediately stored in a thermos containing ice cubes. The samples were taken into the lab within 2-4 hs after collection and deep frozen (–25°C) until they were processed for the determination of nitrate, nitrite and total phosphorus. The nutrient analysis was conducted using conventional colorimetric methods at the Hydrobiological station of the Furnas Reservoir.

Zooplankton samples were collected with a conical net (90 μm mesh size, 40 cm diameter) covering the entire water column. The filtrate was transferred into a polyethylene vial of 250 ml, diluted to 200 ml with lake-filtered cold water (4°C). A few drops of the dye Bengal Rose were added and each sample was allowed to rest in a box covered with ice some 5-10 minutes prior to fixation. A buffered (pH = 7.0) solution of Sucrose-Formaline (250 g sucrose to 1.0 liter of formaline 37%) was added to achieve a final concentration of 4% of formaline.

Prior to counting, the samples were examined on a Leica WilD M3C stereo-microscope for identification of the organisms. The following literature was used: Smith & Fernando, 1980; Matsumura-Tundisi & Rocha, 1983; Van de Velde, 1984; Reid, 1985; Matsumura-Tundisi, 1986; Reid, 1987; Elmoor-Loureiro, 1990; Sendacz & Kubo, 1992; Elmoor-Loureiro, 1997. The counting and measuring of zooplankton was carried out using a Sedgewick-Rafter counting chamber on a Leica DMLB microscope equipped with a CCD Sony camera. The subsampling was done using a 1 ml Hensen pipette. A minimum of 700 individuals was always enumerated. The determination of biomass was done using length × weight regressions provided by McCauley (1984).

Nado Reservoir

During each sampling period, the water temperature and the dissolved oxygen were measured through vertical profiles with 50 cm intervals in the central region of the reservoir utilizing a YSI model 55 oxymeter - thermistor.

To investigate the effect of oxygen in the vertical positions of zooplankton, we estimated the oxygen deficit in the reservoir was estimated. The oxygen actual deficit (OAD) was calculated by the difference between the real oxygen concentration in a given depth and the saturated concentration for the prevailing water temperature in the same conditions of pressure (Wetzel, 1983). All calculations used in the present study can be found in Henry et al. (1989).

Zooplankton was collected every month between October 1999 and September 2000 at a central sampling point (z = 6.5 m). The samples were taken every meter, covering the whole column. All samplings included a daylight period set of samples (09:00 and 11:00 hs a.m) and a night period samples (22:00 and 24:00 hs p.m.). A large (17 liter) modified Patalas trap was used. The trap was constructed using a box made of pexiglas inside a frame of thin steel rigid tubes. This design provided a good transparency of the device. A previous comparative study (Bezerra-Neto and Pinto-Coelho, submitted) demonstrated the higher efficiency of this trap for collecting all instars of *Chaoborus* and other large zooplankters. Chemical fixation with formaline sucrose (4% final concentration of formaline) was done immediately after sampling.

At the laboratory, every single *Chaoborus* individual was counted and the total length and well as the length of the cephalic carapace were measured. The determination of the instar was performed according to the head length. The measurements were done using a Leica DM3 stereomicroscope equipped with a SONY-CCD video camera. A Scion Image software (Scion Corporation) of digital image analysis was used to obtain the measurements.

The mean weighted depth of each instar population for day and night hours was calculated according to the formula (Dini et al. 1993):

$$\text{Mean depth} = \frac{(\Sigma XiZi)}{(\Sigma Xi)}$$

Where Xi is the i-instar density and Zi is the depth of the i-sample.

Statistical (Systat 8.0) and graphic (Sigma Plot 6.0) packages were used for statistical analysis and graphics, respectively.

LONG-TERM RESPONSES OF ZOOPLANKTON TO EUTROPHICATION

The water quality of Pampulha Reservoir reflected an increase of eutrophication during the years 1993-1998 (Fig. 9.4). Despite seasonal variation, a clear long-term pattern of nutrient increase can be observed in most variables such as total phosphorus, conductivity and particulate carbon. The bottom values (6 m) of total phosphorus, ammonium and conductivity (Fig. 9.4) always increased their maximum annual values in the study period. The mean annual value total phosphorus concentrations passed from 53.8 to 234.0 $\mu g.P\text{-}PO_4.l^{-1}$ in the study period. Like TP, electric conductivity values also suffered a major increase and the mean annual values passed from 267 $\mu S.cm^{-1}$ to 317 $\mu S.cm^{-1}$ (Fig. 9.4).

The amount of sestonic carbon as well as chlorophyll-*a* also exhibited clear trends of long-term increase during the study period (Fig. 9.4). After the complete macrophyte removal in 1995, there was a pulse in the sestonic POC concentrations, reaching a maximum of 7.8 $mgC.l^{-1}$. In 1996, the POC concentrations were somewhat lower (annual maximum of 7.0 $mgC.l^{-1}$) but this variable again showed an increase in 1997 and 1998 with annual maxima of 12.1 and 15.0 $mgC.l^{-1}$, respectively.

The chemical composition of the reservoir water column has a typical seasonal cycle with higher nutrients values being found during the dry season, especially between July and October (Fig. 9.4). During the rainy season, which usually begins in November, a sharp decrease in the concentrations of conductivity and inorganic nutrients was observed in most years.

The temporal evolution of crustacean zooplankton showed a complex pattern of long-term succession characterized first by the disappearance of calanoids in late 1995, followed by a large development of cladoceran biomass in 1995 and 1996, especially daphnids (Fig. 9.5). The third and last event covered in this study refers to the dominance of cyclopoids (*Thermocyclops decipiens*) from 1997 and during the whole year of 1998.

Cladocerans were the dominant zooplankters until late 1996. Between 1993 and 1995, this taxocenosis consisted of two species of *Daphnia* (*D. gessneri* and *D. laevis*), *Diaphanosoma birgei* and the small cladocerans *Bosmina longirostris, B. hagmanni* and *Ceriodaphnia cornuta*. A significant regression was found between the size of cladocerans and the total phosphorus (Fig. 9.6). This regression probably reflects the disappearance of these small cladocerans between 1993 and 1995.

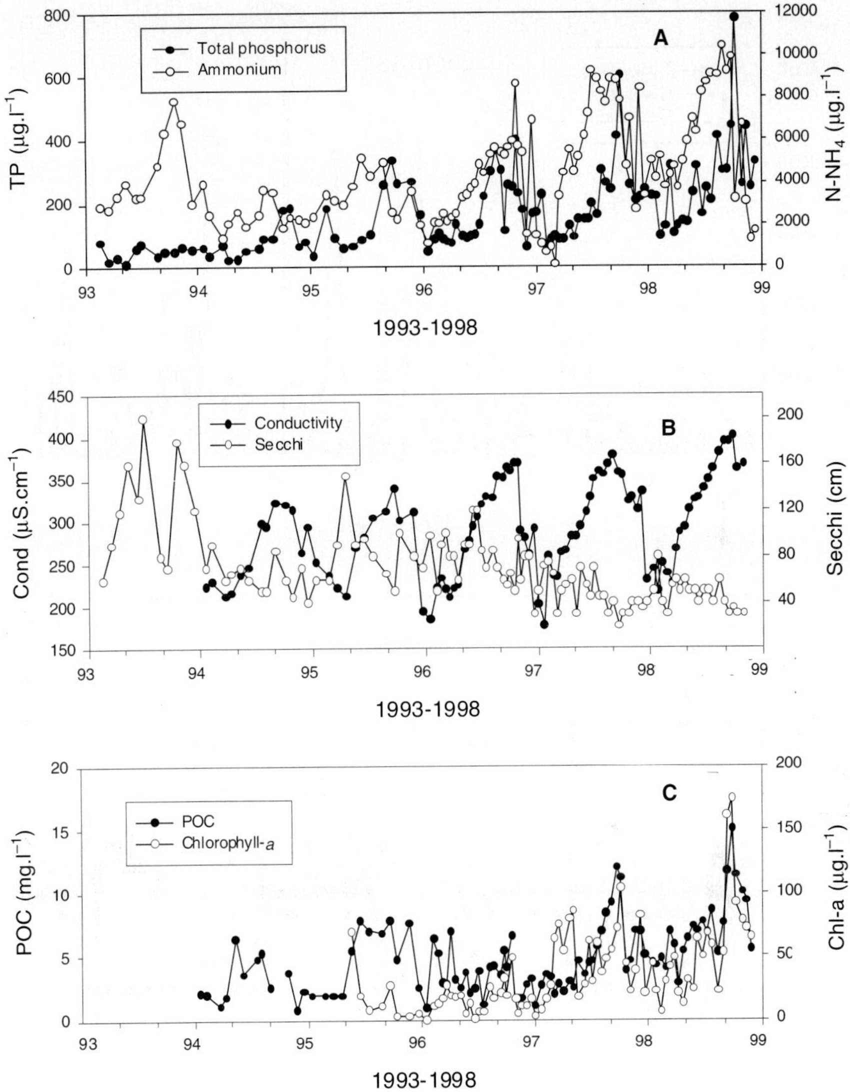

Fig. 9.4 Long-term (1993-1998) evolution of eutrophication in Pampulha Reservoir. Top: Total phosphorus (bottom); middle: conductivity (bottom) and bottom: particulate organic carbon, POC (at 0.5 m)

After the removal of macrophyte beds in 1995, *D. laevis* alone practically dominated the cladoceran community showing several population picks in 1995 and 1996. During this time, the biomass of

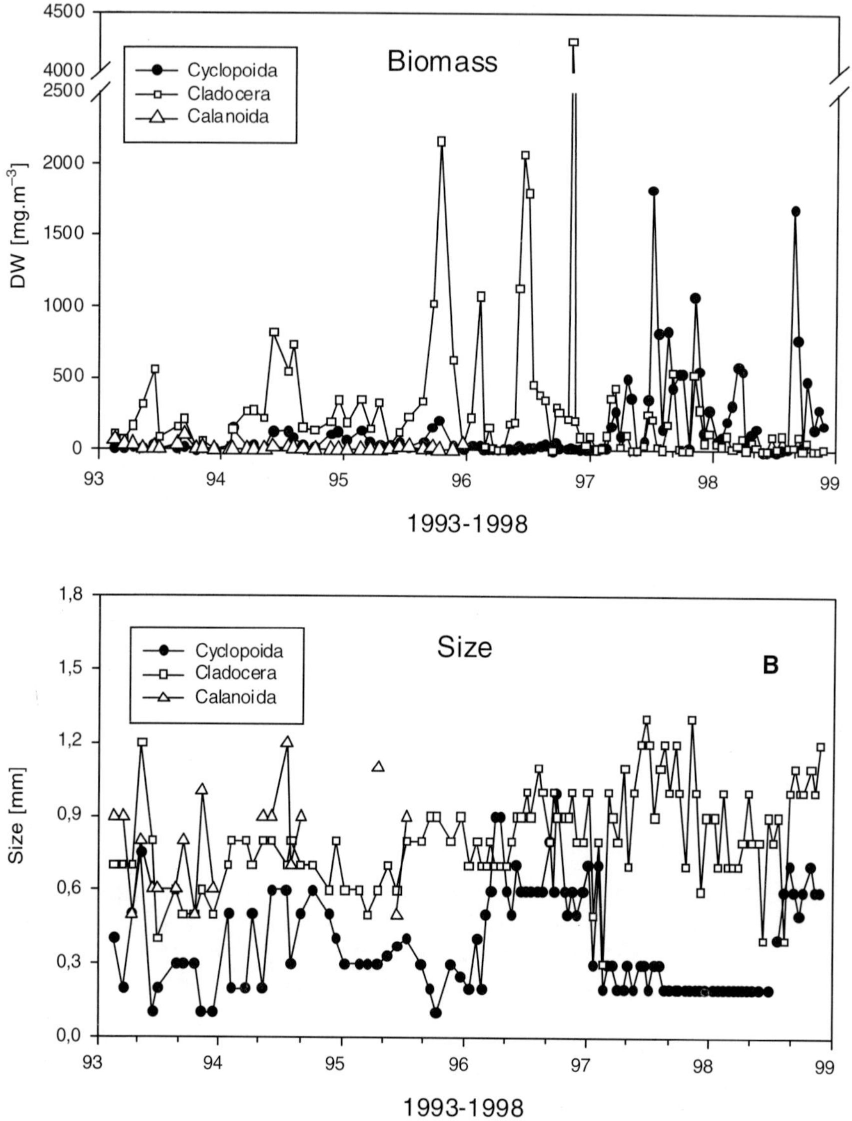

Fig. 9.5 Temporal evolution of the biomass (top) and size (bottom) of crustacean zooplankton at Pampulha Reservoir (1993-1998). Data refer to the 0.5 m samples

cladocerans reached the maximum of 4263 mgDW.m^{-3} (Table 9.3). After the disappearance of *Daphnia* in 1997, the cladoceran group was dominated by *Moina micrura*. In the 1997-1998 period, the maximum biomass reached by cladocerans was just 545 mgDW.m^{-3} (Fig. 9.5).

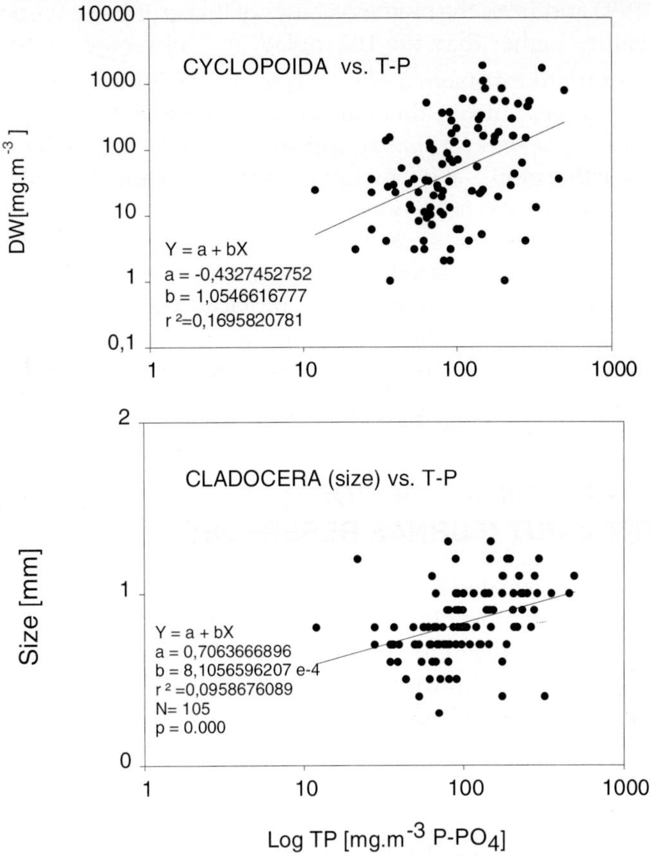

Fig. 9.6 Linear regressions between biomass of cyclopoida and total phosphorus and size of clacocerans and total phosphorus. Data of zooplankton refer to the 0.5 m samples and data of total phosphorus refers to the bottom samples

Cyclopoids in Pampulha showed distinct temporal patterns of biomass (Fig. 9.5). The biomass pattern of cyclopoids can be described by two phases: one period of lower biomass (1993-1996) and a second period with higher biomass (1997-1998). The maximum biomass during the first period did not surpass 204 mgDW.m^{-3}. The second phase occurred after the end of dominance period of daphnids in late 1996, when the biomass of cyclopoids responded with a vigorous increase. Values as high as 1829 mgDW.m^{-3} (Fig. 9.2-B, Table 9.3) were found in this *Daphnia*-free period. A significant linear regression was found between the biomass of cyclopids and total phosphorus (Fig. 9.6)

A single species of calanoid, *Scolodiaptomus corderoi* was present in the reservoir. This calanoid never built up large populations but when present, they were the largest organisms in the samples (Fig. 9.5).

Between 1993 and 1996, the biomass values of this group of copepods never reached values higher than the 102 $mgDW.m^{-3}$, observed in 1993.

Mesozooplankton (sensu Sieburth et al., 1978) have also suffered important changes during this period. The calanoid *S. corderoi* and the cladoceran *Ceriodaphnia cornuta* disappeared from the lake in the middle of nineties. Furthermore, the eutrophication of Pampulha affected the temporal dynamics of most species of this community (Pinto-Coelho, 1998). Some large cladocerans, such as the dominant *Daphnia laevis* and *D. gessneri*, have also disappeared from the lake from 1997 up to the present. The dominance of this organism was replaced by the small cyclopoid *Thermocyclops decipiens*. The disappearance of daphnids is probably associated with a sharp increase of an exotic fish population, the African cichlid *Oreochromis niloticus*, which occurred during this period in the reservoir (Álvares et al. 2000).

SPATIAL RESPONSES OF ZOOPLANKTON TO EXTERNAL NUTRIENT INPUT (FURNAS RESERVOIR)

Furnas reservoir revealed a clear trend for lower transparency and higher concentrations of Chlorophyll-a in the eutrophic regions, F1, F2 and F3 (Fig. 9.7). The concentrations of nitrate (Fig. 9.8) and total phosphorus (Fig. 9.9) also reflected in trophy gradient), especially during the months of dry season (August, October and July). The non-parametric statistics (Table 9.1) showed that chlorophyll, Secchi disk and ammonium reflected well the spatial gradient in the trophy.

Twenty crustacean species were identified in the (meso)zooplankton community of Furnas Reservoir. Twelve species belonged to the order Cladocera and eight to the Copepoda group. Cladocerans were always present in all sampling stations, excepting *Moina minuta* and *Bosminopsis deitersi* that were only observed in more eutrophic areas (I1, I2, I3 and F1, F2 and F3). They reached the highest figures of biomass (32.85 $\mu gPS.l^{-1}$ for *M. minuta* and 26.62 $\mu gPS.l^{-1}$ for *B. deitersi*) in the eutrophic regions.

Four species of calanoids were found in the reservoir with the highest biomass always being found in the oligotrophic regions (M points region). *Notodiaptomus henseni* was the most important species, reaching a biomass maximum of 5.47 $\mu gPS.l^{-1}$. Other calanoid species such as *N. cearensis* and *N. henseni* were not even collected in the eutrophic part of Furnas.

Cyclopoids were represented by the following species: *Metacyclops mendocinus*, *Mesocyclops ogunnus*, *T. decipiens* and *T. minutus*. The biomass proportion of this crustacean group increased gradually from

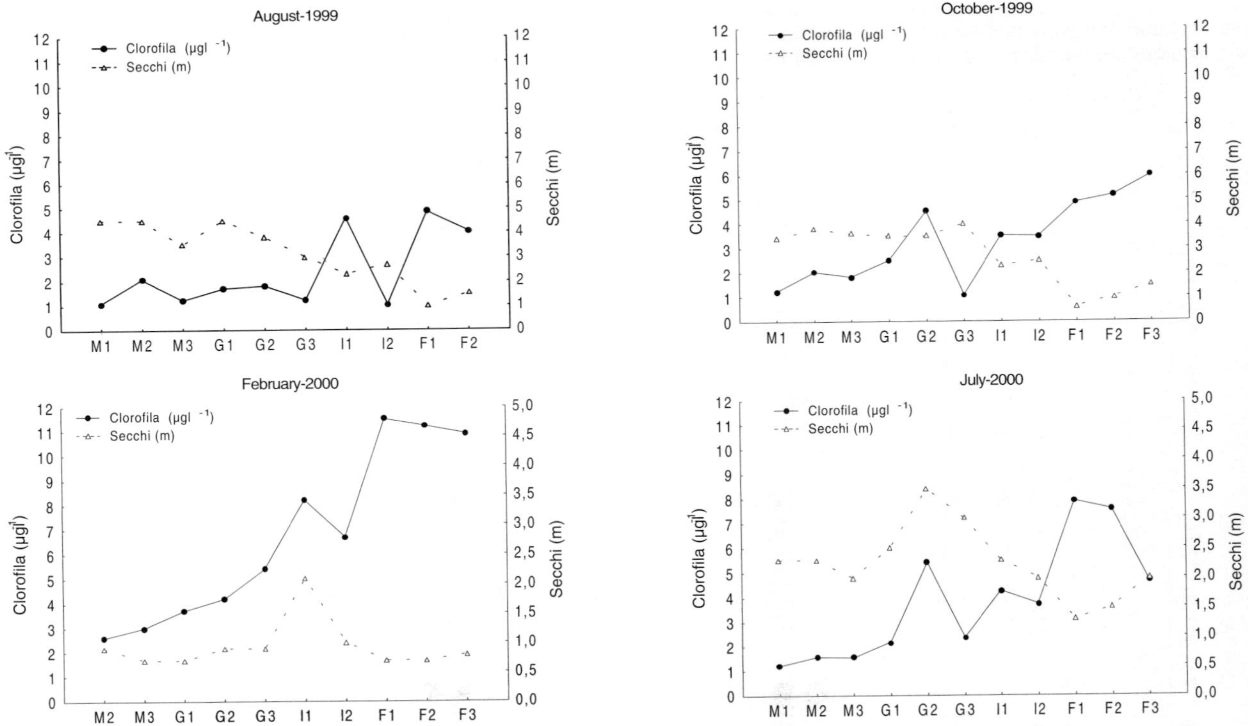

Fig. 9.7 Concentrations of chlorophyll-a ($\mu g.l^{-1}$) in samples taken at 0.5 m and Secchi disk readings in Furnas Reservoir in four different occasions covering the annual cycle of 1999-2000. (M-points: oligotrophic region, G-points: mesotrophic region and I- and F-points : eutrophic region. Charts refer to the following sampling times: (a) August 1999 (top, left); (b) October 1999 (top, right); (c) February 2000 (bottom, left); and (d) July 2000 (bottom right)

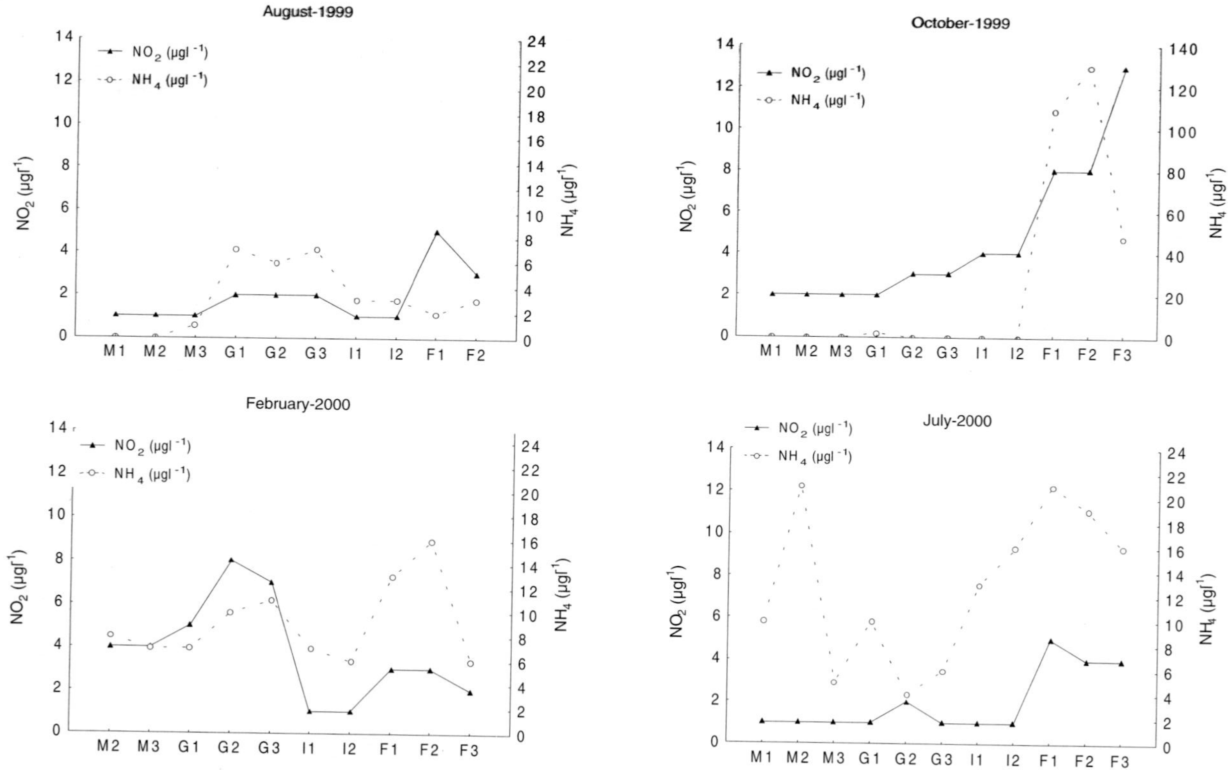

Fig. 9.8 Concentrations of nitrate NO_3 (μgl^{-1}) and ammonium NH_4 (μgl^{-1}) in Furnas Reservoir. (M-points: oligotrophic region, G-points: mesotrophic region and I- and F-points: eutrophic region. Charts refer to the following sampling times: (a) August 1999 (top, left); (b) October, 1999 (top, right); (c) February 2000 (bottom, left); and (d) July 2000 (bottom right)

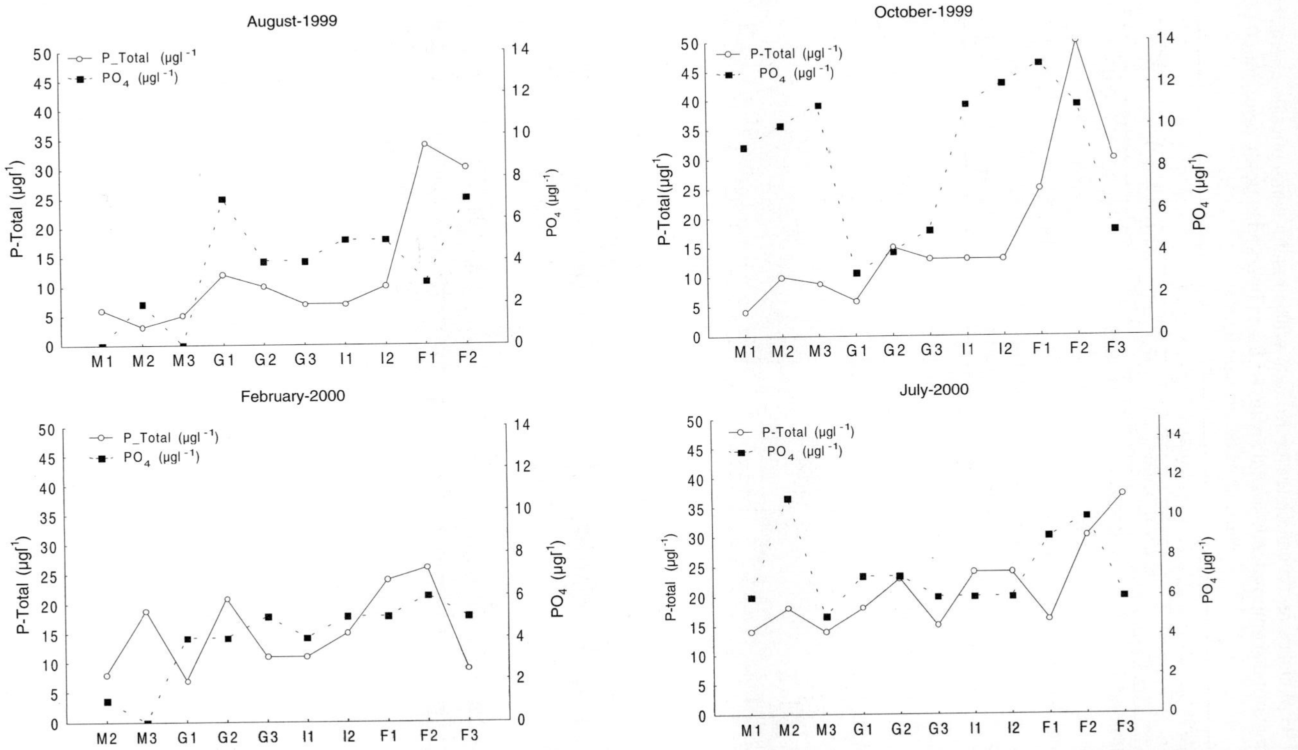

Fig. 9.9 Concentrations of soluble orthophosphate PO_4 ($\mu g l^{-1}$) and total phosphorus ($\mu g l^{-1}$) in Furnas Reservoir in four different occasions covering the annual cycle of 1999-2000. (M- points: oligotrophic region, G-points: mesotrophic region and I- and F-points: eutrophic region. Charts refer to the following sampling times: (a) August 1999 (top, left); (b) October 1999 (top, right); (c) February 2000 (bottom, left); and (d) July 2000 (bottom right)

Table 9.1 H statistics and p values for the Kruskal-Wallis test for some physico-chemical variables testing the temporal and spatial scale. P values (marked in bold) lower than 0.05 were considered significant.

Variables	*Spatial*		*Temporal*	
	H (3; n = 42)	*P*	*H (3; n = 42)*	*P*
Condutividade	6.69024	0.0825	22.35620	0.0001
Clorofila-a	22.87510	0.0000	11.57560	0.0090
Disco de Secchi	13.8760	0.0031	19.53040	0.0002
$N\text{-}NO_3$	2.824495	0.4194	18.59321	0.0003
$N\text{-}NH_4$	10.6889	0.0135	13.53040	0.0043

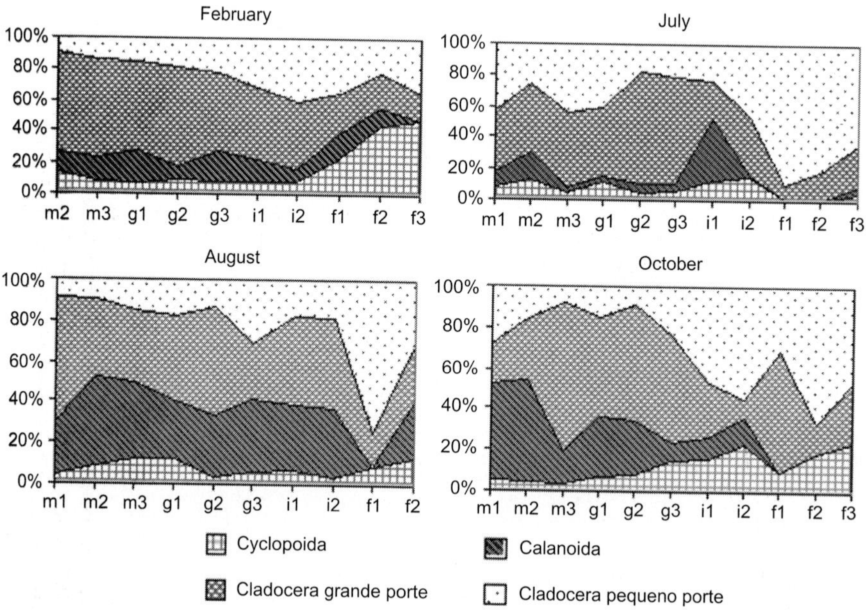

Fig. 9.10 Relative biomass distribution of major zooplankton groups in Furnas Reservoir. (M- points: oligotrophic region, G-points: mesotrophic region and I- and F-points: eutrophic region Charts refer to different sampling times: (d) February 2000 (top, left); (b) July 2000 (top, right); (c) August, 1999 (bottom, left); and (d) October, 1999 (bottom, right)

oligotrophic to eutrophic regions (Fig. 9.10). The cyclopoid *Metacyclops mendocinus* was only found in the eutrophic region. *T. decipiens* was the most important species in terms of biomass reaching 5.8 $\mu gPS.l^{-1}$, i.e. 5.5% of the total zooplankton in the distal and eutrophic regions (F points).

Large cladocerans such as *Daphnia guessneri, Diaphanosoma fluviatille, D. spinulosum, D. brevireme, M. minuta* and *M. micrura* appeared with higher biomass in the oligotrophic regions (M points and G points). In the eutrophic regions, smaller cladocerans such as *B. deitersi, B. longirostris, B. hagmanni, C. cornuta, C. rigaudi, Ceriodaphnia sp.* were more abundant. *B. longirostris* and *B. deitersi* reached 23.1 % and 28.8 % of the total biomass in these eutrophic regions.

Zooplankton community also suffered modifications during the seasonal cycle. Cyclopoida decreased their biomass in the dry season (June and August), while Calanoids exhibited an inverted trend. A non-parametric test was run to test the effects of seasonality (each month was considered a different treatment) on the biomass distribution of different zooplankton groups. The statistics H (and the p values) demonstrated that the seasonal effect 44 was not so important as it was shown for spatial effects (Tables 9.2 and 9.3). Only cyclopoid nauplii showed significant differences (Table 9.3) when the four different periods were compared.

Table 9.2 H statistics and p values for the Kruskal-Wallis test for biomass of major zooplankton groups testing the spatial scale. P values (marked in bold) lower than 0.05 were considered significant. Cyclopoida—sum of biomass of copepodites and adults, Calanoida—sum of biomass of copepodites and adults; large Cladocera—organisms of the genus: *Daphnia, Diaphanosoma* and *Moina*, small Cladocera—organisms of the genus, *Bosmina, Bosminopsis* and *Ceriodaphnia*.

Group	*H (3; n = 42)*	*p*
Cyclopoida	13.9300	0.0030
Calanoida	9.125594	0.0277
Nauplii (Cyclopoida)	8.204491	0.0420
Nauplii (Calanoida)	10.58182	0.0142
Large Cladocerans	6.306712	0.0976
Small Cladocerans	24.99699	0.0000

Table 9.3 H statistics and p values for the Kruskal-Wallis test for of major zooplankton groups testing the temporal scale. P values (marked in bold) lower than 0.05 were considered significant. Cyclopoida: sum of biomass of copepodites and adults, Calanoida: sum of biomass of copepodites and adults, large Cladocera: organisms of the genus: *Daphnia, Diaphanosoma* and *Moina,* small Cladocera: organisms of the genus: *Bosmina, Bosminopsis* and *Ceriodaphnia*.

Group	*H (3; n = 42)*	*p*
Cyclopoida	7.30897	0.0627
Calanoida	5.082243	0.1659
Nauplii (Cyclopoida)	17.05925	**0.0007**
Nauplii (Calanoida)	6.670774	0.0832
Large Cladocerans	2.574504	0.4620
Small Cladocerans	3.352458	0.3404

Canonical analysis

The eutrophic gradient in the reservoir was reflected in the CCA multivariate analysis (Figs. 9.11 and 9.12). The F-stations formed a distinct group in the positive region of axis 1 (Fig. 9.11). The oligotrophic region near the dam (M-points) points are all plotted in the negative part of axis 1. It is interesting to note that the intermediary regions such as G or I appeared between these two extremes.

The biplot diagram of the biotic and abiotic explicative variables (Fig. 9.12) justify the stations' distribution illustrated above (Fig. 9.11). There is a clear relation between the observed spatial patterns and the eutrophication. The biotic explicative variables were: clorophyll-a, biomass of calanoids (adult + copepodites and nauplii), cyclopiods (adult + copepodites and nauplii), large cladocerans and small cladocerans biomass of zooplankton. Cyclopoids including their nauplii stages and cladocerans were plotted in the positive region of axis 1. All vectors of environmental factors which are also good descriptors of eutrophic conditions such as chlorophyll-a, ammonium, nitrite, nitrate and P-total are pointed to this region. The calanoids were plotted isolated in the negative region of axis 1 together with the vector of Secchi disk.

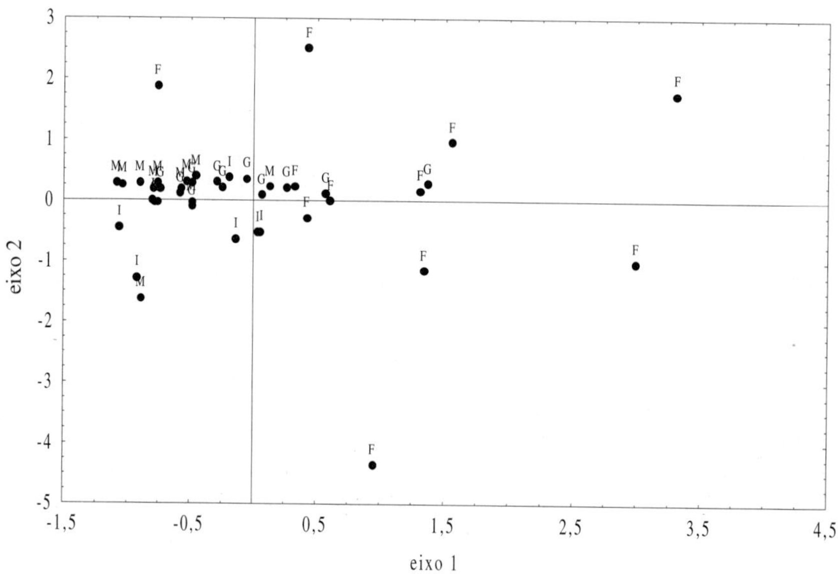

Fig. 9.11 CCA biplot diagram containing the spatial ordenation of sampling points. (M-points: oligotrophic region, G-points: mesotrophic region and I- and F-points: eutrophic Only axis 1 (horizontal) was significant (p = 0,0006)

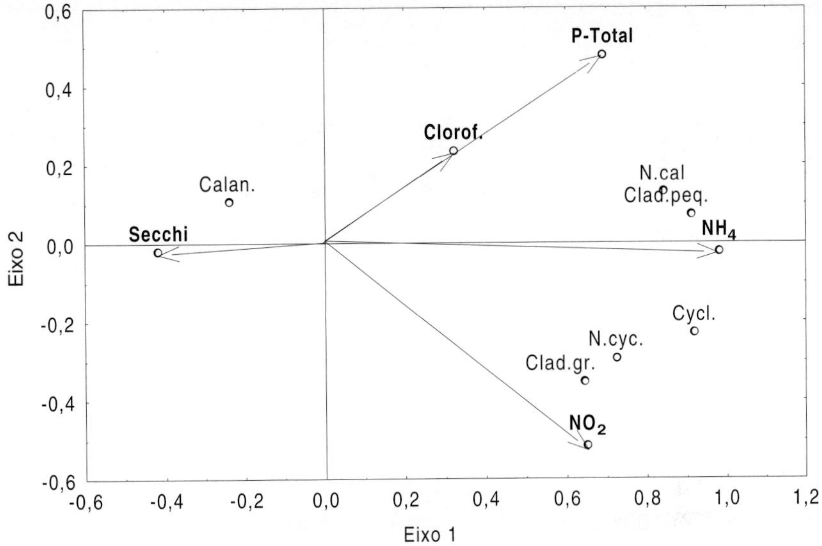

Fig. 9.12 CCA biplot diagram showing the abiotic explicative variables: P-total, Chlorophyll-a, ammonium -NH_4, nitrate - NO_3 (arrows) and the major crustacean zooplankton groups: calanoida (copepodite + adults), calanoida (nauplii), small cladocerans, large cladocerans, cyclopoida (nauplii) and cyclopoida (copepodites + adults)

The two first axis showed the eigenvalues of 0.63 and 0.29, respectively. These numbers indicate that these two first axis of the multivariate model explain 66.3% of the total data variance. Only axis 1 was significant (Chi^2 (30) = 115.2953; p = 0.0000). The correlation coefficient between axis 1 and abiotic variables (NH_4, P-Total, NO_2, chlorophyll-a and Secchi disk) were, respectively, 0.98, 0.69, 0.65, 0.32 and −0.42.

EFFECTS OF OXYGEN DEFICIT ON VERTICAL MIGRATION OF ZOOPLANKTON

The water temperature of surface and bottom of Nado Reservoir is represented in Fig. 9.13. The reservoir had a stable thermal stratification that began in August 1999 and finished at the beginning of the dry season (winter) in May 2000. A gradual increase of surface temperatures was observed from September through February. The hypolimnion remained as a rather stable region. The temperatures of the bottom stratum oscillated in a narrow range between 18.9 and 22.3°C in the annual cycle.

The Nado Reservoir has a permanent oxygen deficit, probably due to the stable thermal structure. The oxygen deficit (OAD) expressed by area

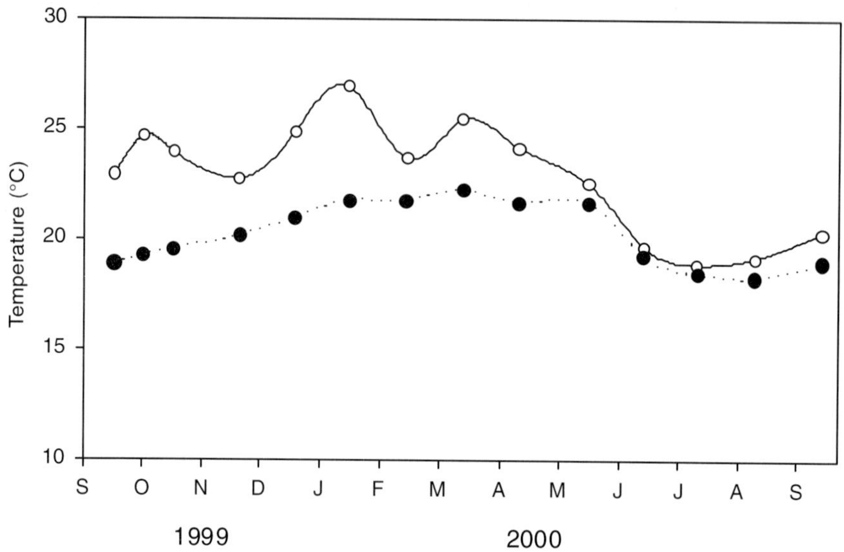

Fig. 9.13 Water temperature at surface and bottom (6 m) at Nado Reservoir, Belo Horizonte city between September 1999 and September 2000

unit of the lake reservoir is represented in Fig. 9.14. The oxygen deficit was lower during winter (dry seaon) and higher in summer months (rainy season). It varied from 0.53 mg.O_2.cm^{-2} (August, 1999) to 1.86 mg.O_2.cm^{-2} (December, 2000). These oxygen deficit (OAD) values are typical for tropical eutrophic lakes (Bezerra-Neto & Pinto-Coelho, 2001).

The four instars of *C. brasiliensis* showed diel vertical migration during the study period. They stayed in deeper layers during the day and moved upwards at night. The density-weigthed mean depth (Pm_i) during day hours and at night along the annual cycle is represented in Fig. 9.15. Between November 1999 and April 2000, no significant difference in the range of vertical migration was found between the different instars (pairwise comparison t-test, $p > 0.05$). No significant difference between these mean depth was found for instar I, II and III. However, instar IV individuals showed a significant greater depth not only during the day but also at night (parwise *t-test*, $p < 0.05$). Thus, the partition of habitats (water column) as found by López and Cressa (1996) and Voss and Mumm (1999) was not found. These authors found younger instars in the upper layers and older instars living in deeper strata. A possible reason for the lack of such a rigid vertical segregation pattern may be related to the fact that the Nado Reservoir is a rather shallow water body (Z_{max} = 7.3 m). The only pattern observed is the avoidance of the 0-2 m depth

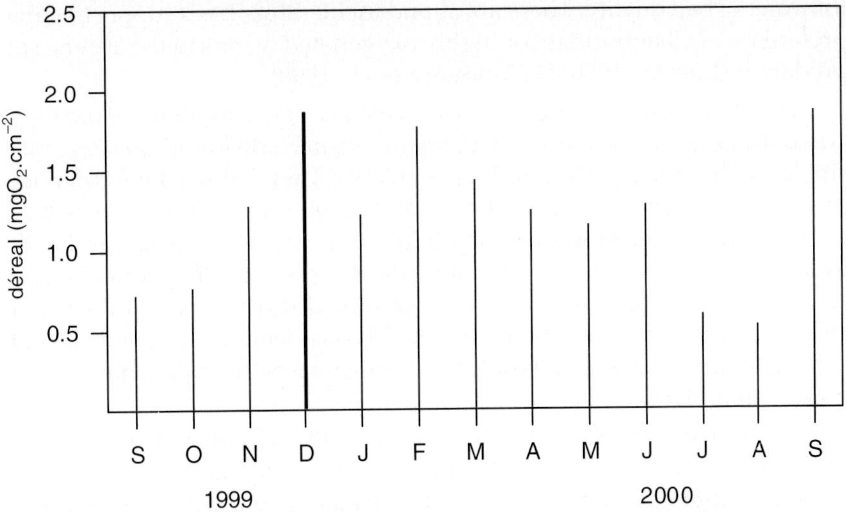

Fig. 9.14 Dissolved Oxygen actual deficit expressed by surface units in the Nado Reservoir for the annual cycle between September 1999 and September 2000

strata of all larval instars during most parts of the diel cycle (97% and 73% of all counted larvae avoid surface during day and night, respectively).

According to Davis (1975), the critical limit of dissolved oxygen for fish is the 1.0-2.0 mg $O_2.l^{-1}$, boundary. This depends, however, on the considered species and the prevailing temperature. The present study showed that the mean depth (Pm_i) of the instars of *C. brasiliensis* are always correlated with the depth at which the dissolved oxygen is 1.0 mg.l^{-1} ($n = 7$; $r^2 = 0.29$; $p = 0.05$, for instar I; $n = 10$; $r^2 = 0.80$; $p < 0.05$, for instar II; $n = 11$; $r^2 = 0.68$; $p < 0.05$, for III; $n = 11$; $r^2 = 0.34$; $p < 0.05$, for instar IV).

The mean depth (Pm_i) of *Chaoborus* larval stages in Nado Reservoir in the night hours also were positively correlated with the critical oxygen concentration as defined above ($n = 7$; $r^2 = 0.73$; $p = 0.05$, for instar I; $n = 8$; $r^2 = 0.63$; $p < 0.05$, for instar II; $n = 10$; $r^2 = 0.76$; $p < 0.05$, for III; $n = 11$; $r^2 = 0.50$; $p < 0.05$, for instar IV).

The vertical distribution of *C. brasiliensis* in the Nado Reservoir is clearly associated with the widely accepted hypothesis that vertical migration is a predation avoidance mechanism. Besides, the vertical distribution of *C. brasiliensis* in Nado Reservoir is also affected by oxygen. It is not acceptable that the preference of chaoboridae in this reservoir for low oxygen layers may be attribute to some odd physiological preference for water of bad quality. There are abundant literature basis supporting the preference of Chaoboridae for highly oxygenated waters in the absence of predation (Luecke, 1986; Dawidowicz et al., 1990).

The relationship between the critical oxygen concentration (1.0 $mg.O_2.l^{-1}$) and the vertical distribution of *C. brasiliensis* in Nado Reservoir was quite similar to the study of Rine and Kesler (2001). They followed the behavior of *C. punctipennis* in Poplar Tree Lake, USA and found a consistent relationship between the mean depth (i) of *C. punctipennis* instars and with the critical oxygen concentration along the diel cycle in a long term data set (13 years). This correlation could be observed during the fall (begin of October). However, since their data set did consider the different instars of *C. punctipennis*, it was not possible to investigate whether this behavior was present at all larval Instars.

Other investigations, however, recognize that Chaoboridae can evade predators using other factors such as light intensity, and fish-produced chemical stimuli (kairomones) to modulate their displacement in the water column in order to decrease the predation risk. In a set of laboratory experiments, Tjossen (1990) observed that *C. albatus* population tend to remain in the sediments if there is fish exudates in the water. They also observed that the vertical displacement of *C. flavicans* is reduced by adding kairomones in mesocosm experiments. Similar results were also found by Dawidowicz et al. (1990) with *C. flavicans*. Abrupt changes in the behavior of vertical migration were also observed in mesocosms biomanipulated with different fish biomass (McQueen et al. 1999). It was observed that the IV instar of *C. flavicans* changed their behavior on leaving the water column and remaining the entire day in the mesocosm sediment if the fish densities was higher that 2400/ha.

The larger zooplankton of Nado Reservoir is a simple array of just a few species. The major components of large zooplankton (mesozooplankton, larger than 200 um) were the copepod *T. minutus* and the cladoceran *M. micrura*. The vertical positioning of these organisms was clearly influenced by the presence of chaoboridae (Fig. 9.16).

The nauplii of *T. minutus* showed a diel mean depth of 2.7 m and a mean night depth of 2.5 m for the period under study. A vertical migration was detected only in five months (Fig. 9.15). In some occasions (October—November 1999 and July 2000), the vertical migration can be seen as a clear avoidance of the invertebrate predador. Nevertheless, the mean depths indicate that the nauplii and copepod always remained in a upper layer on being compared to Chaoboridae.

The mean depth of copepodites of *T. minutus* were 3.0 m and 2.2 m for daily and night hours, respectively. Except for December 1999, the

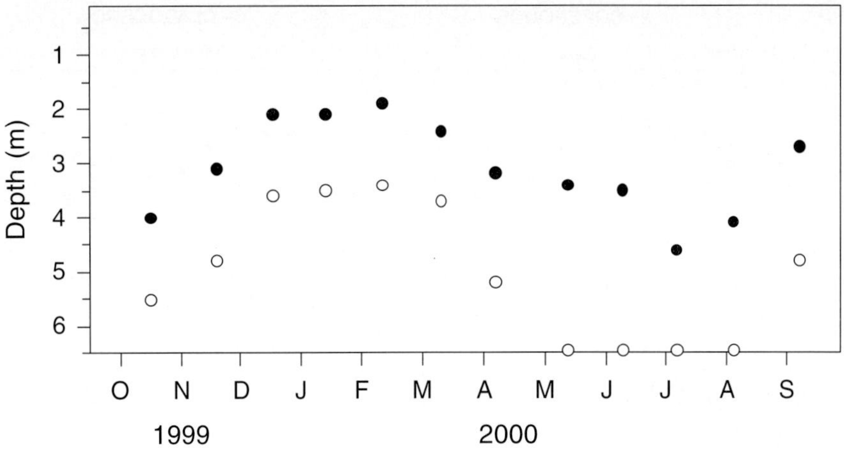

Fig. 9.15 Density-weigthed mean depth (*Pm*$_i$) (see page 10 for additional details) for instars III and IV of *Chaborus brasiliensis* at Nado Reservoir in the annual cycle. Between October 1999 and September 2000. Black dots: night values and white dots: daily sampling

copepodites had a clear pattern of diel vertical migration. The range of migration varied from 0.4 m in January 2000 to 1.1 m in May and August 2000 (Fig. 9.4). In nine sampling days, the copepodites of *T. minutus* showed a clear avoidance of predators (Fig. 9.15, middle chart).

The adults of *T. minutus* also had a typical pattern of diel vertical migration (Fig. 9.15, bottom). The range of vertical displacement was greater than those observed for nauplii and copepodites. Considering the study period, the diel mean depth was 3.3 m and the night mean value was only 1.9 m. There was a significant difference in the ranges of vertical migration between adults and copepodites and anuplii of *T. minutus*. The adults had a significant higher vertical displacement than the copepodites (paired t-test, $p < 0.005$) as well as with nauplii (teste *t* paired, $p < 0.005$).

The cladoceran *Moina minuta* had the highest ranges of vertical migration in the zooplankton community of Nado Reservoir (Fig. 9.15, top, right). A clear spatial segregation was also found in the vertical positions of this cladoceran and its potential predator, *C. brasiliensis*. The ranges of migration varied from 0.4 m (September 2000) to 3.0 m (June and August 2000), with an annual mean of 1.5 m.

CONCLUSION

The present study demonstrated that similar effects of eutrophication were found in both temporal and spatial scales. Eutrophication causes a reduction in biomass and species richness of the Calanoida group. These

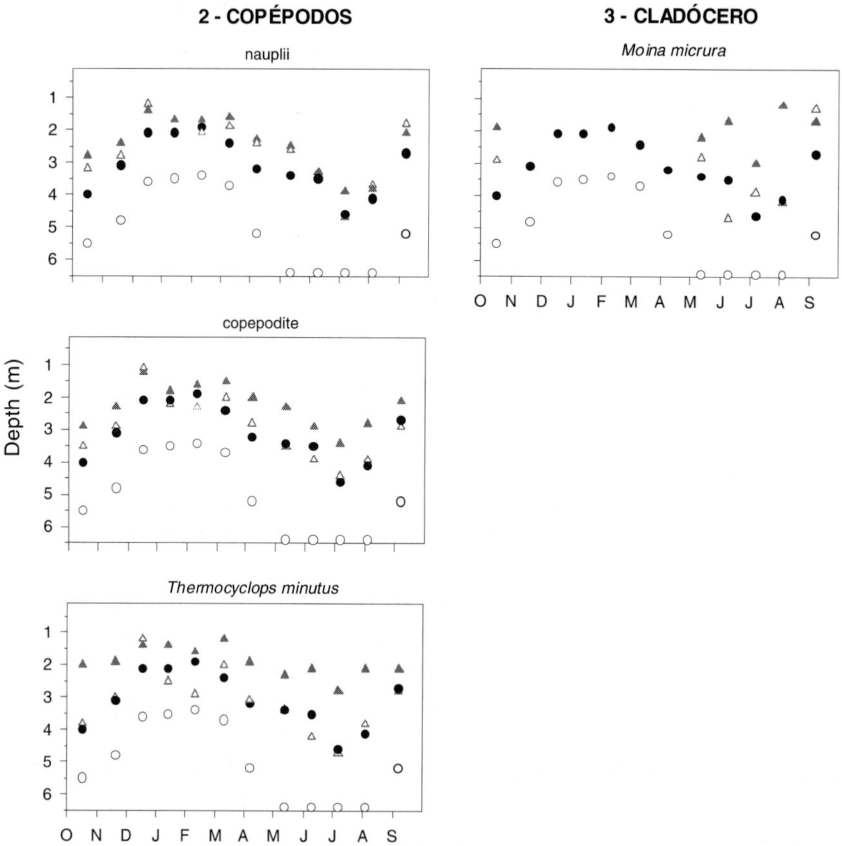

Fig. 9.16 Density-weigthed mean depth (*Pm*$_i$) during the daytime (White dots) and during the night hours (black dots) of the instars III and IV of *C. brasiliensis* and its potential preys (hollow triangles: day; solid black triangles: night) in Nado Reservoir. Reservoir in the annual cycle October 1999 and September 2000

organisms disappeared from Pampulha Lake during the last decade. In Furnas Reservoir, they are restricted only to the less eutrophicc regions.

On a long-term temporal scale, eutrophication is also associated with the biomass increase of cyclopoids. The small cyclopoid *Thermocyclops* spp. seems to be specially adapted to dominate the zooplankton in eutrophic conditions. A similar effect was also found at the spatial gradient of trophy in the Furnas Reservoir.

A trend for smaller size of cladocerans were also found in the more eutrophic stations of the Furnas Reservoir. This pattern, however, could not be observed in the Pampulha Reservoir.

The oxygen is a key factor, regulating the vertical structure of zooplankton in tropical reservoirs. It limits the vertical distribution not only of small but also larger species of zooplankton. Finally, the invertebrate predator, *C. brasiliensis* is able to limit the vertical migration of its potential prey species in the Nado Reservoir.

REFERENCES

Alvares, E., Canelas, M. A. S., Costa, R. G., Dias, L. G., Ribeiro-Mendes, H. N. & Pinto-Coelho, R.M. 2000. Aspectos socioeconômicos e estimativa da atividade de pesca no reservatório da Pampulha, Belo Horizonte, MG. *Anais V Simpósio de Ecossistemas Brasileiros, Vitória, Brasil.* pp. 375-382.

Bezerra-Neto, J. 2001. F. *A influência da larva de Chaoborus (Insecta: Diptera) na distribuição espacial da comunidade zooplanctônica na Lagoa do Nado, Belo Horizonte – MG.* 2001. Dissertação (Mestrado) – Instituto de Ciências Biológicas, Universidade Federal de Minas Gerais, Belo Horizonte, 2001.

Bezerra-Neto, J. F. & Pinto-Coelho, R. M. Avaliação da eficiência de dois amostradores na estimativa da abundância e biomassa de organismos planctônicos (submitted).

Bezerra-Neto, J. F. & Pinto-Coelho, R.M. 2002. A estrutura térmica, o trabalho do vento e a estabilidade de um reservatório tropical raso: Lagoa do Nado, BH – MG. Bios, **10:** 19-33.

Bottrell, H. H., Duncan, A., Gliwicz, Z. M., Grygierek, E., Herzig, A. & Hillbricht-Ilkows, A. 1976. A review of some problems in zooplankton production studies. Norw. J. Zool., **24:** 419-456.

Carpenter S. R., Z. Kitchell J. F. 1992. Trophic cascade and biomanipulation-interface of research and management–reply. Limnol. Oceanogr., **37:** 208-213.

Culver, D. A., Bouchete, M. M., Bean, D. J. & Fletcher, J.W. 1985. Biomass of freshwater crustacean zooplankton from length-weight regressions. Can. J. Fish. Aquat. *Sci.,* **42:** 1380-1390.

Davis, J.C. 1975. Minimal dissolved oxygen requirements of aquatic life with emphasis on Canadian species: a review. J. Fish. Res. Board Can., **32:** 2295-2332.

Dawidowicz, P., Pijanowska, J. & Ciechomski, K. 1990. Vertical migration of *Chaoborus* larvae is induced by the presence of fish. Limnol. Oceanogr., **35:** 1631-1637.

Dini, M.L., Soranno, P.A., Scheuerell, M.D. & Carpenter, S.R. 1993. Effects of predators and food supply on diel vertical migration of *Daphnia.* In: S.R. Carpenter, Kitchell, J.F. (eds) The Trophic Cascades In Lakes. New York: University of Cambridge, cap. 9, pp. 153-171.

Elmoor-Loureiro, L.M.A. 1990. *Diaphanosoma birguei* e *Diaphanosoma barchyurum.* Possível necessidade de revisão das identificações no Brasil. Acta Limnologica Brasiliensia **3:** 157-160.

Elmoor-Loureiro, L. M. A. 1997. Manual de identificação de cladóceros límnicos do Brasil. Editora Universa. Universidade Católica de Brasília. Brasilía. pp. 156.

Gannon, J. E. & Stemberger, R. S. 1978. Zooplankton (especially crustaceans and rotifers) as indicators of water quality. Trans. Amer. Micros. Soc., **97:** 16-35.

Henry, R. 2003. *Ecotonos nas Interfaces dos Ecossistemas Aquaticos*. Ryma Editora, Sao Carlos, SP., p. 349.

Henry. R. 1999. Ecologia de reservatórios: estrutura, função e aspectos sociais. Fundbio-Fapesp, Botucatu. 800p.

Henry R. Pontes, M.C. & Tundisi, J.G. 1989. O déficit de oxigénio do lago dom Helvécio (Parque Florestal do Rio D'oce). Revta. Bras. Biol. **49:** 251-260.

Lazzaro, X. 1987. A review of planktivorous fishes: Their evolution, feeding behaviors, selectivities and impacts. Hydrobiologia, **146:** 97-167.

López, C.; Cressa, C. 1996. Ecological studies on a *Chaoborus* larvae population in a tropical reservoir (Socuy Reservoir, Venezuela). Arch. Hydrobiol., **136:** 421-431.

Lorenzen, C. J. 1967. Determination of chlorophyll and phaeo-pigments: spectrophotometric equations. Limnol. Oceanogr., **12:** 343-346.

Luecke, C. A. 1986. Change in the pattern of vertical migration of *Chaoborus flavicans* after introduction of trout. J. Plankton Res., **8:** 649-657.

Mackereth, F. J. H., Heron, J. & Talling, J. F. 1978. Water analysis: some revised methods for limnologists. Freshwater Biol. Assoc. Scientific. Publication *36*, Cumbria, UK, p. 120.

Malley, D. F., Lawrence, S. G., Maciver, M. A. & Findlay, W. J. 1989. Range of variation in the estimates of dry weight for planktonic crustacea and rotifera from temperate North American lakes. Can. Tech. Rep. of Fish. Aquat. Sci., **1666:** 1-49.

Masundire, H. M. 1994. Mean individual dry weight and length-weight regressions of some zooplankton of Lake Kariba. Hydrobiologia, **272:** 231-238.

Matsumura-Tundisi, T. 1983. Occurence of copepod (Calanoida Cyclopoida and Harpacticoida) from Broa Reservoir (Sao Carlos, Sao Paulo, Brazil). Revta. bras. Biol. **43**(1): 1-17.

Matsumura-Tundisi, T. 1986. Latitudinal distribution of Calanoida copepods in freshwater aquatic systems of Brasil. Revta. Bras. Biol. **46(3):** 527-553.

Matsumura-Tundisi, T. & Rocha, O. 1983. On the mechanisms for the coexistence of congeneric species of zooplankton in Amazonian lakes. Tropical Ecology **24**(1): 43-47.

McCauley, E. 1984. The estimation of abundance and biomass of zooplankton in samples. In: J.A. Downing, & F.H. Rigler, (eds) A manual on methods for the assessment of secondary productivity in fresh waters. IBP Handbook 17, Blackwell Sci. Publ. pp. 228-265.

McQueen, D. J. et al. 1999. *Chaoborus* behavioral responses to changes in fish density. Arch. Hydrobiol., **145:** 165-179.

Murphy, J. & Riley, J.P. 1962. A modified single solution method for the determination of phosphate in natural waters. Analytica Chimica Acta, **27:** 31-36.

Pinto-Coelho, R.M. 1998. Effects of eutrophication on seasonal patterns of mesozooplankton in a tropical reservoir: A four-years' study in Pampulha Lake, Brazil. Freshwat. Biol., **40:** 159-174.

Pinto-Coelho, R.M. 1994. Limnological perspectives for the control of eutrophication in large tropical reservoirs. In: R.M. Pinto-Coelho, et al., (eds) Ecology and Human Impact on Lakes and Reservoirs in Minas Gerais. SEGRAC. Belo Horizonte, pp. 11-26.

Pinto-Coelho, R.M. & Corgosinho, P. H. 1998. Alterações na estrutura do zooplâncton em resposta a um gradiente de trofia no reservatório de Furnas, Minas Gerais. *Anais do VIII Seminário de Ecologia Vol.VII, 1173-1188.*

Ravera, O. 1996. Zooplankton and trophic state relationships in temperate lakes. Mem. Ist. Idrobiol., **54:** 195-212.

Reid, J.A. 1985. Chave de identificação e lista de referências bibliográficas para as espécies continentais sul-Americanas de vida libre da ordem Cyclopoida (Crustacea, Copepoda). Bolm. Zool., **9:** 17-143

Reid, J.A. 1987. *Scolodiaptomus,* a new genus proposed for *Diaptomus* (sensu latu) *corderoi* Wright, and description of *Notodiaptomus brandorffi,* new species (Copepoda: Calanoida), from Brazil. Journal of Crustacean Biology, **7:** 364-379.

Rine, H.E. & Kesler, D.H. 2001. A long-term study of larval Chaoborus *punctipennis* (Diptera : Insecta) vertical migration and the role of critical oxygen concentration. J. Fresh. Ecol., **16:** 145-150.

Rosen, R.A. 1981. Length-dry weight relationships of some freshwater zooplankton. J. Freshwat. Ecol., **1:** 225-227.

Sendacz, S. & Kubo, E. 1982. Copepoda (Calanoida e Cyclopoida) de reservatórios do Estado de São Paulo. *Boletim Instituto de Pesca (único):* 51-89.

Siebuith, J.N., V., Smetacek & Lenz, J. 1978. Pelagic ecosystem structure: heterotrophic compartiments of the plankton and their relationships to plankton size fractions. Limnol. Oceanogi. **23(6):** 1256-1263.

Smith, K.E. & Fernando, C.H. 1980. Guia para los copépodos (Calanoida y Cylopoida) de las aguas dulces de Cuba. Editora de la Academia de Ciencias de Cuba. La Habana. p. 28.

Tjossem, S.F. 1990. Effects of chemical cues on vertical migration behavior of *Chaoborus.* Limnol. Oceanogr., **35:** 1456-1468.

Thomaz, S.M. & Bini, L.M. 2003. *Ecologia e Manejo de Macrofitas Aquaticas* Editora da Universidade Estadual de Maringa, Parana, p. 341.

Torres, I.C., Araújo, M.A. & Pinto-Coelho, R.M. 1998, Variação sazonal e diurna do carbono orgânico particulado no reservatório da Pampulha, Belo Horizonte, MG. Rev. Bras. Biol., **58:** 131-141.

Tundisi, J.G. 2003. Água no século XXI: enfrentando a escassez. São Carlos: Rima, IIE, p. **248.**

Van De Velde, I. 1984. Revision of the African species of the genus *Mesocyclops* Sars, 1914 (Copepoda: Cyclopoidae). Hydrobiologia, **109:** 3-66.

Voss, S.; Mumm, H. 1999. Where to stay by night and day: size-especific and seasonal differences in horizontal and vertical distribution of *Chaoborus flavicans* larvae. Freshwater Biology, **42:** 201-213.

Wetzel, R.G. 1983. Limnology. 2 ed. Saunders College Publishing. New York. p. 767.

Wetzel, R. & Gopal, B. 2001. Limnology in Developing Countries, Volume 3 Int. Assoc. Limnology, SIL, North Carolina Chapel Hill, USA, p. 189.

CHAPTER

10

Eutrophication and its Influences on the Fish Fauna of Lake Victoria

F. Witte, J.H. Wanink, H.A. Rutjes, H.J. van der Meer and G.E.E.J.M. van den Thillart

Institute of Biology Leiden, University of Leiden, P.O.Box 9516, 2300 RA Leiden, The Netherlands

ABSTRACT

During the past decades, dramatic changes were observed in the ecosystem of Lake Victoria. The introduced Nile perch (*Lates niloticus*) boomed in the 1980s and many other fish species disappeared. The more than 500 haplochromine cichlid species, that originally made up more than 80% of the demersal fish mass, were most severely affected. Within 10 years, they had almost vanished from the catches in the sublittoral and offshore waters, and lakewide, some 200 species may have disappeared. In the same period, the native zooplanktivorous cyprinid dagaa (*Rastrineobola argentea*) and the introduced Nile tilapia (*Oreochromis niloticus*) strongly increased in numbers. Concomitant with the changes in fish species composition, other changes in the ecosystem were also observed. Most prominent was the increase in eutrophication that had started already in the 1930s when a strong increase of the human population was also observed. As a result of soil runoff, eutrophication and phytoplankton blooms, dissolved oxygen concentrations and water transparency in the lake declined in the 1980s.

In the 1990s, after a decline of the Nile perch due to over-fishing, some 20 haplochromine species recovered in the Mwanza Gulf. Apart from predation by Nile perch—which is still high when compared to predation in general before the ecological changes—the recovering species have to cope with environmental changes, such as increased hypoxia and decreased water clarity.

Laboratory experiments revealed that most haplochromine species, including those from the littoral zone, are tolerant to low dissolved oxygen concentrations (0.8 mg l^{-1}). They seem to be more tolerant than the Nile perch and dagaa that are flourishing in the current ecosystem. Several haplochromine species—both in captivity and in the field—revealed adaptive responses in gill morphology to long-term hypoxia. Negative impacts of hypoxia are, however, still possible, as the scope for activity may have decreased in hypoxic areas and because severe hypoxia might hamper embryonic development.

Decreased water clarity seems to pose a greater problem on haplochromine cichlids, as they heavily depend on visual clues for foraging and social interactions. Species diversity appears to be lower in areas with low visibility than in clear areas. A decrease in water clarity seems to hamper mate recognition and may lead to hybridization, or even may completely frustrate breeding. Further, low visibility may decrease prey selectivity and result in increased interspecific competition, which will have a negative impact on species coexistence.

Measures to stop the increasing eutrophication of Lake Victoria are urgently needed. A decrease of the eutrophication might enhance the recovery of haplochromine species and thus the fish diversity. An increased biodiversity in turn, might result in a more predictable, more stable and in the long term more productive ecosystem.

LAKE VICTORIA AND ITS FISH FAUNA

Lake Victoria is the world's largest tropical lake by area. Bound by Tanzania, Uganda and Kenya, it straddles the equator in eastern Africa. Lake Victoria has a surface area of ca. 68,800 km^2, a maximum depth of about 70 m and is situated at an altitude of 1134 m above sea level (Fig. 10.1). The lake originated less than a million years ago and probably even dried up completely ca. 14,000 years ago (Johnson et al., 1996). In spite of its young age, Lake Victoria has had an extremely rich fish fauna, comprising over 500 cichlid species (Greenwood, 1974; Witte & van Oijen, 1995; Seehausen, 1996; Kaufman et al., 1997), making up more than 80%

of the total demersal fish mass (Kudhongania & Cordone, 1974). The remaining 20% comprised ca. 45 species divided over 11 families (Table 10.1; Greenwood, 1974; van Oijen, 1995). The cichlids in Lake Victoria belong to two tribes: the Tilapiini, with only two herbivorous species that grow to an adult size of more than 25 cm standard length (SL), and the Haplochromini with more than 500 species. Haplochromine cichlids generally do not grow larger than 15 cm SL and comprise specialists to virtually every possible food source in the lake (Greenwood, 1981; Witte & van Oijen, 1995; Seehausen, 1996). In the 1970s, detritivores/phytoplanktivores made up ca. 50% of the total fish mass and zooplanktivores ca. 25%. Piscivorous species made up less than 10% of the total ichthyomass (Goldschmidt et al., 1993). The total annual fish catch in Lake Victoria in the 1960s and 1970s approximated 100,000 tonnes (Reynolds et al., 1995).

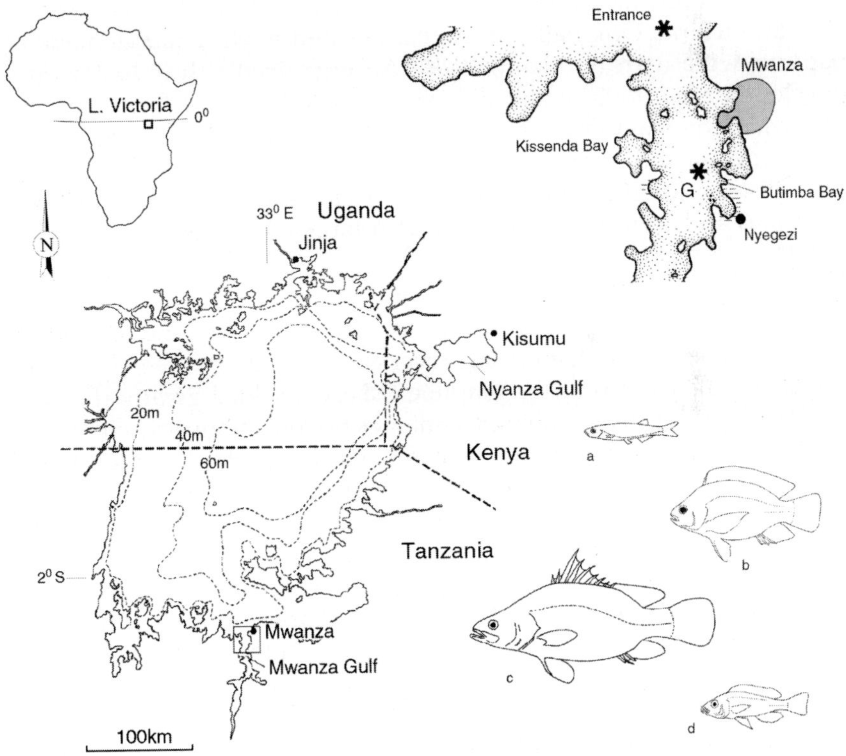

Fig. 10.1 Map of Lake Victoria and the northern part of the Mwanza Gulf where the main field research was carried out. Sampling stations "G" and "Entrance" have depths of, respectively, 14 m and 20 m. Major fish species mentioned in the text: (a) dagaa (*Rastrineobola argentea*); (b) Nile tilapia (*Oreochromis niloticus*); (c) Nile perch (*Lates niloticus*); (d) a haplochromine cichlid (zooplanktivore).

Table 10.1 Fish species composition in Lake Victoria before the Nile perch boom in the 1980s

Taxon	*nr of species*
Haplochromine cichlids	500 +
Tilapiine cichlids	2
Cyprinids	17
Catfishes (4 families)	10
Rest (6 families)	19

HUMAN ACTIVITIES AND THEIR IMPACTS DURING THE PAST CENTURY

Fishery and Species Introductions

From time, immemorial, fishing has been carried out in the coastal areas of Lake Victoria. Originally, traps, baskets and moving fences made of papyrus were used (Graham, 1929). An organized fish industry only developed after 1905, when gill nets were introduced from Europe. Lakewide, the main target species was the tilapiine cichlid *Oreochromis esculentus*, but other large growing fish species such as *Oreochromis variabilis*, the lungfish *Protopterus aethiopicus*, and the catfishes *Clarias gariepinus* and *Bagrus docmak*, were also targeted for fishing (Graham, 1929).

During the first half of the twentieth century, fishing pressure increased at a steady pace. Railway lines reached the lake and opened up new market areas (Graham, 1929; Fryer & Iles, 1972). Between the 1930s and 1940s, the catches of the tilapiine fishery declined greatly (Fryer & Iles, 1972). In the 1950s, fishing techniques became more efficient due to the introduction of nylon nets and outboard engines, but the catches dropped further. Subsequently, fishermen started to use illegal mesh sizes of less than 5" (12.7 cm). A couple of years later, the minimum mesh of 5" was officially repealed (Fryer & Iles, 1972). As a result of the decreasing catches, exotic tilapiine cichlids were introduced into the lake in the 1950s, in order, to fill the supposed empty niches (Beauchamp, 1958; Welcomme, 1967). In the same period, a huge predator, the Nile perch (*Lates niloticus*), was introduced illegally. Later, between 1961 and 1963, Nile perches from Lakes Albert and Turkana were officially released (Arunga, 1981, Welcomme, 1988), in spite of fierce opposition (Fryer, 1960).

During the first few decades after these introductions, the alien species did not boost fishery. Nile perch catches remained low (mean catch rate in bottom trawls ≤ 2 kg h^{-1}; Kudhongania & Cordone, 1974), and the introduced tilapiines appeared to compete for spawning sites and nurseries with the native species (Welcomme, 1967; Lowe-McConnell, 2000; Goudswaard et al., 2002b).

Because of the continuing decline in the catches, fishermen increasingly used small mesh sizes and so, smaller, thus far under-exploited, species were caught (Balirwa et al., 2003). In the middle of the 1960s, a light-fishery developed for the small cyprinid *Rastrineobola argentea*, locally called dagaa (Okedi, 1981). A trawl fishery on haplochromine cichlids was started in 1976 near Mwanza in the south of the lake, where a fish meal factory converted 10-15 tonnes of haplochromines per day into fodder. Due to the heavy fishing pressure, haplochromine catches declined locally in the Mwanza Gulf (Witte & Goudswaard, 1985). When the Nile perch boomed at the beginning of the 1980s, the decline of the haplochromines increased and became a lakewide phenomenon (Barel et al., 1985, 1991; Ogutu Ohwayo, 1990a). By 1987, the haplochromines had disappeared almost completely in waters of more than 6 m deep and Nile perch made up more than 90% of the bottom dwelling fish mass (Barel et al., 1991; Witte et al., 1992b). In the sublittoral and offshore areas of the lake, a decline was also observed for other fish groups, such as the catfishes (Ogutu-Ohwayo, 1990a; Goudswaard & Witte, 1997). However, the population of the zooplanktivorous cyprinid dagaa increased substantially, possibly as a result of the release of competition by haplochromines (Wanink, 1991, 1999; Wanink & Witte, 2000a; Witte et al., 1992a). Catch rates of the introduced Nile tilapia, *Oreochromis niloticus*, also increased from the middle of the 1980s (Ogutu-Ohwayo, 1990a; Goudswaard et al., 2002b). This species replaced the native tilapiines. The limited overlap in distribution between piscivorous Nile perch and Nile tilapia of consumable sizes is probably an important factor in explaining the co-existence of the two species (Goudswaard et al., 2002b). Fish catches lakewide rose from some 100,000 tonnes in the 1970s to over 450,000 tonnes in 1989, at that time mainly comprising Nile perch, Nile tilapia and dagaa (Reynolds et al., 1995).

Concomitant with the decline of the haplochromines, increases in densities of invertebrate species such as midge larvae and molluscs were observed. Most striking was the increase of the shrimp *Caridina nilotica* (Goldschmidt et al., 1993; Lehman et al., 1996). These shrimps became important prey of the Nile perch after the disappearance of the haplochromines (Ogari & Dadzie, 1988; Ogutu-Ohwayo, 1990b; Hughes, 1992; Mkumbo & Ligtvoet, 1992). For the pelagic zone, a decrease in the relative abundance of the large-bodied herbivorous calanoids and cladocerans and an increase of small-bodied predatory cyclopoid copepods was reported (Mwebaza-Ndawula, 1994; Gophen et al., 1995).

The decrease in herbivorous calanoids and cladocerans, which had commenced between the 1920s and 1950s, may have been caused by concurrent changes in the phytoplankton composition (Wanink et al., 2002) that shifted in predominance from diatoms to cyanobacteria (Gophen et al., 1993, 1995; Hecky, 1993; Mugidde, 1993)

In 1989, the South American water hyacinth, *Eichhornia crassipes*, entered Lake Victoria. This new intruder quickly established itself (Njuguna, 1991). For almost a decade, extensive mats of this weed covered large areas along the lake's shores, depriving the waters of light and oxygen. To eradicate the weed, South American weevils (*Neochetina eichhorniae* and *N. bruchi*) were introduced into the lake, and in 1998, the water hyacinth strongly declined again. However, it is not certain whether this decline was due to the weevils only or to other factors as well (Twongo & Odongkara, 2000).

During the second half of the 1990s, after a decline of the Nile perch due to over-fishing (e.g. Pitcher & Bundy, 1995; Ntiba et al., 2001), some haplochromine species recovered in the Mwanza Gulf (Seehausen et al., 1997b; Witte et al., 2000). A similar pattern of faunal loss and recovery in relation to Nile perch upsurge and decline was observed in Lake Nabugabo (Chapman et al., 2003).

Eutrophication

During the 1980s, concomitant with the Nile perch boom, an increase in algal blooms, particularly cyanobacteria was observed in several areas of Lake Victoria (Ochumba & Kibaara, 1989; Witte et al., 1992a; Mugidde, 1993). Data derived from sediment cores in the deepest part of the lake revealed that eutrophication must have started much earlier, between the 1920s and 1930s (Hecky, 1993; Verschuren et al., 1998, 2002). The human population of the Lake Victoria drainage-basin has grown from 4.6 million in 1932 to 27.7 million in 1995, and eutrophication ran parallel with human-population growth and agricultural activity (Verschuren et al., 2002). The increase in eutrophication is most probably due to an increase in the nutrient input from atmospheric deposition, and land runoff as a result of forest burning and land exploitation for agriculture (Scheren et al., 2000). The wetland zone that regulates material transport to the lake has also been under intense pressure from human activities (Balirwa, 1995; Kairu, 2001).

The Nile perch explosion in the 1980s has contributed through a bottom-up and a top-down effect to this increasing eutrophication. Unlike the small haplochromines, the relatively fat and large Nile perch could not be dried in the sun. Therefore, the fish were chopped into pieces and subsequently fried in the fat that had been removed from the intestines (Ligtvoet et al., 1995). The smaller Nile perch were mainly smoked. As a consequence, fish smoking became more common in the area. For instance, at Wichlum Beach (Kenya), the number of smoking kilns increased between 1984 and 1991 from about 10 to over 50 (Riedmiller, 1994). For frying and smoking purposes, firewood was needed, which enhanced the deforestation along the lake's shore, thus increasing land erosion and eutrophication of the lake.

Apart from this bottom-up effect, there was also a top-down effect. Prior to the ecological changes, phytoplanktivorous and detritivorous fish (mainly haplochromines, which could digest cyanobacteria) made up respectively ca. 17% and 30% of the total fish mass in the sublittoral area of the Mwanza Gulf (Goldschmidt et al., 1993). Based on the ingestion rates calculated by Moriarty et al., (1973) for similar fish from Lake George, the total phytoplankton intake by fish prior to the changes in the Mwanza Gulf was estimated at a dry weight of 230 mg m^{-2} day^{-1}. After the decline of the haplochromines, the impact of phytoplankton grazing by fish in the Mwanza Gulf was negligible. However, the precise impact of the disappearance of phytoplanktivorous fish on the phytoplankton densities and composition is not clear. For a proper estimation of this impact, data are needed on phytoplankton production and on phytoplankton intake by other organisms before and after the ecological changes.

High rates of photosynthesis due to phytoplankton blooms created nearly continuous super-saturation in surface waters in the 1990s, and oxygen concentrations in the upper twenty metres of the offshore waters became higher than in the past (Hecky et al., 1994). Sedimentation and decomposition of the increased amount of phytoplankton consumed oxygen in the hypolimnetic waters. As a result, oxygen concentrations near the bottom were lower than in the 1960s. Moreover, these low oxygen concentrations were present for a longer part of the year and occurred higher into the water column (Table 10.2; Hecky et al., 1994). Severe hypoxic conditions (< 1.0 mg l^{-1}) persisted from October to March at depths below 40-53 m (Table 10.2; Hecky et al., 1994), which cover ca. 35% of the lake's total bottom area. Even in the shallow (< 20 m) Mwanza Gulf, an extension of the hypoxic period was observed between 1979 and 1988, and the isopleths of both 1 and 5 mg O_2 l^{-1} were situated further from the bottom in 1988 (Fig. 10.2; Wanink et al., 2001). Measurements in the upper 5 m at the entrance of the gulf (20 m deep), in 1987, revealed super-saturation, depending on the time of the day (Wanink & Kashindye, 1998). Oxygen levels in these top 5 metres increased strongly between 9.45 h and 14.00 h, at the peak of pelagic photosynthesis, after which they dropped again.

Table 10.2 Mean dissolved oxygen levels at 40-60 m depth in Lake Victoria and depth at which DO level was 1 mg l^{-1}, during 1960/61 and 1990/91 (from Talling, 1966 and Hecky et al., 1994)

Month	*mg l^{-1} DO*		*1 mg l^{-1} depth*	
	1960/61	*1990/91*	*1960/61*	*1990/91*
Sep.	3.48	3.25	> 60	> 60
Oct.	3.99	1.46	> 60	52
Nov.	3.48	0.52	> 60	40
Dec.	2.91	0.87	> 60	45
Jan.	4.34	2.20	60	54
Febr.	-	-	-	-
Mar.	6.29	2.21	> 60	53
Apr.	1.79	3.76	51	> 60
May.	2.75	6.74	53	> 60
Jun.	-	-	-	-
Jul.	6.34	7.72	> 60	> 60
Aug.	5.66	4.16	> 60	> 60

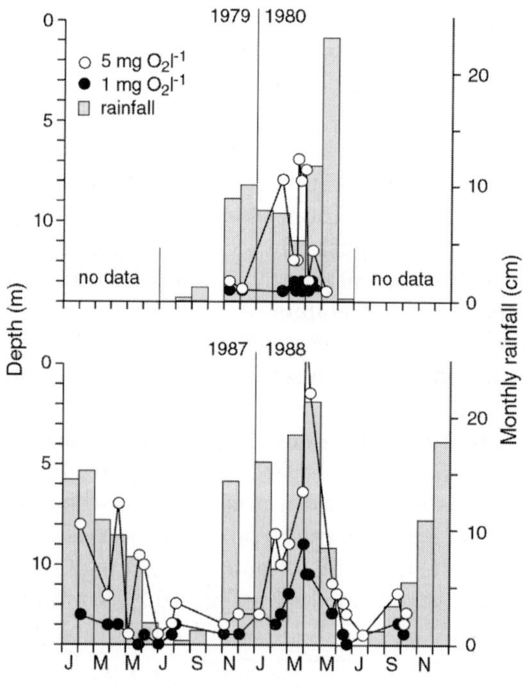

Fig. 10.2 Seasonal fluctuations in dissolved oxygen at station G (14 m deep) in the Mwanza Gulf and rainfall in Nyegezi during 1987 and 1988. The plotted isopleths represent the distance from the bottom at which the oxygen concentrations reached values of 1 and 5 mg l^{-1}, respectively. After Wanink et al. (2001).

Increased water temperatures in Lake Victoria since the 1960s, as a result of climate change, may have increased the thermal stability of the water column (Hecky, 1993; Hecky et al., 1994). This may have contributed to the persistence of hypoxia in the deeper layers as it would hamper the effective mixing of the lake (Ogutu-Ohwayo et al., 1997). The decreased levels of dissolved oxygen (DO) in the hypolimnetic waters seemed to coincide with an increase in mass fish kills due to sudden upwelling of deoxygenated water (Ochumba & Kibaara, 1989; Ochumba, 1990, 1995; Kaufman, 1992; Kudhongania & Chitamwebwa, 1995). In the past, such upwellings and massive kills used to occur only occasionally in Lake Victoria (Fish, 1956, 1957; Kitaka, 1972).

The water transparency in Lake Victoria has always been low compared to that in the deep rift lakes, Malawi and Tanganyika. As early as 1927-28, Secchi depths were measured in Lake Victoria (Worthington, 1930), ranging from 0.6-1.8 m in gulfs and bays and from 6.4-8.2 m in open waters (Table 10.3; Worthington, 1930). Increasing eutrophication had a strong impact on water transparency. Maximum Secchi values in the offshore waters in the northern part of the lake decreased to 1.3-3 m in 1989-91 (Table 10.3; Mugidde, 1993). In the sublittoral waters of the Mwanza Gulf, a decrease was observed from 1.1-2.5 m in 1973-80 to 0.7-1.2 m in 1986-87 (Table 10.3; Akiyama et al., 1977; van Oijen et al., 1981). A similar decline was also observed in the Nyanza Gulf (Lung'ayia et al., 2001).

Table 10.3 Water transparency in the Mwanza Gulf and in offshore waters in the northern part of the lake. Transparency was measured (in m) with a Secchi disc of 20 or 30 cm

Date	*depth (m)*	*Secchi depth (m)*	*source*
Mwanza Gulf before 1984			
Apr. - Dec. 1973	8	1.4 - 1.8	Akiyama et al., 1977
Jan. - Dec. 1974	8	1.1 - 1.9	Akiyama et al., 1977
Feb. - Apr. 1980	3 - 14	1.7 - 2.5	van Oijen et al., 1981
Mwanza Gulf after 1984			
Nov. 1986	7 - 14	0.7 - 0.9	J.H. Wanink, unpublished
Feb. - Jun., Aug., Nov. 1987	5	0.7 - 0.9	J.H. Wanink, unpublished
Feb. - Jun., Aug., Dec. 1987	14	1.0 -1.2	J.H. Wanink, unpublished
Open water before 1984			
Sep. 1927	60 - 70	7.3 - 7.9	Worthington, 1930
Jan. 1928	38 - 61	6.4 - 8.2	Worthington, 1930
Open water after 1984			
Sep. - Nov. 1985	25 - 40	1.3 - 2.5	Ochumba & Kibaara, 1989
Feb. - May. 1986	25 - 40	0.2 - 1.7	Ochumba & Kibaara, 1989
1989 - 1991	65	1.3 - 3	Mugidde, 1993
1990 - 1991	28 - 66	2.1 (mean)	Gophen et al., 1993

We will now offer a survey of the potential and observed effects of the increased hypoxia and decreased water clarity on the fish fauna of Lake Victoria. Emphasis will be on the haplochromine cichlids as they were by far the dominant group in the undisturbed system, and also because our studies mainly concentrated on these fishes.

THE EFFECTS OF INCREASED HYPOXIA ON THE FISH FAUNA

Behavioural and Physiological Responses

In this section, some general behavioural and physiological responses of fish to hypoxia will be discussed. Subsequent sections will focus on the tolerance for and responses to hypoxia in Lake Victoria fish species.

Immediate responses of fish to hypoxia occur at two different levels, i.e. behavioural and physiological. Fish exposed to hypoxia may either try to flee from the area, which increases their energy demand, or they may become inactive and reduce their energy demand. Each of these two responses has a survival value, as the first brings the fish to a more acceptable environment, while the second allows the animal to survive longer under hypoxic conditions. Which response will be chosen by a particular fish depends on the environmental situation (how severe is the hypoxia) and on the species involved (what is its capacity to cope with the conditions). The major physiological effect of hypoxia is a reduction of the oxygen extraction capacity of the gills, because oxygen diffusion is directly proportional to the oxygen gradient. This can hardly be compensated for, since the diffusion distance, the surface area, and the oxygen gradient determine the net oxygen transport over the gills. When these are optimized, the extraction capacity is reached. While the diffusion distance is fixed, the surface area of the gills can be controlled by shunting of the blood flow through the gills, thus regulating the effective exchange surface. If there is more oxygen available than needed, the entire gill surface area is not used. At maximal oxygen uptake, all the blood flowing through the gills comes in contact with the respiratory surface.

The extraction capacity decreases with decreasing oxygen level, reducing the maximal performance (swimming speed) of the fish. The balance between energy expenditure and oxygen uptake determines the anaerobic threshold, the boundary between aerobic and anaerobic metabolism. This is a critical point, since the chances of survival drop rapidly once anaerobic metabolism is activated. Though anaerobic capacity may vary, it is always limited. Fish may survive below the anaerobic threshold somewhere between minutes to days, depending on the species and the severity of the condition (van den Thillart & van

Waarde, 1985). Some species of cichlids such as the South American *Astronotus ocellatus*, have extreme capacities to survive hypoxia. This species survived more than 16 hours of severe hypoxia (DO ≤ 0.4 mg l^{-1}) and even four hours of complete anoxia at 28°C (Muusze et al., 1998).

Short-term hypoxia exposure (less than one day) does not change the oxygen-binding characteristics of haemoglobin. However, the affinity of the haemoglobin of a particular species to oxygen (P_{50} value) has a large effect on its critical oxygen level. The P_{50} of cichlids is rather low (Weber & Wells, 1989), allowing for high extraction efficiencies at low oxygen values. Haemoglobin affinity and, consequently, hypoxia sensitivity vary widely among fish species. Fast swimmers require high P_{50} values, which make them susceptible to hypoxia. In contrast, sedentary fish—like many haplochromine cichlids—are better off with lower P_{50} values, which allow them to live in hypoxic areas. Most fish species have several types of haemoglobin, which allow them to live in a variety of habitats (Weber, 1990). Long-term hypoxia may change both the oxygen-binding characteristics as well as the haemoglobin composition, thus adapting the fish to the hypoxic environment.

Oxygen uptake on the blood side of the gills is mainly determined by shunting of blood flow, and on the water side by the total water flow. At the exchange site, water and blood flow have to remain at a fixed ratio which is determined by the total oxygen-carrying capacities (Jones et al., 1970). It is a physical requirement for optimal oxygen transport that the total oxygen transport is equal at the water and the blood side. The oxygen-carrying capacity of the blood depends on the haemoglobin content. Since the oxygen content of blood is about 20 times higher than that of normoxic water, the water flow must be around 20 times as high as the blood flow. When the oxygen content of the water declines, the water flow must increase in order to maintain the same uptake. For example, at 10% air saturation level of the water, the water flow over the gills must increase at least tenfold. On the other hand, under assumption of a constant energy demand, the cardiac output will remain the same, because the high affinity of haemoglobin results in a constant carrying capacity of blood over a wide range of oxygen concentrations. This is the reason why fish mainly increase their expirational flow rate when exposed to hypoxia. In Nile tilapia, an approximately tenfold decrease in oxygen tension of inspired water (towards the critical oxygen tension) resulted in a gill ventilation flow that was 6-12 times higher than under normoxia. Ventilation volume increased much more than ventilation frequency (Fernandes & Rantin, 1989). In haplochomines that were acutely exposed to hypoxia (0.8 mg l^{-1}), the ventilation frequency increased from around 45-80 to 100-130 times per minute. The ventilation volume also increased strongly, but was not measured (H.A. Rutjes, G.E.E.J.M. van den Thillart & F. Witte, unpublished data).

At any oxygen level, a fish has a particular minimal and a maximal oxygen consumption rate. The minimal rate is required for maintenance; below this level, the animal will die. The maximal oxygen consumption rate allows extra activity (e.g. swimming). The space between the two is called the scope for activity. This scope for activity decreases when the oxygen level decreases. Thus, at each activity level, there is an oxygen concentration that limits the extraction and, therefore, the activity. This becomes clear when the activity level is fixed such as with swimming in a swimming tunnel. Smit (1965) found that when goldfish swam, the oxygen content in the tunnel decreased slowly; the collapse point at which the fish stopped swimming was each time at exactly the same oxygen level. The collapse point of a fish is determined by the oxygen extraction capacity. The enormous increase of ventilation flow in fish that are acutely exposed to hypoxia increases the energy demand of the animal, which obviously has a negative effect on its scope for activity.

Hypoxia Tolerance in Lake Victoria Fish Species

Several studies have been done—and are currently carried out—to establish the hypoxia tolerance of Lake Victoria fish species. Depending on dissolved oxygen (DO) availability in the habitat, Chapman et al. (2002) found differences in behaviour, gill morphology, haematological characters and metabolic rates in fish species from the Lake Victoria region. Species living in swamps of satellite lakes of Lake Victoria had higher haematocrit and haemoglobin concentrations, a larger gill surface area and a lower critical oxygen tension than fishes from the open water in the satellite lakes. Similar differences seem to be present between species from different habitats *within* Lake Victoria. Of the closely related species *Haplochromis (Gaurochromis) hiatus* and *H. (G.) iris*, the former lived at depths between 3 and 9 m and the latter between 8 and 15 m (Hoogerhoud et al., 1983). During the rainy season, stratification and low DO levels (2-3 mg l^{-1}) were occasionally observed in the habitat of *H. iris* (van Oijen et al., 1981). A comparison of the gill morphology of the two species revealed that the total gill area in *H. iris* was ca. 1.6 times that of *H. hiatus* (Hoogerhoud et al., 1983). Similar differences in gill morphology were found in other cichlid species from Lake Victoria (Galis & Barel, 1980).

A swimming experiment was carried out with *H. (Astatotilapia) elegans* and *H. (Gaurochromis) angustifrons* from Lake George (Galis & Smit, 1979). This experiment proved that the collapse point of *H. elegans* increased from about 1.5 to 2.0 mg l^{-1} when the swimming speed increased from 16 to 25 cm s^{-1}. Over the same range, *H. angustifrons* had collapse points that were about 0.5 mg l^{-1} lower than in *H. elegans*. This difference may be related to differences, both, in habitats and in feeding habits. *H. elegans* is a dweller of shallow sandy bottoms near papyrus fringes and feeds on chironomid larvae, whereas *H. angustifrons* is essentially an offshore species that seems to prefer mud substrates. It mainly feeds on *Chaoborus* larvae that are buried in the soft anoxic mud (Greenwood, 1973).

Verheyen et al. (1986) measured relatively high haemoglobin-oxygen affinities in five haplochromine species from Lake Victoria. The P_{50} values ranged from 1.41 to 3.14 mmHg. Although the value 1.41 of the rock dwelling *H. (Neochromis) greenwoodi* (H. "velvet black" in Verheyen et al., 1986) is rather low, it is not as low as the P_{50} value of Nile tilapia (1 mmHg), which is a very hypoxia-tolerant fish species (Verheyen et al. 1985). It should be noted that the five haplochromine species studied by Verheyen et al. (1986) were all restricted to relatively well-oxygenated, shallow (< 6 m) sandy and rocky habitats (see Witte et al., 1992b, for data on distribution). In contrast, Nile tilapia also occurs in deeper water and over mud bottoms (e.g. Goudswaard et al., 2002b). A study of Fish (1956) revealed a P_{50} value of 17 mmHg for Nile perch (referred to as *Lates albertianus*) and a P_{50} value of ca. 2.5 mmHg for *Oreochromis esculentus*, which is closely related to Nile tilapia. The mormyrid *Mormyrus kannume* and the catfish *Bagrus docmak*, which both have been caught up to depths of more than 60 m (Kudhongania & Cordone, 1974), had P_{50} values of ca. 1 mmHg.

Experiments testing the response to short-term low DO regimes of nine cichlid species from Lake Victoria (seven haplochromines, Nile tilapia and *O. esculentus*) and three species from Lake Tanganyika, revealed interspecific differences (Chapman et al., 1995). In general, Lake Victoria cichlids showed higher levels of hypoxia tolerance than those of Lake Tanganyika. *Astatoreochromis alluaudi*, which beyond Lake Victoria is widespread in a variety of aquatic habitats including satellite lakes and affluent streams, revealed a well-developed aquatic surface respiration and no loss of equilibrium (i.e. the fish could control their position) during acute hypoxia (30 minutes at < 0.2 mg l^{-1} DO). In contrast, the rock-dwelling H. (*Neochromis) rufocaudalis* (= *N. nigricans* in Chapman et al., 1995), which normally lives in shallow well-oxygenated water, lost equilibrium more quickly than other Lake Victoria species. This species also had the highest P_{50} value (3.14 mmHg) in the study of Verheyen et al. (1986). When there was no access to the surface, only the native tilapiine *Oreochromis esculentus* and the introduced Nile tilapia did not lose equilibrium during acute hypoxia (30 minutes at < 0.2 mg l^{-1} DO; Chapman et al., 1995), thus revealing an extreme hypoxia tolerance.

In spite of this tolerance, Nile tilapia was a victim of mass fish kills observed in the 1980s (Ochumba, 1990). Ochumba suggested that apart from low DO levels, low pH and high levels of suspended material clogging gills may have played a role in these fish kills. Nile perch was the main victim of these massive kills. Considering its sensitivity to low DO levels, this is not remarkable (Fish, 1956; Kaufman, 1992; Chapman et al., 2002). However, the sensitivity may be less extreme than sometimes suggested. Kaufman (1992) used 5 mg l^{-1} DO as a critical field value, but, based on trawl catches at different DO levels, Wanink et al. (2001) estimated that the critical field level amounted to 2-2.5 mg l^{-1}.

In long-term hypoxia experiments, the DO level in fish tanks was gradually decreased in a couple of days (or weeks) and thereafter maintained for months at this low level. These experiments showed that juveniles of several Lake Victoria cichlids—both from the littoral (< 6 m deep) and sublittoral areas (6-20 m deep)—could be raised successfully at 10% air saturation (ca. 0.8 mg l^{-1} DO) for more than a year until adulthood (H.A. Rutjes, G.E.E.J.M. v.d. Thillart, J. Visee & F. Witte, unpublished data). Preliminary results of these experiments indicated increases in gill size, as well as in haematocrit and haemoglobin levels in the hypoxia-reared fish. Chapman et al. (2000) raised a brood of the cichlid *Pseudocrenilabrus multicolor victoriae* from the Lake Victoria region for five months under normoxia and hypoxia (1.0 mg l^{-1} DO). The total gill surface area in their hypoxia group became 18% greater than in the normoxia group.

Chapman et al. (2000), also compared field populations of *P. multicolor victoriae* from a stable hypoxic habitat (a swamp) and a stable well-oxygenated habitat (a lake margin). In this case, the swamp-dwelling fish had a 29% greater gill surface than those from the well-oxygenated habitat. According to Chapman et al. (2000) this may be the result of a combination of inherited changes and phenotypic plasticity.

We observed changes in gill morphology in individuals of *Haplochromis (Yssichromis) pyrrhocephalus* and dagaa, two zooplanktivorous fish species that both had been collected before and after the ecological changes in Lake Victoria. At the end of the 1990s, *H. pyrrhocephalus* became the dominant species among the recovering haplochromines (Witte et al., 2000). An increase in the gill surface area of 73% was observed when fish from 1999 were compared with those from 1978-79 (M. Welten, J. Wanink & F. Witte, unpublished data). In dagaa (the cyprinid *Rastrineobola argentea*), which boomed shortly after the Nile perch in the 1980s, the number of gill filaments on the first gill arch increased with 3.6% between 1983 and 1988 (in this case, the gill filament length was not measured; Wanink & Witte, 2000b). In spite of this increase in filament numbers, dagaa did not move to the bottom during periods of stratification, but dwelled at the oxycline (Fig. 10.3). The critical level for this species in the wild apparently amounted to 2-2.5 mg l^{-1} DO, which is approximately the same as that for Nile perch (Wanink et al., 2001), and considerably higher than the DO level (0.8 mg l^{-1}) under which haplochromines could live in our long-term laboratory experiments. It should be mentioned, however, that well-developed hypoxia tolerance in captivity does not automatically imply survival under similar conditions in nature. The scope for activity needed to survive in the field will probably be greater than that in captivity.

Hypoxia Tolerance of Cichlid Embryos

Though juvenile and adult cichlids can apparently cope with extreme hypoxia, we do not yet know whether earlier stages—like embryos and larvae—can survive under these low DO levels. So far, two out of five cichlid species kept under long-term hypoxia produced vital juveniles at 0.8 mg l^{-1} DO. These two species were Nile tilapia and *A. alluaudi*, of which the adults also appeared to be extremely hypoxia tolerant. One of the other species produced brood several times that died within three days after spawning (H.A. Rutjes, G.E.E.J.M. van den Thillart & F. Witte, unpublished data).

In early fish embryos, viz. embryos before the development of heart and blood vessels, oxygen transport depends on diffusion. This may

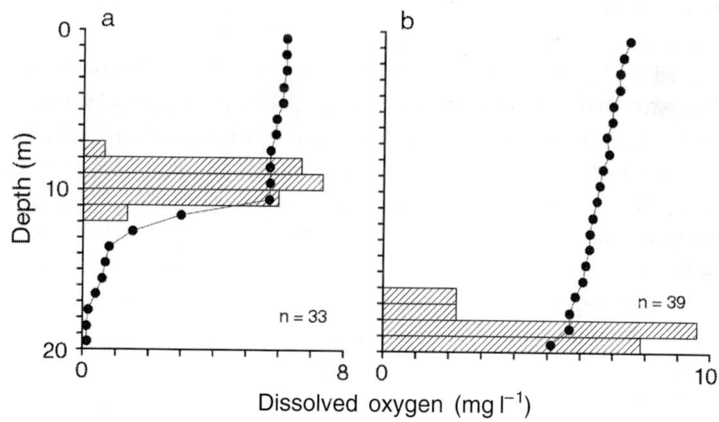

Fig. 10.3 Daytime depth distribution of dagaa (hatched bars) in relation to the depth of the oxycline (black dots) at the entrance of the Mwanza Gulf (20 m deep) on: (a) 28 April and (b) 14 October 1988. Plotted sample sizes refer to the total number of dagaa on which the frequency distributions were based. After Wanink et al. (2001).

hamper successful development under hypoxia in species with large embryos (Kranenbarg et al., 2000). Indeed a high correlation (Pearson correlation coefficient: $R = 0.9$, $P < 0.01$) was found between the mean DO levels in spawning habitats and the mean egg size of nine species of Lake Victoria haplochromines (T. Goldschmidt, pers. com.). *A. alluaudi,* that successfully spawned in our hypoxia experiments, is among the Lake Victoria haplochromine species with the smallest egg sizes (Goldschmidt & Goudswaard, 1989).

Hypoxic Areas: Refugia or a Cause of Mortality or Migration?

As shown above, Nile perch has a lower hypoxia tolerance than many other fish species, including haplochromine cichlids (Fish, 1956; Chapman et al., 2002). There are indications that Nile perch migrates upward into the column when DO levels near the bottom become less than 2 mg l^{-1} (Wanink et al., 2001; P.C. Goudswaard, E.F.B. Katunzi, J. H. Wanink & F. Witte, unpublished data). Therefore, several authors have suggested that hypoxic areas might function as refugia for haplochromines against Nile perch predation (e.g. Kaufman & Ochumba, 1993; Hecky et al., 1994; Chapman et al., 1995). In contrast, Wanink et al. (2001) suggested that Nile perch might dive beneath the oxycline in order to forage, which would limit the use of this habitat as a refugium. Low-oxygen areas in wetlands are more likely to function as refugia for some haplochromine cichlids (Chapman et al., 1995, 2002, 2003; Rosenberger & Chapman, 1999), because they are large areas that, except for the margins, probably cannot be penetrated by short raids of foraging Nile perch. The fact that these habitats are structurally complex, thus providing hiding places and making it difficult to pursue a prey, also contributes to their importance as refugia (Chapman et al., 1996).

It was also suggested that the decrease in DO levels in Lake Victoria might have played a major role in the decline of the haplochromines in the 1980s (e.g. Hecky et al. 1994; Bundy & Pitcher, 1995; Ochumba, 1995). According to Hecky et al. (1994) the hypoxic deep-water area may have originally functioned as a refugium for the haplochromines that had lower critical DO levels than Nile perch. The increasing deoxygenation (< 1 mg l^{-1}) of the deep water might have forced the demersal populations to the shallower waters, where they were exposed to Nile perch predation. In the 1970s, several haplochromine species living at the entrance of the Mwanza Gulf (> 20 m deep) migrated into the gulf during the second half of the rainy season (March-May; van Oijen et al., 1981). This was the period during which low DO levels, due to stratification, were common in deeper waters (Fig. 10.2; Talling, 1966; van Oijen et al., 1981), but it was also the main spawning period in the Mwanza Gulf

(Witte, 1981). Therefore, it is not clear whether this seasonal migration was triggered either by hypoxia or by the need for suitable spawning grounds or a combination of the two. *Haplochromis (Yssichromis) laparogramma*, one of the species with its main distribution at the entrance of the gulf (> 20 m deep), became more abundant year round at our main sampling station G (14 m deep) within the gulf during the first years of the Nile perch boom (Witte et al., 1995). This migration into shallower waters might indeed have been caused by deoxygenation at greater depths. Another explanation suggests migration away from areas with a high predation pressure (Witte et al., 1995), as, in spite of their sensitivity to low DO levels, densities of large piscivorous Nile perch (> 25 cm) were higher at the entrance rather than inside the gulf (Goudswaard et al., 2002a). Remarkably, *H. laparogramma* was the only deep-water species displaying an inshore migration.

Apart from migration, severe hypoxia may cause heavy mortalities. Fish (1956) reported a local mortality near Entebbe on 31 July 1953. The affected fish consisted almost entirely of tilapiine and haplochromine cichlids. During the fieldwork of the Haplochromis Ecology Survey Team, from 1977 until 1990, large mortalities of haplochromines that could be related to hypoxia have never been observed in the Mwanza Gulf and its entrance. However, after 1987, a mass mortality of Nile perch was observed several times (Wanink et al., 2001; P.C. Goudswaard, pers. com.). For water of less than 30 m deep, where hypoxia is generally less severe than in deeper areas, we have no indications that the decline of cichlids in the 1980s was caused by mass mortalities due to hypoxia. This is supported by the observation that other fish species which are extremely tolerant to hypoxia, such as catfishes, mormyrids and lungfish (e.g. Fish, 1956), also declined in the 1980s (Ogutu-Ohwayo, 1990a; Goudswaard & Witte, 1997; Goudswaard et al., 2002a), whereas the relatively hypoxia-sensitive Nile perch and dagaa both flourished after the ecological changes. In our opinion, Nile perch was the major cause of the decline of the haplochromine cichlids during the 1980s in the sublittoral waters, though fishery also contributed in some areas (Witte et al., 1995).

EFFECTS OF DECREASED WATER TRANSPARENCY

Light Transmission and Underwater Vision

Transmission of light by water is dependent on the wavelength. Bodies of water that do not contain much organic matter absorb violet and red light much more than light of intermediate wavelengths (Levine & MacNichol, 1982; Loew & McFarland, 1990). Hence, Lakes Malawi and Tanganyika get bluer with depth. Waterbodies, such as Lake Victoria, containing yellow-green phytoplankton and dissolved organic matter, absorb light of all wavelengths more strongly than clear water. However, the short wavelengths (violet and blue) are most strongly absorbed and yellow-green light becomes increasingly dominant with depth (Levine & MacNichol, 1982; Loew & McFarland, 1990). The increasing eutrophication in Lake Victoria may have caused further shifts towards longer wavelengths. Light measurements in March and April 1987 revealed that mainly orange-red light penetrated over mud bottoms at depths of more than 10 m (Fig. 10.4; de Beer, 1989). Transmission spectra measured in 1996, at a water depth of 2 m at 13 research stations near rocky shores, showed a decrease of the width of the transmission spectrum with decreasing transparency (Seehausen et al., 1997a). The peak of the transmission spectrum was near the yellow-orange light.

Most freshwater fish are thought to have trichromatic vision, with visual pigments having absorption peaks around 455 nm (blue), 530 nm (green) and 625 nm (red) (Levine & MacNichol, 1982; Jobling, 1995). Apart from enabling colour vision, this facilitates visual contrast. Dark objects are best observed with a pigment that is matched to the ambient light, as the object will be dark against a bright background. In contrast, bright objects are best observed by visual pigments that are set off from the colour of the ambient light (Levine & MacNichol, 1982).

Light and Dark Adaptations

Like in other vertebrates, fish retinas contain cones for photopic vision and rods for scotopic vision. In the majority of fish species, these receptor cells lie internally to a pigment epithelium. Most teleosts cannot control

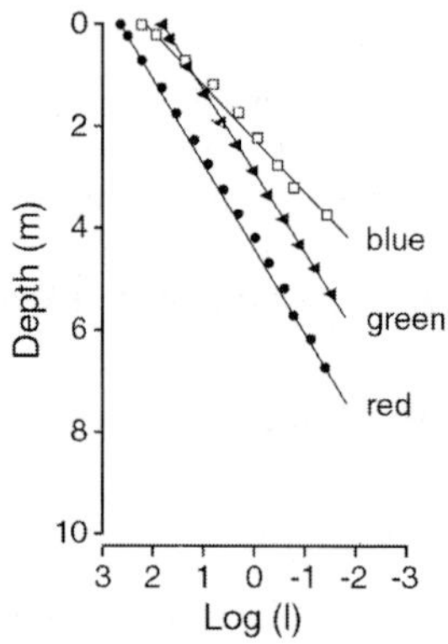

Fig. 10.4 Spectral absorption of the downwelling light in the Mwanza Gulf. The light intensity ($\mu E\, m^{-2} s^{-1}$) of three components (red, green, and blue light) is plotted against depth on a semi-log scale. After van der Meer & Bowmaker (1995).

the amount of light that reaches the retina by changing the size of their pupil. Adaptation of the eye to light and dark conditions occurs via retino-motor movements. These are movements of the receptor cells and the pigments in the epithelium (Jobling, 1995). In the light-adapted eye, the extremely light-sensitive rods are shielded by the pigment. In the dark-adapted eye, the rods are fully exposed to the incident light. In some fish species, visual sensitivity may be increased by the presence of a tapetum lucidum, a mirror, at the back of the eye. Apart from retino-motor movements, which tend to follow a diel cycle, temporal changes in concentrations and in relative proportions of different visual pigments may occur in response to seasonal changes in incident light intensities (Jobling, 1995).

Cone size and density can be related to photon-catching ability and retinal resolution, respectively. Cones with a large diameter receive more light energy per second than smaller cones, thus lowering the sensitivity threshold. If the same surface area of the retina is occupied by a high density of small cones, it will provide a better detail discrimination due to a higher resolution (van der Meer & Anker, 1984; van der Meer, 1994, 1995; Kröger et al., 1999).

Retina Structure in Cichlids

A comparison of retinas of haplochromine cichlid species from different habitats revealed the differences in retina structure and visual pigments. As in all Perciformes, the cones in the retinas of haplochromine cichlids are mostly arranged in a square pattern, viz., units of four double cones surrounding a central single cone (Fig. 10.5; Ali & Anctil, 1976; Wagner, 1978; Otten, 1981; van der Meer, 1992). According to van der Meer (1992), this pattern can be easily transformed into a row pattern consisting of double cones only. Microspectrophotometry of the photopigments in cichlids showed that in all the studied species, the double cones were non-identical, with long wavelength (red) sensitive and middle-wavelength (green) sensitive members. The single cones were all short-wavelength (blue) sensitive (Bowmaker, 1990, 1995; van der Meer & Bowmaker, 1995).

In the Mwanza Gulf, species living near the surface or in shallow water generally had square patterns with large single cones. In species dwelling at greater depths, where short wavelengths did not penetrate, the blue-sensitive single cones were either small in size and number or even completely absent (Fig. 10.5; van der Meer et al., 1995). At these greater depths, the double cones were often larger in size than in species dwelling closer to the surface and they were arranged in a row pattern (Fig. 10.5; van der Meer et al., 1995). A row pattern provides maximal

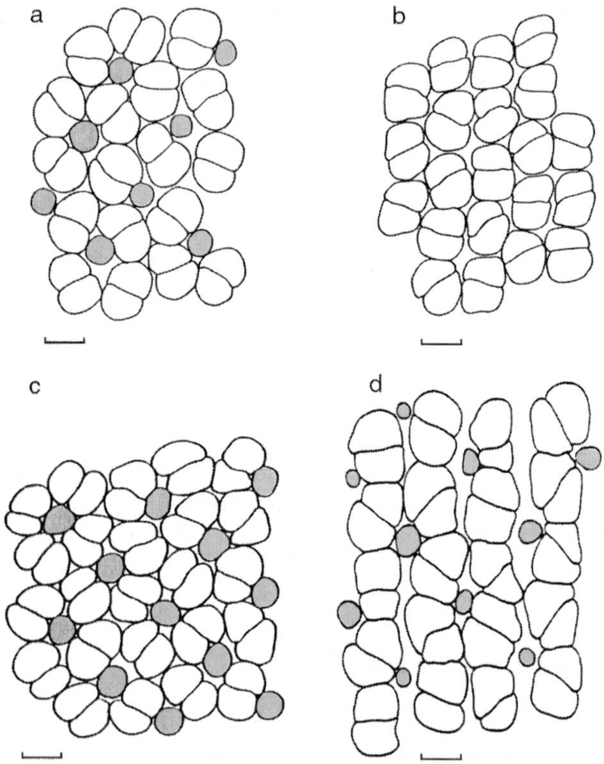

Fig. 10.5 Cone patterns in retinas of haplochromine cichlids: (a) A square pattern of four double cones, surrounding a single cone in the surface dwelling zooplanktivore *H.* 'argens'; (b) A row pattern of double cones in the bottom dwelling zooplanktivore *H.* 'reginus'; (c) A square pattern in the oral shelling molluscivore *H. sauvagei* raised under a broad light spectrum; (d) A blue light deprived specimen of *H. sauvagei* with (occasional) row pattens of double cones and small single cones. Scale bar = 5 μm. (a) and (b) after van der Meer et al. (1995); (c) and (d) after van der Meer (1993).

packing when only double cones are needed, thus maintaining the resolution as good as possible (van der Meer, 1992; Barel, 1993). It seemed that the species that lived in deeper waters were dichromatic, probably with one pigment matching the ambient light and the other mismatching it.

A study of the retinas of nine zooplanktivorous haplochromine cichlids from the Mwanza Gulf indicated distinct interspecific differences in photon-catching abilities of the cones and a slight differentiation in retinal resolution (van der Meer et al., 1995). The authors concluded that the differences in these optical functions correlated poorly with photic conditions of the species-specific habitats, and suggested that they may be explained by different modes of prey detection. This may indeed be the case, but it should also be kept in mind that foraging is only one of the ecological roles to which eyes are important; intraspecific communication and predator avoidance are other major roles (Guthrie & Muntz, 1993). Different ecological roles may require incompatible functional demands. Which of these functional demands will dominate, or how a compromise will be reached, may depend on ecological differences between the species (differences in diet and/or in spawning place and period).

As mentioned, haplochromine species living in deeper waters generally had tiny or no single cones and enlarged double cones in a row pattern. Similar retina configurations were induced by rearing individuals of the shallow water species *Haplochromis (Ptyochromis) sauvagei* in darkness or in absence of short-wave light (Fig. 10.5; van der Meer, 1993). The possibility to induce such changes within a species does not imply that the above-mentioned *interspecific* differences are merely the result of phenotypic plasticity. Species that had been reared under standardized conditions in our laboratory for several generations still showed interspecific differences in arrangement and spectral sensitivity of their photoreceptors (van der Meer & Bowmaker, 1995; Smit & Anker, 1997).

Light Conditions and Cichlid Species Diversity

Cichlid behaviour, such as feeding and breeding, depends highly on visual clues (Fryer & Iles, 1972). Female haplochromine cichlids choose mates on the basis of male colouration. Laboratory experiments have demonstrated that females under monochromatic light cannot discriminate between a conspecific male and that of a sibling species with a different colouration, while under broad spectrum illumination, they do prefer conspecific over heterospecific males (Seehausen & van Alphen, 1998). Comparing rock-dwelling cichlid communities at different rocky islands in Lake Victoria, Seehausen et al. (1997a) discovered that the clearer the water and the broader the spectrum of transmitted light, the more colour morphs of a species and the more species of a genus coexisted. Moreover, the distinctiveness of nuptial hue difference between sympatric species was also greater in areas with a great spectral width. The high diversity of closely related species in clear water could be explained by strong assortative mating, leading to sexual isolation of colour morphs. In contrast, water with a narrow spectral width poses constraints on the functional diversity of colour signals and limits the number of species that can be sexually isolated (Seehausen et al., 1997a). Indeed, the frequency of phenotypes that were intermediate between sympatric forms was negatively correlated with water transparency at different sampling stations in the lake (Seehausen, 1997, 2000).

The need for fairly clear water for reproduction in Lake Victoria haplochromines is confirmed by observations on spawning periods. Before the ecological changes, most cichlids living over mud bottoms in the sublittoral area (> 6 m deep) of the Mwanza Gulf spawned during the dry season, when water transparency in this area became temporarily high (Fig. 10.6; Witte et al., 1999). In contrast, in the relatively clear shallow sandy and rocky habitats, most haplochromines spawned year-round (Witte, 1981; Seehausen et al., 1998). The restricted spawning period at deeper mud bottoms will increase competition for spawning sites. This will have another negative impact on species coexistence, added to the above-mentioned impact of the loss of sexual isolation.

Apart from the explanations related to breeding, impacts of water transparency on feeding may also explain the higher species diversity in clear water. Based on models from foraging theory, Seehausen et al. (2003) suggested that a decline in the visual range decreases the 'encounter rates' with prey. This will have negative effects on selectivity and, thus, on the degree of specialization on food. Loss of specialization,

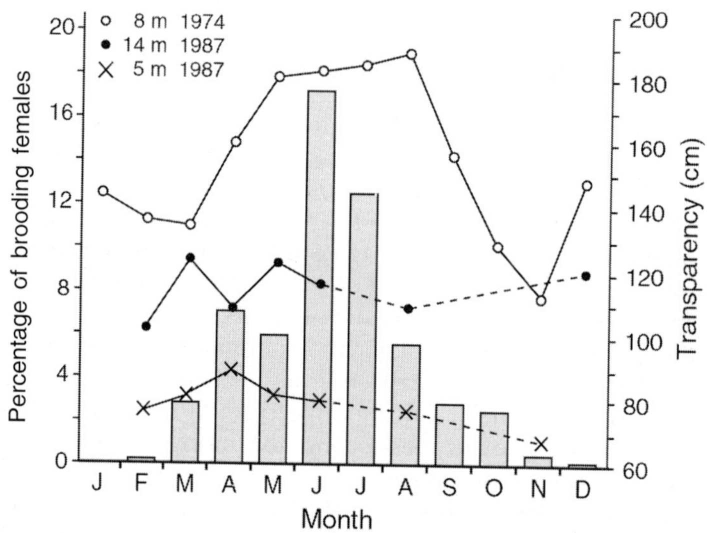

Fig. 10.6 Percentage of brooding individuals (bars) among 16,098 female haplochromines caught at depths between 8 and 15 m before the ecological changes (1979/80) in the Mwanza Gulf, and water transparency (Secchi values) before (1974; white dots) and after the ecological changes (1987; black dots and crosses). In 1974, transparency was measured at an 8-m-deep station (data from Akiyama et al., 1977), in 1987 at the 14-m-deep station G and at a 5-m-deep station more southwards. Note that the transparency increases with water depth. This implies that the transparency at 8 m depth in 1987 was somewhere in between that given for 5 m and 14 m. After Witte et al. (1999; 2000).

will in turn have a negative impact on species coexistence as niche overlap, and thus competition, will increase. Finally, some feeding habits, like oral shelling of molluscs seem only possible if visibility is sufficient. Grasping a snail at its foot, as performed by the so-called oral shellers, appears to require careful observation of the prey in order to successfully catch it (Slootweg, 1987). Experiments in our laboratory with two oral shelling species revealed a strong decrease in successful attacks with decreasing light intensity (M. Rensing & F. Witte, unpublished data). This may explain why oral shelling species mainly occur over sand bottoms (Witte, 1981).

In conclusion, the impacts of water transparency on breeding and feeding may, to a large extent, explain the diversity of haplochromine cichlids. Apart from the correlation between water transparency and species diversity at rocky islands (Seehausen et al., 1997a) other examples were also found. For instance, at several localities in Lake Victoria, the number of species over relatively clear sand bottoms was approximately 1.5 times higher than over mud bottoms of the same depth (Fig. 10.7: Greenwood, 1974; F. Witte, unpublished data). Finally, as noted by

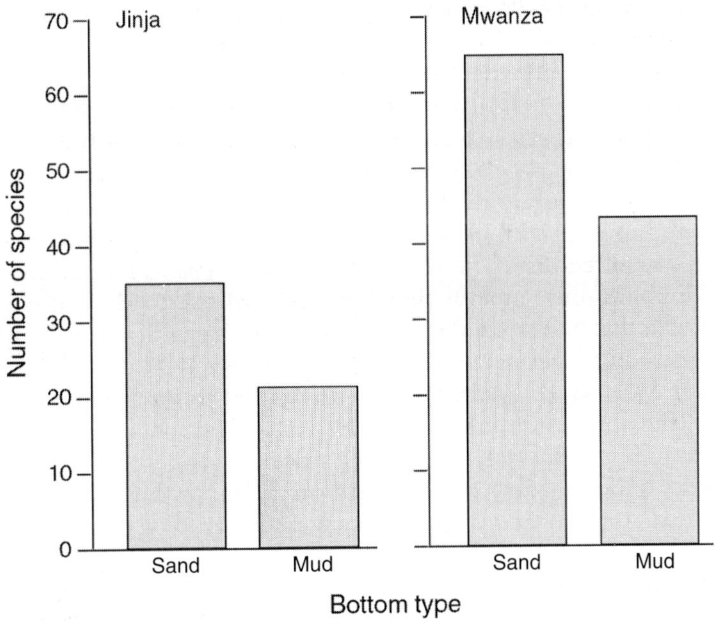

Fig. 10.7 Total number of haplochromine species caught over a sandy beach and in a muddy bay near Jinja between 1950 and 1956, and over a sand and mud bottom of the same depth in Butimba Bay, near Mwanza in 1979-80. Data from Jinja are based on Greenwood (1974). Similar results were found during a survey west of the Mwanza Gulf in June 1985.

Seehausen et al. (1997a) "the seven Great Lake basins of tropical Africa in which haplochromine cichlids formed endemic species flocks, have distinctly clearer water than the five in which they did not".

Effects of the Decrease in Water Clarity

The decrease in water transparency since the 1970s (Fig. 10.6) narrowed the light spectrum. This may have changed the way female haplochromines perceived the colours of the males, and could mask certain differences between the species (Seehausen et al., 1997a). Following the decrease in water clarity in the Mwanza Gulf, the brightness and distinctiveness in a population of *Haplochromis (Pundamilia) nyererei* decreased. This has been explained by Seehausen et al. (1997a) as a result of relaxation in sexual selection for costly male colouration. It might finally lead to hybridization between species that were mainly isolated by premating barriers.

Another possible effect of decreased water clarity concerns the selective recovery of cichlid species. At the end of the 1990s, after the decline of the Nile perch due to over-fishing, about 20 cichlid species recovered in the northern part of the Mwanza Gulf (Seehausen et al., 1997b; Witte et al., 2000). The zooplanktivorous *H. (Yssichromis) pyrrhocephalus* became the most common cichlid in the area, whereas its sister species *H. (Y.) heusinkveldi* had disappeared. Prior to the ecological changes, both species had nearly identical distribution patterns (Goldschmidt et al., 1990) and were equally abundant (Witte et al., 2000). Apart from male colouration—that was grey-blue in *H. heusinkveldi* and orange-red in *H. pyrrhocephalus*—the two species differed in retinal, and thus, in visual features. The retina of *H. pyrrhocephalus* mainly contained large double cones in low densities in contrast to the retina of *H. heusinkveldi* that had small double cones in high densities (Fig. 10.8b). The rod density in *H. pyrrhocephalus* was 15% higher than in *H. heusinkveldi* (Van der Meer et al. 1995). Further, compared to other haplochromine species, the absorption maxima of the visual pigments in double cones and rods in *H. pyrrhocepalus* were at significantly longer wavelengths (van der Meer & Bowmaker, 1995). Therefore, compared to the eyes of *H. heusinkveldi*, those of *H. pyrrhocephalus* were likely to be more sensitive to light in general and to red light in particular. The higher number of smaller double cones in *H. heusinkveldi* could result in a higher visual resolution at the cost of its photopic threshold. This has been interpreted as an adaptation to a different feeding strategy in combination with the detection of small food items (van der Meer et al., 1995). Before the occurrence of ecological changes in the lake, the diet of *H. heusinkveldi* consisted for 20% to 40% of relatively small phytoplankton, that of *H.*

pyrrhocephalus for less than 5% (Fig. 10.8a; Goldschmidt et al., 1990). Spawning in *H. heusinkveldi* mainly occurred in months with a high water transparency at depths of 8-14 m, but *H. pyrrhocephalus* spawned all-year-round at depths of less than 9 m (Fig. 10.8c; Goldschmidt & Witte, 1990). As has been suggested by Witte et al. (2000) the adaptation to feed on phytoplankton may have made it impossible for *H. heusinkveldi* to spawn during murky periods. Since, after the ecological changes water transparency in the habitat of the two species was low throughout the year (Fig. 10.6), this may have hampered the recovery of *H. heusinkveldi*.

Preliminary results of a study on the eyes of the recovering zooplanktivore *H. tanaos* revealed an adaptive response of its retina. This

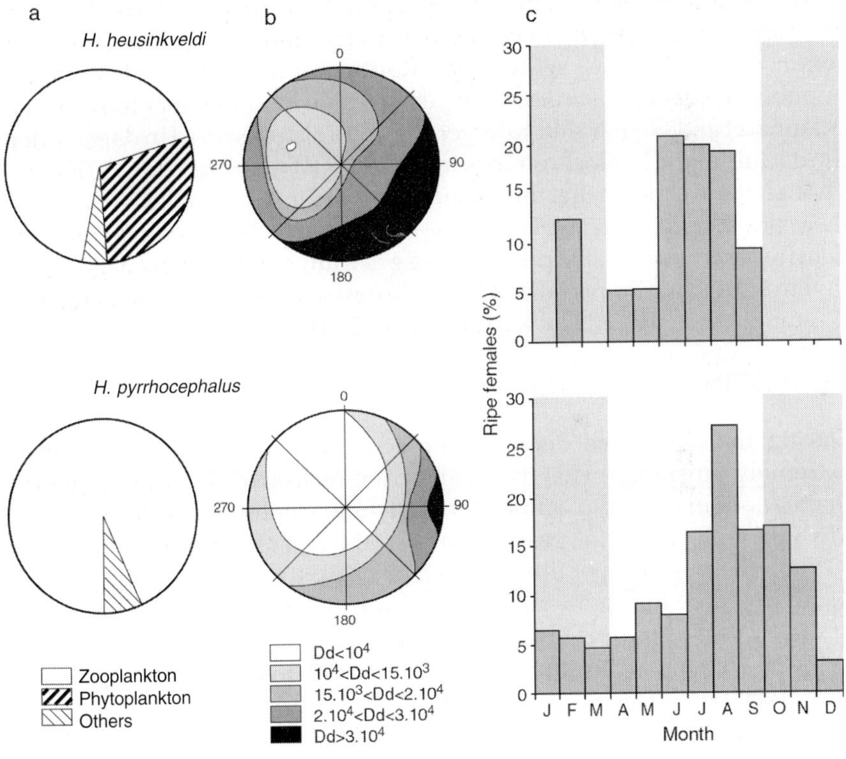

Fig. 10.8 (a) Diet composition of *H. heusinkveldi* and *H. pyrrhocephalus* in 1981-82. (b) Patterns of double cone density (Dd in numbers per mm^2) in *H. heusinkveldi* and *H. pyrrhocephalus*. Patterns of cone size are reversed to those of cone density. (c) Percentage of ripe females of *H. heusinkveldi* and *H. pyrrhocephalus* collected at station G (14 m deep) between March 1983 and October 1984. Grey bars indicate periods with relatively murky water before the ecological changes in the Mwanza Gulf (see also Fig. 10.7). (a) After Goldschmidt et al. (1990) (b) after van der Meer et al. (1995); (c) after Witte et al. (2000).

species, which used to live over sand bottoms at depths upto 6 m, virtually disappeared in the 1980s. After recovery in the 1990s, it extended its habitat and currently, it is also caught over mud bottoms upto a depth of 10 m (van Oijen & Witte, 1996; Seehausen et al., 1997b; J.H. Wanink unpublished data). Compared to those of the fishes of the 1970s and early 1980s, the blue-sensitive single cones in the recovering population decreased in size or even vanished; in certain areas of the retina, the square pattern was transformed into a row pattern (L. Wagenaar, H.J. van der Meer, J.H. Wanink & F. Witte, unpublished data). It is not known whether these changes were caused by natural selection, or whether they represent environmentally induced phenotypic responses.

The recovering zooplanktivores *H. pyrrhocephalus*, *H. laparogramma* and *H. tanaos* all changed their diet. The proportion of zooplankton in the diet decreased and larger preys such as insects, fish and shrimps became more common (van Oijen & Witte, 1996; Katunzi et al., 2003; J. H. Wanink, unpublished data). *H. pyrrhocephalus* even fed on molluscs after its recovery (Katunzi et al., 2003). A shift to larger prey was also observed in dagaa, after the decline of the haplochromines in the 1980s (Wanink & Witte, 2000a, b). Though these shifts in diet might be a result of the decreased water clarity, there are several other explanations—for instance, the strong increase in density of some profitable prey types (e.g. shrimps and juvenile dagaa) and the strong decline in specialized haplochromines that used to feed on them (Wanink and Witte, 2000a; Katunzi et al., 2003).

RESTORING THE BALANCE

During the past three decades, the ecosystem of Lake Victoria was extremely unstable. This is clearly demonstrated by the dramatic fluctuations in the composition of the fish fauna in the northern part of the Mwanza Gulf within the past 25 years. At the end of the 1970s, more than 80% of the bottom-dwelling fish stock consisted of haplochromines, including more than a hundred ecologically different species. By the end of the 1980s, this pattern changed into a system dominated by Nile perch (> 90 %), with haplochromines making up less than 1%. Another ten years later, Nile perch catches declined and haplochromines increased again, making up more than 20% of the demersal fish mass (Witte et al., 2000; J.H. Wanink, unpublished data). However, the current total catches are much lower than in the past, and only about 20 of the more than 100 species in the northern part of the Mwanza Gulf have recovered. The trophic diversity of the recovering haplochromine species is smaller than it used to be. Also, the trophic composition was transformed.

Zooplanktivores are now far more common than detritivores/ phytoplanktivores, whereas the reverse was the case in the past (Witte et al., 2000). One of the major causes of the recovery of the haplochromines seems to be the decline of the Nile perch due to the heavy fishing pressure. The switch to a Nile perch fishery with large-meshed gear in the 1980s and the ban on the trawl and beach seine fishery at the end of the 1990s may have also played a distinct role (Bundy & Pitcher, 1995). Some haplochromine species that may have survived the Nile perch boom in refugia (Kaufman & Ochumba, 1993; Chapman et al., 1995, 2002) currently seem to have a chance to successfully recolonize the lake. However, the resurging haplochromines have to cope with a dramatically changed environment, including decreased DO levels and water transparency.

Though haplochromine cichlids appeared to be hypoxia tolerant in captivity, extreme hypoxia and periodic anoxia in waters of more than 50 m deep may make these areas uninhabitable to most haplochromines. In shallower areas, a decreased scope for activity and an impaired embryonic development due to hypoxia may have negative impacts on the recovering haplochromines. Decreased water clarity may even more seriously affect the recovering haplochromines, as visual information seems to be important in cichlid fish for social interactions and foraging. Breeding of certain species may be hampered in insufficient light conditions. Further, a reduced width of the transmission spectrum increases the chance of hybridization, which may erode species diversity by amalgamation (Seehausen et al., 1997a). Decreased water clarity has probably less impact on non-cichlid fishes in the lake, as most of these species do not seem to depend as much on visual cues for reproduction and feeding as cichlids. Irrespective of the fact whether hypoxia or low water transparency has a dominant impact, it is obvious that eutrophication negatively affects fish diversity in Lake Victoria. Moreover, there is a possibility of a synergetic effect when the impact of two or more adverse features are combined (Beverton, 1992).

The effects on the ecosystem of the recovery of predominantly zooplanktivorous haplochromines are still unknown. Witte et al. (2000) mentioned two possibilities: (1) a decline of the zooplanktivorous dagaa population, which had increased substantially after the decline of the haplochromine cichlids in the 1980s; or (2) an increase of the total zooplanktivorous ichthyomass. Through cascading effects, the latter might further enhance phytoplankton growth (Ogutu-Ohwayo & Hecky, 1991) which, in turn, may influence water turbidity and oxygen concentrations. However, as suggested by Witte et al. (2000) the current dominance of zooplanktivores may be only an intermediate stage in the process of resurgence of haplochromine cichlids. The zooplanktivores were the last haplochromines to disappear from the sublittoral waters in the Mwanza Gulf. This was supposed to be the result of their partly pelagic lifestyle that reduced habitat overlap with Nile perch (Witte & Goudswaard, 1985; Wanink, 1991; Witte et al., 1992b). If their semi-pelagic lifestyle did lead to a relatively high survival of zooplanktivores in refugia, it seems plausible that they are the first to recover with the decline of Nile perch pressure. Nevertheless, only part of the zooplanktivorous species have as yet recovered, while other pelagic haplochromines such as the phytoplanktivores (Goldschmidt et al., 1993) are still absent. It is expected that the recovery of haplochromines will be more successful if the negative impacts of eutrophication can be reduced. Though the return of all the species is unlikely, the diversity of species and trophic types may still increase, depending on the number of species that survived in refugia.

The effects of reductions in diversity of ecosystems have been a controversial issue during the past 50 years (Rozdilsky & Stone, 2001). However, ecological studies during the past decade showed that a greater diversity may lead to greater ecosystem predictability, temporal stability and, in the long term, to greater ecosystem productivity (Naeem & Li, 1997; McGrady-Steed, 1997; Norberg et al., 2001; Tilman 2001; Rozdilsky & Stone, 2001; Bolam et al., 2002). Moreover, Nile perch seems to grow fastest on a haplochromine prey base (Kaufman & Schwarz, 2002). Thus, besides the aesthetic quality of the lake, a further recovery of haplochromine species may be of great importance for a healthy and productive ecosystem (Balirwa et al., 2003).

The catchment area of Lake Victoria covers 195,000 km^2 and has a human population of ca. 30 million people with a growth rate of 6% per annum in urban centres and over 3% in rural areas (Ntiba et al., 2001). The people largely depend on the agriculture and cattle breeding and the use of firewood as fuel for cooking. Deforestation, coupled with bad agricultural practices, strongly increases land erosion. Rapid industrial development in the major urban centres contributes to pollution of the lake (Scheren et al., 2000; Ntiba et al., 2001; Kishe & Machiwa, 2003). Unless adequate measures are taken, eutrophication and pollution of Lake Victoria is bound to increase rapidly. Such protective and preventive measures comprise reforestation of the riparian zones of the lake and its tributaries; restoration of the wetlands, which function as natural filter systems; building of sewage plants near densely populated areas; and other actions reducing the influx of materials and contaminants into the lake. Since the 1990s, there has been a growing awareness of the regional and the international scientific community, and of the riparian governments regarding the need for effective measures urgently. The governments of Kenya, Tanzania and Uganda have established the Lake Victoria Fisheries Organization (LVFO) in 1994 as a permanent institution to protect and restore the lake. A Global Environment Facility (GEF) project is providing funding through the Worldbank to the LVFO and addresses land use management, catchment forestation, fisheries research and management, water hyacinth control, industrial effluent treatment and municipal waste treatment (Ntiba et al., 2001). Other regional projects are: (1) the Lake Victoria Fisheries Research Project (LVFRP), funded by the European Union, which is addressing the status of the fish stocks in the lake; and (2) an FAO-funded Lake Victoria Water Resources Project (LVWRP) addressing issues about lakewater balances (Ntiba et al., 2001). These are important initiatives, but much work remains to be done before measures can be taken that will prove to be really effective in restoring the balance of the ecosystem.

Finally, it should be noted that eutrophication is not only a serious threat to Lake Victoria but to all of the great African lakes. Only with great regional and international efforts will it be possible to conserve these precious freshwater bodies that supply millions of people with water and food and within them contain a unique biodiversity.

ACKNOWLEDGEMENTS

The first two authors thank their colleagues from the Haplochromis Ecology Survey Team (HEST) and the Tanzania Fisheries Research Institute (TAFIRI) for support and cooperation during the fieldwork. Patrick Niemantsverdriet, Peter Snelderwaard and their colleagues are acknowledged for taking care of the laboratory-reared fishes and their assistance with the hypoxia conditioning experiments. We are indebted to Kees Barel for his comments on an earlier draft of this chapter and also to Martin Brittijn for making of the figures. The research of HEST was financially supported by The Netherlands Foundation for the Advancement of Tropical Research (WOTRO; grants W87-129, W87-161, W87-189, W84-488), the Section for Research and Technology of The Netherlands Minister of Development Cooperation, the Schure Beijerinck-Popping Fonds, the van Tienhoven Stichting and by Yellow Springs Instruments.

REFERENCES

Akiyama, T., Kajumulo, A. A. & Olsen, S. 1977. Seasonal variations of plankton and physico-chemical condition in Mwanza Gulf, Lake Victoria. Bull. Freshwat. Fish. Res. Lab. 27: 49-61.

Ali, M. A. & Anctil, M. 1976. Retinas of Fishes: An Atlas. Springer Verlag. Berlin.

Arunga, J. O. 1981. A case study of the Lake Victoria Nile perch *Lates niloticus* (Mbuta) fishery. In: Proceedings of the Workshop of the Kenya Marine and

Fisheries Research Institution on Aquatic Resources of Kenya, Mombasa, 13-19 July 1981. KMFRI & Ken. Nat. Ac. Adv. Arts Sci., Nairobi: 165-184.

Balirwa, J. S. 1995. The Lake Victoria environment: Its fisheries and its wetlands—a review. Wetl. Ecol. Managem. 3: 209-224.

Balirwa, J. S., Chapman, C. A., Chapman, L. J., Cowx, I. G., Geheb, K., Kaufman, L., Lowe-McConnell, R. H., Seehausen, O., Wanink, J.H., Welcomme, R. L. & Witte, F. 2003. Biodiversity and fishery sustainability in the Lake Victoria basin: An unexpected marriage? BioScience 53: 703-715

Barel, C. D. N. 1993. Concepts of an architectonic approach to transformation morphology. Acta Biotheor. 41: 345-381.

Barel, C. D. N., Ligtvoet, W., Goldschmidt, T., Witte, F. & Goudswaard, P. C. 1991. The haplochromine cichlids of Lake Victoria: An assessment of biological and fisheries interest. In: Keenleyside, M.H.A. (ed.), Cichlid Fishes: Behaviour, Ecology and Evolution. Chapman & Hall, London: 258-279.

Barel, C. D. N., Dorit, R., Greenwood, P. H., Fryer, G., Hughes, N., Jackson, P. B. N., Kanawabe, H., Lowe-McConnel, R. H., Nagoshi, M., Ribbink, A. J., Trewavas, E., Witte, F. & Yamaoka, K. 1985. Destruction of fisheries in Africa's lakes. Nature 315: 19-20.

Beauchamp, R. S. A. 1958. Utilising the natural resources of Lake Victoria for the benefit of fisheries and agriculture. Nature 181: 1634-1636.

Beverton, R. J. H. 1992. Fish resources; threats and protection. Neth. J. Zool. 42: 139-175.

Bolam, S. G., Fernandes, T. F. & Huxham, M. 2002. Diversity, biomass and ecosystem processes in the marine benthos. Ecol. Monogr. 72: 599-615.

Bowmaker, J. K. 1990. Visual pigments of fishes. In: Douglas R. H. & M. B. A. Djamgoz (eds), The Visual System of Fish. Chapman & Hall, London: 81-108.

Bowmaker, J. K. 1995. The visual pigments of fish. Prog. Retina Eye Res. 15: 1-31.

Bundy, A. & Pitcher, T. J. 1995. An analysis of species changes in Lake Victoria: Did the Nile perch act alone? In: Pitcher, T. J. & P. J. B. Hart (eds), The Impact of Species Changes in African Lakes. Chapman & Hall, London: 111-135.

Chapman, L. J., Galis, F. & Shinn, J. 2000. Phenotypic plasticity and the possible role of genetic assimilation: Hypoxia-induced trade-offs in the morphological traits of an African cichlid. Ecol. Lett. 3: 387-393.

Chapman, L. J., Chapman, C. A., Nordlie, F. G. & Rosenberger, A. E. 2002. Physiological refugia: swamps, hypoxia tolerance and maintenance of fish diversity in the Lake Victoria region. Comp. Biochem. Physiol. A 133: 421-437.

Chapman, L. J., Kaufman, L. S., Chapman, C. A. & McKenzie, F. E. 1995. Hypoxia tolerance in twelve species of East African cichlids: Potential for low oxygen refugia in Lake Victoria. Conserv. Biol. 9: 1274-1288.

Chapman, L. J., Chapman, C. A., Ogutu-Ohwayo, R., Chandler, M., Kaufman, L. & Keiter, A. E. 1996. Refugia for endangered fishes from an introduced predator in Lake Nabugabo, Uganda. Conserv. Biol. 10: 554-561.

Chapman, L. J., Chapman, C. A., Schofield, P. J., Olowo, J. P., Kaufman, L., Seehausen, O. & Ogutu-Ohwayo, R. 2003. Fish faunal resurgence in Lake Nabugabo, East Africa. Conserv. Biol. 17: 500-511.

de Beer, M. 1989. Light measurements in Lake Victoria, Tanzania. Ann. Mus. r. Afr. c., Zool. 257: 57-60.

Fernandes, M. N. & Rantin, F. T. 1989. Respiratory responses of *Oreochromis niloticus* (Pisces, Cichlidae) to environmental hypoxia under different thermal conditions. J. Fish Biol. 35: 509-519.

Fish, G. R. 1956. Some aspects of the respiration of six species of fish from Uganda. J. Exp. Biol. 33: 186-195.

Fish, G. R. 1957. A seiche movement and its effect on the hydrology of Lake Victoria. Fish. Publ. Lond. 10: 1-68.

Fryer, G. 1960. Concerning the proposed introduction of Nile perch into Lake Victoria. E. Afr. J. Agric. 25: 267-270.

Fryer, G. & Iles, T. D. 1972. The Cichlid Fishes of the Great Lakes of Africa: Their Biology and Evolution. Oliver & Boyd, Edinburgh.

Galis, F. & Barel, C. D. N. 1980. Comparative functional morphology of the gills of African lacustrine Cichlidae (Pisces, Teleostei): An eco-morphological approach. Neth. J. Zool. 30 : 392-430.

Galis, F. & Smit, H. 1979. Hypoxia tolerance of two closely related *Haplochromis* species (Pisces: Cichlidae): *Haplochromis elegans* Trewavas, 1933 and *H. angustifrons* Boulenger, 1914. Comp. Biochem. Physiol. 64A: 137-139.

Goldschmidt, T. & Goudswaard, P. C. 1989. Reproductive strategies in haplochromines (Pisces, Cichlidae) from Lake Victoria. In: Goldschmidt T. An Ecological And Morphological Field Study on the haplochromine cichlid fishes (Pisces, Cichlidae) of Lake Victoria. PhD thesis Leiden: 93-119.

Goldschmidt, T. & Witte, F. 1990. Reproductive strategies of zooplanktivorous haplochromine cichlids (Pisces) from Lake Victoria before the Nile perch boom. Oikos 58: 356-368.

Goldschmidt, T., Witte, F. & de Visser, J. 1990. Ecological segregation in zooplanktivorous haplochromine species (Pisces: Cichlidae) from Lake Victoria. Oikos 58: 343-355.

Goldschmidt, T., F. Witte & Wanink, J. 1993. Cascading effects of the introduced Nile perch on the detritivorous/phytoplanktivorous species in the sublittoral areas of Lake Victoria. Conserv. Biol. 7: 686-700.

Gophen, M., Ochumba, P. B. O. & Kaufman, L. S. 1995. Some aspects of perturbation in the structure and biodiversity of the ecosystem of Lake Victoria (East Africa). Aquat. Living Resour. 8: 27-41.

Gophen, M., Ochumba, P. B. O., Pollingher, U. & Kaufman, L. S. 1993. Nile perch (*Lates niloticus*) invasion in Lake Victoria (East Africa). Verh. Int. Ver. Limnol. 25: 856-859.

Goudswaard, P. C. & Witte, F. 1997. The catfish fauna of Lake Victoria after the Nile perch upsurge. Env. Biol. Fish. 49: 21-43.

Goudswaard, P. C., Witte, F. & Chapman, L. J. 2002a. Decline of the African lungfish (*Protopterus aethiopicus*) in Lake Victoria (East Africa). Afr. J. Ecol. 40: 42-52.

Goudswaard, P. C., Witte, F. & Katunzi, E. F. B. 2002b. The tilapiine fish stock of Lake Victoria (East Africa) before and after the Nile perch upsurge. J. Fish Biol. 60: 838-856.

Graham, M. 1929. The Victoria Nyanza and its fisheries: a report on the fishing survey of Lake Victoria (1927-1928). Crown Agents for the Colonies, London.

Greenwood, P. H. 1973. A revision of the *Haplochromis* and related species (Pisces, Cichlidae) from Lake George, Uganda. Bull. Br. Mus. Nat. Hist. (Zool.) 25: 139-242.

Greenwood, P. H. 1974. The cichlid fishes of Lake Victoria, East Africa: the biology and evolution of a species flock. Bull. Br. Mus. Nat. Hist. (Zool.) Suppl. 6: 1-134.

Greenwood, P. H. 1981. The Haplochromine Fishes of the East African Lakes: Collected Papers on their Taxonomy, Biology and Evolution. Kraus International Publications, Munich.

Guthrie D. M. & Muntz, W. R. A. 1993. Role of vision in behaviour. In Pitcher, T. J. (ed.), Behaviour of Teleost Fishes 2nd edn. Chapman & Hall, London: 112-115.

Hecky, R. E. 1993. The eutrophication of Lake Victoria. Verh. int. Ver. Limnol. 25: 39-48.

Hecky, R. E., Bugenyi, F. W. B., Ochumba, P., Talling, J. F., Mugidde, R., Gophen, M. & Kaufman, L. 1994. Deoxygenation of the deep water of Lake Victoria, East Africa. Limnol. Oceanogr. 39: 1476-1481.

Hoogerhoud, R.J.C., Witte, F. & Barel, C.D.N. 1983. The ecological differentiation of two closely resembling *Haplochromis* species from Lake Victoria (*H. iris* and *H. hiatus*; Pisces, Cichlidae). Neth. J. Zool. 33: 283-305.

Hughes, N. F. 1992. Nile perch, *Lates niloticus*, predation on the freshwater prawn, *Caridina nilotica*, in the Nyanza Gulf, Lake Victoria, East Africa. Env. Biol. Fish. 33: 307-309.

Jobling, M. 1995. Environmental Biology of Fishes. Chapman & Hall, London.

Johnson, T. C., Scholz, C. A., Talbot, M. R., Kelts, K., Ricketts, R. D., Ngobi, G., Beuning, K., Ssemmanda, I. & McGill, J. W. 1996. Late Pleistocene desiccation of Lake Victoria and rapid evolution of cichlid fishes. Science 273: 1091-1093.

Jones D. R., Randall, D. J. & Jarman, G. M. 1970. A graphical analysis of oxygen transfer in fish. Resp. Physiol. 10: 285-298.

Kairu, J. K. 2001. Wetland use and impact on Lake Victoria, Kenya region. Lakes Reserv.: Res. Manage. 6: 117-125.

Katunzi, E. F. B., Zoutendijk, J., Goldschmidt, T., Wanink, J. H. & Witte, F. 2003. Lost zooplanktivorous cichlid from Lake Victoria reappears with a new trade. Ecol. Freshw. Fish. 12: 237-240.

Kaufman, L. 1992. Catastrophic change in species-rich freshwater ecosystems: The lessons of Lake Victoria. BioScience 42: 846-858.

Kaufman, L. & Ochumba, P. 1993. Evolutionary and conservation biology of cichlid fishes as revealed by faunal remnants in northern Lake Victoria. Conserv. Biol. 7: 719-730.

Kaufman, L. & Schwarz, J. 2002. Nile perch population dynamics in Lake Victoria: implications for management and conservation. In: M. Ruth, & J. Lindholm (eds), Dynamic Modelling for Marine Conservation. Springer-Verlag, New York: 257-313.

Kaufman, L. S., Chapman L. J. & Chapman, C. A. 1997. Evolution in fast forward: Haplochromine fishes of the Lake Victoria region. Endeavour 21: 23-30.

Kishe, M. A. & Machiwa, J. F. 2003. Distribution of heavy metals in sediments of Mwanza Gulf of Lake Victoria, Tanzania. Environ. Int. 28: 619-625.

Kitaka, G. E. B. 1972. An instance of cyclonic upwelling in the southern offshore waters of Lake Victoria. Afr. J. Trop. Hydrobiol. Fish. 1: 85-92.

Kranenbarg, S., Muller, M., Gielen, J. L. W. & Verhagen, J. H. G. 2000. Physical constraints on body size in teleost embryos. J. Theor. Biol. 204: 113-133.

Kröger, R. H. H., Bowmaker, J. K. & Wagner, H. J. 1999. Morphological changes in the retina of *Aequidens pulcher* (Cichlidae) after rearing in monochromatic light. Vis. Res. 39: 2441-2448.

Kudhongania, A. W. & Chitamwebwa, D. B. R. 1995. Impact of environmental change, species introductions and ecological interactions on the fish stocks of Lake Victoria. In: T. J. Pitcher, & P. J. B. Hart (eds), The Impact of Species Changes in African Lakes. Chapman & Hall, London: 19-32.

Kudhongania, A. W. & Cordone, A. J. 1974. Batho-spatial distribution patterns and biomass estimates of the major demersal fishes in Lake Victoria. Afr. J. Trop. Hydrobiol. Fish. 3: 15-31.

Lehman, J. T., G. B. Mbahinzireki & Mwebaza-Ndawula, L. 1996. *Caridina nilotica* in Lake Victoria: Abundance, biomass, and diel vertical migration. Hydrobiologia 317: 177-182.

Levine, J. S. & MacNichol, E. F. 1982. Color vision in fishes. Sci. Am. 246: 140-149.

Ligtvoet, W., Mous, P. J., Mkumbo, O. C., Budeba, Y. L., Goudswaard, P. C., Katunzi, E. F. B., Temu, M. M., Wanink, J. H. & Witte, F. 1995. The Lake Victoria fish stocks and fisheries. In: F. Witte, & W. L. T. van Densen (eds), Fish Stocks and Fisheries of Lake Victoria: A Handbook for Field Observations. Samara Publishing Ltd., Cardigan, UK: 12-53.

Loew, E. R. & McFarland, W. N. 1990. The underwater visual environment. In: R.H. Douglas, & M. B. A. Djaamgoz (eds), The Visual System of Fish. Chapman & Hall, London: 1-40.

Lowe-McConnell, R. H. 2000. The roles of tilapias in ecosystems. In: M. C. M. Beveridge, & B. J. McAndrew (eds), Tilapias: Biology and Exploitation. Kluwer Academic Publishers, Dordrecht, The Netherlands: 129-162.

Lung'ayia, H., Sitoki, L. & Kenyanya, M. 2001. The nutrient enrichment of Lake Victoria (Kenyan waters). Hydrobiologia 458: 75-82.

McGrady-Steed, J., Harris, P. M. & Morin, P. J. 1997. Biodversity regulates ecosystem predictability. Nature 390: 162-165.

Mkumbo, O. C. & Ligtvoet, W. 1992. Changes in the diet of Nile perch, *Lates niloticus* (L), in the Mwanza Gulf, Lake Victoria. Hydrobiologia 232: 79-83.

Moriarty, D. J. W., Darlington, J. P. E. C., Dunn, I. G., Moriarty, C. M. M. & Tevlin, M. P. 1973. Feeding and grazing in Lake George, Uganda. Proc. Roy. Soc. Lond. 184: 299-319.

Mugidde, R. 1993. The increase in phytoplankton primary productivity and biomass in Lake Victoria (Uganda). Ver. int. Ver. Limnol. 25: 846-849.

Muusze B., Marcon, J., van den Thillart, G. & Almeida-Val, V. 1998. Hypoxia tolerance of amazon fish. Respiratory and energy metabolism of the cichlid *Astronotus ocellatus*. Comp. Biochem. Physiol. 120A: 151-156.

Mwebaza-Ndawula, L. 1994. Changes in relative abundance of zooplankton in northern Lake Victoria, East Africa. Hydrobiologia 272: 259-264.

Naeem, S. & Li, S. 1997. Biodiversity enhances ecosystem reliability. Nature 390: 507-509.

Njuguna, S. G. 1991. Water hyacinth: The world's worst aquatic weed infests lakes Naivasha and Victoria. Swara 14 (6): 8-10.

Norberg, J., Swaney, D. P., Dushoff, J., Lin, J., Casagrandi, R. & Levin, S. A. 2001. Phenotypic diversity and ecosystem functioning in changing environments: a theoretical framework. Proc. Natl. Acad. Sci.USA, 98: 11376-11381.

Ntiba, M. J., Kudoja, W. M. & Mukasa, C. T. 2001. Management issues in the Lake Victoria watershed. Lakes Reserv.: Res. Manage. 6: 211-216.

Ochumba, P. B. O. 1990. Massive fish kills within the Nyanza Gulf of Lake Victoria, Kenya. Hydrobiologia 208: 93-99.

Ochumba, P. B. O. 1995. Limnological changes in Lake Victoria since the Nile perch introduction. In: T. J. Pitcher, & P. J. B. Hart (eds), The Impact of Species Changes in African Lakes. Chapman & Hall, London: 33-43.

Ochumba, P. B. O. & Kibaara, D. I. 1989. Observations on blue-green algal blooms in the open waters of Lake Victoria, Kenya. Afr. J. Ecol. 27: 23-34.

Ogari, J. & Dadzie, S. 1988. The food of the Nile perch, *Lates niloticus* (L.) after the disappearance of the haplochromine cichlids in the Nyanza Gulf of Lake Victoria. J. Fish Biol. 32: 571-577.

Ogutu-Ohwayo, R. 1990a. The decline of the native fishes of Lakes Victoria and Kyoga (East Africa) and the impact of introduced species, especially the Nile perch, *Lates niloticus* and the Nile tilapia, *Oreochromis niloticus*. Env. Biol. Fish. 27: 81-96.

Ogutu-Ohwayo, R. 1990b. Changes in the prey ingested and the variations in the Nile perch and other fish stocks of Lake Kyoga and the northern waters of Lake Victoria (Uganda). J. Fish Biol. 37: 55-63.

Ogutu-Ohwayo, R. & Hecky, R. E. 1991. Fish introductions in Africa and some of their implications. Can. J. Fish. Aquat. Sci. 48, Suppl. 1: 8-12

Ogutu-Ohwayo, R., Hecky, R. E., Cohen, A. S. & Kaufman, L. 1997. Human impacts on the African Great Lakes. Env. Biol. Fish. 50: 117-131.

Okedi, J. 1981. The *Engraulicypris* "Dagaa" fishery of Lake Victoria: with special reference to the southern waters of the lake. In Proceedings of the Workshop of the Kenya Marine and Fisheries Research Institution on Aquatic Resources of Kenya, Mombasa, 13-19 July 1981. KMFRI & Ken. Nat. Ac. Adv. Arts Sci., Nairobi: 445-484.

Otten, E. 1981. Vision during growth of a generalized *Haplochromis* species: *H. elegans* Trewavas 1933 (Pisces, Cichlidae). Neth. J. Zool. 31: 650-700.

Pitcher, T. J. & Bundy, A. 1995. Assessment of the Nile perch fishery in Lake Victoria. In: T. J. Pitcher, & P. J. B. Hart (eds), The Impact of Species Changes in African Lakes. Chapman & Hall, London: 163-180.

Reynolds, J. E., Gréboval, D. F. & Mannini, P. 1995. Thirty years on: the development of the Nile perch fishery in Lake Victoria. In: T. J. Pitcher, & P. J. B. Hart (eds), The Impact of Species Changes in African Lakes. Chapman & Hall, London: 181-214.

Riedmiller, S. 1994. Lake Victoria fisheries: the Kenyan reality and environmental implications. Env. Biol. Fish. 39: 329-338.

Rosenberger A. E. & Chapman, L. J. 1999. Hypoxic wetland tributaries as faunal refugia from an introduced predator. Ecol. Freshw. Fish 8: 22-34.

Rozdilsky, I. D. & Stone, L. 2001. Complexitity can enhance stability in competitive systems. Ecol. Lett. 4: 397-400.

Scheren, P. A. G. M., Zanting, H. A. & Lemmens, A. M. C. 2000. Estimation of water pollution sources in Lake Victoria, East Africa: Application and elaboration of the rapid assessment methodology. J. Env. Managem. 58: 235-248.

Seehausen, O. 1996. Lake Victoria Rock Cichlids: Taxonomy, Ecology, and Distribution. Verduyn Cichlids, Zevenhuizen, The Netherlands.

Seehausen, O. 1997. Distribution of and reproductive isolation among color morphs of a rock-dwelling Lake Victoria cichlid (*Haplochromis nyererei*). Ecol. Freshw. Fish 6: 59-66.

Seehausen, O. 2000. Explosive speciation rates and unusual species richness in haplochromine cichlid fishes: effects of sexual selection. Adv. Ecol. Res. 31: 237-274.

Seehausen, O. & van Alphen, J. J. M. 1998. The effect of male coloration on female mate choice in closely related Lake Victoria cichlids (*Haplochromis nyererei* complex). Behav. Ecol. Sociobiol. 42: 1-8.

Seehausen, O., van Alphen, J. J. M. & Witte, F. 1997a. Cichlid fish diversity threatened by eutrophication that curbs sexual selection. Science 277: 1808-1811.

Seehausen, O., van Alphen, J. J. M. & Witte, F. 2003. Implications of eutrophication for fish vision, behevioral ecology and species coexistence. In: T. L. Crisman, L. Chapman & C. Chapman (eds), Conservation, Ecology, and Management of African Freshwaters. University Press of Florida, Florida. : 268-287.

Seehausen, O., Witte, F., van Alphen, J. J. M. & Bouton, N. 1998. Direct mate choice is the mechanism that maintains diversity among sympatric cichlids in Lake Victoria. J. Fish Biol. 53: 37-55.

Seehausen, O., Witte, F., Katunzi, E. F., Smits, J. & Bouton, N. 1997b. Patterns of the remnant cichlid fauna in southern Lake Victoria Conser. Biol. 11: 890-904.

Slootweg, R. 1987. Prey selection by molluscivorous cichlids foraging on a schistosomatic vector snail, *Biomphalaria glabrata*. Oecologia (Berlin) 74: 193-202.

Smit, H. 1965. Some experiments on the oxygen consumption of goldfish (*Carassius auratus* L.) in relation to swimming speed. Can. J. Zool. 43: 623-633.

Smit, S. A. & Ch.Anker, G. 1997. Photopic sensitivity to red and blue light related to retinal differences in two zooplanktivorous haplochromine species (Teleostei, Cichlidae). Neth. J. Zool. 47: 9-20.

Talling, J. F. 1966. The annual cycle of stratification and phytoplankton growth in Lake Victoria (East Africa). Int. Rev. ges. Hydrobiol. 51: 545-621.

Tilman, D. 2001. An evolutionary approach to ecosystem functioning. Proc. Natl. Acad. Sci. USA, 98: 10979-10980.

Twongo, T. & Odongkara, O. K. 2000. Invasive water weeds in Lake Victoria basin: proliferation, impacts and control. In: Lake Victoria 2000: A New Beginning, International Conference 16–19 May, 2000, Jinja, Uganda, Book of Abstracts. Lake Victoria Fisheries Organization, Jinja, Uganda: 52-53.

van der Meer, H. J. 1992. Constructional morphology of photoreceptor patterns in percomorph fish. Acta biotheor. 40: 51-85.

van der Meer, H. J. 1993. Light-induced modulation of retinal development in the cichlid fish *Haplochromis sauvagei* (Pfeffer, 1896). Zool. J. Linnean Soc. 108: 271-285.

van der Meer, H. J. 1994. Ontogenetic change of visual tresholds in the cichlid fish *Haplochromis sauvagei*. Brain Beh. Evol. 44: 40-49.

van der Meer, H. J. 1995. Visual resolution during growth in a cichlid fish: A morphological and behavioural case study. Brain Behav. Evol. 45: 25-33.

van der Meer, H. J. & Anker, G. C. 1984. Retinal resolving power and sensitivity of the photopic system in seven haplochromine species (Cichlidae, Teleostei). Neth. J. Zool. 34: 197-209.

van der Meer, H. J. & Bowmaker, J. K. 1995. Interspecific variation of photoreceptors in four co-existing haplochromine cichlid fishes. Brain Behav. Evol. 45: 232-240.

van der Meer, H. J., Anker, C. Ch. & Barel, C. D. N. 1995. Ecomorphology of retinal structures in zooplanktivorous haplochromine cichlids (Pisces) from Lake Victoria. Env. Biol. Fish. 44: 115-132.

van Oijen, M. J. P. 1995. Key to Lake Victoria fishes other than haplochromine cichlids. In: Witte, F. & W. L. T. van Densen (eds), Fish Stocks and Fisheries of Lake Victoria: A Handbook for Field Observations. Samara Publishing Ltd., Cardigan, UK: 209-300.

van Oijen, M. J. P. & Witte, F. 1996. Taxonomical and ecological description of a species complex of zooplanktivorous and insectivorous cichlids from Lake Victoria. Zool. Verh. Leiden 302: 1-56.

van Oijen, M. J. P., Witte, F. & Witte-Maas, E. L. M. 1981. An introduction to ecological and taxonomic investigations on the haplochromine cichlids from the Mwanza Gulf of Lake Victoria. Neth. J. Zool. 31: 149-174.

van den Thillart G. & van Waarde, A. 1985. Teleosts in hypoxia: Aspects of anaerobic metabolism. Mol. Physiol. 8: 393-410.

Verheyen E., Blust, R. & Decleir, W. 1986. Hemoglobin heterogeneity and the oxygen affinity of hemolysate of some Victoria cichlids. Comp. Biochem. Physiol. 84A: 315-318.

Verheyen E., Blust, R. & Doumen, C. 1985. The oxygen uptake of *Sarothereodon niloticus* L. and the oxygen binding properties of its blood and hemolysate (Pisces: cichlidae). Comp. Biochem. Physiol. 81A: 423-426.

Verschuren, D., Edgington, D. N., Kling, H. J. & Johnson, T. C. 1998. Silica depletion in Lake Victoria: sedimentary signals at offshore stations. J. Great Lakes Res. 24: 118-130.

Verschuren, D., Johnson, T. C., Kling, H. J., Edgington, D. N., Leavitt, P. R., Brown, E. T., Talbot, M. R. & Hecky, R. E. 2002. History and timing of human impact on Lake Victoria, East Africa. Proc. R. Soc. Lond. B 269: 289-294.

Wagner, H. J. 1978. Cell types and connectivity patterns in mosaic retinas. Adv. Anat. Embryol. Cell Biol. 55 (3): 1-81.

Wanink, J. H. 1991. Survival in a perturbed environment: the effects of Nile perch introduction on the zooplanktivorous fish community of Lake Victoria. In: O. Ravera, (ed.), Terrestrial and Aquatic Ecosystems: Perturbation and Recovery. Ellis Horwood Ltd., Chichester, UK: 269-275.

Wanink, J. H. 1999. Prospects for the fishery on the small pelagic *Rastrineobola argentea* in Lake Victoria. Hydrobiologia 407: 183-189.

Wanink, J. H. & Kashindye, J. J. 1998. Short-term variations in pelagic photosynthesis demand well-timed sampling to monitor long-term limnological changes in Lake Victoria. Hydrobiologia 377: 177-181.

Wanink, J. H. & Witte, F. 2000a. The use of perturbation as a natural experiment: Effects of predator introduction on the community structure of zooplanktivorous fish in Lake Victoria. Adv. Ecol. Res. 31: 553-570.

Wanink, J. H. & Witte, F. 2000b. Rapid morphological changes following niche shift in the zooplanktivorous cyprinid *Rastrineobola argentea* from Lake Victoria. Neth. J. Zool. 50: 365-372.

Wanink, J. H., Kashindye, J. J., Goudswaard, P. C. & Witte, F. 2001. Dwelling at the oxycline: does increased stratification provide a predation refugium for the Lake Victoria sardine *Rastrineobola argentea*? Freshw. Biol. 46: 75-85.

Wanink, J. H., Katunzi, E. F. B., Goudswaard, P. C., Witte, F. & van Densen, W. L. T. 2002. The shift to smaller zooplankton in Lake Victoria cannot be attributed to the 'sardine'*Rastrineobola argentea* (Cyprinidae). Aquat. Living Resour. 15: 37-43.

Weber, R. E. 1990. Functional significance and structural basis of multiple hemoglobins with special reference to ectothermic vertebrates. In: J-P. Truchot & B. Lahlou (eds), Animal Nutrition and Transport Processes. 2. Transport, Respiration and Excretion: Comparative and Environmental Aspects. Comp. Physiol. Karger, Basel, 6: 58-75.

Weber, R. E., & Wells, R. M. G. 1989. Hemoglobin structure and function. In: S. C. Wood, (ed.), Comparative Pulmonary Physiology. Marcel Dekker inc., New York: 279-309.

Welcomme, R. L. 1967. Observations on the biology of the introduced species of Tilapia in Lake Victoria. Rev. Zool. Bot. Afr. 76: 249-279.

Welcomme, R. L. 1988. International introductions of inland fish species. FAO Fish. Tech. Pap. 294: 1-318.

Witte, F. 1981. Initial results of the ecological survey of the haplochromine cichlid fishes from the Mwanza Gulf of Lake Victoria, Tanzania: breeding patterns, trophic and species distribution. Neth. J. Zool. 31: 175-202.

Witte, F. & Goudswaard, P. C. 1985. Aspects of the haplochromine fishery in southern Lake Victoria. FAO Fish. Rep. 335: 81-88.

Witte, F. & van Oijen, M. J. P. 1995. Biology of haplochromine trophic groups. In: F. Witte, & W. L. T. van Densen (eds), Fish Stocks and Fisheries of Lake Victoria: A Handbook for Field Observations. Samara Publishing Ltd., Cardigan, UK: 321-335.

Witte, F., T. Goldschmidt & Wanink, J. H. 1995. Dynamics of the haplochromine cichlid fauna and other ecological changes in the Mwanza Gulf of Lake Victoria. In: T. J. Pitcher, & P. J. B. Hart (eds), The Impact of Species Changes in African Lakes. Chapman & Hall, London: 83-110.

Witte, F., T. Goldschmidt, Goudswaard, P. C., Ligtvoet, W., van Oijen, M. J. P. & Wanink, J. H. 1992a. Species extinction and concomitant ecological changes in Lake Victoria. Neth. J. Zool. 42: 214-232.

Witte, F., Goudswaard, P. C., Katunzi, E. F. B., Mkumbo, O. C., Seehausen, O. & Wanink, J. H. 1999. Lake Victoria's ecological changes and their relationships with the riparian societies. In: H. Kawanabe, G. W. Coulter & A. C. Roosevelt (eds), Ancient Lakes: Their Cultural and Biological Diversity. Kenobi Productions, Ghent, Belgium: 189-202.

Witte, F., Goldschmidt, T., Wanink, J., van Oijen, M., Goudswaard, K., Witte-Maas, E. & Bouton, N. 1992b. The destruction of an endemic species flock: quantitative data on the decline of the haplochromine cichlids of Lake Victoria. Env. Biol. Fish. 34: 1-28.

Witte, F., Msuku, B. S., Wanink, J. H. Seehausen, O., Katunzi, E. F. B., Goudswaard, P. C. & Goldschmidt, T. 2000. Recovery of cichlid species in Lake Victoria: An examination of factors leading to differential extinction. Rev. Fish Biol. Fish. 10: 233-241.

Worthington, E. B. 1930. Observations on the temperature, hydrogen-ion concentration, and other physical conditions of the Victoria and Albert Nyanzas. Int. Rev. Gesamt. Hydrobiol. 24: 328-357.

PART 3

Restoration and Management

CHAPTER

11

Lake Restoration and Biomanipulation in Temperate Lakes: Relevance for Subtropical and Tropical Lakes

Erik Jeppesen[1,2*], Martin Søndergaard[1], Nestor Mazzeo[3], Mariana Meerhoff[1], Christina C. Branco[4], Vera Huszar[5] and Flavio Scasso[3]

[1]National Environmental Research Institute, Dept. of Freshwater Ecology, Vejlsøvej 25, DK-8600 Silkeborg, Denmark

[2]Dept. of Botanical Ecology, University of Aarhus, Nordlandsvej 68, DK-8230 Risskov, Denmark

[3]Sección Limnología, Departamento de Ecología, Facultad de Ciencias, Iguá 4225, Montevideo, Uruguay

[4]Universidade do Rio de Janeiro, Depto. Ciências Naturais, Av. Pasteur 458, 22290-240 Rio de Janeiro, Brazil

[5]Universidade Federal do Rio de Janeiro, Museu Nacional, Departamento de Botânica, Laboratório de Ficologia. Quinta da Boa Vista, São Cristóvão, Rio de Janeiro, Brazil, 20940-040

ABSTRACT

This chapter gives a broad survey of the responses of temperate lakes to increased nutrient loading (the eutrophication process) and during remediation. Chemically and biologically conditioned resistance may cause a delayed response to the nutrient loading reduction and different methods of overcoming this resistance have been briefly described. Biological restoration methods (termed biomanipulation) are promising new tools. However, a significant difference in biological interactions in temperate versus subtropical—tropical lakes renders it difficult to directly apply the biological restoration methods. These include often higher dominance and abundance of small fish, higher aggregation of fish in vegetation, higher number of fish cohorts per year, higher degree of omnivorous feeding by fish and less piscivory in subtropical and tropical lakes than in temperate lakes. Unfortunately, very little is known about the trophic dynamics and the role of fishes in warm lakes. Since many subtropical and tropical lakes are heavily eutrophicated, today there is a great need for gaining new knowledge of trophic interactions and possible lake restoration methods for these regions. This is amplified by the fact that the eutrophication problem is expected to increase in the near future due to both the economical development and global warming.

EUTROPHICATION OF LAKES

During the past centuries, increased urbanization and sewage disposal, regulation of wetlands and streams and more intensive farming practices have increased the nutrient loading to many shallow lakes worldwide, not least in the industrialized part of the world. This has resulted in major changes in the biological structure and dynamics of the lakes and often in a noticeable shift from a clear to a turbid state.

A typical succession from temperate lakes in Northern Europe is detailed in Fig. 11.1. At the top of the food web, major changes have occurred in the fish community. At low nutrient concentrations, predatory perch (*Perca fluviatilis*) and pike (*Esox lucius*) dominate the fish community. When the nutrient loading increases, the biomass of fish increases as well. However, a shift occurs to dominance of cyprinids, especially roach (*Rutilus rutilus*) and bream (*Abramis brama*). Roach and bream are partly zooplanktivorous and the predation pressure on zooplankton, therefore, increases, which, in turn, results in a lower grazing pressure on phytoplankton. Changes in the size structure of zooplanktivorous fish towards dominance of small specimens further enhance the predation pressure on zooplankton. In addition, due to competition, perch rarely reach the size of piscivory and, therefore, mainly predate on zooplankton and benthic invertebrates instead. Primarily due to the high fish predation pressure, the biomass ratio of zooplankton to phytoplankton decreases from 0.5-0.8 in mesotrophic lakes to less than 0.2 when phosphorus concentrations are above 0.1-0.15 mg P l^{-1}, the latter figure being so low that zooplankton are incapable of

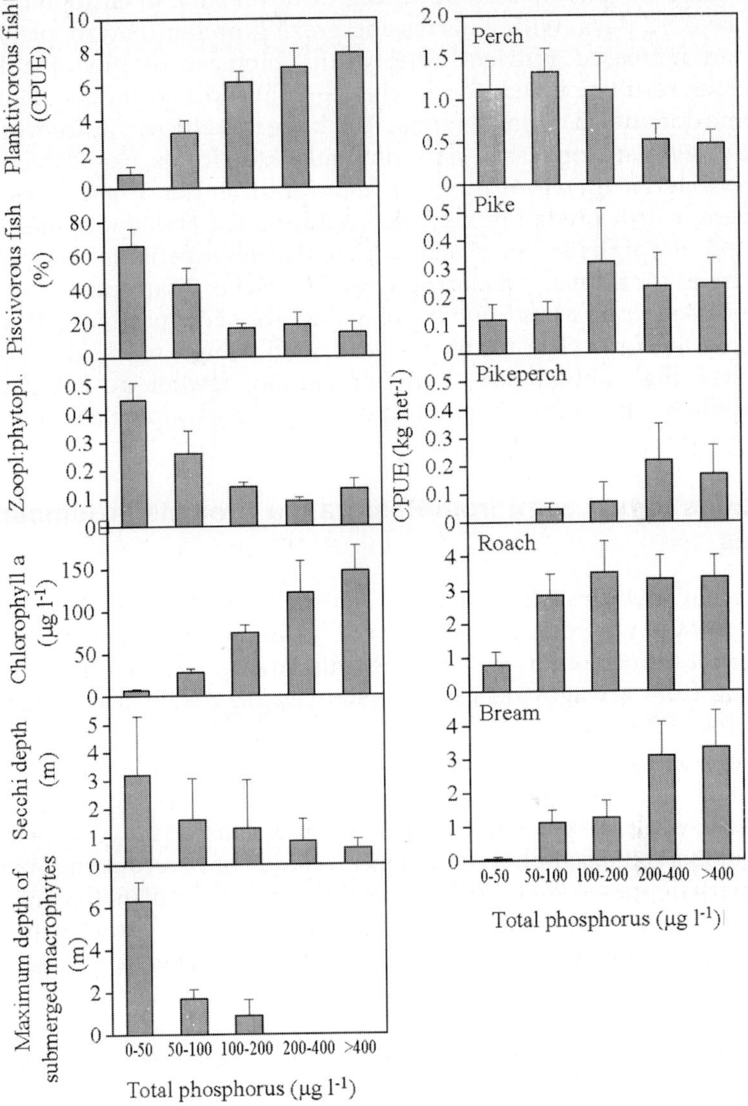

Fig. 11.1 (Left) August biomass of zooplanktivorous fish (measured as CPUE; catch in multiple mesh size gill nets, 14 different mesh sizes 6.25-75 mm) versus summer mean lake water total phosphorus (TP). Also shown are the percentage of piscivorous fish, summer mean (1 May-1 October) of zooplankton : phytoplankton biomass ratio, chlorophyll *a*, Secchi depth and the maximum depth of submerged macrophytes versus the lake water TP-concentration. Mean ± SD of the five TP groups is shown. From Jeppesen et al. (1999). (Right) Biomass (CPUE) of various quantitatively important fish species in Danish lakes versus summer mean TP. The first three species are piscivorous, while the last two are plankti-benthivorous. Modified from Jeppesen et al. (2000)

controlling the phytoplankton, whose turnover time in eutrophic lakes may be 0.5-2 days. With a decreasing grazing pressure by zooplankton and an increased nutrient supply, the biomass of phytoplankton increases, resulting in reduced Secchi depth. In addition, the zooplankton become dominated by small forms, which are less efficient grazers on the large phytoplankton. Large-sized phytoplankton forms, like filamentous cyanobacteria, therefore thrive in many nutrient-enriched lakes. An increase in fish predation may also reduce snail abundance and thus grazing of epiphytes on plants, which also impoverishes the growth conditions for submerged macrophytes. The plants disappear and with it the food source for a large number of herbivorous and macroinvertebrate-eating waterfowl. The result is a lake with a high roach and bream biomass, high abundance of phytoplankton, few or no submerged macrophytes and a greatly reduced density of birds dominated by fish-eating species (Fig. 11.2).

Lessons to be Learnt from Restoration Projects in Temperate Lakes

In recent decades, large efforts have been devoted to combat eutrophication by reducing the external loading of phosphorus in many Western European countries and in North America and, in consequence, loading from sewage and industry sources has declined significantly since the 1970s.

Several lakes respond rapidly and positively to loading reductions; nuisance algal blooming and plankti-benthivorous fish abundance decrease, while the percentage of piscivores increases as do water clarity and submerged macrophyte abundance. A study of 14 recovering lakes in Denmark (Jeppesen et al., 2002) showed that the phytoplankton and fish biomass generally declined, leading to an overall higher zooplankton:phytoplankton ratio, which suggests that enhanced grazing pressure on phytoplankton results in clearer water. Moreover, the biomass of planktivorous fish declined and the share of potential piscivores increased, most likely resulting in a stronger control by piscivores of planktivores. Accordingly, in most lakes, the share of the large-bodied zooplankton *Daphnia* spp., which is particularly sensitive to predation by zooplanktivorous fish, and the body weight of *Daphnia* spp. and other cladocerans generally increased, further enhancing the grazer control of phytoplankton. The improvements in lakewater clarity are, therefore, a result of higher resource control of phytoplankton (lower availability of nutrients) as well as of enhanced predator control of planktivorous fish, cascading to the phytoplankton level via increased grazer control by large-bodied zooplankton. Several other lakes have

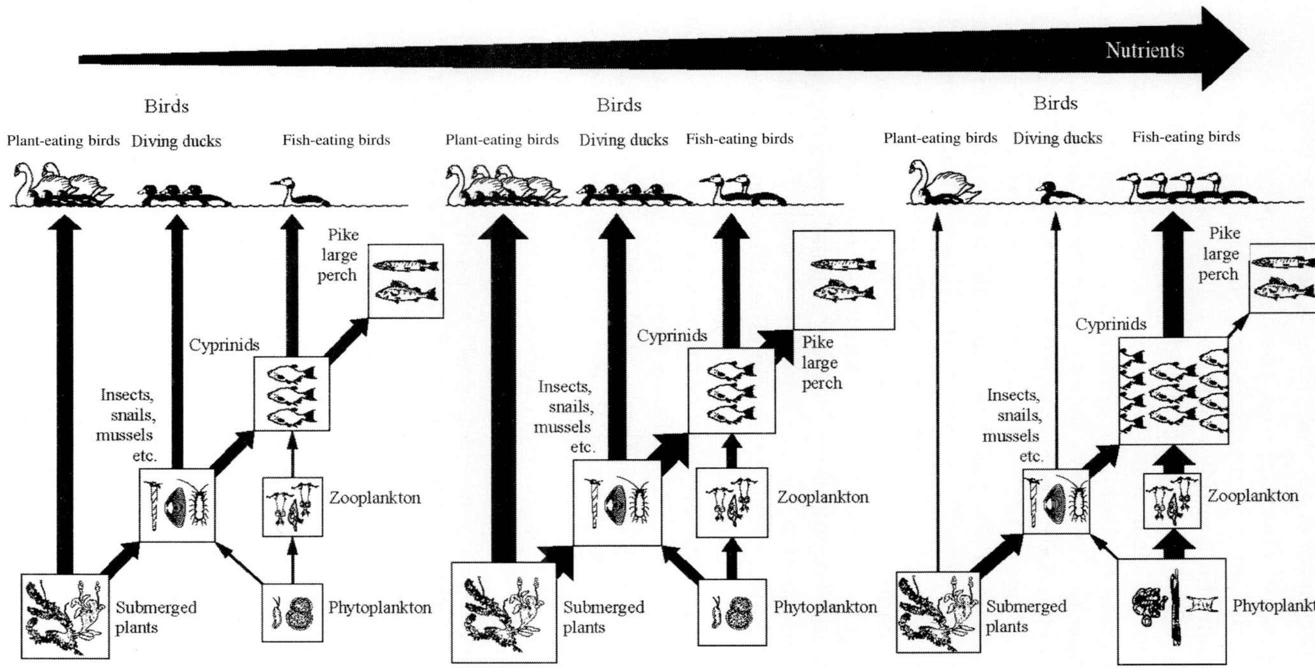

Fig. 11.2 Picture showing how the trophic structure changes along a gradient in nutrient loading in shallow temperate lakes. Low-nutrient lakes are clear with submerged vegetation and a well-balanced pelagic food web. With increasing loading, the lakes first remain clear with higher abundance and production at all trophic levels and a high diversity of waterfowl. At a certain threshold, a shift occurs to a turbid state, with high abundance of zooplanktivorous cyprinids, few piscivores, few zooplankton, high phytoplankton biomass (often with blooms of cyanobacteria in summer), few submerged plants and few waterfowl dominated by piscivores (from Jeppesen et al., 1999, modified from Andersson et al. 1990)

exhibited a swift response to nutrient loading reduction (Edmondson & Lehman, 1981; Bernhardt et al., 1985; Bäuerle & Gaedke, 1998).

However, positive effects cannot always be expected to occur, as many lakes have proven to be highly resistant to loading reductions and have exhibited only little improvement (Sas, 1989, Marsden, 1989). For some lakes, this reflects the insufficient reduction of the nutrient input to trigger a shift to the clear water state. For example, significant and sustaining changes in the biological community and water transparency of shallow temperate freshwater lakes cannot be expected to appear unless the TP concentration has been reduced to a level below 0.05-0.1 mg P l^{-1} (Jeppesen et al., 2000), or for deep lakes below 0.02-0.03 mg P l^{-1} (Sas, 1989). Even when the P loading has been sufficiently reduced, often resistance to improvement is observed. This resistance may be "chemical": P concentrations remain high because of P release from the sediment pool accumulated when loading was high (Søndergaard et al., 2002). Many years may pass before the surplus pool is either released or permanently buried. The duration of this transitional period depends on, for instance, the duration of the period with high TP loading, the residence time and the phosphorus-binding sites (like iron) supplied from the surroundings. In some cases, this transient phase towards a new equilibrium has spanned several decades (Søndergaard et al., 2001; Jeppesen et al., 2003). Various methods have been used to reduce the internal loading of phosphorus (Cooke et al., 1993; Søndergaard et al., 2002), including sediment removal and chemical treatment of the sediment with aluminum or iron salts. In stratified lakes, injections with oxygen or nitrate to the bottom layer or destabilization of the thermocline have also been used.

The resistance may also be "biological". In particular, planktivorous and benthivorous fish contribute to biological resistance in shallow eutrophic lakes (Moss, 1990; Scheffer et al., 1993). A continuously high fish predation pressure prevents both the appearance of large herbivorous zooplankton that would otherwise clear the water as well as diminishes the number of benthic animals stabilizing and oxidizing the sediment. Moreover, excretion of nutrients to overlaying waters from benthic-feeding fish or fish bioturbation of surface sediment may also play a role (Breukelaar et al., 1994; Persson, 1997).

To overcome the biological resistance by fish, various fish manipulation methods have been developed (Benndorf, 1995; Drenner & Hambright, 1999; Søndergaard et al., 2001). One such method is the enhancement of top-down control of phytoplankton by selective removal of planktivorous fish; a method employed worldwide in the temperate zone. Removal of 75-80% of the planktivorous and benthivorous fish stock during a 1-2 year period is recommended to avoid regrowth and to stimulate the growth of potentially piscivorous perch (Perrow et al., 1997; Hansson et al., 1998; Meijer et al., 1999; Mehner et al., 2002). An alternative or supplementary method to fish removal is ample stocking of 0^+ pike (> 1000 h.$^{-1}$) to control newly hatched plankti-benthivorous roach and bream (Preijs et al., 1994; Berg et al., 1997). Others have used stocking of pike perch (*Stizostedion lucioperca*), walleye (*Stizostedion vitreum*) and largemouth bass (*Micropterus salmoides*) (Benndorf, 1995; Lathrop et al., 2002). Opposite to the above-mentioned physico-chemical methods, fish manipulation is often cheap (Jeppesen & Sammalkorpi, 2002) and therefore attractive, though its long-term stability is uncertain. The findings to date indicate that fish manipulation may have a long-term effect in shallow temperate lakes provided that the nutrient loading is reduced to a TP level below 0.05-0.1 mg P l^{-1} in the future state of equilibrium. However, if the nitrogen loading is low, fish manipulation may sometimes have a positive impact at higher TP concentrations (Moss et al., 1994; Jeppesen et al., 1999). The 0.05-0.1 mg P l^{-1} threshold is in accordance with the empirical data appearing in this range (Fig. 11.1). However, temporary effects of fish manipulation can be obtained in lakes with high nutrient concentrations, but it seems unlikely that the effect will prevail in such lakes in the long term unless the abundance of planktivorous fish is repeatedly reduced. The TP threshold for positive effects to occur is most likely lower for deep lakes, but an accurate threshold has to be defined.

CAN THE KNOWLEDGE OF BIOMANIPULATION BE TRANSFERRED TO SUBTROPICAL AND TROPICAL LAKES?

Fish Community

Several factors indicate that fish stock manipulations would not have the same positive effect on the environmental state in tropical lakes as in temperate lakes. First, the richness in fish species is often higher in both tropical and subtropical lakes. South America and Africa have an extraordinarily rich freshwater fish fauna (Sunaga & Verani, 1997; Lévêque & Paugy, 1999). For example, 273 native fish species were found in aquatic environments in 44 × 10^3 km^2 Rio de Janeiro State (Bizerril & Primo, 2001), and in Uruguay (171 × 10^3 km^2), for instance, 51 characid species are found (Reichert, 2002). Many of the fish species show partial niche overlap, which expectedly increases the predator control of prey items (Lazzaro, 1997; Aguiaro & Caramaschi, 1998).

Second, the fish stock in tropical and subtropical lakes is often dominated by omnivorous species that feed on zooplankton but also

nutrient-rich lakes, not the least submerged plants often act as daytime refuges for zooplankton against fish predators (Timms & Moss, 1984; Lauridsen et al., 1996; Burks et al., 2002). At night, when the risk of predation is lower, zooplankton migrate to the open water for feeding and thereby, contribute to maintaining clearwater conditions in lakes with high macrophyte coverage (Jeppesen et al., 1997a). However, in the tropics and subtropics, the effects of macrophytes on trophic interactions are more complex, as all life forms (emergent, submerged, floating-leaved and large free-floating species) can be extremely prominent. The few studies conducted so far in the subtropics and tropics indicate that fish, particularly the smallest species and individuals, aggregate in high numbers in the vegetation (Conrow et al., 1990; Meerhoff et al., 2003; Branco et al., unpublished data). Studies of the composition of tropical ichthyofauna associated to macrophytes have shown that the life cycle of some species is completely connected to this biotope, examples being small-sized tetragonopterins, characiforms and some cichlids (Sazima & Zamprogno, 1985; Delariva et al., 1994). Many fish species exhibit spatial distribution patterns often connected with zooplankton and predator densities, and also with different macrophyte life forms (Agostinho et al., 1994; Sunaga & Verani, 1997; Meerhoff et al., 2003). Besides, tropical aquatic vegetation can be intensively colonized by periphyton and provide microhabitats for many invertebrates, including chironomids, trichopterans and molluscs (Callisto et al., 1996; Masifwa et al., 2001).

One might, therefore, expect that the vegetation is a poor refuge for large-bodied zooplankton in warm lakes, which seems to be supported by the few field and experimental studies conducted so far (Meerhoff et al., 2003; Meerhoff, unpublished data). Fish aggregation in the vegetation and lack of zooplankton aggregated among plants have also been observed in brackish lakes dominated by small-sized sticklebacks in Denmark (Jeppesen et al., 1997b) and, accordingly, nutrient-rich brackish lakes are turbid, even when macrophyte coverage is high. Likewise, from subtropical freshwater lakes in Florida, Bachmann et al. (2002) showed no positive effect of plants on water clarity and nutrient-rich lakes with high plant biomass were often turbid. These results may well be of general importance for subtropical and subtropical lakes, although more research is needed within this field. Also, contrasting management approaches can alter the role of macrophytes in lake functioning and trophic interactions. While in temperate systems the introduction and development of aquatic plants are considered a key step in a restoration process (Moss et al., 1996), many aquatic plants in the tropics and subtropics are often considered a nuisance and subject to severe eradication measures.

LIMITATIONS OF BIOMANIPULATION IN THE SUBTROPICS AND TROPICS

As can be deduced from above, it seems more difficult to provoke and not least maintain a trophic cascade effect in subtropical and tropical lakes than in temperate lakes, for which the concept of biomanipulation as a restoration tool was developed. Supporting this view, Nagdali & Gupta (2002) found positive, but only short-term cascading effects of a massive (> 80%) kill (due to fungal infection) of the most abundant planktivorous mosquito fish (*Gambusia affinis*) in Lake Naini Tal, India. Zooplankton abundance increased significantly, phytoplankton biomass and productivity declined as did nutrient concentrations, resulting in higher water transparency. However, only four months later, the abundance of mosquito fish, plankton and nutrients had returned to the level recorded in the previous year (Fig. 11.3). In a study performed by Scasso et al. (2001), a slowly increasing abundance of larger species was observed after two years of biomanipulation involving removal of small planktivorous fish, without, however, inducing cascading effects leading to clear-water conditions.

Yet, only few studies have investigated the applicability of the biomanipulation theory to tropical and subtropical freshwater ecosystems, and most of the existing ones have examined food interactions in eutrophic lakes and reservoirs with the aim to control cyanobacterial blooms via enhanced grazing by omnivorous fish such as silver carp (Arcifa et al., 1986; Northcote et al., 1990; Starling, 1993; Jones & Poplawski, 1998; Saha & Jana, 1998). The results obtained indicate that omnivory and strong shifts in fish diet and in the fish and zooplankton composition jeopardize successful biomanipulation (Boon et al., 1994; Arcifa et al., 1995; Starling & Lazarro, 1997; Starling et al., 1998; Boulton & Brock, 2001), but more information on phytoplankton-zooplankton interactions in the tropics is needed before its potential can be fully elucidated. Moreover, the absence of a native piscivorous fish culture in many tropical countries precludes the application of biomanipulation. There is a huge richness of fish species potentially useful for this purpose, but mass production for biomanipulation purposes has not yet been considered. Generally, aquaculture has so far focused on exotic species (i.e. common carp and grass carp) that have negative effects on the water quality and biodiversity.

In many of the experiments undertaken in the temperate zone, an improvement in environmental state has, however, been recorded without the occurrence of a trophic cascade, i.e. without changes in the zooplankton species composition towards higher dominance of large-sized individuals and with it a higher grazing pressure on phytoplankton

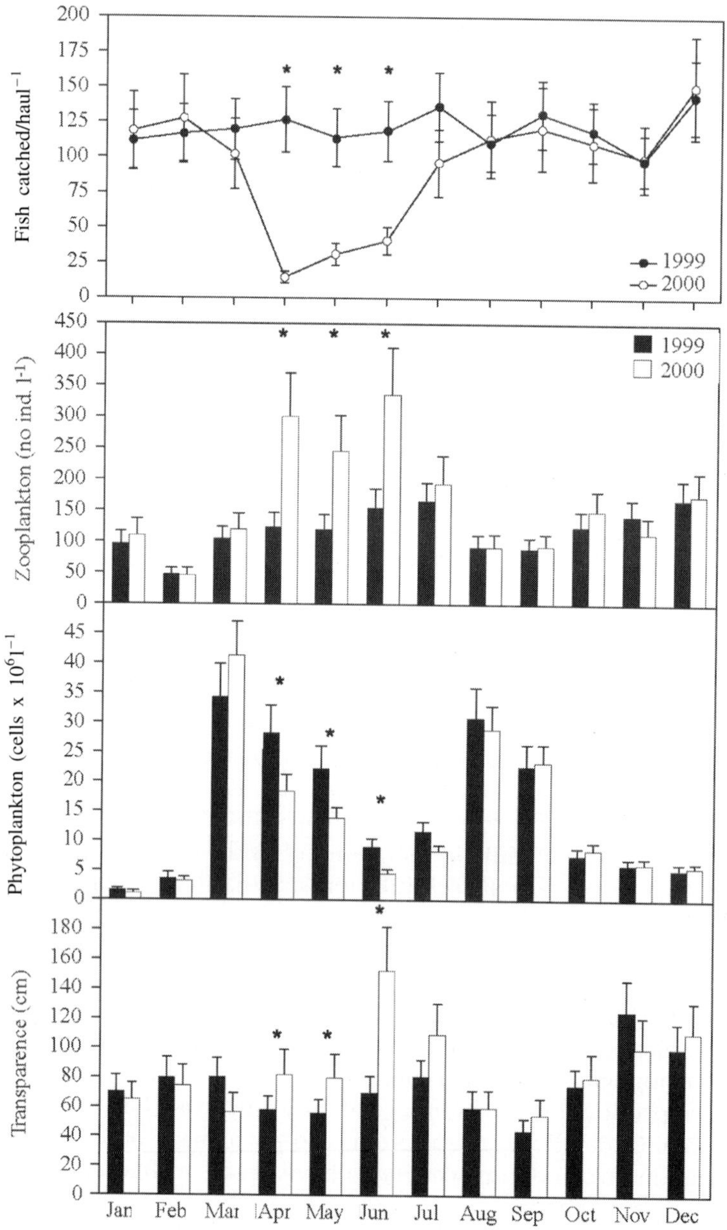

Fig. 11.3 Seasonal variation in fish catch, zooplankton, phytoplankton and Secchi depth in Lake Naini Tal, India, before (1999) and after a major fish kill in April 2000. Note that the fish stock recovered four months after the fish kill and that a short-term cascading effect on zooplankton, phytoplankton and transparency was recorded. Stars indicate significant differences (modified from Nagdali & Gupta, 2002)

(Horppila et al., 1998). Even without such a trophic cascade, a significant reduction has been observed in the occurrence of cyanobacteria, total phosphorus has declined and water clarity increased. These phenomena have been ascribed to reduced release of phosphorus from the sediment, not least due to the lower rate of fish foraging in the sediment following biomanipulation (Horppila et al., 1998). Biomanipulation may, therefore, potentially also reduce the nutrient release from the sediment in tropical lakes, but the dominance of small fish species and the improved growth conditions for cyanobacteria suggest that the effect will not be long lasting. Therefore, a drastic reduction of the external nutrient loading seems to be the best way forward for restoring lakes also in the tropics, but clearly the scientific basis (e.g. nutrient threshold levels) on which to make decisions is yet too limited.

STRONG NEED FOR MORE RESEARCH

As the situation is today, many subtropical and tropical lakes are heavily eutrophied and biodiversity has declined, and in the years to come, it is to be feared that many other lakes in these regions, especially in coastal areas, will follow the same negative developmental pattern both as a consequence of the future economical development and climate change. According to Vitousek et al. (1997), around 60% of the world's population is concentrated in the outer 100 km coastal zone, and disordered urban expansion and agricultural activities have heavily impacted aquatic ecosystems in the developing countries located in tropical and subtropical regions. These already disturbed environments will expectedly face more stressful challenges as a result of increasing human activities and changing meteorological events associated to global warming.

One may hope that the developing countries will learn from the mistakes made in Europe and North America and adopt a farsighted policy and take actions to avoid the eutrophication of present-day high quality lakes and thus the need for initiating expensive restoration projects in the future. However, whether they do so or not, more knowledge is needed to initiate suitable restoration projects of already eutrophied lakes in these regions. This book is an interesting step on the long road to the procurement of such knowledge.

ACKNOWLEDGEMENTS

We thank the counties for access to data from the Survey Programme of Danish Lakes. The work was supported by BIOMAN (EVK2-CT-1999-00046), and the Danish Natural Science Research Council funded research project "Consequences of weather and climate changes for marine and freshwater eco-systems. Conceptual and operational forecasting of the aquatic environment" (SWF: 2052-01-0034) and the Eu Eurdimpacs project (GOCE-CT-2003-505540) . We also thank Kathe Møgelvang and Anne Mette Poulsen for technical assistance and Prof. Vikram Reddy for inviting EJ to write this chapter.

REFERENCES

Agostinho, A.A., Benedito-Cecilio, E., Gomes, L.C. & Sampaio, A.A. 1994. Spatial and temporal distribuition of sadela, *Hypophthalmus edentatus* (Pisces, Siluroidei), in the area of influence of the Itaipu reservoir (Paraná, Brazil). Revista UNIMAR 16(3): 27-40.

Aguiaro, T. & Caramaschi, E.P. 1998. Trophic guilds in fish assemblages in three coastal lagoons of Rio de Janeiro State (Brazil). Verh. Internat. Verein. Limnol. 26: 2166-2169.

Aguiaro, T., Branco, C.W.C., Verani, J.R. & Caramaschi, E.P. 2003: Diet of the clupeid fish *Platanichthys platana* (Regan, 1917) in two different Brazilian coastal lagoons. Brazilian Archives of Biology and Technology 46 : 215-222.

Arcifa, M.S., Northcote, T.G. & Froelich, O. 1986. Fish-zooplankton interactions and their effects on water quality of a tropical Brazilian reservoir. Hydrobiologia 139: 49-58.

Arcifa, M.S., Starling, F.L.R.M., Sipaúba-Tavares, L. & Lazzaro, X. 1995. Experimental Limnology, pp.257-280. In: J.G. Tundisi, C.E.M. Bicudo & T. Matsumura-Tundisi (eds). Limnology in Brazil. Brazilian Acad. of Sci. and Brazilian Limnol. Soc., São Paulo.

Andersson, G., Blindow, I., Hargeby, A. & Johansson, S. 1990. Det våras för Krankesjön. Anser 29: 53-62.

Araújo Lima, C.A.R.M., Agostinho, A.A. & Fabré, N.N. 1995. Trophic aspects of fish communities in Brazilian rivers and reservoirs, pp. 105-136. In: J.G. Tundisi, C.E.M. Bicudo & T. Matsumura-Tundisi (eds). Limnology in Brazil. Brazilian Acad. of Sci. and Brazilian Limnol. Soc., São Paulo.

Bachmann, R.W., Horsburgh, C.A., Hoyer, M.V., Mataraza, L.K. & Canfield, Jr. D.E. 2002. Relations between trophic state indicators and plant biomass in Florida lakes. Hydrobiologia 470: 219-234.

Benndorf, J. 1995. Possibilities and limits for controlling eutrophication by biomanipulation. Int. Rev. Ges. Hydrobio. 80: 519-534.

Berg, S., Jeppesen, E. & Søndergaard, M. 1997. Pike (*Esox Lucius* L.) stocking as a biomanipulation tool. 1. Effects on the fish population in Lake Lyng (Denmark). Hydrobiologia 342/343: 311-318.

Bernhardt, Von H., Clasen, J., Hoyer, O. & Wilhelms, W. 1985. Oligotrophication in lakes by means of chemical nutrient removal from the tributaries. Its demonstration with the Wahnbach Reservoir. Arch. Hydrobiol. Suppl. 70: 481-533.

Bistoni, M.A., Hued, A., Videla, M. & Sagretti, L. 1999. Water quality effects on fish communities of the central part of Argentina. Revista Chilena de Historia Natural 72: 325-335.

Bizerril, C.R.S.F. & Primo, P.B.S. 2001. Fish of the continental waters of the State of Rio de Janeiro (in Portuguese). Environment Secretary (SEMADS) and Marine Studies Foundation (FEMAR), p. 417.

Blanco, R., Romo, S., Villena, M.J. & Martinez, S. 2003. Fish communities and food web interaction in some shallow Mediterranean Lakes. Hydrobiologia 506/509 : 473-480.

Boon, P.I., Bunn, S.E., Green, J.D. & Shiel, R.J. 1994. Consumption of cyanobacteria by freshwater zooplankton – implications for the success of top-down control of cyanobacterial blooms in Australia. Aust. J. Mar. Freshwat. Res. 45: 875-887.

Boulton, A.J. & Brock, M.A. 2001. Australian freshwater ecology: processes and management. Coop. Res. Centre for Freshw. Ecol. & SIL. p. 300.

Branco, C.W.C., Aguiaro, T., Esteves, F.A. & Caramaschi, E.P. 1997. Food sources of the Teleost *Eucinostomus argenteus* in two coastal lagoons of Brazil. Stud. Neotrop. Fauna & Environm. 32, 33-40.

Branco, C.W.C., Esteves, F.A. & Kozlowsky-Suzuki, B. 2000. The zooplankton and other limnological features of a humic coastal lagoon (Lagoa Comprida, Macaé, R.J.) in Brazil. Hydrobiologia 437: 71-81.

Branco, C.W.C., Rocha, M-I., Pinto, G.F.S. & Gômara, Filippo, R.D. 2002. Limnological features of Funil Reservoir (R.J., Brazil) and indicator properties of rotifers and cladocerans of the zooplankton community. Lakes & Reservoirs: Research and Management 7: 87-92.

Breukelaar, A. W., Lammens, E. H. R. R., Klein Breteler, J. P. G. & Tatrai, I. 1994. Effects of benthivorous bream (*Abramis brama* L.) and carp (*Cyprinus caprio* L.) on sediment resuspension and concentration of nutrients and chlorophyll *a*. Freshwat. Biol. 32: 113-121.

Burks, R.L., Jeppesen, E., Lodge, D.M. & Lauridsen, T. 2002. Diel Horizontal Migration of Zooplankton: Costs and benefits of inhabiting littoral zones. Freshw. Biol. 47: 343-365.

Bäuerle, E. & Gaedke, U. (eds). Advances in Limnology 53. Lake Constance: Characterization of an ecosytem in transition. E. Schwizerbart'sche Verlagsbuchhandlung, pp. 610.

Callisto, M., Serpa-Filho, A., Oliveira, S.J. & Esteves, F.A. 1996. Chironomids on leaves of *Typha domingensis* in a lagoon of Rio de Janeiro State (Brazil). Stud. Neotrop. Fauna & Environm. 31: 51-53.

Conrow, R., Zale, A.V. & Gregory, R.W. 1990. Distributions and abundances of early stages of fishes in a Florida lake dominated by aquatic macrophytes. Trans. Am. Fish. Soc. 119: 521-528.

Cooke, G. D., Welch, E. B., Peterson, S. A. & Newroth, P. R. 1993. *Restoration And Management of Lakes and Reservoirs.* Boca Raton, Florida: Lewis Publishers.

Crisman, T.L. & Beaver, J.R. 1990. Applicability of planktonic biomasipulation for managing eutrophication in the subtropics. Hydrobiologia 200/201: 177-185.

Delariva, R.L., Agostinho, A.A., Nakatani, K. & Baumgartner, G. 1994. Ichthyofauna associated to aquatic macrophytes in the upper Parana River Floodplain. Revta. UNIMAR 16 (suppl. 3): 41-60.

Drenner, R. & Hambright, D. 1999. Review: Biomanipulation of fish assemblages as a lake restoration technique. Arch. Hydrobiol. 146: 129-165.

Dumont, H.J. 1994. On the diversity of the Cladocera in the tropics. Hydrobiologia 272: 27-38.

Edmondson, W.T. & Lehman, J.T. 1981. The effect of changes in the nutrient income on the condition of Lake Washington. Limnol. Oceanogr. 26: 1-29.

Fernando, C.H. 1994. Zooplankton, fish and fisheries in tropical freshwaters. Hydrobiologia 272: 105-123.

Garcia, P.R., Nandini, S., Sarma, S.S.S., Valderrama, E.R., Cuesta, I. & Hurtado, M.D. 2002. Seasonal variations of zooplankton abundance in the freshwater reservoir Valle de Bravo (Mexico). Hydrobiologia 467: 99-108.

Hansson, L.-A., Annadotter. H., Bergman. E., Hamrin, S. F., Jeppesen, E., Kairesalo, T., Luokkanen, E., Nilsson, P.-Å., Søndergaard, M. & Strand, J. 1998. Biomanipulation as an application of food chain theory: Constraints, synthesis and recommendations for temperate lakes. Ecosystems 1: 558-574.

Horppila, J., Peltonen, H., Malinen, T., Luokkanen, E. & Kairesalo, T. 1998. Top-down or bottom-up effects by fish—Issues of concern in biomanipulation of lakes. Restor. Ecol. 6: 1-10.

Jeppesen, E., Søndergaard, M., Kanstrup, E., Petersen, B., Henriksen, R.B., Hammershøj, M., Mortensen, E., Jensen, J.P. & Have, A. 1994. Does the impact of nutrients on the biological structure and function of brackish and freshwater lakes differ? Hydrobiologia 275/276: 15-30.

Jeppesen, E., Jensen, J.P., Søndergaard, M., Lauridsen, T.L., Pedersen, L.J. & Jensen, L. 1997a. Top-down control in freshwater lakes: The role of nutrient state, submerged macrophytes and water depth. Hydrobiologia 342/343: 151-164.

Jeppesen, E., Søndergaard, M., Jensen, J.P., Kanstrup, E. Petersen, B. 1997b. Macrophytes and turbidity in brackish lakes with special emphasis on the role of top-down control. In: E. Jeppesen, Ma. Søndergaard, Mo. Søndergaard & K. Christoffersen (Eds), The structuring role of submerged macrophytes in lakes. Ecological Studies, Vol. 131. Springer Verlag, 1997: 369-377.

Jeppesen, E., Søndergaard, M., Kronvang, B., Jensen, J.P., Svendsen, L.M. & Lauridsen, T. 1999. Lake and catchment management in Denmark. Hydrobiologia 395/396: 419-432.

Jeppesen, E., Jensen, J.P., Søndergaard, M., Lauridsen, T. & Landkildehus, F. 2000. Trophic structure, species richness and biodiversity in Danish lakes: Changes along a phosphorus gradient. Freshwat. Biol. 45: 201-213.

Jeppesen, E. & Sammalkorpi, I. 2002. Chapter 14 "Lakes". In: M. Perrow & T. Davy (eds). Handbook of Restoration Ecology, Cambridge University Press: 297-324.

Jeppesen, E., Jensen, J.P. & Søndergaard, M. 2002. Response of phytoplankton, zooplankton and fish to re-oligotrophication: an 11-year study of 23 Danish lakes. – Aquat. Ecosys. Health & Manage. 5: 31-43.

Jeppesen, E., Søndergaard, M., Jensen, J.P. & Lauridsen, T.L. 2003. Recovery from eutrophication – global perspective. In: Kumagai, M. and W.F. Vincent (eds). Freshwater Management—Global versus Local Perspectives, Springer Verlag, Tokyo: 135-152.

Jones, G.J. & Poplawski, W. 1998. Understanding and management of cynobacterial blooms in subtropical reservoirs of Queensland, Australia. Wat. Sci. Tech. 37: 161-168.

Jones, J.I. & Waldron, S. 2003. Combined stable isotope and gut contents analysis of food webs in plant-dominated, shallow lakes. Freshw. Biol. 48: 1396-1407

Kalff, J. 2002. Limnology—Inland Water Ecosystems. Prentice-Hall Inc.

Lathrop, R.C., Johnson, B.M., Johnson, T.B., Vogelsang, M.T., , Hrabik, T.R., Carpenter, S.R., Kitchell, J.F,.Magnuson, J.J.,Rudstam L.G. & Stewart R.S. 2002. Stocking piscivores to improve fishing and water clarity: A synthesis of the Lake Mendota biomanipulation project. Freshw. Biol. 47:2410-2424

Lauridsen, T.L., Pedersen, L.J., Jeppesen, E. & Søndergaard, M. 1996. The importance of macrophyte bed size for cladoceran composition and horizontal migration in a shallow lake. J. Plank. Res. 18: 2283-2294.

Lazzaro, X. 1997. Do the trophic cascade hypothesis and classical biomanipulation approaches apply to tropical lakes and reservoirs? Ver. Int. Verein. Limnol. 26: 719-730.

Lévêque, C. & Paugy, D. 1999. Caractéristiques générales de la faune ichtyologique. In: Lévêque, C. & Paugy, D. 1999 (eds) *Less poisons des eaux continentales africaines: Diversité, écologie, utilisation par l'homme*. IRD Editions, Paris, pp. 43-54.

Lewis, W.M. Jr. 1996. Tropical lakes: how latitude makes a different. In: F. Schiemer & K.T. Boland (eds) Perspectives in tropical limnology, pp. 43-64. SPB Academic Publishing bv, Amsterdam, The Netherlands.

Masifwa, W.F., Twongo, T. & Denny, P. 2001. The impact of water hyacinth, *Eichhornia crassipes* (Mart) Solms on the abundance and diversity of aquatic macroinvertebrates along the shores of northern Lake Victoria, Uganda. Hydrobiologia 452: 79-88.

Marsden, S. 1989. Lake restoration by reducing external phosphorus loading: The influence of sediment phosphorus release. Freshw. Biol. 21: 139-162.

Mazzeo, N., Rodríguez-Gallego, L., Kruk, C., Meerhoff, M., Gorga, J., Lacerot, C., Quintans, F., Loureiro, M., Larrea, D. & García-Rodrígues, F. 2003. Effect of *Egeria densa Planch.* beds on a shallow lake without piscivorous fish. Hydrobiologia 506/509 : 591-602.

Meerhoff, M., Mazzeo, N., Moss, B. & Rodríguez-Gallego, L. 2003. The structuring role of free-floating versus submerged plants in a subtropical shallow lake. Aquat. Ecol. 37: 377-391.

Mehncr, T., Benndorf, J., Kasprzak, P. & Koschel, R. 2002. Biomanipulation of lake ecosystems: Successful applications and expanding complexity in the underlying science. Freshw Biol. 47 : 2453-2466.

Meijer, M.-L., de Boois, I., Scheffer, M., Portielje, R. & Hosper, H. 1999. Biomanipulation in the Netherlands: an evaluation of 18 case studies in shallow lakes. Hydrobiologia 408/409: 13-30.

Moss, B. 1990. Engineering and biological approaches to the restoration from eutrophication of shallow lakes in which aquatic plant communities are important components. Hydrobiologia 200/201: 367-377.

Moss, B., McGowan, S. & Carvalho, L. 1994. Determination of phytoplankton crops by top-down and bottom-up mechanisms in a group of English lakes, the West Midland Meres. Limnol. Oceanogr. 39: 1020-1030.

Moss, B., Madgwick, J. & Phillips, G.L. 1996. A guide to the restoration of nutrient-enriched shallow lakes. Broads Authority & Environment Agency, Norwich, 180 pp.

Nagdali, S.S. & Gupta, P.K. 2002. Impact of mass mortality of a mosquito fish, *Gambusia affinis* on the ecology of a fresh water eutrophic lake (Lake Naini Tal, India). Hydrobiologia 468: 45-52.

Northcote, T.G., Arcifa, M.S. & Munro, K.A. 1990. An experimental study of the effects of fish zooplanktivory on the phytoplankton of a Brazilian reservoir. Hydrobiologia 194: 31-45.

Paugy, D. & Lévêque, C. 1999. La Reproduction. In: Lévêque, C. and Paugy, D. (eds) Less poisons des eaux continentales africaines: Diversité, écologie, utilisation par l´homme. IRD Editions, Paris, pp. 129-152.

Persson, A. 1997. Phosphorus release by fish in relation to external and internal load in a eutrophic lake. Limnol. Oceanogr. 43: 577-583.

Preijs, A., A. Martyniak, S. Boron, Hliwa, P. & Koperski, P. 1994. Food web manipulation in small, eutrophic Lake Wirbel, Poland: effect of stocking with juvenile pike on planktivorous fish. Hydrobiologia 275/276: 65-70.

Perrow, M. R., Meijer, M. L., Dawidowicz, P. & Coops, H. 1997. Biomanipulation in shallow lakes: state of the art. Hydrobiologia 342/343: 355-365.

Quintans, F., Scasso, F., Loureiro, M. & Yafe, A. submitted. Diet of *Cnesterodon decemmaculatus* (Poecilidae) and *Jenynsia multidentata* (Anablepidae) in a manmade lake.

Quirós, R. 1998. Fish effects on trophic relationships in the pelagic zone of lakes. Hydrobiologia 361: 101-111.

Reichert, J.J. 2002. Freshwater fish of Uruguay (In Spanish). European Union, UNDP, GEF. PROBIDES, Uruguay, p. 327.

Roche, K.F., Smapaio, D., Teixeira, T., Matusumura-Tundisi, T., Tundisi, J.G. & Dumont, H.J. 1993. Impact of *Holoshestes heterodon* Eingenmann (Pisces: Characidae) on the plankton community of a subtropical reservoir: The importance of predation by *Chaoborus* larvae. Hydrobiologia 254: 7-20.

Saha, S.D. & Jana, B.B. 1998. Control of bloom in a tropical lake: grazing efficiency of some herbivorous fish. J. Fish Biol. 53: 12-34.

Sas, H. (Ed.). 1989. *Lake restoration by reduction of nutrient loading. Expectation, experiences, extrapolation.* Acad. Ver. Richardz Gmbh.,

Sazima, I. & Zamprogno C. 1985. Use of water hyacinth as shelter, foraging place, and transport by young piranhas, *Serrasalmus spilopleura*. Env. Biol. Fishes 12: 237-240.

Scasso, F., Mazzeo, N., Gorga, J., Kruk, C., Lacerot, G., Clemente, J., Fabián, D. & Bonilla, S. 2001. Limnological changes of a subtropical shallow hypertrophic lake during its restoration. Two years of whole-lake experiments. Aquat. Conserv: Mar. Freshwater Ecosys. 11: 31-44.

Scheffer, M., Hosper, S.H., Meijer, M.-L., Moss, B. & Jeppesen, E. 1993. Alternative equilibria in shallow lakes. Trends Ecol. Evol. 8: 275-279.

Starling, F.L.R.M. 1993. Control of eutrophication by silver carp (*Hypophthalmichthys molitrix*) in the tropical Paranoá Reservoir (Brasília, Brazil): A mesocosm experiment. Hydrobiologia 257: 143-152.

Starling, F.L.R.M. & Lazzaro, X. 1997. Experimental investigation of the feasibility of improving water quality by controlling exotic planktivore overpopulation in the eutrophic Paranoá Reservoir. Verh. Internat. Verein. Limnol. 26: 789-794.

Starling, F.L.R.M., Beveridge, M., Lazzaro, X. & Baird, D. 1998. Silver carp biomass effects on plankton community in Paranoá Reservoir (Brazil) and an assessment of its potential for improving water quality in lacustrine environments. Internat. Rev. Hydrobiol. 83: 499-508.

Sunaga, T. & Verani, J.R. 1997. The fish community of four lakes. In: J.G. Tundisi & Y. Saijo (eds), Limnological Studies on the Rio Doce Valley Lakes, Brazil, pp. 359-369. Brazilian Acad. of Sci., Univ. of S. Paulo and Center for Water Res. and Appl. Ecol., São Carlos, SP.

Søndergaard, M., Jensen, J.P. & Jeppesen, E. 2001. Retention and internal loading of phosphorus in shallow, eutrophic lakes. The Scientific World 1: 427-442.

Søndergaard, M., Wolter, K.-D. & Ripl, W. 2002. Chapter 10, "Chemical treatment of water and sediments with special reference to lakes". In: M. Perrow & T. Davy (eds): Handbook of Restoration Ecology, Cambridge University Press: 184-205.

Timms, R.M. & Moss, B. 1984: Prevention of growth of potentially dense phytoplankton populations by zooplankton grazing in the presence of zooplanktivorous fish, in a shallow wetland ecosystem. Limnol. Oceanogr. 29: 472-486.

Thomaz. S.M. & Bini, L.M. 2003. Ecologia e manejo de macrófitas aquáticas. EDUEM, Maringá, pp. 341.

Vadeboncoeur Y., Vander Zanden M.J. & Lodge D.M. 2002. Putting the lake back together: Reintegrating benthic pathways into food web models. BioScience 52: 44-55.

Vander Zanden M.J. & Vadebonceur, Y. 2002. Fish as integrators of benthic and pelagic food webs in lakes. Ecology 83: 2151-2161.

Vitousek, P.M., Mooney, H.A., Lubchenco, J. & Melillo, J.M. 1997. Human domination of Earth's ecosystems. Science 277: 494-499.

Yafe, A., Loureiro, M., Scasso, F. & Quintans, F. 2002. Feeding of two Cichlidae species in a hypertrophic urban lake. Iheringia. Sér. Zool., Porto Alegre, 92(4): 73-79.

CHAPTER

12

Using Biomanipulation to Control Eutrophication in a Shallow Tropical Urban Reservoir (Lago Paranoá, Brazil)

Xavier Lazzaro[1] and Fernando L. R. M. Starling[2]

[1]Institut de Recherche pour le Développement (IRD), UR098/FLAG, Bel-Air, BP 1386, Dakar, Senegal

[2]Universidade Católica de Brasília (UCB)/Companhia de Saneamento do Distrito Federal (CAESB), SAIN - R1 ETA Brasília, CEP 70620-000, Brasília-DF, Brazil

ABSTRACT

Biomanipulations are almost exclusively performed on temperate North American and European lakes to reduce excessive phytoplankton abundance. They use piscivore stocking and/or zooplanktivore removal to enhance zooplankton grazing on algae. Their success depend on intense trophic cascade interactions along a linear food chain from piscivores to visual zooplanktivores, herbivorous zooplankton and phytoplankton. Piscivore cascading effects down to phytoplankton requires a substantial grazing pressure by large *Daphnia*, the key invertebrate herbivores, when relaxed from zooplanktivory. In contrast, tropical lakes have a weak, zooplankton-phytoplankton link due to the lack of *Daphnia* and prevalence of large algae, mostly inedible by zooplankton, particularly cyanobacteria that may form efflorescences. Their more complex pelagic food webs make the fish-algae relationships less predictable. Dominant planktivores are filter-feeding omnivores that consume large algae in addition to zooplankton and detritus. Most omnivores have benthic habits (e.g., Nile tilapia and common carp), whereas only few are pelagic feeders (e.g., silver carp). Not limited by resources, filter-feeding planktivores often reach high biomasses and enhance internal nutrient cycling, thus promoting algal growth. Because of these peculiarities, biomanipulation in the Tropics should target the control of cyanobacteria by enhancing filter feeding by pelagic planktivores and/ or reducing nutrient loading by benthic omnivores. We present results of laboratory trials (10-50-L aquaria), *in situ* experiments (2.5-6.5-m^3 mesocosms, 1,000-m^2 littoral lilmnocorrals), an overall acoustic fish stock assessment, and a massive accidental fish kill (300 tons) to demonstrate the applicability of fish manipulation within CAESB Restoration Program of Lago Paranoá, the reservoir of Brasilía. We use them to test a biomanipulation approach based on the reduction of the benthic tilapia overpopulatioon (*Oreochromis niloticus* plus *Tilapia rendalli*) and stocking of pelagic silver carp (*Hypophthalmichthys molitrix*) to reduce phosphorus loading and thus control the occurrence of cyanobacteria blooms (*Microcystis aeruginosa, Botryococcus* spp., *Cylindrospermopsis raciborskii*). We also explore the benefits of stocking herbivorous grass carp (Ctenopharyngodon idella) to control excessive macrophytes growth, as well as implementing environment-friendly silver carp cage culture and promoting sport fishery of native piscivores (tucunaré *Cichla ocellaris,* dourado *Salminus maxillosus,* surubim, *Pseudoplatystoma coruscans*) to improve water quality. We discuss the potential and effectiveness of food-web manipulations for the restauration of shallow eutrophic tropical lakes, emphasizing their peculiarities, e.g. the importance of omnivory and internal nutrient cycling favoring nitrogen limitation.

BIOMANIPULATION OF NORTH TEMPERATE LAKES

Lake biomanipulation, synonymous with biological control in lake restoration or top-down manipulation of any lake food web, can be an efficient ecotechnological approach to control lake eutrophication. Nevertheless, it is only a complementary measure to the necessary first step of reducing external nutrient loadings. During the last two decades, biomanipulation has been widely applied to temperate lakes of Europe and North America, with variable successes. To decrease phytoplankton biomass and improve under-water light penetration, biomanipulation has mainly focussed on the limitation of excessive growth of undesirable algae, mainly cyanobacteria (Shapiro et al. 1975). The classical biomanipulation approach is based on the selective reduction of zooplanktivorous and/or benthivorous fish abundance through fisheries, use of fish poisoning substances, and/or stocking of piscivorous fish. Zooplanktivores and benthivores indirectly favour phytoplankton growth. Zooplanktivores also eliminate large-bodied herbivorous zooplankton (cladocerans, mainly *Daphnia*), whereas benthivores recycle nutrients internally to the water column, either excreted or added by bioturbation when feeding on the bottom sediments (Lammens et al., 1990; Perrow et al., 1997).

Another biomanipulation strategy involves the introduction or development of macrophyte beds in the littoral area. Macrophytes create a refuge for herbivorous zooplankton and promote a clear water state by sequestering nutrients and increasing shading, thus preventing phytoplankton development (Moss et al., 1996). Two alternative states occur in shallow lake systems across a wide range of trophic states—a clear state dominated by plants and a turbid state dominated by phytoplankton (Scheffer et al., 1993). By favouring the development of *Daphnia* that increases the grazing pressure on algae, fish manipulation can increase the water clarity. Due to the increase in light penetration, macrophyte seeds develop and promote buffer mechanisms, called *reverse switches* by Moss (1998). This buffering is necessary in order to maintain the state of plant dominance, allowing the ecosystem to self-sustain and maintain the established new fish community.

Lazzaro's (*unpubl. data*) analysis of case studies reveals that 93 temperate lakes or reservoirs have been biomanipulated during the past 25 years—68% located in Europe and 31% in North America—with only one lake in Asia (Russia). From 49 ecosystems where morphometric data were supplied, 73% have a mean depth < 4 m and 69% a surface area < 50 ha. (range 0.04-4,000 ha.). In a recent literature review, Drenner & Hambright (1999) analyzed the successes and failures of 41 published biomanipulation experiments in 39 lakes (79% in Europe, 80% < 100 ha. 78% < 4 m in mean depth, 88% with < 5 years of post-implementation data). Among the 61% of successes, piscivore introduction was the least successful approach for algal control, whereas partial fish removal had the highest success, as compared with these two strategies combined. The authors concluded that biomanipulation is a promising ecotechnology tool for the restoration of small shallow lakes. However, long-term success cannot be guaranteed through this method due to the peculiarities of each system, lack of observations beyond 5 years, and possible return to a turbid state (high chlorophyll concentration) favoured by fish presence and progressive loss of vegetation.

POTENTIAL FOR LAKE BIOMANIPULATION IN THE TROPICS AND SUBTROPICS

Contrasting with those achievements in Europe and North America, lake biomanipulation represents a poorly investigated and implemented strategy in the tropics and subtropics, despite its extremely promising potential.

The concept of efficiently controlling cyanobacterial blooms and increasing the water clarity via fish management of food webs has drawn the attention of developing countries in both tropical and subtropical regions because of expected long-lasting improvements in water quality and a comparatively low cost of implementation (Rast & Thornton, 1996). Uncertainty regarding the efficiency and improvement duration observed for biomanipulations performed in temperate regions, plus some peculiarities in the food-web components and structure of low latitude lakes, have however slowed down the development of a tropical approach to biomanipulation.

Lazzaro (1997) emphasized that, in shallow tropical lakes, food webs may present a series of distinctive features, e.g.: (i) generalist carnivores replace obligate piscivores; (ii) omnivores dominate over planktivores; (iii) rotifers and small Cladoceran species represent the dominant herbivorous zooplankton with *Daphnia* virtually absent, and (iv) high densities of large algae that can form efflorescences are mostly inedible by zooplankton. The bottleneck that results from this weak zooplankton-phytoplankton link favours detritivory and omnivory by planktivorous fishes. This may partially explain the apparently higher complexity of tropical lake food webs, as compared with the frequently depicted temperate food chains. In the tropics and subtropics, the most feasible alternative biomanipulation approaches target the control of cyanobacteria, either by directly using phytoplanktivorous filter-feeding fishes, or indirectly, reducing the internal nutrient loading induced by fish.

In low latitude lakes, the potential for biomanipulation is exemplified by the frequent herbivory by fish (e.g. phytoplankton and macrophyte consumption by tilapia and Chinese carps), high-carrying capacity of omnivorous fish (biomass often = 1,000 kg ha.$^{-1}$) and the key role of fish as a vital source of N and P (internal loading via excretion, egestion and bioturbation) or sink (exportation via fisheries and aquaculture). To our knowledge, published works on lake food web manipulations in the tropics and subtropics are lacking. However, some experimental studies that evaluate planktivorous fish impacts or fish management practices conducted in Asia, Africa, Oceania and the Americas provide evidences for water quality alterations mediated by the presence of fish (Table 12.1).

Table 12.1 Summary of experimental initiatives relevant to biomanipulation of tropical and subtropical lakes

Country	*Management approaches and their results*	*References*
China	Cyanobacteria reduction by silver carp in limnocorrals	Bangxi et al., 1993
	Cyanobacteria reduction by silver carp in lakes	Miura, 1990
	Extensive aquaculture of Chinese carps in reservoirs	Sifa & Senlin, 1995
Singapore	Extensive removal of tilapia and bighead carp cage culture to reduce eutrophication	Ling, 1982
	Use of silver carp to control eutrophication in reservoirs	Buck, 1977
New Zealand	Experimental control of eutrophication in lake using silver carp	Carruthers, 1986
	Control of macrophytes using grass carp	Rowe, 1984
Israel	Polyculture of carps and tilapia using silver carp in tanks	Hepher, 1977 (In: Buck, 1977)
	Stocking silver carp into reservoirs to control microphytoplankton	Leventer & Teltsch, 1990; Teltsch et al., 1991; Sagi, 1992
	Use of native *Tilapia galilea* for fisheries production and control of *Peridinium* in Lake Kinneret	Gophen,1995
India	Use of silver carp to improve sewage effluent	Rajan & Raj, 1994
	Experimental study in limnocorrals showing better algal control by silver carp than by tilapia	Datta & Jana, 1998
Morocco	Use of silver carp to control eutrophication in reservoirs	ONEP, 1991
South Africa	Extensive removal of tilapia and stocking of silver carp to control eutrophication in reservoir	Ferreira & Schoonbee, 1983
East Africa	Ecological damage of Nile perch introduction into Lake Victoria: *Haplochromis* extinction and enhanced eutrophication	Lowe-McConnell, 1999
Zimbabwe	Sequence of events in Lake Kariba: invasion of macrophytes (*Salvinia*) biologically controlled by insects, introduction of planktivorous fish and shimp, increasing fisheries production and nutrient sink	Magadza, 1995
United States	Extensive fisheries removal of omnivorous fish resulting in water quality improvements in Florida lakes.	Godwin et al., 1994a,b
	Use of grass carp to control macrophytes in Texas and Florida lakes	Blackwell & Murphy, 1996 Sutton, 1977, Leslie *et al.*, 1983

(Table 12.1 Contd.)

(Table 12.1 Contd.)

Uruguay	Macrophytes management to control eutrophication in urban lake	Mazzeo et al., 1999
Chile	Use of cultured sea weeds to improve effluents from cage aquaculture	Buschman et al., 1996
Peru	Re-use of effuents to culture tilapia	Moscoso, 1998
Brazil	Experimental studies of zooplanktivores top-down effects	Arcifa et al., 1986; Roche et al., 1993;
	Experimental studies on omnivorous fish impacts	Starling & Rocha, 1990; Lazzaro, 1991; Starling, 1993a; Figueiredo, 2000
	Re-use of effluents in aquaculture	Matheus, 1986, 1993; Felizatto et al., 2000
	Biomanipulation of urban reservoirs using extensive removal of tilapias	Starling & Lazzaro, 1997 Starling et al., 2000
	Macrophytes control by grass carp in irrigation channels and reservoirs	CODEVASF, unpubl. data; Starling, in prep.

Most studies involve the stocking/removal of fish to control cyanobacteria or macrophytes and generate beneficial results. It must be emphasized that the stocking of Chinese silver carp (*Hypophthalmichthys molitrix*), bighead carp (*Hypophthalmichthys nobilis*) and grass carp (*Ctenopharyngodon idella*) is frequently associated with the removal of omnivorous fishes, such as tilapias. Management strategies consisting of stocking some omnivorous fish species while removing others to control phytoplankton could appear contradictory. It is justified, however, by the major role played by fish excretion in the eutrophication process (Lamarra, 1975; Brabrand et al., 1990; Vanni & Findlay, 1990). The capacity of fish to influence nutrient dynamics is well known. It is either direct, via liquid excretion and egestion of N and P, or indirect via alterations in the nutrient recycling of other components of the food web, e.g. bacteria and zooplankton (Attayde, 2000). In tropical lakes where omnivorous fish naturally reach an extremely high biomass, herbivory by fish may be offset by their antagonistic stimulation of phytoplankton growth via nutrient recycling.

A biomanipulation approach fitted for the tropics should, therefore, balance the two antagonistic effects of onmivorous fishes. Our experience gained through a decade of experiments on Lago Paranoá (Brazil) suggests that a biomanipulation based on the management of omnivorous fishes is an efficient means to control eutrophication.

THE *BIOMANIPULATION PROGRAM* OF LAGO PARANOÁ (BRASÍLIA-BRAZIL)

Evolution of the Eutrophication Process

Lago Paranoá is an urban reservoir (area 3,800 ha., mean depth 14 m, residence time 299 days) created in 1959 in order to improve climatic conditions, promote leisure and receive urban drainage and sewage of Brasilia, the newly constructed Brazilian capital. Since then, within the scope of Lago Paranoá Restoration Program, CAESB (Brasilia Water and Sewage Corporation) has been operating the Northern and Southern wastewater treatment plants, located upstream of the Bananal and Riacho Fundo branches, respectively (Fig. 12.1a). Since early 1970s to 1990s, the water quality deteriorated, particularly in those two branches, with low transparency (Secchi depth ≤ 0.5 m), excessive algal biomass (chlorophyll-*a* concentrations $\geq 40\ \mu g\ L^{-1}$) and high primary productivity (c.a., 900 gC $m^{-2}\ yr^{-1}$, oxygen method using light and dark bottles; CAESB data base), as a result of an increasing loading of raw or inadequately treated domestic sewage. In 1978, an ecological disaster occurred in the surroundings of the southern plant located in the Riacho Fundo Branch. An intense *Microcystis aeruginosa* bloom caused an extensive fish kill, affecting the entire water mass and spreading out a pungent foul odour all over the city. Since then, until the early 1990s, eutrophication intensified in Lago Paranoá. Occurrences of blooms have been prevented using periodic copper sulphate applications (upto several tonnes per year). During 1980s and 1990s, permanent blooms of filamentous cyanobacteria *Cylindrospermopsis raciborskii* have always been associated with reduced grazing by rotifers and small cladocerans, low Secchi depth (< 50 cm) and high nutrient levels (total phosphorus: 30-80 g TP L^{-1}, total nitrogen: 1.8-2.5 mg TN L^{-1}; Altafin et al. 1995). Stocked since early 1960s, exotic omnivorous Nile and Congo tilapias (*Tilapia rendalli* and *Oreochromis niloticus*, respectively) and common carp (*Cyprinus carpio*) responded to nutrient enrichment by reaching extremely high densities and biomasses, especially in Riacho Fundo and Bananal, the two most eutrophic branches.

As a final step of the Restoration Program, starting in 1993-1994, CAESB implemented two new units to their wastewater treatment plants located in the vicinity of the old ones. They were built to treat the sewage from the watersheds at advanced tertiary level (i.e. 90% removal of N and P, 70% reduction in external P-loading). Nevertheless, the water quality

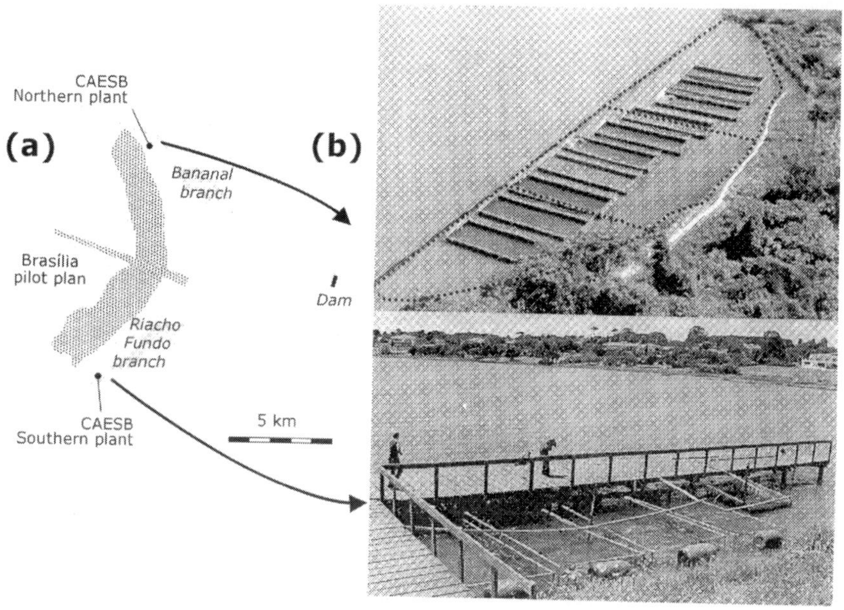

Fig. 12.1 (a) Map of Brasilia pilot plan and Lago Paranoá with location of Northern and Southern wastewater treatment plants on the hypertrophic Bananal and Riacho Fundo branches. (b) Net-cages (10-20 m^3) and large littoral limnocorrals (ten 50 m^3 units in place, and location of three 100 m^3 units shown as dashed line) installed nearby CAESB plants

improved less than expected until a complementary ecotechnological approach was implemented by the end of the dry season (August-October) of 1998. At this time, following the winter circulation period when water level was lowest and cyanobacteria reached their highest concentrations CAESB and CEB, the Brasília hydroelectric company that operates the dam, coordinated their efforts and performed an ecohydrology operation called "flushing" by opening the dam gates. This operation was impressively successful due to an extreme manipulation of residence time (shift from 200 to 900 days, and back to 200 days within a few months). Consequently, Lago Paranoá exhibited a clear reduction in trophic status with overall mean values of chlorophyll-*a* decreasing from 90 μg L^{-1} in 1992 to 9 μg L^{-1} in 1999, remaining below 10 μg L^{-1} until 2002 (Fig. 12.2), and transparency (Secchi depth) increasing to 2.5 m (Burnett et al., 2001). Simultaneously, the nearly permanent monoculture of filamentous cyanobacteria *Cylindrospermopsis raciborskii* was replaced by a dominance of green algal species. In the Riacho Fundo branch, net mean primary productivity dropped from 969 gC m^{-2} yr^{-1} in 1988 to 467 gC m^{-2} yr^{-1} in 2002, remaining three times higher than that of the Bananal branch (from 600 gC m^{-2} yr^{-1} in 1988 to 161 gC m^{-2} yr^{-1} in 2002) due to its higher phosphorus loading (41 kg TP day^{-1} against 8 kg TP day^{-1}, respectively) (Starling, unpubl.). Consequently, the phytoplankton biomass remained proportionally higher in the Riacho Fundo branch (32 μg Chl-*a* L^{-1}) than in the Bananal branch (10 μg Chl-*a* L^{-1}). However, since the beginning of 2002, chlorophyll-*a* concentration has started increasing again, currently reaching 16 μg L^{-1} in the Riacho Fundo Branch, with reappearance of some locally distributed cyanobacteria blooms being treated by algaecide (copper sulphate). Recent application of ecological modelling approaches revealed that Riacho Fundo Branch is the only area that is still over its carrying capacity for external phosphorous loadings (Starling et al. in prep.)

Assessment of Fish Stocks in Lago Paranoá

During the accelerated eutrophication period, two experimental fishing campaigns (SUDEPE, 1980; Grando, 1989)—the first one coupled with an acoustic survey—have been performed to evaluate the fish stock of Lago Paranoá. Total fish stock was estimated as being over 3,000 tons or 750 kg ha.$^{-1}$ on average, predominantly tilapias and common carp.

Prior to our 1980's biomanipulation studies, preliminary estimates of fish abundance were performed in Lago Paranoá to define the maximum tilapia biomass to be stocked into bag-type enclosures (2.5 m^3). Using cast nets—a highly selective gear for tilapia—Starling (1989) may have overestimated tilapia biomass at 3,000 kg ha^{-1}. A more conservative assessment was obtained in an hypertrophic region of the Bananal branch, by quickly isolating a 80-m^2 area with netting (5-mm mesh size). The overall fish biomass removed after rotenone application (1 mg L^{-1}) reached 1,592 kg ha.$^{-1}$, including 1,312 kg ha.$^{-1}$ (88%) represented mainly by *Oreochromis niloticus*, with few *Tilapia rendalli* (Starling, 1998). This estimate was implemented as tilapia typical stocking density in the 1990's experiments using large littoral limnocorrals (Fig. 12.1b).

The observed overpopulation of tilapia in shallow areas added to the risk of fish kills necessitated a reservoir-wide campaign in 1998, coupling experimental fishing with acoustic echo-integration (Lebourges-Dhaussy et al., 1999), to assess the overall fish stock of Lago Paranoá. Gillnets (50 m × 2 m; 10 mesh sizes, 30-240 mm opposite knots) were set at 3-4 stations in each of the four branches and two central regions. Echo-integration during the day and at night was based on (100-m apart) transversal transects using a 120 KHz vertically-oriented Simrad EY 500 split-beam portable transducer, equipped with a GPS, onboard a 4-m rubber boat propelled at 7 km h^{-1}.

Despite the removal of ca. 300 tonnes of fish following two major fish kills during the 20-month period prior to the acoustic survey, the lake fish stock recovered quickly. The fish stock estimated from the acoustic survey reached > 1,400 tonnes for the entire reservoir area. This is an underestimate as fishes occupying areas < 1 m in depth could not be detected by the transducer. The highest fish biomasses were detected in Riacho Fundo (355 tons, average 791 kg $ha.^{-1}$) and Bananal (304 tons, 564 kg $ha.^{-1}$) branches, especially in the vicinity of the Southern and Northern sewage plants.

To provide an independent estimate of fish abundance in areas susceptible to fish kill and also prevent further ecological damages in the Riacho Fundo branch, a large-scale selective tilapia removal campaign was conducted in May 1999. This was the preliminary step to implementing a fisheries management plan at the reservoir scale. A 78 ha. area upstream of the Riacho Fundo Branch was isolated using a net (300 m × 10 m, 5-mm mesh size) to estimate tilapia stock using the mark-recapture procedure. During two days, five cast-net fishermen captured, marked and released 2,083 fishes (94% tilapias). During the following five days, a group of 20 cast-net fishermen captured a total of 14,454 fishes (3 tons), comprising 90% tilapias. With 66 of the initially marked fishes recaptured, the total exploitable stock was estimated using Petersen's modified method as a half million adult fishes, or approximately 100 tons (Starling et al., in prep.). Extending for three more days, the period of capture increased the total capture by 5 tons. This estimate of tilapia stock is consistent with the 200 tons (including fish larvae and fingerlings) estimate from the 1998 acoustic campaign for the same 78 ha. area, assuming that adults represent 30-70% of the total fish abundance (Starling, 1998).

Experimental Studies on the Potential of Biomanipulation to Restore Lago Paranoá

Owing to the association between cyanobacteria dominance and high planktivorous fish biomass, the eutrophication context of Lago Paranoá appeared favourable for testing scenarios of food web manipulations. Hence, in 1987, within the CAESB Restoration Program of Lago Paranoá, we initiated a Biomanipulation Program supported by CAESB and various research agencies (e.g. IFS, CNPq, CAPES) based on experimental studies focusing on fish effects. These studies were performed along various spatio-temporal scales at the Northern plant site in the Bananal branch (Starling, 1989). Series of laboratory trials evaluated fish foraging behaviour and measured their feeding rates in aquaria (10-50 L; duration hours-days). They were coupled with *in situ* experiments that evaluated fish impacts in bag-type (2.5 m^3) and tube-type enclosures (6.5 m^3; duration 1-3 months). These experiments demonstrated the stimulation of

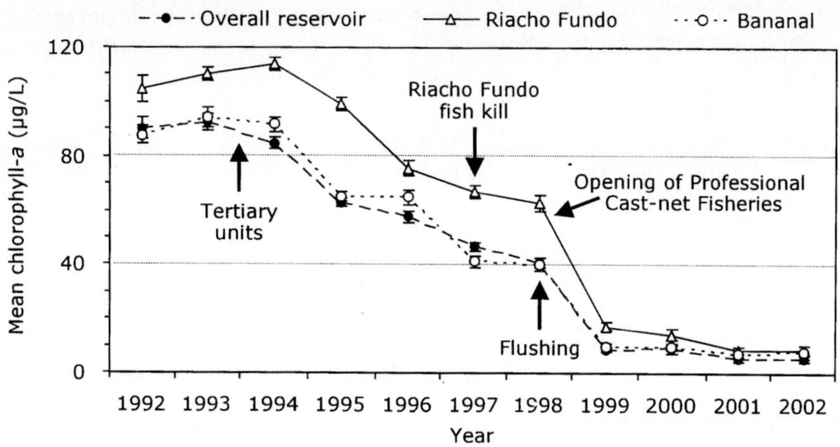

Fig. 12.2 Evolution of mean annual chlorophyll-*a* concentrations in the Bananal and Riacho Fundo branches and for the overall Lago Paranoá, during the 1992-2002 period (CAESB database). Dates of implementation of tertiary treatment units (1993-1994), Riacho Fundo fish kill (1997), opening of professional cast-net fisheries and major water flushing (1998) indicated by arrows along the time axis

eutrophication by Congo tilapia (Starling & Rocha, 1990), as well as the potential for using silver carp, a pelagic filter feeder, for direct control over cyanobacteria blooms (Starling, 1993a, 1993b). Further experiments conducted in large isolated littoral areas (1000 m^3) demonstrated the occurrence of synergistic effects between tilapia overpopulation and the intensification of eutrophication. Cyanobacteria were favoured at high biomass of tilapia because of the direct supply of nutrients to the system by fish excretion/egestion plus stirring up of sediments (bioturbation) caused by foraging activities (Starling & Lazzaro, 1997; Fig. 12.3).

The applicability of biomanipulation to Lago Paranoá was evaluated through various experiments in large limnocorrals (80-1,000 m^2 or 50-100 m^3).

(1) In two large unreplicated limnocorrals (1,000 m^2, 100 m^3), the control of tilapia overpopulation to reduce internal nutrient loading was subdued to a practical test (Starling & Lazzaro, 1997).

(2) The alternative approach of stocking pelagic filter-feeding sterile silver carps to selectively consume cyanobacteria in the water column was also evaluated.

As pointed out in a review concerning the use of silver carp to control cyanobacteria (Starling et al., 1998), the large majority of successful examples of suppressions came from low latitude

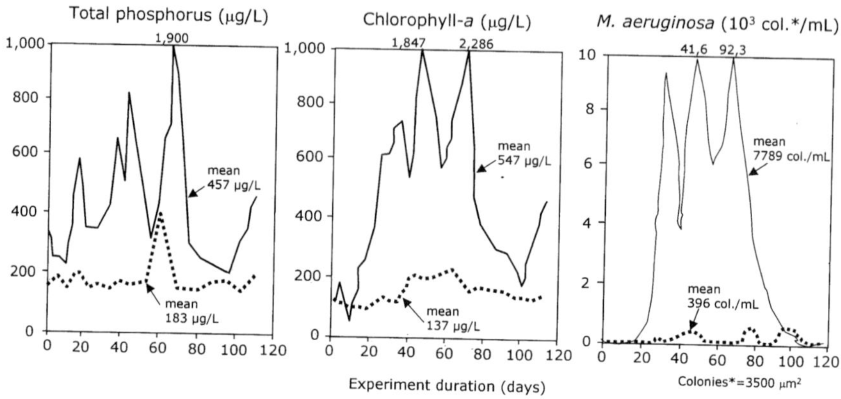

Fig. 12.3 Evolution of total phosphorus, chlorophyll-*a* and cyanobacteria *Microcystis aeruginosa* responses to high (initial 547 kg ha.$^{-1}$, continuous lines) and low tilapia biomass (initial 206 kg ha.$^{-1}$, dotted lines) in two large littoral limnocorrals (834 m^2/619 m^3 and 1,065 m^2/1,015 m^3, respectively) during 113 days of the dry season (modified from Starling & Lazzaro, 1997). The responses between the two treatment biomass levels differed significantly (p = 0.001 for either variable, repeated-measures ANOVA without replication)

ecosystems where large inedible algae dominance was contained using a moderate fish biomass (< 1,200 fish ha.$^{-1}$, < 70 g m^{-3} or < 900 kg ha.$^{-1}$). Limiting the initial stocking density of silver carp plus removing (through fisheries) excess fish biomass produced through time are key points to the success of this type of management. Indeed, at high biomass levels, silver carp stimulate algal growth indirectly through both nutrient excretion and zooplanktivory. A gradient experiment was carried out to evaluate the optimum range of silver carp biomass capable of suppressing cyanobacteria in Lago Paranoá. Adult silver carps were stocked at five biomass levels (0, 10, 20, 40 and 60 g m^{-3} or 0, 60, 120, 240 and 360 kg ha.$^{-1}$) replicated into ten large limnocorrals (80 m^2, 50 m^3) to evaluate fish effects on water quality during a 35-day-experiment (Starling et al. 1998). The presence of silver carp significantly reduced the biomass of colonial algae, mainly *Microcystis*, regardless of fish stocking density (repeated-measures ANOVA (RMA), $p = 0.093$). Only at moderate-high biomass levels (40-60 g m^{-3}), silver carp decreased both the percentage of large-sized algae in the overall phytoplankton biomass (RMA, $p = 0.015$) and zooplankton abundance (RMA, $p = 0.021$; Fig. 12.4).

(3) Experimental studies performed in large littoral limnocorrals (80 m^2, 100 m^3) compared the two management strategies—implemented either alone or in combination. One experiment was

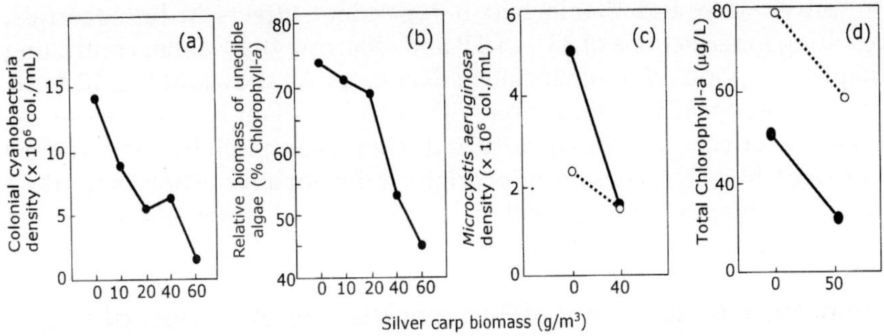

Fig. 12.4 Mean responses of: (a) colonial cyanobacteria density (*Microcystis* and *Botryococcus* spp.); (b) relative biomass of unedible algae (% chlorophyll-*a* ≥ 30 μm) to a biomass gradient of silver carp (60-360 kg ha.$^{-1}$) in ten large littoral limnocorrals (80 m^2/ 50 m^3) during 35 days of the wet season (modified from Starling et al., 1998); (c) mean response of *Microcystis aeruginosa* density to high (120 g m^{-3}, mixture of *O. niloticus* and *T. rendalli*, continuous line) vs low tilapia biomass (40 g m^{-3}, dotted line) in absence or presence of silver carp (40 g m^{-3}), in eight of these limnocorrals during 84 days of the wet season; and (d) mean response of total chlorophyll-*a* to high (100 g m^{-3}, mixture of *O. niloticus* and *T. rendalli*, continuous line) vs low tilapia biomass (25 g m^{-3}, dotted line) in absence or presence of silver carp (50 g m^{-3}), in eight of these limnocorrals during 56 days of the dry season (modified from Starling, 1998)

performed during the rainy season, the other during the dry season, with four replicated treatments comprising: (i) a high tilapia biomass, typical for the reservoir (control: 100-120 g m^{-3} or 1,200-1,440 kg ha.$^{-1}$); (ii) a low tilapia biomass (reduction down to 25-40 g m^{-3} or 300-480 kg ha.$^{-1}$); (iii) a high tilapia biomass combined with the presence of silver carp (40-50 g m^{-3} or 480-600 kg ha.$^{-1}$); and (iv) a low tilapia biomass combined with the presence of silver carp (Starling, 1998).

Results from the rainy-season experiment showed that silver carp presence, irrespective of tilapia biomass, provoked an average 65% reduction in the biomass of colonial floating algae (*Microcystis* and *Botryococcus*) as compared with the control situation (RMA, p = 0.054). During the dry-season experiment, however, important water quality improvements were associated with the reduction of tilapia biomass. As compared with the control situation (high tilapia biomass typical from the reservoir), transparency increased by 31% (RMA, p = 0.058), turbidity decreased by 44% (RMA, p = 0.019), total phosphorus (TP) decreased by 26% (RMA, p = 0.074) and chlorophyll-*a* dropped by 37% (RMA, p = 0.003). Adding silver carp alone promoted a 25% reduction in chlorophyll-*a* (RMA, p = 0.027) as compared with the control situation (Fig. 12.4).

Silver carp and tilapia had independent effects in limnocorrals, resulting in reductions of 35% of TP and chlorophyll by either controlling tilapia overpopulation or stocking silver carp. As shown in Fig. 12.4, the combination of tilapia reduction with stocking silver carp increased the beneficial effects on water quality, reducing chlorophyll by 70% (from 80 to 25 μg Chl-*a* L^{-1}). These results originated from a reduction in nutrient supply by tilapia removal combined with an algal suppression by silver carp filtration (Starling, 1998).

Laboratory Assessment of Phosphorus Excretion Rate of Tilapia

To estimate the impact of removing tilapia on the reduction of internal phosphorus cycling of Lago Paranoá, Starling (1998) used Braband et al., (1990) protocol to measure P-excretion rates of tilapia. These rates were based on monitoring changes in soluble reactive phosphorus (SRP or PO_4) and total phosphorus (TP) at 4-hour intervals during 24 hours in 50-L aquaria stocked with sixteen 16-g or six 40-g juvenile tilapias (half *Oreochromis niloticus*, half *Tilapia rendalli*). The aquaria were filled with water from the Bananal branch filtered through a 10-μm plankton net. Maximum excretion rates were 1.58 μg SRP g^{-1} h^{-1} and 1.12 μg TP g^{-1} h^{-1} for 16-g fish and 0.53 μg SRP g^{-1} h^{-1} and 1.73 μg TP g^{-1} h^{-1} for 40-g fish. Extrapolated to a stocking biomass of 100 kg $ha.^{-1}$, daily minimum-maximum excretion rates were 0.11-0.30 μg SRP L^{-1} d^{-1} for 16-g fish and 0.10-0.29 μg SRP L^{-1} d^{-1} for 40-g fish, which were consistent with published rates for other species.

Nevertheless, potential errors may have arisen in relation to underestimates due to P-loss (excreted) during fish capture, transport or acclimation, and/or overestimates due to P-released by dying fish during long trials. P-excretion rates of tilapias (*O. niloticus* plus *T. rendalli*) were thus re-assessed for small (6.7 cm SL, 30 g) and medium-sized fish (12.2 cm SL, 63 g) during 4-h trials in 50-L aquaria. Rates were only slightly higher, ranging from 3.09 ± 0.75 μg PO_4 g^{-1} h^{-1} or 7.36 ± 1.14 μg TP g^{-1} h^{-1} for 30-g fish to 1.16 ± 0.29 μg PO_4 g^{-1} h^{-1} or 3.56 ± 0.32 μg TP g^{-1} h^{-1} for 63-g fish (Larangeira, Teza, Zaparoli, Dayane, Silva, Santana & Starling, unpubl.). When applied to the 350 tons of the tilapia stock, assessed from the 1998 acoustic survey of the Riacho Fundo branch, the overall tilapia contribution to the internal P-cycling of this most eutrophic branch reached 46 kg TP day^{-1} by excretion only (without considering the contribution from bioturbation of P-rich bottom sediment). As a comparison, the mean 1998-2001 external P-loading to the Riacho Fundo branch was 49 kg TP day^{-1} (including 41 kg TP day^{-1} from the tributaries and the plant effluent, and 8 kg TP day^{-1} from diffuse sources).

These rates were used to estimate the potential of controlling the P-internal cycling by removing tilapia through fisheries. Twenty professional fishermen belonging to the Biomanipulation Program removed 120 tons within 20 months that correspond to ca. 21 kg TP day^{-1}, i.e. half the external P-loading to the Riacho Fundo branch. These estimates were confirmed by calculations based on a 1997 massive fish kill (see below).

Accidental Biomanipulation and Contribution of Tilapia to Phosphorus Internal Cycling

An extensive fish kill occurred in the Riacho Fundo Branch during early winter 1997. The air temperature during the night dropped to 12ºC and difference between day and night increased abruptly from 1.5 to 7.4ºC. This rapid cooling provoked a vertical circulation of the water column, called thermal inversion, that brought sediment organic matter and gases up into the epilimnion. Consequently, oxygen demand increased, resulting in a fish kill, mainly tilapias. During this mortality event that lasted two months, a total of 150 tonnes of dead and partially decomposed fish were removed from the Riacho Fundo Branch. This event took place after we experimentally demonstrated that water quality improves by reducing tilapia overpopulation. It provided an opportunity to analyze the consequences of a large-scale accidental biomanipulation.

We applied a statistical approach called BACI (Before and After in Control and Impact areas; Stewart-Oaten et al., 1986) that is suited for detecting effects in non-replicated situations. This approach uses t-tests to compare the differences between the Impact (Riacho Fundo) and Control (Bananal) branches of variables (e.g. TP, chlorophyll-*a*) measured during one year before and after the fish kill (Starling et al. 2002).

Analyses revealed the occurrence of a long-term 34% reduction (difference between average values before and after the fish kill) in TP (from 100 to 66 µg L^{-1}, $p < 0.001$) and chlorophyll-*a* (from 84 to 56 µg L^{-1}, $p = 0.02$) after the extensive removal of dead fish. These changes were mainly related to fish effects because during this period, the external nutrient loadings to the branches remained unaltered ($p = 0.27$).

We combined P-excretion rates of tilapia measured in laboratory trials (Starling, 1998) with data from acoustic echo-integration and gillnet fishing obtained during a campaign in April-May 1998 (Lebourges-Dhaussy et al. 1999). Hence, we assessed the overall TP excreted by the tilapia stock in the Riacho Fundo branch (3.9-9.1 kg TP day^{-1}, day-night). This internal TP-loading by the tilapia stock represented 9-21% of the external TP-loading from tributaries to the branch and the sewage treatment plant (43 kg TP day^{-1}, CAESB, unpubl.).

We assessed the reduction in internal TP-loading resulting from the removal of 150 tons of dead tilapias as the sum of the prevented P-released by fish feeding on the reservoir bottom (excretion plus bioturbation) and the corporal P removed. The total amount of TP prevented from being recycled reached 12% (5.1 kg TP day^{-1}) of the external TP-loading to the branch, whereas the exported TP bounded in fish tissue was equivalent to 20 days of external loading (857 kg TP; Starling et al. 2002).

Productivity of the Tilapia Fishery and Socio-economical Benefits

During the 1999 experimental cast-net removal of tilapia in the Riacho Fundo branch, capture per unit effort (CPUE) ranged between 5-10 kg fish fisherman^{-1} hour^{-1}, with a daily yield of 20-100 kg fish fisherman^{-1} day^{-1}. Fish landings from Lago Paranoá illegal cast-net fishery monitored during 1999-2000 rendered a mean capture of 225 kg fish fisherman^{-1} month^{-1} (Walter, 2000). During a 20-month monitoring (April 2000-November 2001), a total of 158 tons of tilapias and common carp were landed by a group of 20 licensed fishermen (i.e. mean 395 kg fisherman^{-1} month^{-1}). Overall evolution during the 2000-2001 period showed sustained captures (30 kg fisherman^{-1} day^{-1}), fishing effort (513 hours month^{-1}) and fishing yield (6.2 kg fisherman^{-1} hour^{-1}). However, during an extensive monitoring in May 2002, fishing yield dropped down to 3.2 kg fisherman^{-1} hour^{-1} (Paula, Bittencourt, Matsuda, Birbeire, Arcoverde & Starling, unpubl.) in relation with an increase in fishing effort reaching 902 hours month^{-1}, associated with the overall decrease in reservoir trophic state (Fig. 12.2).

Such high experimental yields added to the excellent sanitary quality of Lake Paranoá fishes confirmed by a 1996 assessment of (a lack of) heavy metals, pesticides and pathogenic bacteria in fish flesh (ISDF, 1997; Caldas, 1999) persuaded the Brazilian Environment Agency (IBAMA) to authorize professional cast-net fisheries in the Riacho Fundo and Bananal branches. From February 2000 until January 2002, an initial group of 450 professional fishermen was licensed by IBAMA (decree n° 106/99), with an extended authorization for two more years so as to maintain environmental and economical benefits.

The number of licensed fishermen arises from combining the CPUE range with a mean fish stock turnover of 65 kg ha.$^{-1}$ month^{-1} estimated from recruitment rates based on the various experimental studies in limnocorrals. This turnover value leads to a mean fisheries productivity of 780 kg ha.$^{-1}$ year^{-1} in those hypertrophic areas of the reservoir (Starling,

1998). Such fisheries' productivity is consistent with the 600 kg ha.$^{-1}$ year^{-1} estimated as 1% of phytoplankton net primary productivity, according to Lowe-McConnell's (1999) calculation.

This fisheries productivity (P) of 780 kg ha.$^{-1}$ year^{-1} is quantitatively similar to the mean total fish biomass (B) of 787 kg ha.$^{-1}$ acoustically estimated in those areas (Lebourgues et al. 1999). Therefore, the production per biomass ratio (P/B) approximates 1. Such P/B ratio is lower than values computed by Lowe-McConnell (1999) for tropical African Lake Kariba (P/B = 1.5 where P = 1,224 kg ha.$^{-1}$ year^{-1}, B = 827 kg ha.$^{-1}$) and Lake Tanganyika (P/B = 3.0; P = 700 kg ha.$^{-1}$ yr^{-1}, B = 230 kg ha.$^{-1}$), and Amazonian Lake Luísa (P/B = 1.8; P = 2,800 kg ha.$^{-1}$ yr^{-1}, B = 1,600 kg ha.$^{-1}$).

The first year of monitored landings from the initial 50 licensed fishermen—already organized in a cooperative—returned a mean capture > 5,000 kg month^{-1}, predominantly tilapias. With tilapias (whole fish without gut and scales) sold at 1 US$ kg^{-1}, the mean gross income of an individual fisherman licensed by IBAMA under the Biomanipulation Program exceeds 200 US$ month^{-1} (using an average conservative capture of 200 kg month^{-1}, cf. mean value of 225 kg month^{-1} from Walter 2000), i.e. more than twice the Brazilian minimum salary.

Beyond the environmental benefits of controlling the internal phosphorous loading and preventing new fish kills, the tilapia fishery produces social and economical benefits to (formerly low-income) professionals and their families. As each licensed fisherman brings up to four jobs (fishermen, rower, fish processor and seller), the Biomanipulation Program could generate ≥ 1,000 new direct and indirect jobs (some fishermen work alone or have only 1-2 helpers) related to the ecological exploitation of Lago Paranoá fish production.

FURTHER DEVELOPMENT OF *LAGO PARANOÁ BIOMANIPULATION PROGRAM* AS A MANAGEMENT TOOL TO SUSTAIN OLIGOTROPHICATION

Cyanobacteria Control by Stocking Free Roaming Silver Carp

Except for the Riacho Fundo Branch, the only area of the reservoir already exceeding its carrying capacity for external phosphorus loading, stocking sterile silver carps to biologically control cyanobacteria became unnecessary, at least until the current oligotrophication process of Lago Paranoá is persisting.

Environment-friendly Silver Carp Cage Culture

The great potential to implement an extensive, environment-friendly, silver carp cage culture to convert part of the considerable sestonic production originated in eutrophic areas, such as the Riacho Fundo Branch, into edible fish biomass deserves a major attention. Experiments in floating net-cages (2.5 m × 2.5 m × 1.7 m, 10 m^3) explored growth responses of silver carp juveniles (300 g, 25 cm SL) and adults (1,100 g, 40 cm SL) maintained at two stocking densities (400 and 800 g m^{-3}) for six months without supplementary feeding. A survivorship ≥ 90% demonstrated that silver carp adapts very well to the most eutrophic conditions of the reservoir in the Bananal Branch (Starling et al. 1999). Growth rates of 2.76 g d^{-1} (0.5% wet weight d^{-1}) for juveniles and 4.4 g d^{-1} (0.29% wet weight d^{-1}) for adults silver carp are comparable to values from the literature (Hamada et al., 1983; Bayne et al.,1991). Intense filtration activity by silver carp (0.3 L g $w.w^{-1}$ h^{-1}, Starling, 1989) resulted in high consumption rates of surface colonial bloom-forming algae, cyanobacteria, such as *Microcystis aeruginosa* and *Botryococcus braunni*. Confirmation came from stomach content analyzes of fishes recovered from the experiment. A new experimental study of the potentiality of silver carp cage-culture is currently in progress in the Riacho Fundo branch. Leão & Starling (unpubl.) performed two long-term experiments using fingerlings (4 g, 7 cm SL) and small fish (68 g, 19 cm SL) stocked at 250-800 g m^{-3}, respectively, in small 10 m^3 and large 20 m^3 net-cages. Fingerlings grew upto 70 g after 232 days, i.e. growth rate of 0.34 g day^{-1} or 9.01% day^{-1}, and small fish to 1,132 g in 187 days, i.e. 5.6 g day^{-1} or 9.14% day^{-1}. Those growth rates were higher than those reported for net-cage cultures (Table 12.2), attesting the great potential for environment-friendly silver carp cage culture in eutrophic areas of tropical lakes and reservoirs. Besides major fish losses (30-60%) in the above experiments, total silver carp yields reached 35 and 297 tonnes $ha.^{-1}$ $year^{-1}$, for fingerlings and small fish, respectively. The maximum amount of phosphorus sequestered and exported from those cages as fish flesh would be 467 mg TP m^{-2} $year^{-1}$. Thus, a large-scale silver carp cage culture occupying 8.78 ha. (equivalent to only 1.9% of the surface area) would be able to export yearly the equivalent of the external TP loading to Riacho Fundo branch (549 ha., 8.9 mg TP m^{-2} $year^{-1}$).

Stocking Herbivorous Fish to Control of Excessive Macrophytes Growth

Although a clear water state is promoted by the predominance of aquatic plants (Scheffer et al., 1993), the excessive development of macrophytes can be a problem in tropical lacustrine ecosystems. In upstream areas of

Table 12.2 Comparative growth rates of silver carp obtained in net-cages

		Growth rates	
Reference	*Country*	*(g ww day^{-1})*	*(% ww day^{-1})*
	Young fishes		
Coche (1979)	Hungary	1.19	0.50
Pradhan & Swar (1987)	Nepal	3.85	0.98
Starling (1995)	Brazil	2.87	0.51
Starling (in prep.)	Brazil	0.36	9.48
	Adult fishes		
Hamada et al. (1983)	Japan	4.60	0.36
Adamek & Spittler (1989)	Czech Republic	1.62	0.18
Starling (1995)	Brazil	4.39	0.29
Starling (in prep.)	Brazil	5.27	6.92

the Riacho Fundo and Bananal branches, it is detrimental to water clarity and quality. In an experiment using three 50 m^2 limnocorrals, performed for 120 days in nearby Vale Verde reservoir (28 ha.), characterized by the proliferation of submerged *Utricularia,* sterile grass carp completely eliminated macrophytes irrespective of fish biomass (600 and 2,000 kg ha.$^{-1}$), but increased turbidity (from 2.6 to 11.7 NTU) and chlorophyll-a (from 8.6 to 52.2 µg L^{-1}) at highest grass carp biomass. Grass carp stocked at a lower 400 kg ha.$^{-1}$ in a similar experiment conducted in CAESB Paranoazinho water supply pond (0.4 ha.) eliminated ca. 50% of rooted submerged macrophytes and 60% of undesirable filamentous macro-algae (1-10 mm *Mougeotia, Spirogira*) without promoting such a dramatic side effect of stimulating phytoplankton (Viana & Starling, unpubl.). These results show that besides the potential for herbivorous fish to control macrophytes, special attention should be paid to the critical use of a low fish biomass (stocking < 20 kg ha.$^{-1}$, using 1-kg individuals; with a control after five years when reaching 300 kg ha.$^{-1}$) in order to maintain a relative macrophytes cover preventing phytoplankton from developing and to avoid detrimental bioturbation effects.

Piscivores Stocking and Sport Fishery

Another complementary initiative of the Biomanipulation Program refers to the stocking of piscivores in some areas of Lago Paranoá, to intensify the control of tilapia overpopulation and promote a sport fishery. Current experimental studies in large limnocorrals (80 m^2, 50-100 m^3) involve the comparative assessment of impacts from tucunaré (*Cichla ocellaris*), an Amazonian species already introduced into the reservoir, with other piscivorous fish from the watershed, such as dourado (*Salminus maxillosus*), matrinxã (*Brycon spp.*), and surubim (*Pseudoplatystoma coruscans*). In the first experiment, both tucunaré and surubim had an excellent adaptation to limnocorrals (100% survivorship during 40 days of trial). However, only tucunaré grew well in tanks containing tilapia as the only prey (Ribeiro-Filho, 2002). Final tests involved the stocking of two tucunaré (1.1-1.2 kg; 42-43 cm), two surubim (1.2-1.5 kg; 58-61 cm), two dourado (0.6-0.8 kg; 37-41 cm), and two matrinxã (0.7-1.9 kg; 40-46 cm) into replicated large littoral limnocorrals (5 x 16 m, 80 m^2 and 100 m^3). Each of these eight experimental units plus two predator-free limnocorrals were stocked with the same number of Nile tilapia (27 fish, 7-78 g; 7.5-14.2 cm) and Congo tilapia (27 fish, 7-84 g; 8.0-14.3 cm). After 5 weeks of experiment, there was a small natural reduction of tilapias in the absence of predators (7.3%), but a progressive reduction of tilapia in the presence of surubim (-18.7%), matrinxã (-22.4%), dourado (-33.6%) and tucunaré (-34.3%). These preliminary results showed not only the potential for such piscivore species to contribute in controlling the excessive growth of tilapia but also indirectly improving water quality. As well, they open a new perspective for promoting a sport fishery (catch-and-release) on noble, highly-priced species (Starling et al., in prep.).

PERSPECTIVES OF BIOMANIPULATION FOR THE RESTORATION OF TROPICAL AND SUBTROPICAL LAKES

In summary, these 15 years of experimental research and management initiatives within the framework of Lago Paranoá Biomanipulation Program attest the great potential of implementing this ecotechnological approach in other tropical lake systems suffering from excessive omnivorous fish biomass and cyanobacteria blooms.

The development of a tropical biomanipulation approach still faces some challenges. First of all, it is necessary to integrate and consolidate the theoretical knowledge gathered regarding tropical lake food webs, and verify whether models originally elaborated for temperate systems are general or specific models should be proposed. Extending the use of in situ experimental approaches is vital. Emphasis should be given to testing the currently debated role of key processes or components in the successes and failures of biomanipulation, such as:

(i) Food web manipulations in shallow systems have a greater potential because of the ability of large bottom areas to be colonized by macrophytes. Macrophytes may thus promote the clear-water state by outcompeting algae for nutrients (Van Donk et al., 1990) and/or releasing allelopathic substances against cyanobacteria (Declerck et al., 2000).

(ii) The effectiveness of biomanipulation measures may differ with the trophic state of lakes, which affect food chain length, particularly the upper trophic levels. The "mesotrophic maximum hypothesis" (Elser & Goldman, 1991) according to which mesotrophic lakes—with four trophic levels including piscivores—suffer stronger top-down impact on phytoplankton, has not reach a consensus.

(iii) The importance of internal nutrient recycling by aquatic animals, as compared with the external loadings by the watershed, is still poorly studied. Many effects of fish on phytoplankton may result from indirect effects of nutrient excretion enhanced by microzooplankton (allometry-related), excretion and egestion by fish, nutrient release via bioturbation of bottom sediment by benthivorous fish (Braband et al., 1990; Schindler et al., 1993; Attayde & Hansson, 2001), deterioration of water quality by destruction or uproot of macrophytes by benthivorous fish (Tatrai & Istvanovics, 1986).

(iv) Size-structured interactions and ontogenetic niche shifts are major determinants of trophic interactions. The stronger zooplanktivory of small fish and young-of-the-year (YOY) fish, which have higher biomass-specific food requirements and occur at higher densities (Romare & Bergman, 1999), is generally neglected, as compared to adult fish, which are mostly considered. The size-refuge of prey-fish that may prevent the top-down control by piscivores (Lammens, 1999) occurs between omnivores (e.g. high-bodied and spiny tilapias, only preyed upon at younger stages) and their facultative piscivores (e.g. traíra, *Hoplias malabaricus*) in tropical Brazilian lakes. The unwanted shift from edible to less edible or inedible algae cannot be reverted in eutrophic tropical lakes lacking daphnids without stocking a filter-feeding pelagic planktivore such as silver carp (Starling et al. 1998).

(v) Biomanipulation requires the drastic removal of planktivores and/or benthivores, which is often an unachievable practical goal. On the other hand, piscivore stocking alone returns the lowest rate of success for biomanipulations (Drenner & Hambright, 1999). Wysujack & Mehner (2002) have suggested that an appropriate balance between piscivores and planktivores in fish community should consist of 25-50% of piscivore in order to ensure successful biomanipulations in German lakes. Whether this range is valid for tropical and subtropical lakes needs to be explored. As an example, the shallow eutrophic Northeast Brazilian reservoirs with chlorophyll $< 20\ \mu g\ L^{-1}$ are characterized by fish communities consisting of > 60% of piscivores (in biomass) and biomass ratios of piscivores to omnivores (i.e. tilapias) > 1 (Lazzaro et al., 2003).

Furthermore, the application of biomanipulation as an *ecotechnological tool* requires to be compatible with other uses of aquatic resources. Indeed, water quality restoration conflicts with aquaculture production as well as conservation of natural resources (i.e. native species coexistence may be jeopardized by the introduction of exotic species).

Finally, implementing biomanipulation should not be detrimental to human health. The sanitary quality of organisms living in polluted systems must be checked prior to implementing management procedures. A crucial warning concerns the risk that algal toxins could be incorporated into the food-webs of hypertrophic ecosystems, where fish are used for human consumption (Magalhães et al., 2001).

REFERENCES

Adamek, Z. & Spittler, P. 1984. Particle size selection in the food of silver carp, *Hypophthalmichthys molitrix*. Folia Zoologica **33**(4): 363-370.

Altafin, I. G., Mattos S. P., Cavalcanti C. G. B. & Estuqui V. R. 1995. Paranoá Lake - Limnology and Recovery Program. *In:* Limnology in Brazil. Brazilian Academy of Sciences & Brazilian Limnological Society, pp. 376.

Arcifa, M. S., Northcote T. G. & Froehlich, O. 1986. Fish-zooplankton interactions and their effects on water quality of a tropical Brazilian reservoir. Hydrobiologia **139**: 49-58.

Attayde, J. L. 2000. Direct and indirect effects of fish predation and excretion in aquatic food webs. Ph.D. Thesis. Department of Ecology, Lund University, Sweden, pp. 112.

Attayde, J. L. & Hansson, L. -A. 2001. The relative importance of fish predation and excretion effects on planktonic communities. Limnol. Oceanogr. **46**: 1001-1012.

Bangxi, X., Deshang, L., Qi, L. & Jimin, C. 1993. The impact of filter-feeding fish on the carrying capacity of reservoirs for feeding cage-culture of fish. Acta Hydrobiologica Sinica **17**: 131-143.

Bayne, D. R., Joshi, P. L., Rai, A. R. & Williams, J. C. 1991. Growth and food habits of cage-cultured bighead carp x silver carp hybrids in ponds of varying trophic status. J. Appl. Aquaculture **1**: 45-55.

Blackwell, B. G. & Murphy, B. R. 1996. Low-density triploid grass carp stockings for submersed vegetation control in small impoundments. J. Freshwat. Ecol. **11**: 475-484.

Brabrand, A., Faafeng, B. A. & Nilssen, J. P. M. 1990. Relative importance of phosphorus supply to phytoplankton production: Fish excretion versus external loading. Can. J. Fish. Aquat. Sci., **47**: 364-372.

Buck, D. H. 1977. The integration of aquaculture with agriculture. Fisheries **2**: 11-16.

Burnett, J. A. B., Mattos, S. P. & Azzolin, N. M. P. 2001. Intervenções da Companhia de Saneamento. *In:* F. Fonseca (ed.). *Olhares sobre o Lago Paranoá. SEMARH,* Brasília, Brazil: 199-212.

Buschman, A. H., Troell, M., Kautsky, N. & Kaursky, L. 1996. Integrated tank cultivation of salmonids and *Gracilaria chilensis* (Gracilariales, Rhodophyta). Hydrobiologia **326/327**: 75-82.

Caldas, E. D., Coelho, R., Souza, L. C. K. R. & Silva, S.C. 1999. Organochlorine pesticides in water, sediment, ad fish of Paranoá Lake of Brasília, Brazil. Bull. Environ. Contam. Toxicol. **62**: 199-206.

Carruthers, A. D. 1986. Effect of silver carp on blue-green algal blooms in Lake Orakai. N. Z. Minist. Agric. Fish., Fish. Res. Div. Fish. Envir., Report 68, pp. 63.

Coche, A. G. 1979. A review of cage fish culture and its application in Africa. In: T. V. R. Pillay & W. A. Dill (Eds), Advances in Aquaculture. Londo: Fishing News Books Ltd.: 428-441.

Datta, S. & Jana, B. B. 1998. Control of bloom in a tropical lake: Grazing efficiency of some herbivorous fishes. J. Fish Biol. **53**: 12-24.

Declerck, S., DeMeester, L., De Smedt, P., Rommens, W., Vyverman, W., Geenens, V., Van Wichelen, J., Degans, H. & Decleer, K. 2000. Clear water and charophytes in a hypertrophic pond. Verh. Internat. Verein. Limnol. **27**: 541.

Drenner, R. W. & Hambright, K. D. 1999. Biomanipulation of fish assemblages as a lake restoration technique. Arch. Hydrobiol. **146**: 129-165.

Elser, J. J. & Goldman, C. R. 1991. Zooplankton effects on phytoplankton in lakes of contrasting trophic status. Limnol. Oceanogr. **36**: 64-90.

Felizatto, M., Starling, F. L. R. M. & Souza, M. A. 2000. Estudos Preliminares para Verificação da Possibilidade de Reúso de Água em Piscicultura no Distrito Federal. *Anais do I Simpósio de Recursos Hídricos do Centro-Oeste: valorizando, conservando e preservando uma riqueza regional*. Brasília, 26 a 28 de junho de 2000.

Ferreira, J. T. & Schoonbee, H. J. 1983. Use of heated sewage effluent from a power station for the culture of *Oreochromis mossambicus* and *Hypophthalmichthys molitrix*. Proc. International Symposium on Tilapia in Aquaculture, Nazareth, Israel, 8-13 May 435-445.

Figueiredo, C. C. 2000. Efeitos da tilápia (*Oreochromis niloticus*) nas características físicas e químicas e estrutura da comunidade fitoplanctônica do Reservatório da UHE de Furnas (MG). Dissertação de Mestrado [Master Thesis]. *Instituto de Ciências Biológicas. Universidade Federal de Minas Gerais (UFMG)*, pp. 205.

Godwin, W. F., Crumpton, J. E., Lowe, E. F., Coveney, M. F., Battoe, L. E. & Stites, D. L. 1994a. Fish harvesting for nutrient removal and trophic state effects in Lake Apopka, Florida. Lake & Reserv. Manag. **9**: 76.

Godwin, W. F., Coveney, M. F., Lowe, E. F., Battoe, L. E. & Benton, J. 1994b. Trophic state changes following removal of planktivorous fish from a hypertrophic Florida lake. Lake & Reserv. Manag. **9**: 77.

Gophen, M. 1995. Long term (1970-1990) whole lake biomanipulation experience case study: Lake Kinneret (Israel) In: De Bernardi, R. & Giussani, G. (eds). Biomanipulation in lakes and reservoirs management. Guidelines of lake management. Volume 7. ILEC/UNEP: 171-184.

Grando, J. V. 1989. Considerações sobre o estágio atual da ictiofauna e aspectos alimentares de quatro espécies de peixes do Lago Paranoá, Brasília-DF. M. Sc. Thesis, Univ. of Brasília, Brasília, pp. 98.

Hamada, A., Maeda, W., Iwasaki, J. & Kumamaru, A. 1983. Conversion efficiency of phytoplankton in the growth of silver carp, *Hypophthalmichthys molitrix*. Jap. J. Limnol. **44**: 321-328.

ISDF, UnB, CAESB, IBGE, 1997. Avaliação da Qualidade Sanitária do Pescado do Lago Paranoá. *Relatório Técnico à Fundação de Apoio à Pesquisa do Distrito Federal (FAP-DF)*, pp. 120.

Lamarra, V. A. 1975. Digestive activities of carp as a major contributor to the nutrient loading of lakes. Int. Verein. Theor. Angew. Limnol., Verh. **19**: 2461-2468.

Lammens, E. H. R. R. 1999. The central role of fish in lake restoration and management. Hydrobiologia **395/396**: 619-627.

Lammens E. H. R., Gulati, R. D., Meijer, M. & Van Donk, E. 1990. The first biomanipulation conference: a synthesis. Hydrobiologia **200/201**: 619-627.

Lazzaro, X. 1991. Feeding convergence in South American and African zooplanktivorous cichlids *Geophagus brasiliensis* and *Tilapia rendalli*. Env. Biol. Fish. **31**: 283-293.

Lazzaro, X., 1997. Do the trophic cascade hypothesis and classical biomanipulation approaches apply to tropical lakes/reservoirs? Verh. Internat. Verein. Limnol. **26**: 719-730.

Lazzaro, X., Bouvy, M., Ribeiro-Filho, R. A., Oliveira, V. S., Sales, L. T., Vasconcelos, A. R. M. & Mata, M. R. 2003. Do fish regulate phytoplankton in shallow eutrophic Northeast Brazilian reservoirs? Freshwater Biology **48**: 1-20.

Lazzaro, X., Lebourges-Dhaussy, A., Dhaussy, Ph. & Correia, J.W. (eds) 1998. Serviço de levantamento hidroacústico (ecossondagem) no Lago Paranoá, Brasília-DF – Relatorio final. *Fundação Sustentabilidade e Desenvolvimento*, Brasília-DF, Brasil, pp. 69.

Lebourges-Dhaussy, A., Dhaussy, Ph., Lazzaro, X. & Starling, F. L. R. M. 1999. Acoustical fish biomass assessment of Lago Paranoá – Brasilia – Brazil within the framework of a water quality improvement project. Proc. 2nd International Shallow Water Fisheries Sonar Conference. Seattle, 7-9 September 1999, 22. pp.

Leslie, A. J., Larry, L. E. & Van Dyke, J. M. 1983. Effects of vegetation control by grass carp on selected water quality variables in four Florida Lakes. Trans. Am. Fish. Soc. **112**: 777-787.

Leventer, H. & Teltsch, B. 1990. The contribution of silver carp (*Hypophthalmichthys molitrix*) to the biological control of Netofa reservoirs. Hydrobiologia **191**: 47-55.

Ling, Y. S. 1982. Fish culture and reservoir management in the Republic of Singapore. In: Seminar on Production and Exploitation of Open Waters. Regional Center for Tropical Biology, Bogor, Indonesia, pp. 18.

Lowe-McConnell, R. H. 1999. Estudos ecológicos de comunidades de peixes tropicais. *Editora Universidade de São Paulo (USP)*, pp. 535.

Magadza, C. H. D. 1995. Reservoir management using food chain manipulation. In*:* De Bernardi, R. & Giussani, G. (eds.). Biomanipulation in lakes and reservoirs management. Guidelines of Lake Management. Volume 7. ILEC/ UNEP: 197-205.

Magalhães, V. F., Soares, R. M. & Azevedo, S. M. F. O., 2001. Microcystin contamination in fish from the Jacarepaguá Lagoon (Rio de Janeiro, Brazil): ecological implication and human health risk. Toxicon **39**: 1077-1085.

Matheus, C. E. 1986. Utilização de peixes para melhorar o desempenho de lagoas de estabilização. *Revista DAE* **46**: 87-101.

Matheus, C. E. 1993. Policultivo de peixes em efluentes de Indústrias de Processamento de Frutas Cítricas e efeitos na qualidade da água. Tese de Doutorado, *Departamento de Hidráulica e Saneamento, Universidade de São Paulo,* São Carlos-SP, pp. 375.

Mattos, S. P., Cavalcanti, C. G. B. & Starling, F. L. R. M. 1999. Lago Paranoá: evidências adicionais de recuperação após manipulação do tempo de residência. *VII Congresso Brasileiro de Limnologia "Perspectivas da Limnologia para o Século XXI".* Florianópolis, 18 a 22 de julho de 1999: 65.

Mazzeo, N., Scasso, F., Gorga, J., Kruk, C. & Clemente, J. 1999. Limnological study of a urban hypertrophic lake under restoration process. Proc. 8Th International Conference on the Conservation and Management of Lakes (LAKE 99). S8B-6

Miura, T. 1990. The effects of planktivorous fishes on the plankton community in a eutrophic lake. Hydrobiologia **200/201**: 567-579.

Moscoso, J. 1998. Acuicultura com aguas residuales tratadas en las Lagunas de Estabilizacíon de San Juan, Lima, Perú. *26° Congreso Interamericano de Ingenieria Sanitaria y Ambiental. AIDIS – Associación Interamericana de Ingenieria Sanitaria y Ambiental.* Lima, Peru, Anais do Congresso (CD-Rom), pp. 21.

Moss, B. 1998. Shallow lakes, biomanipulation and eutrophication. SCOPE Newsletter **29**: 1-45.

Moss, B., Madgwick, J. & Phillips, G. 1996. A guide to the restoration of nutrient-enriched shallow lakes. Broads Authority, Environment Agency & EU Life Programme, Norwich, pp. 31.

ONEP. 1991. Eutrophication Control Project of the Sidi Mohammed Ben Abdallah Dam Reservoir: introducing silver carp in the reservoir. Internal Report from National Office of Potable Water, Kingdom of Morocco, pp. 42.

Perrow, M. R., Meijer, M. -J., Dawidowicz, P. & Coops, H. 1997. Biomanipulation in shallow lakes: state of the art. Hydrobiologia **342/343**: 355-365.

Pradhan, B. R. & Swar, D. 1987. Limnology and fishery potential of the Indrasarobar Reservoir at Kulekhani, Nepal. In: Sena S. de Silva (ed.), Reservoir Fishery Management and Development in Asia: 87-93. Workshop held in Kathmandu, Nepal, 23-28 November 1987.

Rajan, M. R. & Raj, S. P. 1994. Sewage recycling through fish culture. Environ. & Ecol. **12**: 247-251.

Rast, W. & Thornton, J.A. 1996. Trends in eutrophication research and control. Hydrological Processes **10**: 295-313.

Roche, K. F., Sampaio, E. V., Teixeira, D., Matsumura-Tundisi, T., Tundisi, J. G. & Dumont, H. 1993. Impact of *Holoshestes heterodon* Eigenmann (Pisces: Characidae) on the plankton community of a subtropical reservoir: the importance of predation by *Chaoborus* llarvae. Hydrobiologia **254**: 7-20.

Romare, P. & Bergman, E. 1999. Juvenile fish expansion following biomanipulation and the resulting effect on the predation pressure on zooplankton. Hydrobiologia **404**: 89-97.

Rowe, D. K. 1984. Some effects of eutrophication and the removal of aquatic plants by grass carp (*Ctenopharyngodon idella*) on rainbow trout (*Salmo gairdnerii*) in Lake Parkinson, New Zealand. New Zealand J. Mar. & Freshwat. Res. **18**: 115-127.

Sagi, G. 1992. The effect of filter feeding fish on water quality in irrigation reservoirs. Agric. Water Manag. **22**: 369-378.

Scheffer, M. 1998. Ecology of Shallow Lakes. Chapman and Hall, London, pp. 357.

Scheffer, M., Hosper, S. H., Meijer, M. -L., Moss, B. & Jeppesen, E. 1993. Alternative equilibria in shallow lakes. Trends Ecol. & Evol. (Tree) **8**: 275-279.

Schindler, D. E., Kitchell, J. F., He, X., Carpenter, S. R., Hodgson, J. R. & Cottingham, K. L. 1993. Food web structure and phosphorus cycling in lakes. Trans. Am. Fish. Soc. **112**: 756-772.

Shapiro, J., Lamarra, V. & Lynch, M. 1975. Biomanipulation: na ecosystem approach to Lake Restoration. In: P. L. Brezonic & J. L. Fox (eds), Symposium on Water Quality Management Through Biological Control, Jan. 20-23, 1975, Univ. of Florida, Gainesville, Dept. Env. Eng. Sci. & USEPA: 85-96.

Sifa, L. & Senlin, X. 1995. Culture and capture of fish in Chinese reservoirs. International Development Research Center and Southbound. China, pp. 128.

Starling, F. L. R. 1989. Estudo experimental dos impactos de peixes planctófagos sobre a comunidade planctônica e a qualidade da água no Lago Paranoá, Brasília-DF. M.Sc. Thesis, Univ. of Brasília, Brasília, pp. 250.

Starling, F. L. R. M. 1993a. Control of eutrophication by silver carp (*Hypophthalmichthys molitrix*) in the tropical Paranoa Reservoir (Brasilia, Brazil): a mesocosm experiment. Hydrobiologia **257**: 143-152.

Starling, F. L. R. M. 1993b. Análise experimental dos efeitos da tilápia do Congo (*Tilapia rendalli*) e carpa prateada (*Hypophthalmichthys molitrix*) sobre a estrutura da comunidade planctônica do Lago Paranoá, Brasília (DF). Acta Limnol. Brasil. **4**: 144-156.

Starling, F. L. R. M. 1998. Development of Biomanipulation Strategies for the Remediation of Eutrophication Problems in an Urban Reservoir, Lago Paranoá – Brazil. PhD Thesis, Institute of Aquaculture, University of Stirling, Scotland, pp. 259.

Starling, F. L. R. M. & Lazzaro, X. 1997. Experimental investigation of the feasibility of improving water quality by controlling exotic planktivore overpopulation in eutrophic Paranoá Reservoir (Brasilia-DF, Brazil). Int. Verein. Theor. Angew. Limnol., Verh. **26**: 789-794.

Starling, F. L. R. M. & Rocha, A. J. A. 1990. Experimental study of the impacts of planktivorous fishes on plankton community and eutrophication of a tropical Brazilian reservoir. Hydrobiologia, **200/201**: 581-591.

Starling, F. L. R. M., Beveridge, M., Lazzaro, X. & Baird, D. 1998. Silver carp biomass effects on the plankton community and its use for improving water quality in Paranoá Reservoir (Brazil). Int. Rev. Hydrobiology **83**: 499-507.

Starling, F. L. R. M., Beveridge, M., Lazzaro, X. & Baird, D. 1999. Silver carp cage culture and the control of algal blooms in tropical man-made lakes. Proc. 8th International Conference on the Conservation and Management of Lakes – LAKE 99. Volume II (S11B-3), 4 pp.

Starling, F. L. R. M., Lazzaro, X., Walter, T. & Ribeiro, M. 2000. Large-scale extensive tilapia production in a tropical eutrophic reservoir (Lago Paranoá, Brasília, Brazil): environmental and socio-economical benefits from biomanipulation. Proc. 5th International Symposum on Tilapia in Aquaculture. Rio de Janeiro, Brazil, 03 a 07 de setembro de 2000.

Starling, F. L. R. M., Lazzaro, X., Cavalcanti, C. & Moreira, R. 2002. Contribution of omnivorous tilapia to eutrophication of a shallow tropical reservoir: evidence from a fish kill. Freshwat. Biol. **47**: 2443-2452.

Stewart-Oaten, A., Murdoch, W. W. & Parker, K. R. 1986. Environmental impact assessment: "Pseudoreplication" in time? Ecology **67**: 929-940.

SUDEPE, 1980. Estudo do potencial pesqueiro do Lago Paranoá. Annual Report from the Brazilian Fisheries Agency (SUDEPE). pp. 58.

Sutton, D. L. 1977. Grass carp *Ctenopharyngodon idella* Val. in North America. Aquatic Botany **3**: 157-164.

Tatrai, I. & Istvanovics, V. 1986. The role of fish in the regulation of nutrient cycling in Lake Balaton. Freshwat. Biol. **16**: 417-424.

Teltsch, B., Juanico, M., Azov, Y., Ben Harim, I. & Shelef, G. 1991. The clogging capacity of reclaimed wastewater: A new quality criterion for drip irrigation. Wat. Sci. Tech. **24**: 123-131.

Van Donk, E., Grimm, M. P., Gulati, R. D. & Klein Breteler, J. P. G. 1990. Whole-lake food-web manipulation as a means to study community interactions in a small ecosystem. Hydrobiologia **200/201**: 275-289.

Vanni, M. & Findlay, D. L. 1990. Trophic cascades and phytoplankton community structure. *Ecology* **71**: 921-937.

Walter, T. 2000. Ecologia da pesca artesanal no Lago Paranoá – Brasília-DF. Dissertação de Mestrado [Master Thesis], *Universidade de São Paulo*, São Carlos, Brasil, pp. 227.

Wysujack, K. & Mehner, T. 2002. Comparison of losses of planktivorous fish by predation and seine-fishing in a lake undergoing long-term manipulation. Freshwat. Biol. **47**: 2425-2434.

CHAPTER

13

Water Quality Amelioration in an Urban Eutrophic Lake by Preventing the Inflow of Municipal Sewage: A Case Study of Husainsagar Lake

M. Vikram Reddy[1] and A. Vijaykumar[2]

[1]Salim Ali School of Ecology and Environmental Sciences, Pondicherry University, Kalapet, Pondicherry 605014, India

[2]Kedriya Vidyalaya, No. 2, Golconda, Hyderabad 500028, India

ABSTRACT

Preventing the inflow of municipal raw sewage by constructing a physical barrier—a necklace road, ameliorated some of the water quality characteristics of surfacewater of the littoral zone of the eutrophic Husainsagar Lake in Hyderabad city (south India). The pH, electrical conductivity, total hardness, magnesium, calcium, chlorides and dissolved oxygen contents of the water have increased and the Biochemical Oxygen Demand (BOD) has decreased in the water significantly. The benthic macroinvertebrate bioindicators of organic pollution have disappeared.

INTRODUCTION

The municipal sewage—the runoff from the houses (domestic sewage) and that of factories and roads being the major source of organic material, a point source of pollution—is discharged directly into different lotic and lentic ecosystems in many developing countries. This malpractice is the unavailability of proper sewage treatment facilities and/or inadequately maintained treatment facilities and sewage systems in most cases, which cause major environmental problems, particularly cultural eutrophication, emanating a foul odour, and leading to the occurrence and spread of various waterborne diseases (Mason, 1961; Reddy & Vijaykumar, 2004). In the Indian subcontinent, there are several cases of discharge of the municipal sewage from different towns and cities into various water bodies like the Ganges, Yamuna and Gomati, Godavari rivers, and the Dal Lake in Kashmir, upper and lower lakes in Bhopal, Powai Lake in Mumbai, Husainsagar Lake in Hyderabad, Ooty Lake in Udhagamandalam, and Pallikarani marsh in Chennai, which grossly pollute these aquatic bodies, in the process, perturbing their water quality characteristics, causing algal blooms, severely depleting the dissolved oxygen, increasing the Biochemical Oxygen Demand (BOD) and rendering the water unfit for any use. Therefore, these water bodies need to be depolluted or prevented from further pollution by stopping the direct inflow of the sewage, which will help in the restoration of these polluted ecosystems. It has been reported that construction of a sewage treatment plant (STP) reduced such nutrient loading of Lake Suwa in Japan (Okino and Kato, 1987; Okino, 1990). Some of the water bodies in Europe were restored to their non-polluted states by diverting and preventing the inflow of allochthonous organic matter, as was the case in Lake Vesijari in Finland (Horppila and Kairesed, 1992). Meeder et al., (1997) reported containment (physical barriers) to prevent exposure of creatures living at ground level to the underlying hazardous waste, and can also be achieved using vertical or horizontal in-ground barriers to prevent the horizontal migration of the contaminants. However, in developing countries like India, very few attempts have been made to prevent the direct inflow of allochthonous organic matter into the aquatic bodies and enough information is not available on such aspects, and on the success stories of their restoration. The present chapter reports a long-term monitoring study on the effects of prevention of direct inflow of municipal raw sewage in ameliorating some of the water quality characteristics of the surfacewater, by constructing a physical barrier in the form of the necklace road on the west side of the eutrophic Husainsagar Lake located in the centre of the twin cities of Hyderabad and Secunderabad, the state capital of Andhra Pradesh (India).

HUSAINSAGAR LAKE

A detailed description of Husainsagar Lake is given in an earlier chapter in this book (Reddy & Vijaykumar, 2005). The lake was in an advanced stage of eutrophication as it received the inflow of municipal raw sewage through different channels such as Picket nallah, Kukatpally nallah, Bajara nallah and the Jeedimetla nallah in between the sites St III and St IV, the former site characterized by washing clothes manually using phosphorus-based detergents, and the latter directly receiving the municipal raw sewage earlier and treated sewage from the STP at present. The municipal wastewater and industrial effluents in the Kukatpalli nallah was estimated to be 55 ld and 15 ld, respectively, and the other three nallahs discharged 17 ld of domestic wastewater into the lake (Kelkar et al. 2001). The municipal raw sewage was discharged into this area of the lake indiscriminately from the catchments (Fig. 13.1). Besides, a large amount of domestic sewage enters the lake directly from the nearby slum areas located in the foreshore of the lake. St-I and St II of the lake were relatively less polluted as there was no direct discharge of municipal sewage in these sites. The water of the lake was green, indicating the profuse growth and abundance of green algae (*Microsystis* sp.) and water hyacinth, *Eichhornia crassipes*, which was rife in between St III and St IV of the lake.

During the years 1993 to 1998, attempts were made to prevent the direct major inflow of municipal raw sewage into the lake at the Site IV of the lake. The adjacent area of the site was reclaimed filling with soil brought from the outside and a necklace road was constructed as a physical barrier that prevented the direct inflow of the municipal raw sewage from the catchment area into the lake (Fig. 13.2). A sewage treatment plant (STP) was set up in May 1998 to treat the raw sewage entering the lake. The treated effluents were released into the lake through a nallah passing underneath a bridge, on the road, in order to maintain the hydrology of the lake. The other nallahs continue to discharge municipal sewage and storm runoff directly into the lake as usual, even today, without any check.

Ameliorating Effects on Water Quality

Analyses of water samples collected from four sites of the littoral zone of the lake, both prior to the construction of the road (when there was direct inflow of municipal raw sewage into the lake, during January 1988 to May 1991), and after the construction of the physical barrier—the necklace road, during January 1999 to May 2000, for the physico-chemical water quality characteristics—pH, electrical conductivity, total hardness,

Fig. 13.1 Direct inflow of municipal raw sewage entering the Husainsagar Lake through nallahs from the catchment area, prior to the connecting necklace road

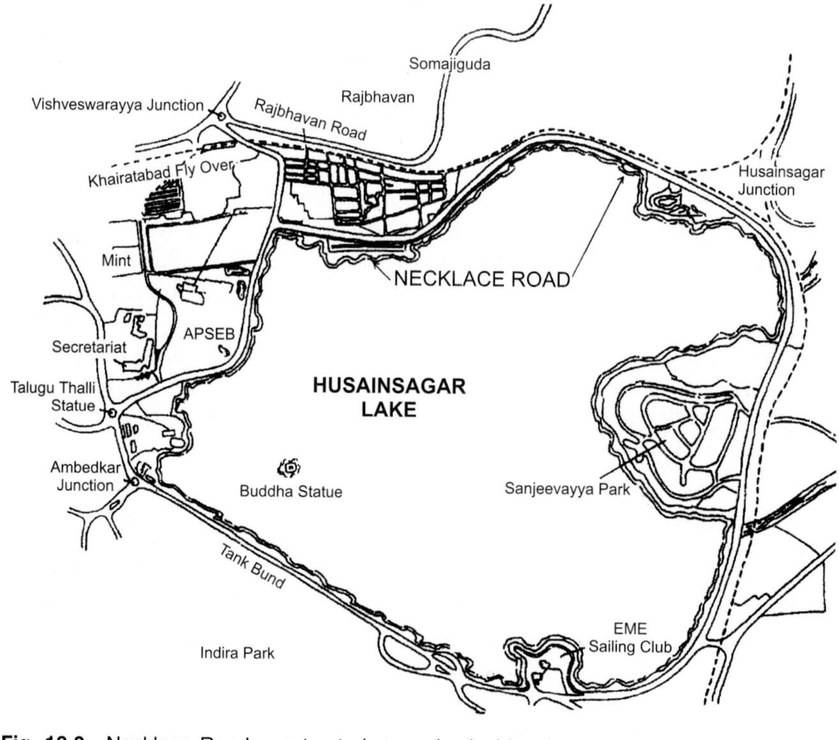

Fig. 13.2 Necklace Road constructed as a physical barrier preventing the direct inflow of municipal raw sewage into the lake (after Huda, MCH)

calcium, magnesium, carbonates, bicarbonates, chlorides, dissolved oxygen (DO) at the biochemical oxygen demand (BOD) following the procedures of APHA (1995), revealed that the water quality of the lake showed a significant improvement (Horpia and Kairesal, 1992), most probably as a result of the prevention of direct inflow of municipal raw sewage into the lake.

The water pH ranging from 7.7 to 9.6 is characteristic of a typical eutrophical lake (APHA, 1995). The mean pH of the surfacewater in the littoral zone of the Husainsagar Lake prior to the construction of the necklace road ranged from 7.5 to 8.3 across the sampling sites, whereas after its construction, it ranged from 8.1 to 8.4, the former in both cases being recorded at St IV (Fig. 13.3). The overall mean pH of the surfacewater of the lake prior to the construction of the road 8.07

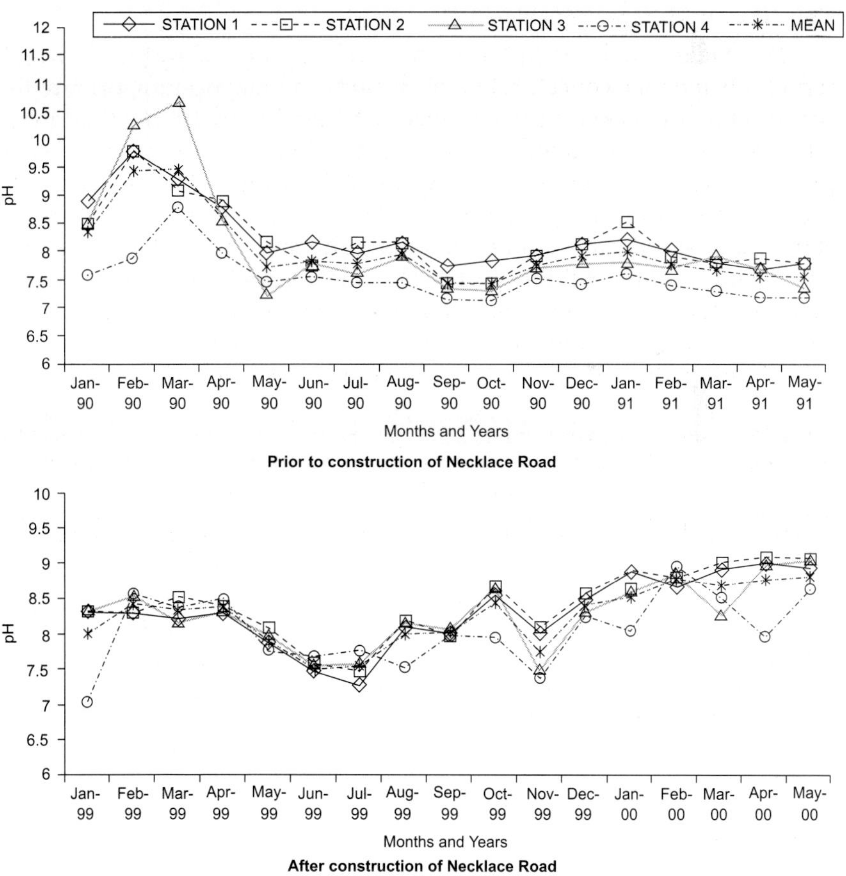

Fig. 13.3 Spatio-temporal variations in pH of surfacewater at four sites (stations) in the littoral zone of the lake

increased to 8.3 after its construction and at the site, St IV of the lake, it has increased form 7.5 to 8.1, these variations across the sites and the years and months being statistically significantly different (Table 13.1).

The mean Electrical Conductivity (EC) of the surfacewater reflects the nutrient status of the lake and its high values indicate a high nutrient load in water. Prior to the construction of the road, this figure ranged 868 to 1998 μmhos/cm, while after the construction of the road, the same ranged 1137 to 1184 μmhos/cm. Its average mean decreased from 1612.17 μmhos/cm prior to the construction to 1157.44 μmhos/cm after the construction of the road. Though the EC of the water at the three sites decreased, there was an increase at the site, St IV from 868 μmhos/cm prior the construction to 1184 μmhos/cm after the construction of the road (Fig. 13.4), probably because of sewage inflow. These variations among the sites and across the years and months are statistically as also significantly different (Table 13.1).

The mean total hardness of the surfacewater of the lake, which represents the total concentration of calcium and magnesium, prior to the construction of necklace road, ranged between 211 to 288 mg/l and that after its construction ranged between 314 to 334 mg/l. Its average mean prior to the construction of the road, 260.48 mg/l has increased to 325.63 mg/l after its construction (Fig. 13.5), the variations being statistically significantly different among the sites, as well as across the years and months (Table 13.1). The mean calcium content of the surfacewater prior to the construction of the road ranged between 42 to 48 mg/l, while that after the construction of the road ranged between 59 to 71 mg/l. Its average mean prior to the construction of the road, 45.71

Table 13.1 The yearly mean values across all the four sites of the lake and ANOVA (Means of Squares) of some of the water quality characteristics before and after the construction of the physical barrier—the Necklace Road

Parameters	*Yearly mean values across the sites*		*Means of Squares*	*F value**
	1990-91	*1999-2000*		
PH	8.07	8.29	1.75	18.06
EC (μmhos/cm)	1612.17	1157.44	7030672.00	73.63
TH (mg/l)	260.48	325.63	144302.00	70.74
Ca (mg/l)	45.71	65.78	13693.16	126.25
Mg (mg/l)	34.44	41.66	928.33	9.54
HCO_3 (mg/l)	388.72	162.52	1739784.00	552.33
Cl (mg/l)	296.38	246.15	85807.00	18.57
DO (mg/l)	1.67	3.95	176.20	72.65
BOD (mg/l)	64.38	16.33	78508.96	229.64

*'F' values of all the parameters except Mg (only across sites) are significant at $P < 0.05$ level.

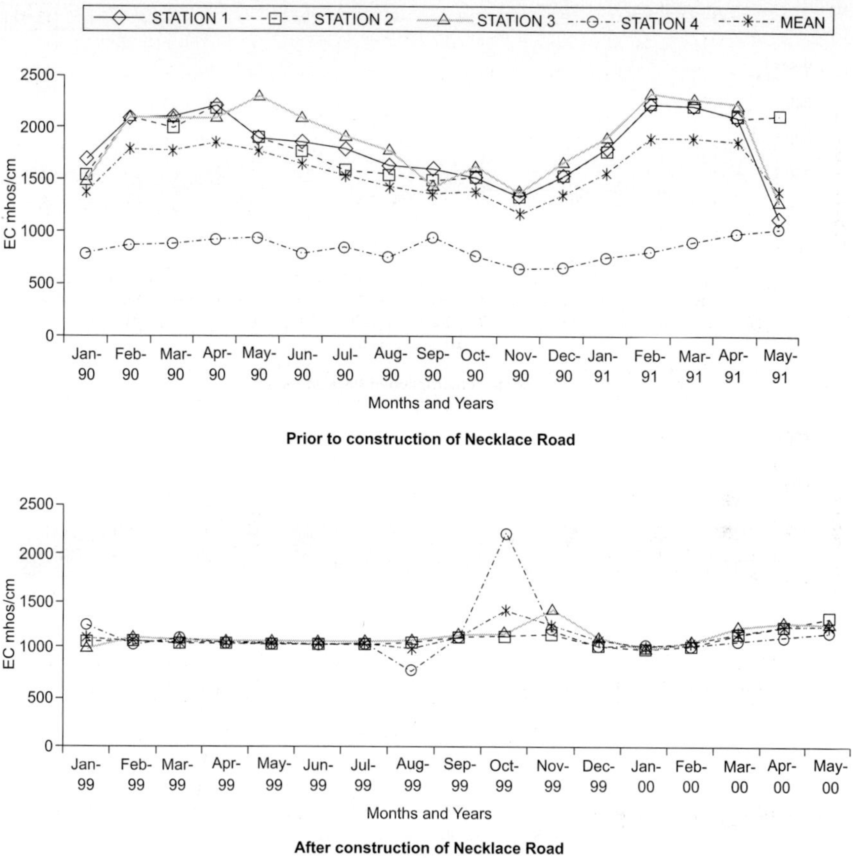

Fig. 13.4 Spatio-temporal variations in Electrical Conductivity (μmhos/cm) of surfacewater at four sites (stations) in the littoral zone of the lake

mg/l, has increased to 65.78 mg/l after its construction (Fig. 13.6), these variations being statistically significantly different among the sites and across the years and months. While the mean magnesium content of the water prior to the construction of the road ranged between 27 to 41 mg/l, and after its construction, it ranged between 40 to 44 mg/l. Its average mean, 36.44 mg/l prior construction of the road has increased to 41.66 mg/l after its construction. Though the increase in magnesium content in the water in between the sites of the lake was not significantly different, it increased from 27 to 43 mg/l at site, St IV of the lake after the construction of the road (Fig. 13.7), the temporal variations being statistically and significantly different (Table 13.1).

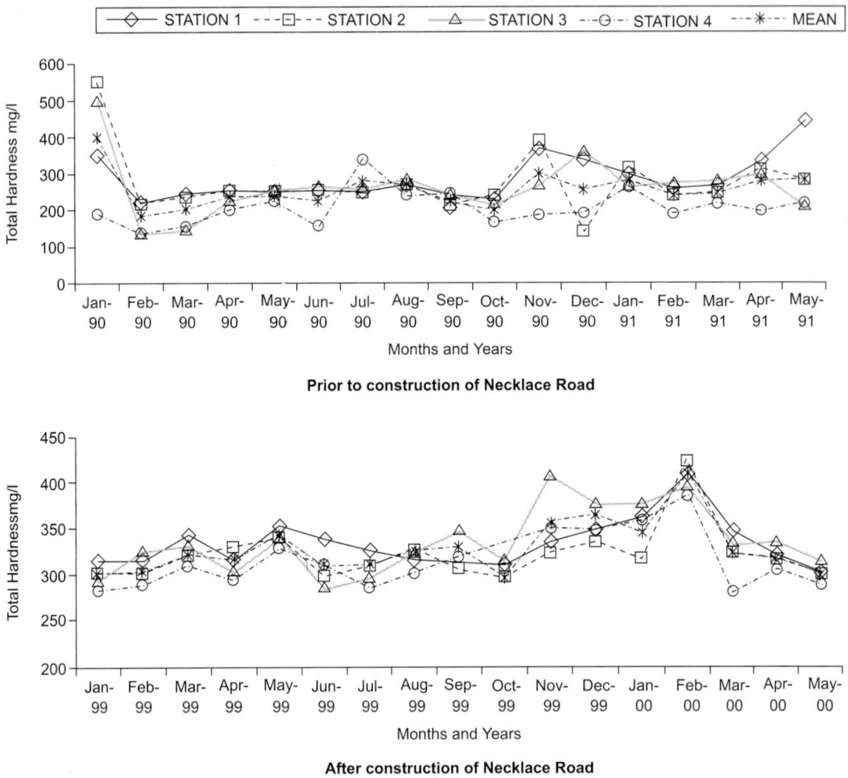

Fig. 13.5 Spatio-temporal variations in total hardness of surfacewater at four sites (stations) in the littoral zone of the lake

The carbonates and bicarbonates define the alkalinity of water. The mean carbonate content of surfacewater prior to the construction of the road ranged between 19 to 57 mg/l, while that after its construction ranged between 20 to 46 mg/l. Its average mean 18 mg/l prior to the construction has increased to 30 mg/l after its construction. However, it decreased at the site, St IV from 57 to 20 mg/l after construction (Fig. 13.8), these variations across the sites being statistically significantly different (Table. 13.1). The mean bicarbonate of the surfacewater prior to the construction of the road ranged between 350 to 446 mg/l, whereas after its construction, it ranged between 145 to 190 mg/l. Its average mean 388.72 mg/l prior to the construction has drastically decreased to 162.62 mg/l after the construction of the road (Fig. 13.9), these variations across the sites and seasons being statistically significantly different (Table 13.1).

The chlorides being the major anions of water, the mean chloride content of the surfacewater of the lake prior to construction of the road

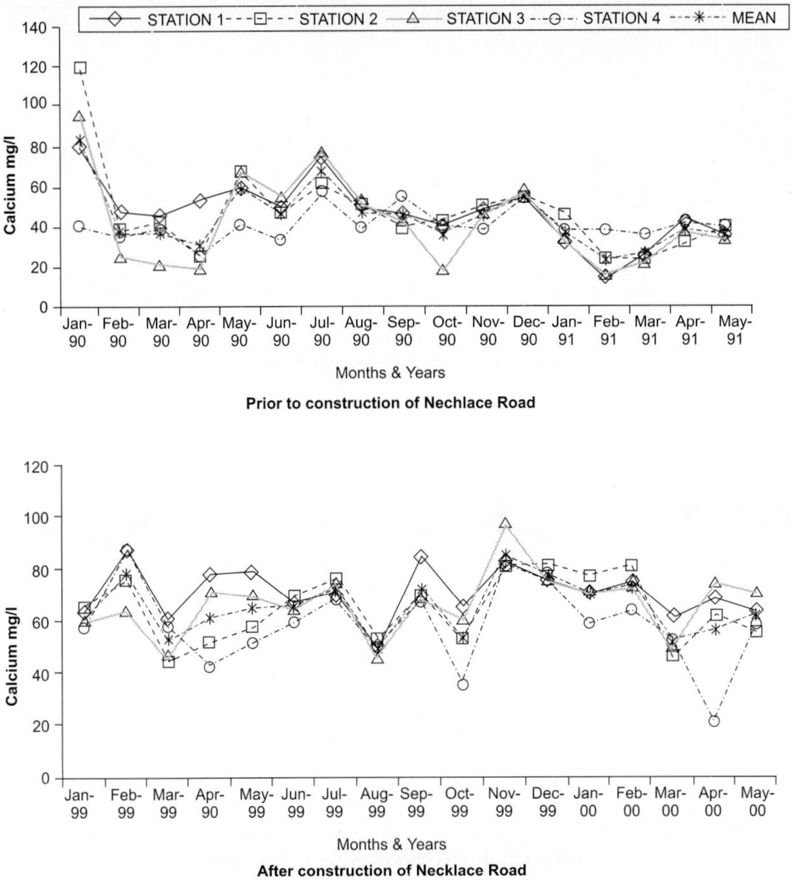

Fig. 13.6 Spatio-temporal variations in calcium of surfacewater at four sites (stations) in the littoral zone of the lake

ranged between 104 to 372 mg/l while that ranged between 237 to 27 mg/l after the construction of the road. Its average mean 296.38 mg/l prior to the construction has decreased to 246.15 mg/l after its construction. Though it decreased in the three sites, it increased at the site, St IV of the lake after the road construction (Fig. 13.10), the differences in variation across the sites as well as across the years and months being statistically significant (Table 13.1).

The dissolved oxygen in the water indicates a water body's water quality, and its depletion is the critical manifestation of eutrophication. At the site St IV and its adjacent area of the lake, anaerobic conditions

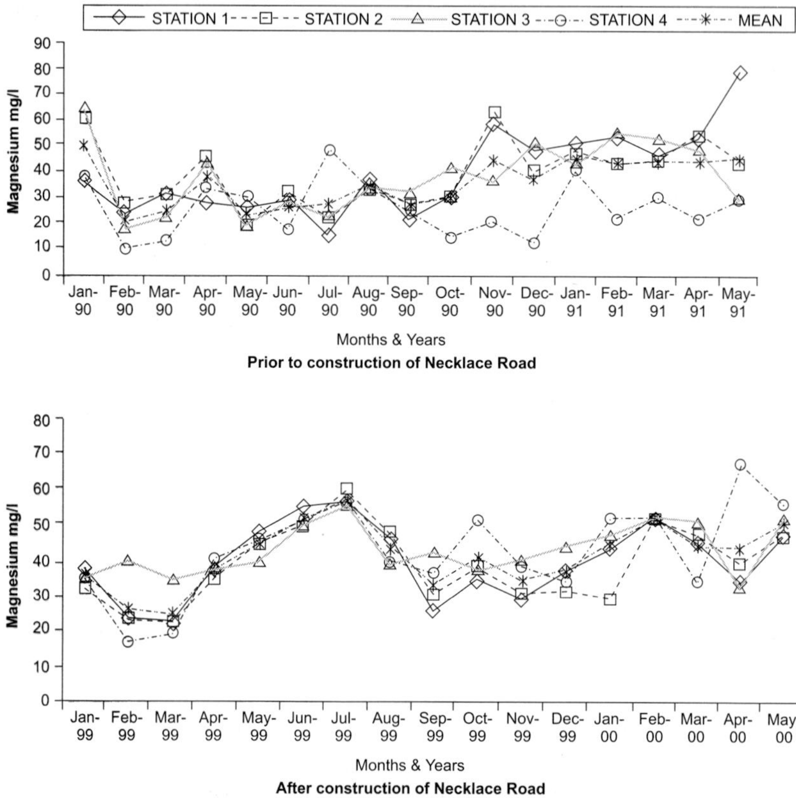

Fig. 13.7 Spatio-temporal variations in magnesium of surfacewater at four sites (stations) in the littoral zone of the lake

prevailed prior to the construction of road with emanation of bad odour, mostly of H_2S. The mean dissolved oxygen of the surfacewater prior to the road construction ranged between zero to 3.1 mg/l, the former being recorded at the site, St IV while that ranged between 1.7 to 5.8 mg/l after the construction. Its average mean 1.7 mg/l prior to the construction of the road has increased to 3.95 mg/l after its construction. Moreover, the increase was observed in all the stations (Fig. 13.11), the differences in variations across the sites and across the years and months being statistically significant (Table 13.1).

The Biochemical Oxygen Demand (BOD) is the most important indicator of organic pollution and water quality. Interestingly the mean BOD of the lake's surfacewater prior to construction of the road ranged between 53 to 96 mg/l, while after its construction, it ranged form 12 to

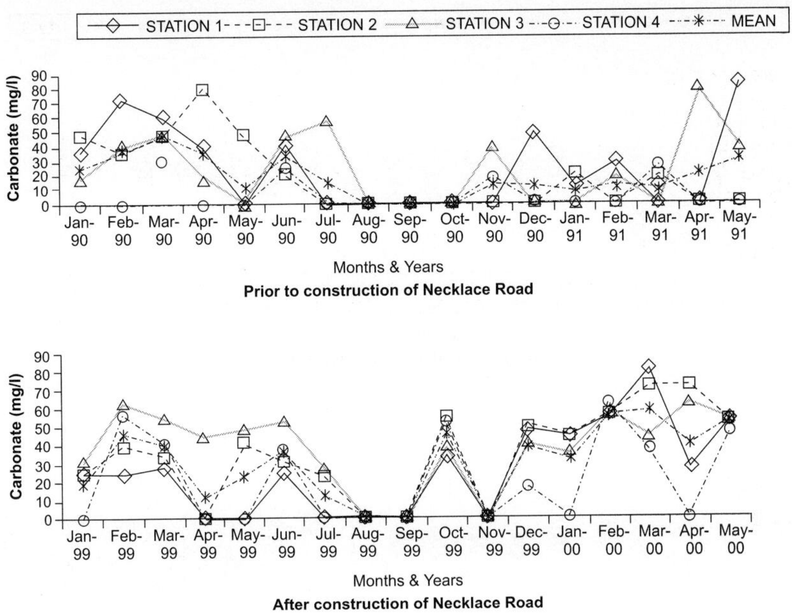

Fig. 13.8 Spatio-temporal variations in carbonates of surface water at four sites (stations) in the littoral zone of the lake.

45 mg/l. Its average mean 64.38 mg/l prior to road construction decreased to 16.33 mg/l after its construction, the decrease being discernible at two sites, i.e. St II and St IV (Fig. 13.12) and the differences in variations across the sites and across the years and months being statistically significant (Table 13.1).

The necklace road constructed, as a physical barrier, prevented the direct inflow of municipal raw sewage from the catchment area of the twin cities into the lake, thus, ameliorating the water quality of the lake. The average pH of water of the lake was within the desirable limits of ISI and WHO acceptable standards for drinking water. The total hardness is often an indicator of water quality, and its average in the surfacewater of the lake was less than the maximum desirable limits of ISI standards (300 mg/l) prior to the construction of the road but has increased slightly after its construction. The average magnesium content of water was slightly higher than the maximum desirable limit of BIS Standards (30 mg/l) prior to the construction of the road, which increased further after its construction. The average calcium content of the surface water has increased from 45.7 to 65. 7 mg/l after the construction of the road, particularly at site, St IV of the lake. Though less than the maximum desirable limits of ISI standards (75 mg/l), the values indicate the eutrophic nature of the water (USEPA, 1975).

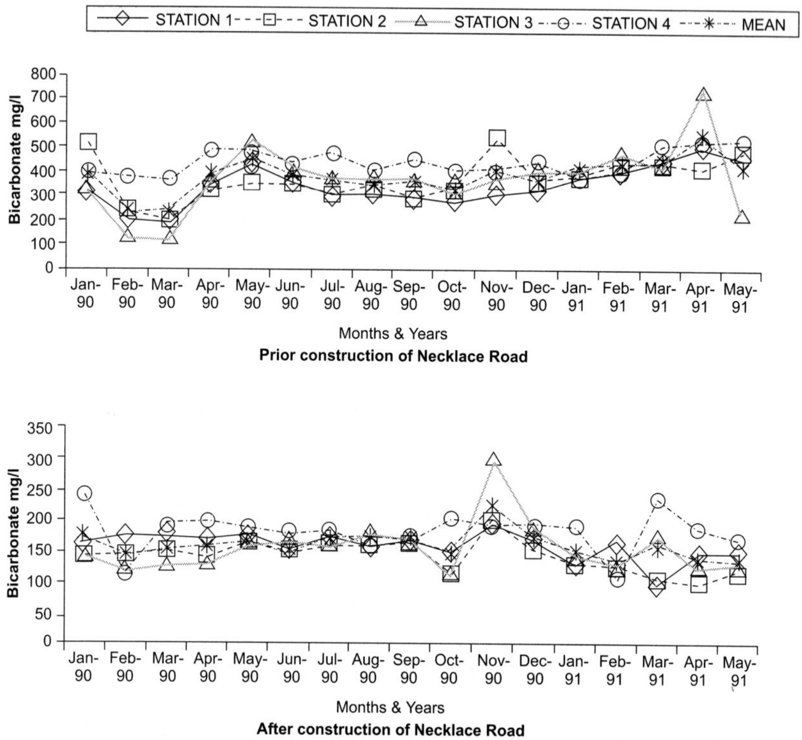

Fig. 13.9 Spatio-temporal variations in bicarbonates of surfacewater at four sites (stations) in the littoral zone of the lake

The chloride content regulates the salinity of water. Its average in the surfacewater prior to the construction of the road was more than the maximum desirable limits of ISI standards (250 mg/l), and was slightly higher than the acceptable limits of WHO standards, which decreased to 246 mg/l after the construction of the road, being less than the maximum desirable limit of ISI standards. However, at the site, St IV, the chloride content was 237 mg/l. These values indicate the nutrient load of water, probably due to municipal sewage.

The average dissolved oxygen content of the water increased significantly across the seasons in all the sampling sites of the lake, particularly at the site, St IV after the road construction, where there was anoxic conditions prior to the construction due to direct inflow of the raw sewage. The biochemical oxygen demand decreased significantly at site, St IV in the lake after the construction of the road, though it is higher when compared that to the ICMR Standards (6 mg/l). The benthic dipteran larvae and pupae of *Psychoda* L., *Tabanus* L., *Limnophora* R & D

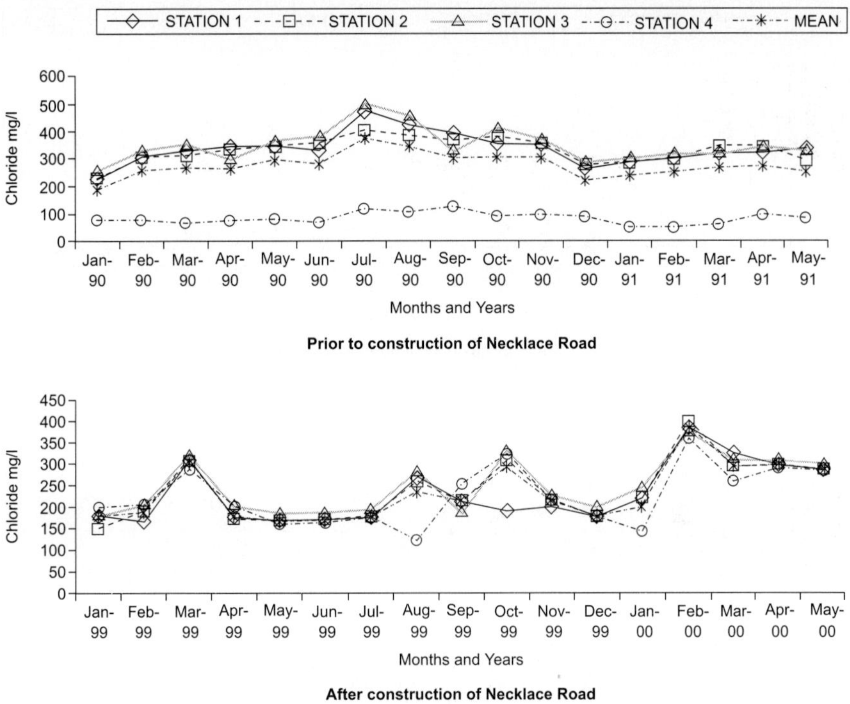

Fig. 13.10 Spatio-temporal variations in chloride of surfacewater at four sites (stations) in the littoral zone of the lake

and *Eristalis tenax,* the bioindicators of organic pollution—prevalent at the site, St IV in the lake prior to the construction of the road—disappeared after its construction. It indicated that prevention of the inflow of municipal raw sewage containing the nutrient load, by constructing the necklace road as a physical barrier, considerably reduced the organic pollution and ameliorated the surfacewater quality of this lake. Similar improvement in water quality in this lake was observed during February 2000 compared to that of October 1998 in terms of BOD, COD and total phosphorus levels (Kelkar et al. 2001). A marked trend in the improvement of water quality in a shallow eutrophic lake, Lake Suwa in central Japan, was recognized in transparency and Chlorophyll-a concentration after the start of the sewage treatment plant (Okino, 1990; Okino and Kato, 1987; Hazanato and Okino, 1998). Prevention of the enrichment of nutrients in lakes is an effective step in tackling the eutrophication problems, which stops the subsequent over growth of phytoplanktons and aquatic weeds (Armstrong and Rohlick, 1976). Lake Vesijarvi in Finland started recovering when the municipal sewage

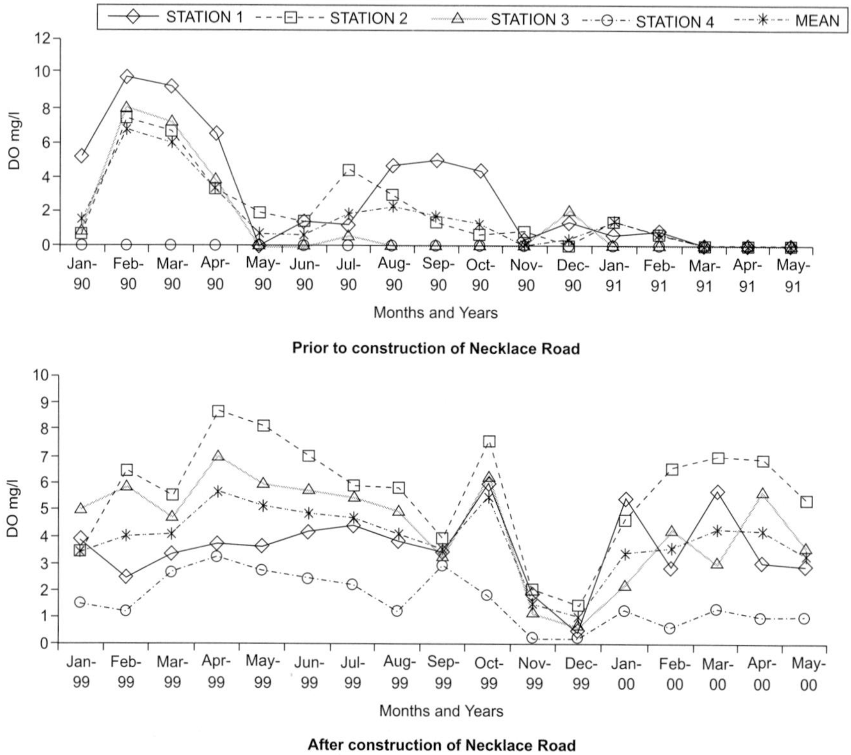

Fig. 13.11 Spatio-temporal variations in dissolved oxygen of surfacewater at four sites (stations) in the littoral zone of the lake

loading was diverted from the lake (Horppila and Kairesalo, 1992). Similarly, the Ooty Lake in south India, polluted mainly due to the inflow of domestic sewage through the main 'Kodappamund channel' running through the Udhagamandalam town, showed improvement in its water quality after the sewage inflow into the lake was diverted.

Presently, there are three more nallahs dumping the municipal raw sewage directly into the Husainsagar Lake. Diversion and outright prevention of such dumping is needed through effective management policies and practices without any further delay. When the inflow of all the raw sewage, including the domestic sewage from the nearby slums entering into the lake is either stopped or completely diverted, there will be further improvement in the water quality of the whole lake. Efforts on these lines are being initiated by upgrading the existing STP and setting up two more STPs for treating domestic sewage and industrial effluent inflowing into the lake through different nallahs. Besides, prevention of

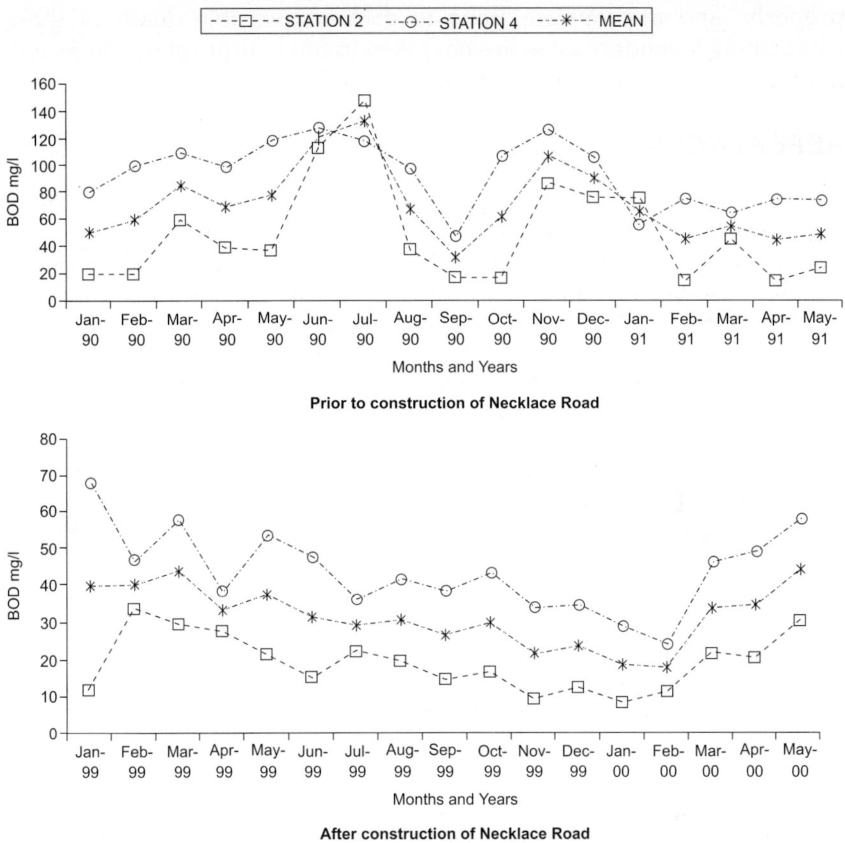

Fig. 13.12 Spatio-temporal variations in biochemical oxygen demand of surfacewater at four sites (stations) in the littoral zone of the lake

the manual washing of clothes with detergents, open defecation on the foreshore, and washing of cattle and vehicles in the littoral zone of the lake will further ameliorate its water quality, and the lake will be in a far better state than what it is now.

The aquatic ecosystems, particularly the urban lakes which are prone to various anthropogenic activities, need to be protected by adapting remedial measures, mainly curbing the entry of external loading of nutrients in the form of municipal sewage wastewater, and dumping garbage and night soil in foreshore of the lakes, which will save these water bodies from being in the list of the endangered water bodies. It is necessary that both abiotic and biotic components of these ecosystems are monitored meticulously for better predictions, in order to maintain the cleanliness of such waters. If these ecosystems are not understood properly, and appropriate steps to prevent possible death of these languishing legendary lakes are not taken in time, future generations will not forgive us for this sort of criminal complacency.

REFERENCES

American Public Health Association, 1995. Standard Methods for the Examination of Water and Wastewater, 19th Edn Wiley., Inc., New York.

Armstrong, D.E., & Rhlich, G.A. 1976. Effects of agricultural Pollution on eutrophication (Pollution and Eutrophicatin). In: Agricultural Practices and Water Quality. (Eds) L. Willrich, and G.E. Smith, Ames Iowa State Univ., Iowa, U.S.A., pp. 314-330.

Hanazato, T. & Okino, T. 1998. Long-term monitoring of water quality after the start of sewage treatment in Lake Suwa, a shallow eutrophic lake in a mountain over. In: T. Iwakuma, (ed.) Proceedings of the 2nd East Asia-Pacific regional conference on long-term ecological research. Natl. Inst. for Environ. Studies, Tsukuba, pp. 166-173.

Horppila, J. & Kairesal, T. 1992. Impacts of bleak (*Alburnus alburnus*) and roach (*Rutilus rutilus*) on water quality, sedimentation and internal nutrient loading. Hydrobiologia, 243-244: 323-331.

Kelkar, P. S., Pande, S.P., & Reddy, R.C., 2001. Pollution status of Hussainsagar lake vis-a-vis rejuvenation measures—Post-implementation scenario. In: Proc. Interm. Conf. on Industrial Pollution and Control Technologies (ICIPACT-2001), JNT University, Hyderabad, A.P. pp. 4.

Mason, C.F. 1996. Biology of Freshwater pollution, 3rd Edn Longman, Essex, England.

Meeder, T., Verskijis, K., Bakker, R., & Soczo, E., 1997. Remedial action techniques for contaminated land. In: The Global Environmental Science, Technology and Management. (Eds) D. Brune, D.V Chapman, M.D. Gwynne, J.M. Pacyna, Vol.2; VCH Willey Comp., S.C. Publ. Weinheim, 887-907.

Okino, T. 1990. Present state of lake environment in Shinshu: Water management and monitoring of Lake Suwa. Jpn. J. Linnol., 51; 17-22.

Okino, T. & Kato, K. 1987. Lake Suwa—Eutrophication and its partial recent recovery, Japan. Geo. J., 14: 373-375.

Reddy, M.V. & Vijaykumar, A. 2005. Seasonal surfacewater quality characteristics of the urban eutrophic Husainsagar Lake in Hyderabad, India. In: Restoration and Management of Tropical Eutrophic lakes, (Ed.) M. Vikram Reddy, Science Publ. Inc., Enfield (NH), USA, pp. 33-85.

U.S.E.P.A. 1975. United States Environmental Protection Agency – Quality Criteria for Water. (Ed.) Train, Caste House Publ. Ltd., Great Britain.

CHAPTER

14

Restoration of Tropical Urban Lakes: Case Study—Lake Pampulha, Brazil

Eduardo von Sperling

Department of Sanitary and Environmental Engineering
Federal University of Minas Gerais, Belo Horizonte, Brazil

ABSTRACT

This chapter describes the restoration of a tropic urban reservoir in Brazil, with emphasis on the engineering processes that have been adopted for this purpose. The main characteristics of tropical lentic systems are discussed in the beginning of the text. Here the most relevant feature is the accelerated dynamics in the functioning of warm water environments. The results of a broad-based research carried out in Brazil, regarding morphological informations of over 600 lakes and reservoirs, are covered next, with the objective of stressing the importance of physical aspects in the management of tropical aquatic systems. Finally, there is the presentation of preventive and corrective measures implemented in order to restore Lake Pampulha, an urban reservoir situated in the city of Belo Horizonte, Brazil (A: 2,6 km^2; V: 14 mi m^3; max. depth: 16 m). The adopted preventive techniques were sewage diversion, slums urbanization, construction of sedimentation basins and of a new sanitary landfill. The corrective processes were mostly directed to the eutrophication control: sediment dredging, deep water discharge, macrophyte removal and hydroseeding for erosion control.

INTRODUCTION

This chapter presents a case study regarding the restoration of a tropical urban lake located in the city of Belo Horizonte, Brazil. Emphasis is given here to the engineering processes adopted in order to rehabilitate this vital tourist atraction of the region. The restoration measures include the implementation of preventive techniques (regulation of the use of the soil, urbanization of slums, erosion control, construction of sedimentation basins, sewage diversion, construction of interceptors and sewage treatment plants) as well as the use of corrective processes (dredging, macrophyte removal, construction of pre-dams, nutrient precipitation). This chapter also deals with two relevant physical influences which are often neglected in the development of lake restoration programs: morphology and hydrodynamics. With respect to morphology, this chapter presents the results of a broad survey carried out by the author involving the determination of the morphogical features of over 600 Brazilian lakes and reservoirs and their relationship to water quality, especially in the evaluation of the trophic degree. The chapter begins with certain considerations about the peculiarities of tropical aquatic systems and their relevant mechanisms of self regulation. The above-mentioned physical aspects (morphology and hydrodynamics) are then discussed, stressing their influence in the structure and function of tropical lakes and reservoirs. The case study about Lake Pampulha is presented, including a general view about the characteristics of the water quality, the origin of the polluting processes (contamination, eutrophication, siltation) and the structure of the restoration programme implemented.

CHARACTERISTICS OF TROPICAL LAKES

The tropic countries, referred to those situated in the warm belt of our planet, present some common features associated with geographical, climatological and also historical aspects. The water degradation problems in tropic countries assume a particular character due to the more intense metabolism observed in warm water aquatic communities. In this sense, the response of the ecosystem is different from that found in temperate water bodies, mainly with respect to the developmemt of eutrophication processes. Moreover, the tropic countries generally present a poor infrastructure for research activities, together with a lack of specialized personnel and severe limitations in the availability of financial resources. Consequently, the activities of lake management in tropic developing countries should be carried out according to the specific features pointed out here. One of the most important aspects related with lake management is well-founded knowledge about structure and function of the aquatic system. This information has a striking influence in the definition of the various steps of the lake management policy. Neverthless, the construction of this informative basis depends on the development of adequate monitoring programmes, which are usually associated with high financial expenditures.

It should be noted that tropical aquatic systems present quite different characteristics in comparison with temperate lakes and reservoirs. These can be summarized as follows (von Sperling, 1997):

- Intense solar radiation and high water temperatures accelerate nutrient uptake by the algae;
- Phytoplankton population peaks are less frequent in comparison with temperate aquatic environments;
- Strong nutrient assimilation associated with high recycling rates leads to the prevalence of intense degrees of productivity;
- As nutrient concentrations in the liquid mass are generally low, many water bodies can be classified as oligotrophic in spite of their high productivity;
- There is a frequent occurrence of low phytoplankton densities which are, however, coupled with high growth rates;
- Due to the high temperature, the water viscosity is low, leading to a higher settling velocity of organisms and nutrients;
- High mineralization rates cause an accelerated oxygen depletion, liberating nutrients which are poor in organic matter; consequently, there is no direct connection between hypolimnetic oxygen deficit or organic matter content in sediment and the water body productivity.

IMPORTANCE OF PHYSICAL ASPECTS IN THE MANAGEMENT OF TROPICAL LAKES

Worldwide, the management of lakes and reservoirs has been strongly influenced by the results of monitoring programmes carried out in order to obtain representative informations concerning the water quality of the lentic environment. Most of the data refers to the chemical and biological conditions of the water body, while the physical parameters—with exception of temperature and turbidity—are usually neglected. This is particularly true in the case of lakes situated in tropic developing countries, where the access to financial resources is quite limited. Here, the most common choice is directed to a simplification of monitoring programmes usually recommended in the technical literature and broadly adopted in temperate regions. As a consequence of this procedure, some parameters of chemical and biological nature can be overevaluated, while the physical aspects of the water body, which frequently are the easiest to be obtained, are shifted to a lower level of priority. In the sequence, some examples are presented vis-a-vis valuable information that can be extracted from a physical approach of the water body.

Hydrology

Water cycle plays a striking role in the configuration of the quality of the aquatic environment. On its own, the simple relationship between precipitation and evaporation (rarely found in limnological studies in tropical regions) can divulge precious information regarding the dilution potential of the water body. The knowledge of the water flow in tributaries to the lake allows for the calculation of the load of pollutants, nutrients, etc., which is undoubtedly more informative than the mere determination of concentrations. This load is part of several mass balance models (e.g. Vollenweider, 1976) which enables the construction of distinct scenarios of water quality according to variations in the input of nutrients. In case of small rivers, this measurement can be easily carried out with the use of appropriate and quite unexpensive equipments. It should be also stressed as to how neglected is the performance of the tributaries inside the framework of many limnological studies, which direct their focus exclusively to the water body characteristics. But maybe the most important hydrological factor for the lake management is the water residence time. Its value regulates the growth of aquatic communities (mainly phytoplankton), the availability of nutrients, the transport of pollutants and many other processes that take place inside the water body. In reservoirs, the control of the water residence time—through the operation of deep discharge—can be used as an efficient technique for their restoration (Ryding and Rast, 1989).

The water residence time also has a striking influence on the quality of recreational waters in lakes, since systems with slow water movements are more susceptible to the effects of wastewater discharges. This characteristic is intensified in tropical lakes, where the nutrient dynamics is far more accelerated, allowing the quick growth of aquatic organisms (algae and macrophytes) that can disturb the intended uses of the water environment.

Hydrodynamics

The transport processes in lakes are responsible for the distribution of particulates, including microorganisms, and dissolved substances. Hence, the **hydrodynamics** of the lake should be carefully analyzed in order to gather useful information about the distribution pattern of the water body. A detailed research about all the transport processes that take place in the lake should also include, besides the typical horizontal and vertical movements measurements in the surface, subsurface and benthic boundary layer (Imberger et al., 1998) and the possible occurrence of Seiches, Ekman spirals, Langmuir currents. Nevertheless, for the objective of lake management, the simple determination of **vertical temperature profiles** at the deepest point can substantially contribute for a reasonable understanding of the circulation pattern of the water body. Without this information, it becomes very difficult to develop a sound management of the lake, since the most important water uses (human and industrial supply, irrigation, energy generation, recreation) are strongly dependent on the water quality that results from the alternation of periods of stratification and circulation.

Morphology

The **morphology** has a strong influence on nearly all physical, chemical and biological parameters of lakes and reservoirs (Wetzel, 1983). It is a widely known fact that the **physical dimensions** (shape and size) of a lake interact with several climatic and surrounding edaphic factors in order to determine the characteristics of the water body as natural (in the case of lakes) or artificial (in the case of reservoirs) environments. In spite of the importance of morphology in defining the quality of the aquatic system, its study is generally not carried out with an adequate level of importance. Some main parameters are surprisingly uncomplicated to be obtained, provided there is a minimum of basic informations from bathymetric maps. These informations can be used in order to evaluate the quality of the aquatic environment, especially with reference to circulation patterns and susceptibility to eutrophication. In a broad survey carried out by the author in the period 1995-97, over 600 Brazilian lentic systems were analyzed. One of the most interesting conclusions is that these water systems generally present low values of **relative depth** (ratio of maximum depth to average diameter of lake surface). Deep water bodies with small areas have high values of relative depth, while large and shallow lakes show low relative depths. In the Brazilian survey, following the distribution of values has been found (Table 14.1).

It can be seen that the absolute majority (about 80%) of surveyed lakes and reservoirs present low relative depths, smaller than 1%. This means that these systems present frequent vertical circulations, which play a very important role on the quality of the environment. This example shows how the utilization of morphometric parameters can be helpful for a deeper understanding of the system and consequently for the establishment of a sound management of water bodies.

Table 14.1

Zr (%)	*Distribution (%)*
0 –0.5	60
0.5 – 1	21
1 – 1.5	8
1.5 – 2	3
2 – 2.5	3
> 2.5	5

Some other morphometric parameters can also be used to estimate the lake water quality. A description of these physical factors following the mean depth is obtained by dividing the lake volume by the lake surface. The relative depth, as pointed to earlier, is defined as the relationship between the lake maximum depth and the mean diameter of the lake (i.e. diameter of a circle that has the same area as the lake). The shoreline development indicates the degree of irregularity of the shoreline. Its value is given by the ratio between shoreline length and the perimeter of a circle that has the same area as the lake. The volume development is a measure used to illustrate the form of the lake basin. It is defined as the quotient between the lake volume and the volume of a cone whose base area is equal to the lake area and whose height is equal to the maximum depth. Finally, there is the areal ratio, the ratio between the area of the drainage basin and the lake area. Lakes with high values of areal ratio generally show a clear trend to eutrophication through a potential intense input of nutrients (originated from diffuse sources in the relatively large drainage area) and due to the enhanced possibility of existence of urban settlements in the drainage basin.

The main characteristic of the distribution pattern of the areal ratio (area of the drainage basin divided by the area of the lake) of the surveyed lakes and reservoirs is the strong concentration of values in the extreme ranges: over 50% of the results are either lower than 50 or higher than 1000. Values ranging from 5 to 300, together with those higher than 1000, represent 88% of the total amount. Small values of areal ratio were associated with natural systems (lakes), while the largest ones were only found in artificial systems (reservoirs). High values of areal ratio were usually linked with small water bodies, which are more susceptible to the effects of diffuse pollution (agricultural activities, cattle breeding). This was particularly true in the case of reservoirs, which are generally situated outside urban areas, thus not receiving discharges of untreated sewage (point sources), which are a common feature in developing countries. On the other hand, small water bodies are also generally shallow. There is a clear inverse relation between the depth and the trophic degree. The assimilation capacity of shallow lakes is generally lower due to the small amount of dissolved oxygen in the bottom layers.

Figure 14.1 presents the box-and-whisker plots for some significant morphological variables, where the association with the trophic status can be clearly seen. It is observed that the lower the depth, the higher is the trophic level, due to the enhanced light penetration. At the beginning of the previous century, when the Aquatic Ecology was moving its first steps, some researchers, such as Thienemann (in Cole, 1983), have already pointed out this inverse relationship between depth and productivity (i.e. trophic degree). During the 50s and 60s, other authors showed inverse curves relating the depth and fish catch or depth and plankton weight (Le Cren and McConnel, 1980). The identification of this relationship represents an efficient tool in the management of lakes and reservoirs.

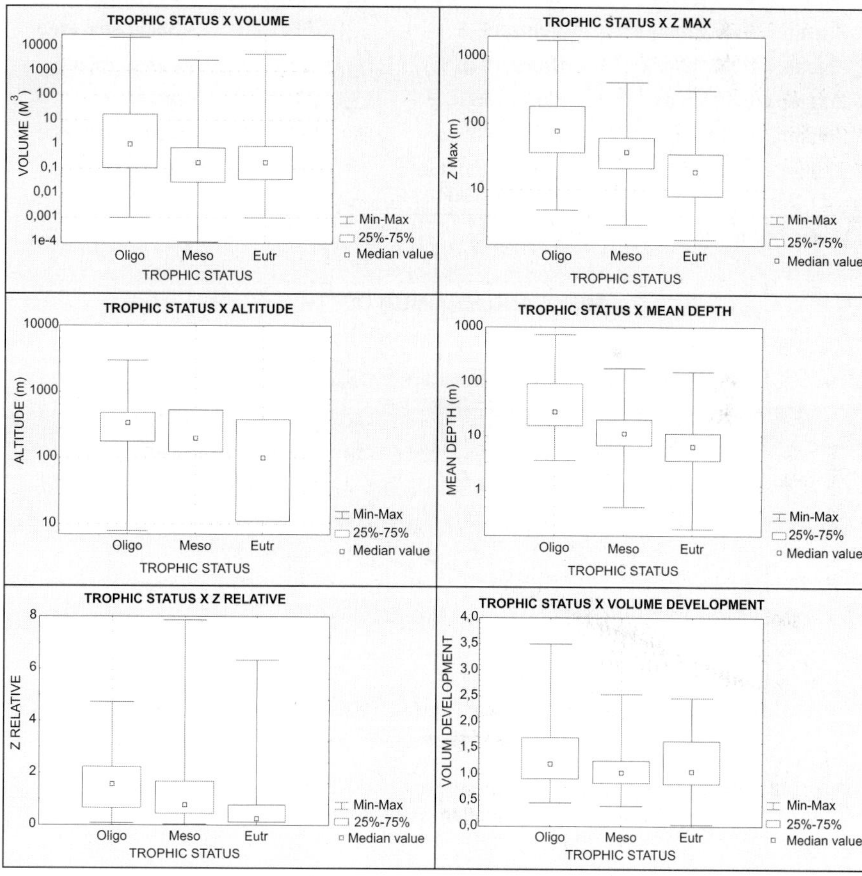

Fig. 14.1 Box-and-whisker plot of the morphological variables with statistical significance, separated according with the trophic status

The relationship between Z_{mean} and Z_{max} are shown in Table 14.2 and are graphically depicted in Fig. 14.2.

Table 14.2 Equations relating Z_{mean} and Z_{max}

Equation	*Reference*	*Number of lakes*	*Location*
Zmean = 0,37 Zmax – 3	Herdendorf,1990	253	World
Zmean = 0,467 Zmax	Neumann,1959	107	World
Zmean = 0,455 Zmax	Gorham,1964	71	World
Zmean = 0,423 Zmax	Hayes,1957	500	World
Zmean = 0,41 Zmax	Koshinsky,1970	68	Canada
Zmean = 0,25 Zmax + 2,34	Von Sperling, 1999	587	Brazil
Zmean = 0,46 $Zmax^{0,95}$	Gorham,1958	262	Scotland
Zmean = 0,75 $Zmax^{0,8}$	Gorham,1958	137	Scotland
Zmean = 0,65 $Zmax^{0,82}$	Ventz,1973	85	Baltic sea area
Zmean = 0,617 $Zmax^{0,95}$	Ventz,1973	39	Alpine region
Zmean = 0,42 $Zmax^{1,03}$	Horie,1962	231	Japan
Zmean = 0,588 $Zmax^{0,93}$	Irwin,1972	39	New Zealand
Zmean = 0,75 $Zmax^{0,83}$	Schwarzbold and Schäfer,1984	-	South of Brazil

Source: von Sperling (1999)

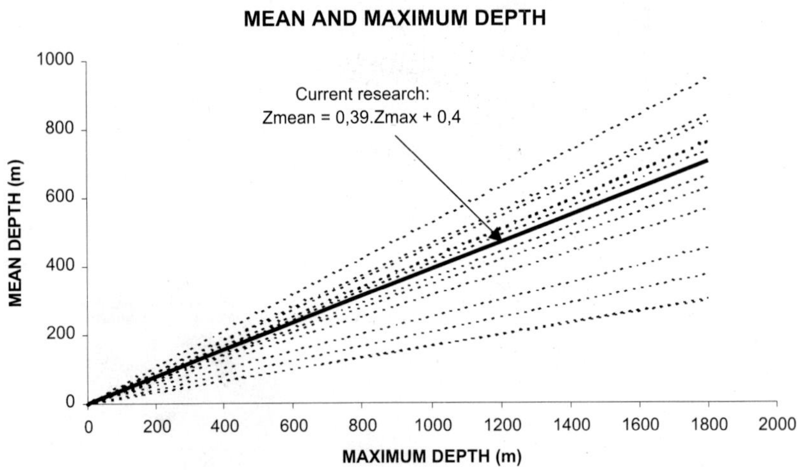

Fig. 14.2 Regression curves of average depth versus maximum depth (equations in Table 14.2 and equation from current research)

CASE STUDY: LAKE PAMPULHA

Lake Pampulha, a manmade reservoir, is located within the urban area of the city of Belo Horizonte, Brazil. The lake was originally constructed with the objective of water supply, but the degradation of the water quality prevented the further use of this source as such. Presently, Lake Pampulha, which is one of the most important tourist attractions of the region, is suitable only for some restricted recreational activities. The lake has an area of approximately 2.6 km^2 and a volume of about 14 million m^3. Its maximum depth is 16 m and the mean depth is 5 m. The drainage area occupies the significant area of 97 km^2. The relationship between these figures indicate that the lake receives a considerable influence (diffuse pollution) from the activities developed in the watershed. The main sources of pollution are the discharge of untreated sewage into the lake, the high degree of erosion of many areas in the drainage basin, leading to a high volume of settlement material in the waterbody and also the presence of diffuse sources of pollution. Even after the diversion of about 80% of the channelized sewage in the year 1997, the lake continued to present eutrophication and silting problems. This fact stresses the key role played by non-point sources in the degradation of the water quality.

Water Quality Characteristics in Lake Pampulha

Physico-chemical aspects

The Secchi depth has been lowering along the last 20 years (from 1.8 m to 0.3 m) as a consequence of frequent episodes of algal blooms. However, it should be stressed that in the case of tropical lakes, the water transparency does not depend exclusively on the intensity of algae growth. Here, the siltation process (transport of mineral and organic soil particles in the rainy period) plays an important role in diminishing the Secchi depth values.

Due to discharge of untreated sewage, nutrient concentrations are high in the waters of Lake Pampulha. The average values for total phosphorus range from 0.05 mg/l to 0.2 mg/l, with peaks of 0.45 mg/l (surface) and 0.4 mg/l (bottom). Nitrate concentrations oscillate between 0.2 and 2 mg/l and ammonium between 1 and 8 mg/l. These numbers stress the prevalence of chemical-reduced compounds in Lake Pampulha. There is a marked seasonal evolution of nutrients, with lower concentrations being detected during the rainy season, from December to March. The N/P ratio is generally over 40, thus indicating that phosphorus is the limiting nutrient for the algae growth. The intense presence of dissolved salts is reflected in high values of electrical conductivity (200-350 μS/cm) (Pinto-Coelho, 1998; Macedo and Pinto-Coelho, 2000).

Some heavy metals are frequently found in Lake Pampulha waters, originated from the discharge of industrial sewage in the neighbouring city of Contagem, which is situated in the drainage basin of the lake. The most common heavy metals are cadmium, lead and chromium.

Bacteriological aspects

Lake Pampulha presents generally very high counts of coliforms (upto 160.000 MPN/100 ml), which are specially concentrated near the main tributaries. It is interesting to point out that due to physical (sedimentation, adsorption in clay particles) and biological (zooplankton filtration) processes, the coliforms density is quite low at the outflow of the lake.

Hydrobiological aspects

The seasonality observed in physical properties and distribution of chemical dissolved substances is also present in the phytoplankton community. While some species are perennial (e.g. *Chorella vulgaris, Pediastrum simplex*), others are periodic in occurrence: *Cylindrospermopsis raciborskii* is quite frequent from October until April, *Peridinium volzii* from June to October and *Cyclotella* sp. in May/July (Gianni, 1994). Other common species are the Cyanobacteria *Microcystis aeruginosa* and *Coelospharium naegelianum.* Chlorophyll a concentrations range usually from 20 to 80 μg/l.

Eutrophication problems in Lake Pampulha lead to the occurrence of blooms of aquatic plants. In many cases, the green carpet, formed mainly by water hyacinths (*Eichhornia crassipes*) and cyanobacteria, covers one third of the surface of the reservoir. Besides water hyacinth, other macrophytes can also be found in Lake Pampulha such as *Salvinia molesta* (floating plant) and *Typha latifolia* (rooted plant). Long-term experiments carried out in Lake Pampulha show the following pattern for the seasonal distribution of algae density:

Rainy season (November – April) lower water residence time	270,000 org/l
Dry season (May – October) higher water residence time	540,000 org/l

It can be clearly seen that algae growth has been directly related with water residence time.

In the zooplankton community, there is a marked presence of rotifers (mainly *Keratella, Brachionus, Polyarthra*) and usually a low density of copepods. On the other hand, the cladocera crustacea are occasionally very strongly represented by the species *Daphnia laevis*.

Restoration Measures in Lake Pampulha

Since the early 90s, the local authorities implemented a broad restoration programme, with the collaboration of specialists from the Federal University of Minas Gerais, including this author. The programme is based on the adoption of preventive and corrective techniques for the rehabilitation of the aquatic system. Some of them have already been applied, while others are still in the project phase.

Some of the most important adopted preventive measures are listed below:

- **sewage diversion** in order to remove direct inputs of wastewater; as a first step, about 9 km of interceptor lines were constructed around the lake;
- **urbanization of slums** situated in the drainage basin;
- construction of a **new sanitary landfill** in the drainage basin;
- construction of **sedimentation basins** in order to avoid siltation problems in the lake; three basins are being constructed upstream the main entrances of the water body.

Besides **sewage diversion**—which is an obvious measure to be adopted for the restoration of eutrophic lakes and probably the most important one—other procedures have been carried out in the last five years. **Erosion control** was achieved through the use of hydroseeding in the gentle slopes where vegetation has been artificially removed. The regrowth of grass vegetation takes place in the period of only a few weeks, thus minimizing the impacts of the very intensive rainfalls that are typical of the region. Regarding **macrophyte removal,** several attempts have been made in order to efficiently remove the considerable mass of aquatic plants.The main problem lies in the very high growth rate of water hyacinths, which are frequently far greater than the removal capacity of the equipments. In 1998, another removal action was undertaken, using appropriate boats and conveyors that transport the water hyacinths to trucks. The final destination of the removed material are islands created within the reservoir as a consequence of the dredging operation. After implementation of this corrective measure, a strong reduction in the total inorganic nitrogen concentrations were observed. However, the most outstanding effect was an abrupt change in the zooplankton community and a remarkable growth of some blue green algae, especially *Microcystis aeruginosa.* The macrophyte and algae populations are supposed to stay permanently in competition for nutrient uptake. The removal of one of them would consequently favour the remaining community. One remediation, that is being currently adopted, is to always leave a certain amount of macrophytes (c. 10-20% in surface area) to act as competitor and thus guarantee a desirable ecological equilibrium. With this procedure, the algal-blooms were significantly reduced. **Sediment dredging** is used to control the release of algae stimulating nutrients from lake sediments by removing the layers of the most highly enriched materials. Examples of the utilization of this technique can be found in Björk (1988) and Moore and Thornton (1988). Among the several types of dredging equipment, the most common choice consists of a hydraulic dredge with a cuttterhead. The sediments are transported as a slurry through a pipeline that traverses the lake from the dredging site to a disposal area. In case of Lake Pampulha, instead of transporting the material to a final destination place, as usually happens in dredging works, the mud has been deposited within the lake, over three islands created as a consequence of the dredging process and that combine gently with the landscape requirements. These islands now function as a recreation site for the population and also as field laboratory for ecological studies. The dredging equipment installed at Lake Pampulha was able to suction 1500 m^3/h of a slurry with c. 80% water content. The suction diameter was 40 cm and the pipeline, of plastic material, had a total length of 800 m. The dredging work consisted in removing the sediment deposited in an area of 0.34 km^2 in order to guarantee a water column of at least 1.5 m. During this action, over 2 million m^3 of sediment were removed from the lake. The water body was monitored biweekly during the period of dredging work. Several sampling points were located in the water column, immediately near the suction site and at the final disposal place. Considering that the potential for severe negative impacts on the lake and surrounding areas due to dredging works is very high, the monitoring programme was conceived to carefully evaluate the effects of this restoration measure on the aquatic environment of Lake Pampulha, specially the possible resuspension of heavy metals from the sediment. The data available from this monitoring action shows that the suction dredging work has been successful in removing the sediment without causing considerable damage to the aquatic environment.

In the year 2005, some further steps will be implemented as part of the restoration programme of Lake Pampulha:

- **long-distance dredging**: construction of a 8-km-long pipeline which will transport about 4 million m^3 of sediment; the destination of the dredged material will be some natural depressions situated close to the lake;

- **installation of a flotation equipment** for rainwater/sewage treatment; this equipment will function inside a gallery situated along the main tributary;
- **installation of deep discharge,** which will allow the removal of the bottom water, usually rich in dissolved nutrients, iron, and manganese and poor in oxygen.

The cost of the implementation of these three measures is approximately US$ 25,000.

FUTURE STRATEGIES OF CONTROL

One very effective management strategy to combat the eutrophication in Lake Pampulha is the **construction of pre-dams**, which are small reservoirs situated at the entrance of the tributaries in the main water body. Although these dams are yet to be built, the project indicates that they will be able to considerably lower the nutrient load to the lake. The retention mechanism is based on biotic (nutrient uptake by the algae, involving biochemical conversion from dissolved to particulate form) and abiotic (phytoplankton sedimentation) processes. Previous experience of the author (von Sperling, 1992) shows that the pre-dams can act very efficiently in reducing the nutrient input in urban lakes. Pre-dams are small reservoirs constructed with the object of reducing the nutrient loading of the main water body. They have an average theoretical water retention time of a few days, which allows the nutrient uptake by the algae. Pre-dams are usually situated immediately upstream of the larger main lake or reservoir, although in some cases, they can be formed by the separation of a small part of the main water body. The dominance of photoautrophic mechanisms, which govern the removal of nutrients, indicates that the mean depth of the pre-dam should not be considerably greater than the depth of the euphotic zone. The water retention time, which is a function of the volume of the water body and of the amount of the inflow, also plays a very important role in the calculation of the depth of the pre-dam. The exercise here is to select an average water retention time which is sufficiently high to ensure efficient nutrient assimilation, but, also low enough to avoid an excessive proliferation of the phytoplankton. In any case, the selected water retention time must be higher than the critical retention time, which is defined as the limit below which the phytoplankton growth is not sufficient to compensate for its loss by hydrodynamic flushing out. The process of nutrient assimilation is directly influenced by three factors: solar radiation, temperature and the nutrient inflow concentration. The relationship between solar radiation and the growth rate of organisms can be expressed by an exponential function (Michaelis-Menten-Monod). This relationship is approximately linear for low values of solar radiation, becoming exponential at higher values. A linear relationship exists between temperature and growth rate, upto a point where an optimum value is reached; then the growth rate falls rapidly with rising temperatures. The relationship between nutrient retention and phytoplankton growth rate follows a pattern which is similar to that for solar radiation. The main parameters for the design of pre-dams are, therefore, the following: depth, water retention time, inflow water discharge, inflow nutrient concentration, solar radiation and water temperature. Some of these parameters, such as solar radiation and water temperature, are beyond human control. On the other hand, the inflow discharge and concentration, depth and volume (and consequently the retention time) of the pre-dam can be calculated and controlled in order to achieve the highest nutrient removal efficiency. Empirical models, developed for a given situation, can be used for data determination (Benndorf and Pütz, 1987). Experiments conducted by the author in a German lake (von Sperling, 1992) show that retention is considerably higher when the nutrient loading (product of flow and concentration) is due to small water discharges and high nutrient concentrations, i.e. the most effective way of achieving a good nutrient retention is by limiting the amount of inflow water. This observation stresses the need, which is often underestimated, of a careful hydrological monitoring of the drainage basin before embarking on the implementation of restoration measures.

Another feasible control strategy is the **nutrient precipitation**. This technique could be applied in order to limit the nutrient input from the non-point sources. The most probable choice is the addition of aluminum sulfate or, eventually, iron chloride. The utilization of zirconium salts, which are very efficient floculants is also possible. The aluminum salts, mainly the sulphates, have the advantage of precipitating independently of the redox potential of the aquatic environment, i.e. the sedimentation happens also in anoxic or hypoxic conditions. The solubility of aluminum, different from that of the iron, does not depend on the oxidation-reduction pattern of the water body. The concentrations to be used are in the range of 0.2 mg/l to 0.15 mg/l (Cooke et al. 1993). The disavantages coupled with the utilization of aluminum salts are the possible clogging of the filtering system of the zooplankton and the toxicity of the metal. Iron can be easily obtained in the region of Lake Pampulha (Iron Quadrangle), which is one of the largest and richest iron ore deposits in the world. Iron is also less toxic for the aquatic communities and can be used as iron chloride in concentrations from 0.3-12 mg/l.

ADMINISTRATIVE ASPECTS

Community participation plays a striking role in the recovery process of Lake Pampulha, since the success of the programme depends on the achievement of several steps of environmental education. It should be noted that the population that lives around urban lakes has not only a very intensive visual contact with the water body but also interacts with some uses such as recreation (swimming) and fishing.

The **administrative structure** of the restoration programme is depicted below. It can be seen that the municipality is responsible for the programme management which includes the participation of universities and research institutes as a technical support. One important feature in the city organizational structure is the existence of 12 regional administrations. One of them corresponds to the area around Lake Pampulha. It is easy to understand that a regional administration can have a far more intense contact with the population, besides being extensively aware of the problems of the region. On the other hand, the citizen has an easier access to the regional administrator, eliminating all formalities and difficulties coupled with audiences with the mayor.

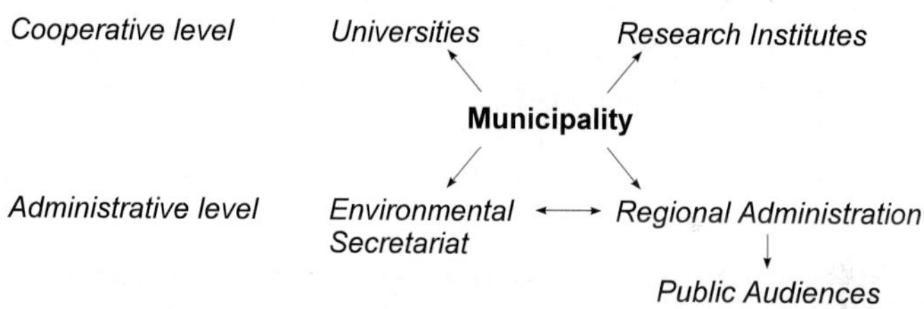

The **public audiences** are the main instrument for the citizen participation. The priority level of public works is defined according to the demands of the population, which are expressed through the convocation of public audiences. This sort of participative budget guarantees the optimization in the use of public fonds. Several meetings have been organized in order to discuss the pros and cons of investing financial resources in the complex task of restoring an urban lake. The majority of the inhabitants of Lake Pampulha district are strongly devoted to the implementation of restoring actions. There is a general consensus that a good environmental quality is closely linked with a good life quality.

One important aspect in the environmental education process is related with **recreational activities.** Since the population is aware of the inconvenience of swimming in a polluted lake, the main leisure action is directed to **fishery.** An enquiry carried out by the Federal University of

Minas Gerais showed that the absolute majority (about 95%) of the fishermen are fishing in Lake Pampulha not for leisure, but simply to obtain food. There is a clear incompatibility between using an aquatic environment as an indirect source of protein (eutrophic conditions) and maintaining its quality for recreational activities such as swimming, sailing and diving (oligotrophic conditions). Moreover, the presence of snails in the shallow bays formed in the reservoir is a serious concern in respect to the transmission of **schistosomiasis.** It is interesting to observe that the reappearance of snails (*Biomphalaria glabrata*) is directly linked with improvements in the water quality, i.e. when the lake was more degraded, the snails had no favourable conditions for their development. Another group exposed to schistosomiasis transmission is formed by people who extract **sand** from the sediments. This sand, in spite of being contamined with coliforms and heavy metals, can be used for some restricted purposes, such as road pavement.

The process of **environmental education** for the community that lives in the drainage area of Pampulha Lake is based not only in prohibitions but rather, in stressing the necessity of change of values. Pollution should be considered not just as an environmental impact but as an issue of social behaviour. A good example of this approach is related with the discharge of solid wastes in the tributaries of the lake, which is a clear responsibility of each citizen. A TV sketch shows that "The same garbage you discharge in the river return to your home after flooding" and "What will your neighbour think about you if he sees how you dispose your wastes?"

Finally, the municipality and the regional administration stress the point that the restoration of Lake Pampulha cannot bring clear results in a short period of time. This **awareness of the existence of a lag time** is a noteworthy issue in order to avoid a future feeling of disappointment in the community.

CONCLUSIONS

The restoration of an aquatic system follows the way back to the situation prior to the existence of the main impacts that degraded its water quality. It is the return to a condition that many citizens, especially children and teenagers, can hardly picture in their minds. In the case of Lake Pampulha, this change in the environment has been accomplished by the adoption of remedial interventions (preventive and corrective techniques) detailed in the paper. Besides the technical aspects, which are managed by the municipality, with support of universities and research centres, efforts are also directed to the informative issues. Thanks to the implementation of a regular monitoring programme, the community is following step by step the first changes that are happening in the lake. It is like a friend or relative that has been ill and fortunately, we can get clear answers to the question "How is he/she doing now ?"

REFERENCES

Benndorf, J. & Pütz, K. 1987. Control of eutrophication of lakes and reservoirs by means of pre-dams. Water Research, Vol. 21: 829-842.

Björk, S. 1988. Development of lake ecosystems: A case study approach. Ambio. 17: 90-98

Cole, G. 1983. Textbook of Limnology. C.V. Mosby Company, Toronto, p. 401.

Cooke, G.D., Welch, E.B, Peterson, S.A. & Newroth, P.R. 1993. Restoration and Management of Lakes and Reservoirs. Lewis Publishers, Boca Raton, 548 p.

Gianni, A 1994. Limnology in Pampulha Reservoir: Some general observations with emphasis on the phytoplankton community.In: (eds), Ecology and Human Impact on Lakes and Reservoirs in Minas Gerais, R.M. Pinto-Coelho, A. Giani, and E. von Sperling, UFMG, p. 193.

Imberger, J. et al. 1998. Dynamics of large stratified lakes. Proceedings of the Kinneret Symposium on Limnology and Lake Management, Ginossar, Israel, p. 16.

Le Cren, E. & McConell, R. 1980. The Functioning of Freshwater Ecosystems. Cambridge University Press, Cambridge, p. 588.

Macedo, C.F. & Pinto-Coelho, R. M. 2000. Diel variations in respiration, excretion rates and nutritional status from the Pampulha Lake. Journal of Experimental Biology 286: 483-494.

Moore, L. & Thornton, K. 1988. Lake and Reservoir Restoration Guidance Manual. United States Environmental Protection Agency, Washington D.C.

Pinto-Coelho, R.M. 1998. Effects of eutrophication on seasonal patterns of mesozzoplankton in a tropical reservoir: a four-year study in Pampulha Lake, Brazil. Freshwater Biology 40: 159-173.

Ryding, S. & Rast, W. 1989. The Control of Eutrophication of Lakes and Reservoirs. UNESCO, Paris.

Vollenweider, R. 1976. Advances in defining critical loading levels for phosphorus in lake eutrophication. Mem. Ist. Ital. Idrobiol. 33, p. 53-83.

von Sperling, E. 1992. Determination of the settling velocity of phosphorus in an aerated lake Wat. Sci. Tech., 26 (7-8): 1973-1979.

von Sperling, E. 1997. The process of biomass formation as the key point in the restoration of tropical eutrophic lakes. Hydrobiologia, v. 342, p. 351-354.

von Sperling, E. 1999. Morfologia de Lagos e Represas. Belo Horizonte, DESA/ UFMG, p. 137. (in Portuguese)

Wetzel, R. 1983. Limnology. Sounders Company, Philadelphia, p. 767.

CHAPTER

15

Dianchi Lake—A Hypereutrophic Lake in China

PU Peimin

Nanjing Institute of Geogrphy &
Limnology, Academia Sinica, China

ABSTRACT

Dianchi Lake has water surface area of 298 km^2, mean depth of 4.4 m and catchment area of 2920 km^2, occupying the sixth largest fresh water lake in China. It is located at the elevation of 1886 meters a.s.l. Getting more intensive sunshine, the highland lake has the higher preliminary production and possesses the characteristics of lakes in the lower latitudes—lakes in the tropical zone. It was rich in aquatic vegetation with more than 100 species, in association with abundant biodiversity in history, The water quality of Dianchi Lake was quite good as the safe drinking water source before the 1970ties. It is a notable recreation site and reputed as "a Sparkling Pearl Imbedded in the Yunnan-Guizhou Plateau". Starting from the last years of 1970ties, the degradation of water quality became more rapidly. A lot of conventional measures have been taking for controlling eutrophication especially since 1980. But the problem for improving water quality, to controlling eutrophication and to reconstructing the health aqua-ecosystem, is still far from its solving. All these aspects, such as the background information on the changes of eoc-environmental situation from healthy to degraded hyper-eutrophic state in Dianchi

Lake, the reasons of rapid development of eutrophication in Dianchi Lake, the various measures and their results performed for controlling eutrophication were introduced in the first paragraphs of this Chapter.

How can we save Dianchi Lake from progressive environmental deterioration? It needs to develop new strategies and technologies based on the laboratory and field experiments. After the event in 1990 (the direct economic lost was about USD 15 million during a week for Wuxi City, Jiangsu Province, after an event of algae bloom in Taihu Lake in summer 1990), a serial research and experiment projects directly towards for reconstruction of the helath aqua-ecosystem in local area, especially in the intake area of drinking water source was performed under the support of the Chinese State and Provincial Government, the Chinese Academy of Sciences, and the International Organizations as well. The small-scale experiments of a couple of thousands or ten thousands square meters were taken place in different environmental conditions. The research results were summarized in the retral paragraphs of this Chapter.

The technology for reconstruction of healthy aqua-ecosystem and improvement of water quality in the surface waters from local to all the surface water systems in the catchment step by step mainly using solar energy was named as "Physic-Ecological Engineering—PEEN. It contains 5 key techniques, namely: i) Soft wall and physic-biological membrane technology, ii) Technology for Removing Algae Plume (TRAP) and Improving Transparency, iii) Mosaic technology and Macrobionics, iv) Technology of Immobilized Nitrogen Cycle Bacteria (INCB), and v) Ecological floating island and 5 M-key taches, namely: a) Macrophyte, b) Microorganism, c) Mollusk, d) Macrobionics, and e) Management.

The possibility of rapid decrease of pollutants loading into the lake in the next a few years is very low. The approach of dredging sediment is a costly one, but the theoretical analysis and the practices in Dianchi Lake and other lakes showd that it is neither the sufficient nor the necessary one for controlling eutrophication in lakes. Controlling Eutrophication by enhancing purification capacity in lakes might be the most cost-effective approach. The purification capacity in water body should to be enhanced up to form the concentration, which would be lower the thresholds of the nutrients for preventing the algae bloom. To use the sunshine and solar energy to improve the preliminary production in lake by high efficiently transforming the nutrients in water into the biomass and taking off them from the water body as usable and economic beneficial resources by using PEEN technology. Therefore, to develop the Bio-Environmental Enterprise (BEE) is the perspective way out for controlling eutrophication. The theoretical basis and technical approaches for controlling eutrophication by enhancing purification capacity in lakes and the technology for utilizing the aquatic plant and other resources were dealt in this Chapter. The perspective for controlling eutrophication in the surface waters appears with optimism. Of course, there are a lot of issues in this field still need to be studied and solved in the further.

INTRODUCTION

Dianchi Lake catchment is located on the Yunnan-Guizhou Plateau between 24°29′ ~25°28′ N, and 102° 29′ ~103° 01′ E, in the southwest of China (Fig. 15.1) With a catchment area of 2920 km^2, Dianchi Lake, based upon a fault, is located at the foot of the Western Hills, southwest of Kunming. It has water surface area of 298 km^2, mean depth of 4.4 m, circumference of 136 km, length of 40 km from north to south, and volume of 12.9×10^8 m^3, occupying the sixth largest freshwater lakes in China. The retention time of the lake is 2.7 year. Being a highland lake with an elevation of 1886 meters above sea level, the lake receives intensive sunshine, has the higher preliminary production and possesses the characteristics of lakes in the lower latitudes—lakes in the tropical zone.

There are other two large lakes in the adjacent region of Dianchi Lake on the Yunnan-Guizhou Plateau, one of them being the Fuxian Lake—the most important deep lake in China (Fig. 15.2). These lakes have similar origin of fault depression, but in different stages of development. Dianchi Lake is in the later stage of its development. Its deposition depth during the Quaternary Period reaches 1 000 m in the northern part of Dianchi Lake. The Fuxian Lake is the deepest big lake in China.

Dianchi Lake is located in the subtropical southwest monsoon climate zone. The mean annual solar radiation reaches 5522 mJ/m^2, the sunlight time—2081-2470 hours, frost-free days—230. The intensive sunshine at plateau benefits the enhancing of the preliminary production. The character of four seasons, which is typical in east China, does not appear to be obvious in the Dianchi Lake catchment. Scene of "no bitter cold in winter, no extreme hot in summer" and "Each season like spring, it becomes winter, while it is raining" is typical in Dianchi catchment. Therefore, Kunming, the capital of Yunnan Province, is named the City of Perpetual Spring. The dry season and wet season can be clearly distinguished. The annual mean precipitation is 1036 mm, 88% of which is concentrated during the period May to October. The dominant wind

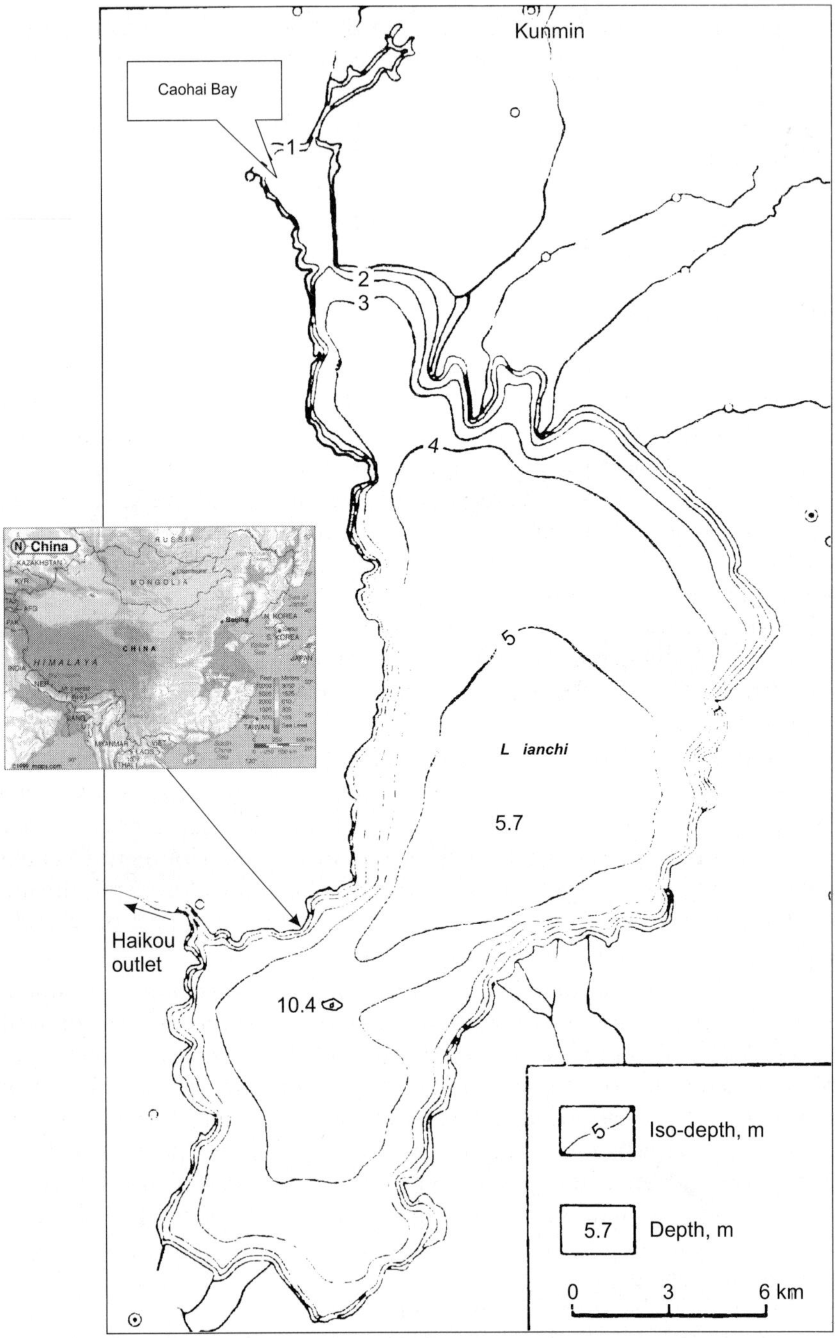

Fig. 15.1 Location of Dianchi in China and its topography map (in southwest of China at elevation of 1886 m above sea level)

Fig. 15.2 Remote sensing image of Dianchi Lake (upper left) basin and its adjacent region (Fuxian Lake, lower right big lake)

directions are southwest and southeast. The annual mean air temperature is 14.4°C the maximum of monthly mean temperature of 19.6°C is observed in July, the minimum of monthly mean temperature of 7.4°C is observed in January (Table 15.1).

The ratio of the catchment area and the watersurface area, namely, the supplement coefficient accounts 8.8. There are more than 20 main inflow rivers (Fig. 15.3), but just one outflow at Haikou in the west bank of the lake (Table 15.2, Table 15.3). The main components of water balance of Dianchi Lake are shown in Table 15.2.

The annual water resource for Dianchi Lake Basin accounts 5.454×10^8 m^3, which is the 1/33 of the world mean volume per capita. Therefore, this is one of the regions suffering from insufficient water resources in China. The population in the catchment accounts for 2, 103, 599 with density of population being 720/ km^2; among them, agricultural residents occupied 32% in 1998. There are more than 10 ethnic minorities living in the catchment. There are also the Yunnan ethnic villages on the southern shore of Dianchi Lake. Each of the 25 recognized ethnic minorities of Yunnan have a village built with a commemorative square, a theatre for the performance of ethnic songs and dances, a folklore museum, and a museum of ethnic waxworks.

The southward Panlong River is the main inflow of Dianchi Lake. This river flows 47.4% of the total annual inflow into Dianchi Lake (Fig. 15.3). Kunming city, the capital of Yunnan Province, is located in the lower reaches of the Panlong River. The intensive pollution issue from Kunming city and its countryside makes a major contribution in the pollution loading to Dianchi Lake, especially to its northern bay—Caohai Bay with a surface area of 7.8 km^2 (Fig. 15.1). The eutrophication of Dianchi Lake developed rapidly, starting from Caohai Bay and spreading from the North basin to the South basin.

At the present, the drinking water source areas had been seriously polluted by nutrients and algae bloom, and most of the drinking water sources have stopped performing their original functions. Controlling eutrophication of Dianchi Lake has become one of the three key projects in controlling eutrophication of lakes in China. The other two lakes are Taihu Lake in Jiangsu Province and Caohu Lake in Anhui Province along the lower reaches of the Changjiang River.

A SPARKLING PEARL IMBEDDED IN THE YUNNAN-GUIZHOU PLATEAU OF CHINA

Being in a pleasant climatic zone, Kunming is named the Spring City. It is the economic, transport, and multi-national cultural center of the

Table 15.1 Basic climatic characters of Dianchi Lake (Kunmin) (1971-1980, Item with *:1951-1980), Unit: Temperature—°C; Pressure—100 Pascal, Relative humidity—%, Precipitation—mm, Evaporation—mm, Sunshine time—hours, Percentage of sunshine time—%, Wind speed—m/s, Cloud cover—bar

Item	*January*	*Feb.*	*March*	*April*	*May*	*June*	*July*	*August*	*Septem.*	*October*	*Novem.*	*Decem.*	*Year*
Mean air temperature	7.4	9.2	12.6	16.1	18.7	19.5	19.6	18.8	17.3	14.7	10.8	7.9	14.4
Mean daily variation of air temp.	13.9	14.3	15.1	14.8	11.1	7.7	7.2	8	8.5	8.8	10.9	12.6	11.1
Mean maximum air temp.*	15.3	17.2	20.8	24	24.9	23.8	24	23.9	22.6	20.1	17.5	15.2	20.8
Mean minimum air temp.*	1.4	2.9	5.7	9.2	13.8	16.1	16.8	15.9	14.4	11.3	6.6	2.6	9.7
External maximum air temp	22.6	23.8	26.6	29.2	29.9	28.6	28.2	29.6	26.9	25.6	23.9	20.6	29.9
External minimum air temp.	-5.4	-0.9	-1.8	1	7	9.9	11.9	9.9	6.7	2.4	-2.9	-4.9	-5.4
Annual variation of air temp.													12.2
Mean water vapour pressure	6.6	6.9	8	10.4	14.5	17.9	18.8	18.1	16.5	13.2	10.1	7.5	12.4
Mean relative humidity	68	64	59	61	69	79	83	84	84	80	79	73	74
Minimum relative humidity	2	8	6	9	11	27	36	16	29	18	10	8	2
Precipitation	14.9	14	18.5	23.9	122.9	187.7	204.2	216.4	112.7	65.1	48.4	7.4	1036.1
Max. daily precipitation	22.1	40.9	46.5	29.2	94.9	71.3	66.4	85.6	56.7	53.6	41.7	10.6	94.9
Total cloud cover	3.4	3.4	3.3	4.2	6.4	8.7	8.8	8.2	7.6	7.1	5	3.8	5.8
Lower layer cloud cover	2.9	2.9	2.3	2.8	5	7.2	7.3	6.9	6.9	6.5	4.5	3.2	4.9
Air pressure	811.3	810.1	810.1	809.4	808.5	807.7	807.4	808.8	811.7	814.1	814.2	813.1	810.5
Sunshine time	239.8	238.5	277.5	260.9	216.3	136.5	152.8	159.2	134.8	161.8	170.7	222.8	2371.7
Percentage of sunshine time	72	75	75	69	52	33	37	40	37	45	52	68	54
Evaporation*	129.2	157.2	222.7	253.5	253.3	151.5	146.6	140.2	119.7	104.9	104.3	105.8	1870.6
Mean wind speed*	2.6	3.1	3.2	3.1	2.7	2.4	2	1.5	1.6	1.7	2	2.1	2.3
Maximum wind speed*	19	18	15	18	13	9	10	10	17	10	11	13	19
Dominant wind direction*	C, SW	C, SW	C, SW	SW	SW	C, SW	C, SW	C, S	NW,C,S	C, SW	C, SW	C, SW	C, SW
Fine days (low cloud)	17.4	14.9	17.9	14.4	6.8	1.7	1.1	1.9	2.6	3.7	9.7	15	106.9
Overcast days (low cloud)	3.5	2.9	1.5	1.3	6.8	13.1	13.5	12.4	11.8	12	6.7	4	89.3

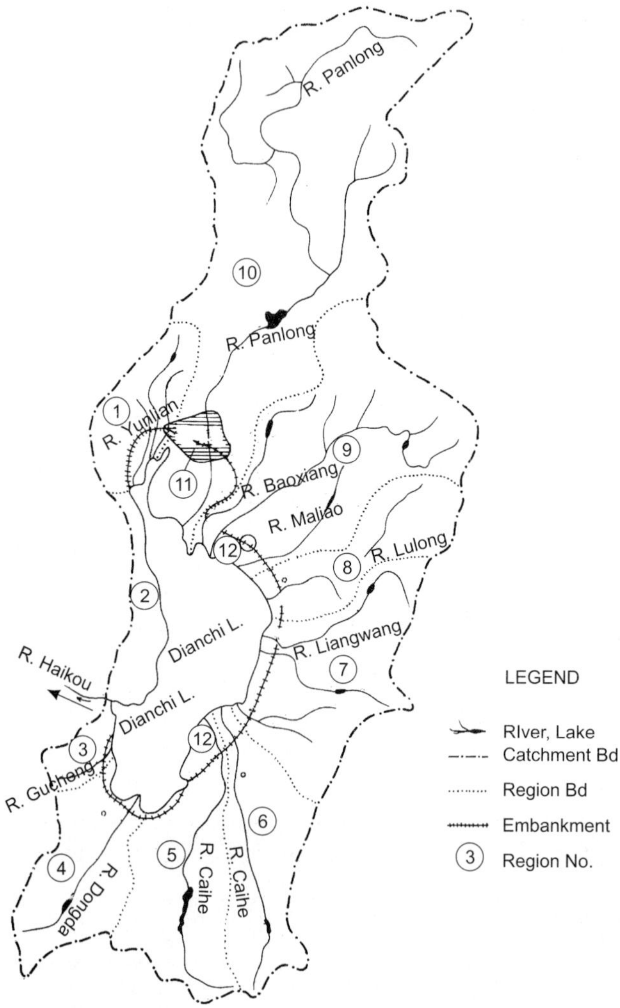

Fig. 15.3 Scheme of the catchment of Dianchi Lake

Table 15.2 Mean water balance components of Dianchi Lake Basin, Unit: 1×10^8 m^3

1 Precipitation over land	24.762	8 Outflow from Haikou	4.071
2 Precipitation over lake	2.772	9 Industry use	0.046
3 Total precipitation	27.534	10 Agriculture use	1.350
4 Evaporation from land	17.800	11 Domestic use	0.025
5 Evaporation from lake	4.279	12 Variation of water level	-0.038
6 Total evaporation	22.080		
7 Net Input to Dianchi Lake	5.454	13 Outout from Dianchi Lake	5.454

Table 15.3 Basic characters of water quality of Dianchi Lake in 1988 and in 1980 (right 2 columns), annual mean, Unit: mg/L

	Caohai Bay	*Dianchi Lake Middle*	*Dianchi Lake Mean*	*Caohai Bay*	*Dianchi Lake Middle*
Water Temperature, °C	17.5	17.6	17.6		
Colour, degree	43.6	15.5	16.5		
pH	8.07	8.69	8.68		
Total hardness German Scale	12.48	9.4	9.51		
Total alkalinity	4.01	2.81	2.85		
K^+	11.95	7.18	7.35	10.97	5.09
Na^+	47.22	30.94	31.5	44.03	15.59
Ca^{2+}	59.99	35.83	36.67	40.37	34.28
Mg^{2+}	17.73	19.02	18.98	16.64	15.93
Cl^-	61.68	35.32	36.23	36.76	17.77
SO_4^{2-}	51.05	30.58	31.29	74.36	22.01
CO_3^{2-}	9.44	10.06	10.04	4.43	4.86
HCO_3^-	230.8	152.24	155.1	176.72	165.92
Total concentration of ions	489.86	321.17	327.16	404.28	281.45
Increasing rate during 1980-1988	21%	14%			

province and southwest China. As a famous tourism lake in Kunming, Dianchi Lake possess abundant biodiversity and high quality water since ancient times. It is a notable recreation was known to site and reputed as "a Sparkling Pearl Imbedded in the Yunnan-Guizhou Plateau".

The water quality in Dianchi Lake was good before the 1960s, with the nutrient state gaining an oligotrophic-mesotrophic status by the end of 1960s. The water quality was classified as grade II, suitable as a drinking source. There were several intake areas for drinking waterworks. A famous park and a natural swimming zone were located around the northern bank and coastal zone of the lake.

The major water quality characters of Dianchi Lake are given in Table 15.3. The total ion concentration was 281 mg/L and 404 mg/L for open water and Caohai Bay of Dianchi Lake, respectively, in 1980. These figures increased to 321 mg/L and 490 mg/L, respectively, in 1988. It means that the total ion concentration increased 14 % and 21 % in open water and Caohai Bay in eight years.

The northern bay of Dianchi Lake is named Caohai Bay—a bay rich in aquatic vegetation (in Chinese). In fact, Caohai Bay was covered by profuse macrophytes. The macrophytes are known to be densely distributed in the coastal zone (Fig. 15.4). Prior to the 1960s, 90% of the

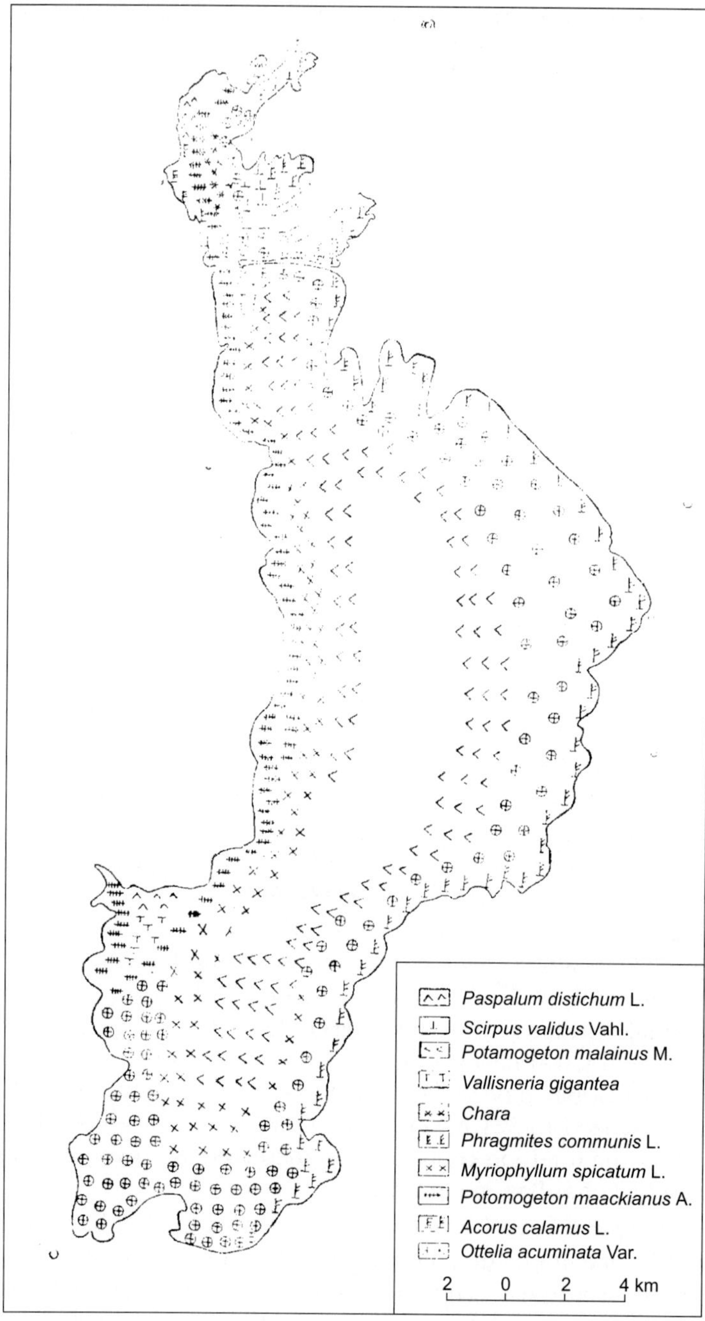

Fig. 15.4 Distribution of macrophytes in Dianchi Lake in 1960s

watersurface in Dianchi Lake was covered by macrophytes, the submerged plant distributed almost to the center of the lake, the transparency was a couple of meters. The total range of macrophytes had more than 100 species. There was community of *Ottelia acuminata*—one of the famous endemic species, which are the typical species for clear waters of grade II.

There were 129 species of phytoplankton and 109 species of zooplankton in Dianchi in 1980. There are also widely distributed water worm and chironomid larva, which are characteristic of the eutrophic state of the lake by 1980.

There is abundant distribution of molluscs in Dianchi Lake, including 29 species of Gastropoda and 21 species of Lamellibranchia (before 1980) (Fig. 15.5).

There are also 27 endemic species of fish in Dianchi Lake, four of them special to this lake.

All these data indicate that Dianchi Lake had a healthy ecosystem with abundant biodiversity, and associated high quality of waters in earlier times (Fig. 15.6).

RAPID DEVELOPMENT OF EUTROPHICATION

The quality of water of Dianchi Lake used to be quite good before the 1970s. The water quality was grade II, suitable for drinking purposes in the 1960s, which became grade III in the 1970s. Beginning from the last years of 1970s, the degradation of water quality became more rapid. This process of degradation developed in the Caohai Bay first and then dispersed to the south, to open waters and finally, to the entire lake.

The increase of nutrients in Caohai Bay lead to a decrease in the transparency of the lake. The submerged plants could not get sufficient sunshine and became weaker, to finally disappear in the deeper part of the lake. The competitive chemical substances released by submerged plants decreased, and the attached algae spread. The submerged plant became much weaker, while the blue-green algae became dominant species of phytoplankton. In suitable climatic conditions (sunshine and temperature), algae bloom appears more and more frequently. The result is a decrease in the vegetation cover in the lake. Figure 15.7 shows the aquatic vegetation in Dianchi Lake during 1981-1983. Obviously, the area with vegetation of submerged plants has decreased significantly in comparison with that in the 1960s (Fig. 15.4).

There were 90% vegetation cover and abundant molluscs prior to the 1960s, and about 13% vegetation and about 0.9 molluscs/m^2 before the

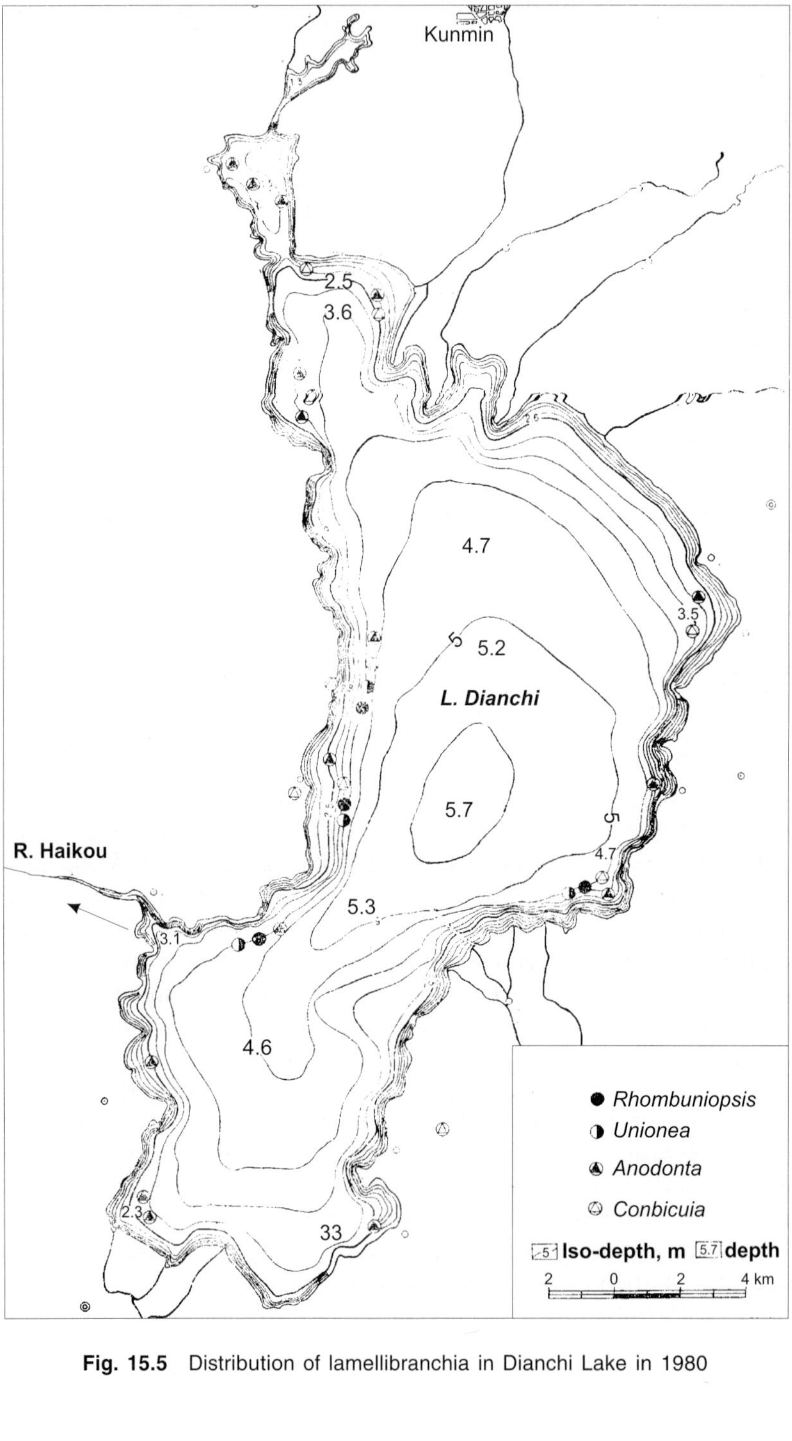

Fig. 15.5 Distribution of lamellibranchia in Dianchi Lake in 1980

Fig. 15.6 Views of Dianchi Lake, Kunming—**City of Perpetual Spring,** China

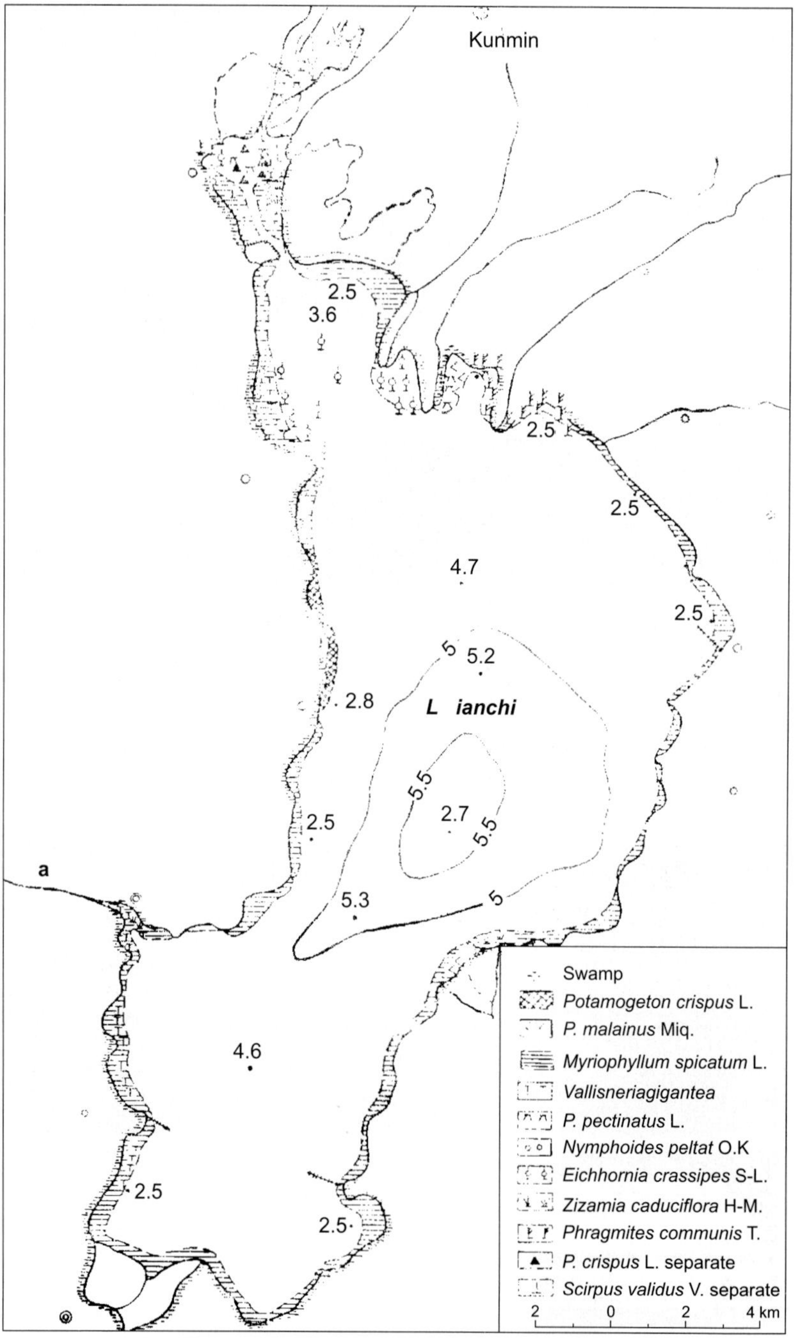

Fig. 15.7 Map of aquatic vegetation in Dianchi Lake during 1981-1983

1980s. There was no trace of submerged plants in Caohai Bay and only about 4 km^2 vegetation in the main part of Dianchi Lake by the end of 1980s. A lot of molluscs died at that time. The algae blooms appear very frequently during most of the year. Transparency is as low as 0.2~0.5 m. Submerged plant and molluscs decreased steadily. Except some species of merged plant existing in the local coastal zone, the abundant water hyacinths and other floating plant become the dominant macrophytes in Dianchi Lake.

Presently, both the wide algae bloom and water hyacinths cover have become very serious issues in the Dianchi Lake (Figs. 15.9 and 15.10).

According to the systematic monitoring data on the water quality indices since 1986, the annual mean value of BOD_5 and COD_{Mn} (Figs. 15.11 and 15.12) in Caohai Bay was about 15 mg/L, which is more than the state standard of $\leq$ 10 mg/L for surfacewater grade V. The annual mean value of the total nitrogen (TN) and total phosphorus (TP) were about 5 mg/L and 0.5 mg/L. This is dangerous for macroorganisms; the lake may be graded as "Desert Aqua-Ecosystem (AES)"— in contrast to its previous "Healthy AES". The annual mean value of TN was more stable at the level of 5 mg/L, while the annual mean value of TP gradually increased from 0.4 mg/L-0.9 mg/L during 1987-1993. There was no obvious relationship between the annual value of BOD_5 and TP during this period. The possible explanation is that in this period, the TP in the organic matters contributing the BOD_5 was not a rich element, and/ or the lake bottom of Caohai had a vast purifying (bounding) capacity, especially for phosphorus. It accumulated a lot of phosphorus in the bottom sediment, when the organic pollutant loading was large and the decomposed phosphate was trapped more intensely by the bottom sediments. Since 1993, there has been a good linear correlation between the TP and BOD_5 (Fig. 15.13). Figure 15.13 included the data for the open water of Dianchi Lake during 1986-2000. The correlation coefficient then was 0.9538. The linear regression equation is as follows:

$$BOD_5 = b + m \times TP;\ b = 1.283;\ m = 15.5169;\ (\text{mg/L}) \qquad (15.1)$$

The average error for estimation of the annual mean BOD_5 according to the annual mean TP value was ± 0.79 mg/L, when the confidence interval was 99%.

The BOD value is mainly related to the concentration of the organic pollutants, and includes natural organic pollutants and synthetic organic pollutants, which are, in turn, related to TP concentration. This is the basis of the above regression equation of (15.1).

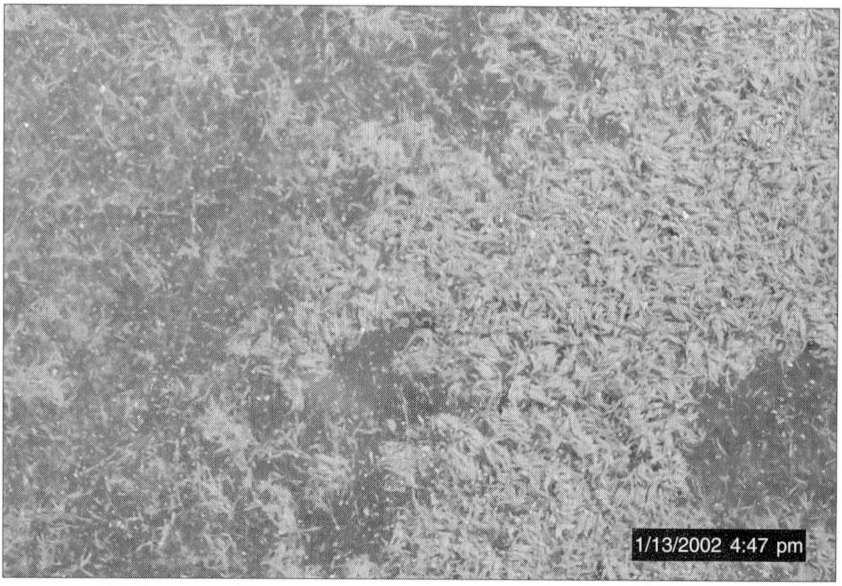

Fig. 15.8 View of algae bloom of *Aphanizomenon flos-aquae* in Dianchi Lake, Jan. 13 2002 (Photo by LI Wenchao)

Fig. 15.9

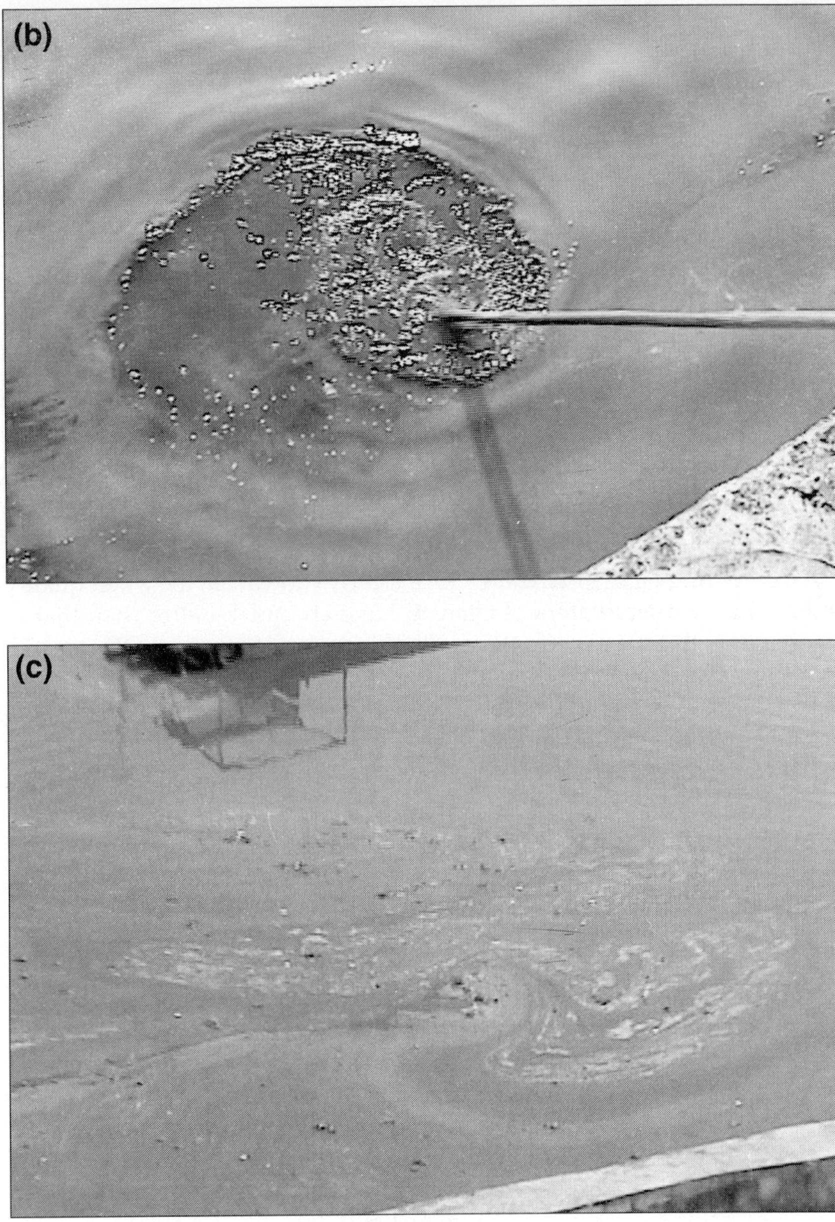

Fig. 15.9 View of algal bloom and water hyacinths detritus in Dianchi Lake (a) June 14, 2001, (b) Nov. 14, 2001, and (c) January 9, 2002

Fig. 15.10 Water hyacinths cover on Dianchi Lake, September 24, 2002

Comparing Figs. 15.12 and 15.13, it clearly shows that the water quality indices in the open waters of Dianchi Lake are much better than that in Caohai Bay. But we can see the gradual increase of annual BOD_5 and TP in the open waters of Dianchi Lake. The BOD_5 and TP reached 6 mg/L and 0.3 mg/L by 1998, which exceeded the grades IV and V for surfacewaters according to the state standard. These indices were worse than that for waters for grades II and III before 1993.

The data in May 2002 (Table 15.4) indicated the process of eutrophication had made new developments during the last few years.

MAIN REASONS OF THE RAPID DEVELOPMENT OF EUTROPHICATION

(1) As was shown in Fig. 15.4, the Caohai Bay was much bigger than its recent pattern. The vast area of the coastal zone with abundant macrophytes and biodiversity disappeared after intensive reclamation. Similar reclamation or reconstruction had been carried out along the other coastal zone. Most parts of coastal wetlands had been reclaimed during the 1960s. A lot of hydrophytes and helophytes in littoral zone disappeared. The self-purifying capacity of Dianchi Lake decreased notably, but the increasing population especially the widespread utilization of flush toilet in urban and countryside led to a rapid increase of nutrients loading into the lake.

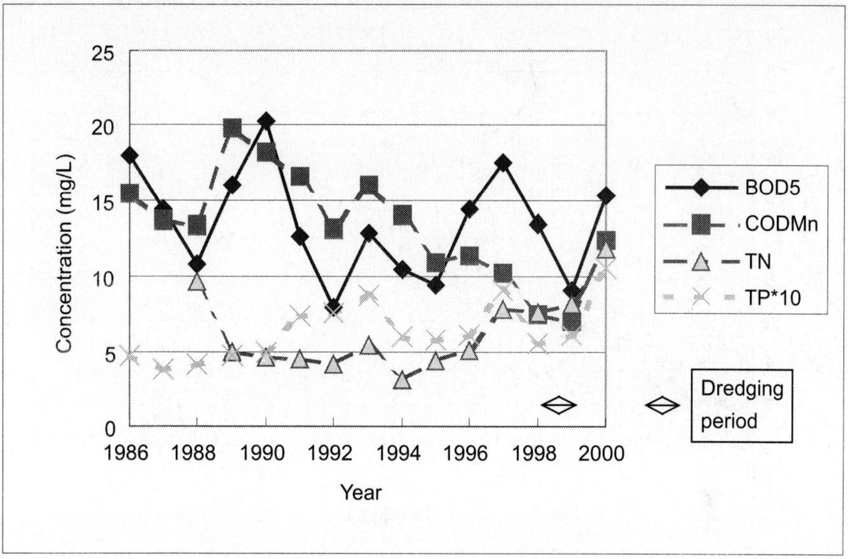

Fig. 15.11 Variation of annual main water quality indices in Caohai Bay, Dianchi Lake, China

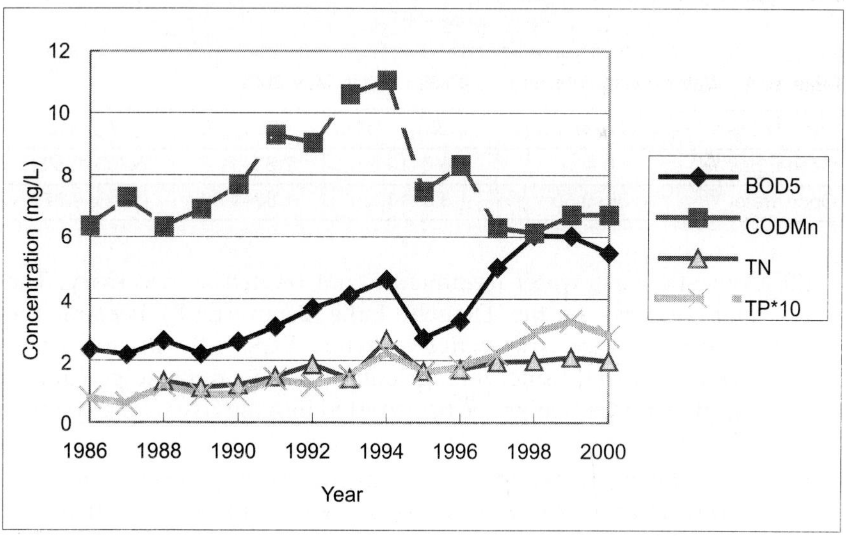

Fig. 15.12 Variation of annual main water quality indices in the open waters of Dianchi Lake, China

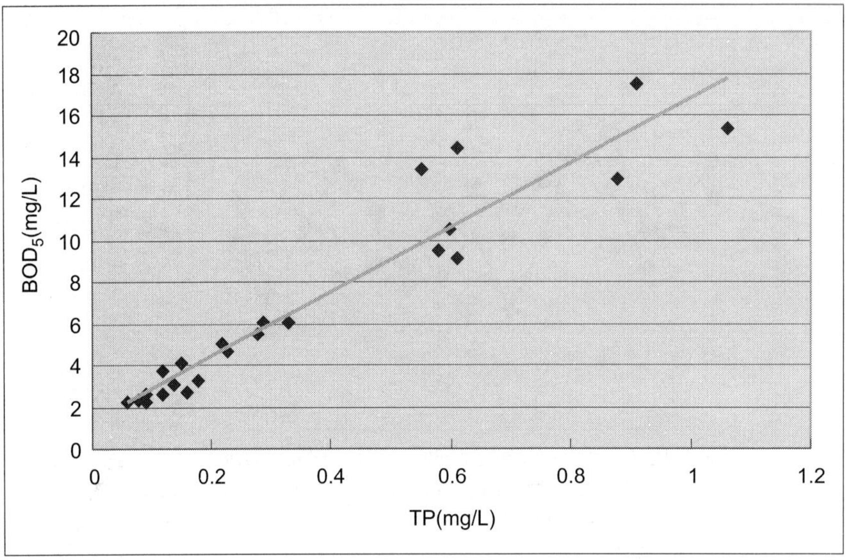

Fig. 15.13 Correlation between the annual mean BOD_5 and TP in Dianchi Lake, China, 1986-2000 (for Caohai Bay 1993-2000); $BOD_5 = b + m*TP$ (mg/L), b = 1.283, m = 15.5169; correlation coefficient = 0.9538. The average error for estimation of the annual mean BOD_5 according to the annual meanTP value was ±0.79 mg/L, when the confidence interval was 99%.

Table 15.4 Water quality indices of Dianchi Lake in May 2002

Dianchi Lake	*Transparency, cm*	*COD_{Mn} mg/L*	*TN, mg/L*	*TP, mg/L*
Caohai Bay	0.4	8.8~10.5	7.89~11.7	0.703~0.809
Open water	0.25~0.50	8.3~10.7	1.67~4.02	0.243~0.770

(2) Domestic wastewater treatment is not treated as necessary. The river systems in the Dianchi Lake Basin (DLB) became the wastewater receiving water systems. Especially during flood periods, all the industrial, agricultural and domestic pollutants stored in the area, might be transported into the river/lake systems in DLB.

(3) We are facing an extremely, serious situation, in the sense that the concentration of the wet and dry precipitation is more than the threshold value for eutrophication by a couple of times (Table 15.5). It means that we cannot expect to control the eutrophication by dilution using precipitation.

Table 15.5 Annual TN, TP loading into Dianchi Lake and Taihu Lake (Kunmin Institute of Environment Sciences, 1999; JIN Xiancai, et al. 1995)

Lake	*Area, km²*	*1988, t/a*		*1995, t/a*		*1988 Rain+Dust,t/a*		*Loading rate, 1998*		*Stand. for*	
		TN	*TP*	*TN*	*TP*	*TN(Concent. mg/L)*	*TP*	*TN(t/km²/a)*	*TP*	*eutro. mg/L*	
Dianchi	298	4197	338	8476	480	412(2.5)	36(0.27)	14.1	1.1,	TN	TP
Taihu	2338	3105	964			3180(1.2)	93(0.04)	1.3	0.4	0.2	0.02

According to the estimation, in comparison with that in 1995, the annual loading of CODcr, TN, TP into Dianchi in 2000 increased by 65.7%, 56.1%, and 66.1%, respectively. The increasing rate would rise to 203.9%, 156.9%, and 187.9%, respectively, in 2010.

The other important fact was that the concentration of pollutants was much bigger (a couple times more) during floods (Kunming Institute of Environment Sciences, 1992) than in the usual cases. Therefore, the estimated pollutants loading were usually underestimated.

All of these statistics suggest that the pollution loading (PL) exceeds the self-purification capacity (SPC) of the waters. All these facts also lead to the rapid expansion of hyper-eutrophic and eutrophic waters. (Figs. 15.14 and 15.15).

The x-axis indicates the percentage of waters with various quality grades in a lake catchment. The y-axis is the time. In history, the most surfacewaters were waters of grades better than grade III in the history. Because of the increase of PL and decrease of SPC, as explained by the above reasons, the water quality first worsened in the waters close to the pollutant source areas and then extended from local to larger waters. The degradation of aquatic ecosystem developed because of the rapid increase of PL as well as the rapid decrease of SPC caused by the degradation of the aquatic ecosystem. The surfacewaters with worse than grade III quality in the lake catchment occupies most parts at the present, as showed in Fig. 15.14. Our purpose is to restore the historical landscape of the once healthy ecosystem, and increase the volume of clear waters. The key approaches are to decrease the PL, enhance SPC, and make the PL<SPC.

(4) This can be explained by Fig. 15.15 and equation 15.2:

In the one-dimensional steady flow issue, the nutrient concentration S changing with distance may be described by the following simplified equation:

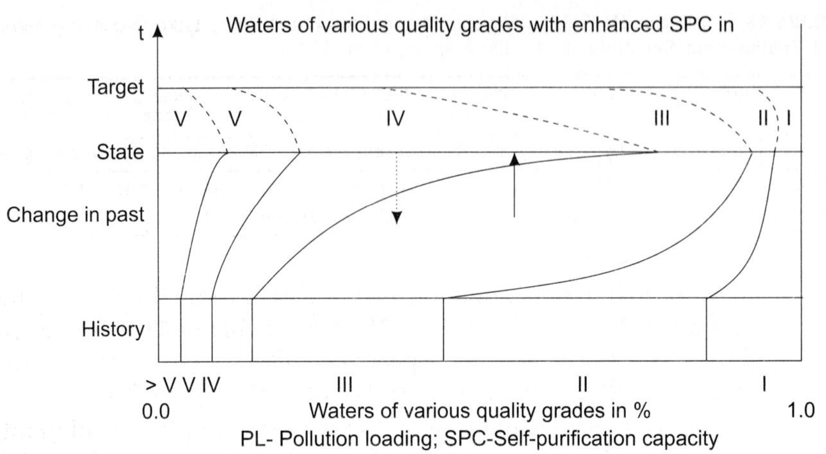

Fig. 15.14 Scheme of changing the state of surfacewaters with various quality grades in relation with pollution loading (PL) and self-purification capacity (SPC)

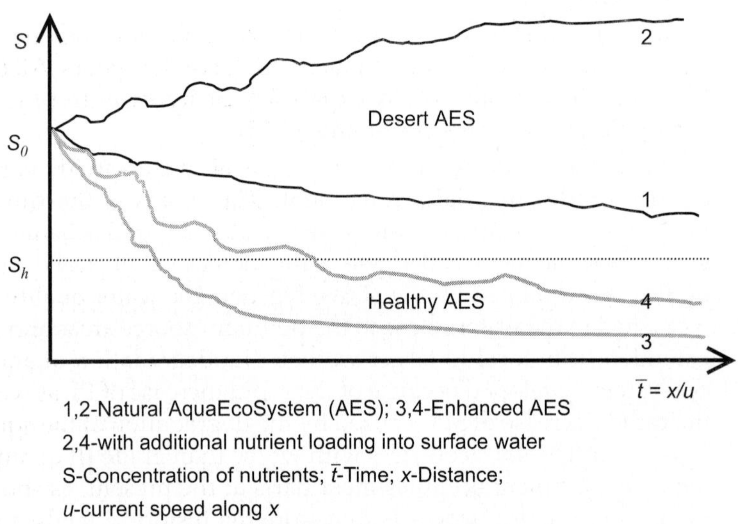

Fig. 15.15 Distribution of nutrient concentration of *S* along current in a steady state in relation with different self-purification capacities

$$\frac{\partial S}{\partial \bar{t}} = PL - \alpha S; \tag{15.2}$$

where $\bar{t} = x/u$; x = distance; u=current speed; $PL = L_s + L_b + L_t + L_i$: pollution loading from water surface (with lower index s), bottom sediment (*b*), transverse (*t*), and internal (*i*); αS-self purification capacity. When $PL < \alpha S$, we have $\partial S / \partial \bar{t} < 0$ (Fig.15.15 curve 1). When $PL > \alpha S$ we have $\partial S / \partial \bar{t} > 0$ (Fig.15.15 curve 2) (Fig. 15.14). Curves 3 and 4 show the enhanced SPC with no additional input and with additional inputs. Because the initial concentration S_0 is much higher than the concentration for a healthy aqua-ecosystem AES (S_h), and the SPC is too small, a large area of the surfacewaters in the basin is occupied by "desert AES", where there are almost only microorganisms. A physico-ecological engineering scheme has been set up to improve the surface water quality. (PEEN, PU, P. et al., 1998). The central tache is to enhance the SPC and the capacity for resisting the possible environmental hazards, even in local areas mainly by using solar energy. All the macrophytes, microorganisms and molluscs might increase SPC greatly (these points will be described in the section 4). Macrobionics is very important for construction the optimum AES. Similarly, management of human activity is the most important element for construction the enhanced AES.

(5) The bound capacity of the bottom sediment for Total Phosphorus (TP) is decreasing. The fact is that the TP was relatively low as 0.5 mg/L during the period before 1990 (Fig. 15.11), while the COD_{Mn} and BOD_5 were relatively as high as about 15 mg/L in Caohai Bay. It seems quite possible that the abundant TP formed by decomposition from abundant organic substances (high concentration of COD and BOD in water) was absorbed and bound by the bottom sediment. The vast binding capacity of the bottom sediment appeared during the start of eutrophication. During the period after 1990, the COD had a general tendency to decrease, while the TP and the BOD had a similar variation curves. Probably, the binding capacity of the bottom sediment for TP decreased after the early period of commencement eutrophication. The fact that the binding capacity decreases with the development of eutrophication was also found in Lake Okeechobee, Florida, USA (Blasland, Bouck & Lee, 2001). During the period after 1993, a close relationship between the TP and BOD appeared clearly in Caohai Bay (Fig. 15.13).

(6) The nutrients TP associated with BOD_5 were steadily increasing in the open waters of Dianchi Lake, while the TN became increasing by weaker (Fig. 15.12). This indicates that the purifying capacity in

the open lake is enhanced by a more strong wind wave mixing process, and the nitrogen released by decomposition of organisms into a gaseous state and then release into the atmosphere.

(7) The speed of development of the trophic state of Dianchi Lake was much faster than in Taihu Lake—the third largest freshwater lake in China with a water surface area of 2338 m^2. The major reasons can be explained by: (1) the TN and TP loading for unit water volume in Dianchi Lake is 1 and 3 times higher than that in Taihu Lake; (2) the mean retention time of Dianchi is 2.7 year, which is 3 times of 0.9 year for Taihu Lake (Table 15.5). It means that the increased eutrophic state remains more stable in Dianchi Lake. The output rate through the outflow is less in Dianchi Lake than in Taihu Lake. The degradation tendency of the aqua-ecosystem is more difficult be stop in the Dianchi Lake.

The SPC in waters decreases gradually after the aqua-ecosystem begins to degrade. The water quality rapidly worsens, because the SPC is less than the pollution loading. This is a vicious circle of the development of aqua-ecosystem. Therefore, the degradation of aqua-ecosystem develops even more rapidly.

(8) There was no abundant output of bio-products from the lake to take off the nutrients from waters. Most inputted nutrients were stored at the lake bottom, which are easy to release into water column in a shallow lake, such as Dianchi Lake, by water-sediment interactions and resuspension.

In conclusion, Dianchi Lake is suffering because of the rapid degradation of water environment. The primary reasons behind this phenomena are:

(1) The ecosystem in the lake has been destroyed (low SPC);
(2) The biomass of preliminary production was not taken off from lake largely (low output of nutrients); therefore, the inputted nutrients accumulated (most parts of that were stored at the bottom;
(3) The nutrient loading did not decrease (high loading of nutrients into lake). The SPC for nutrients is less than the nutrient loading.

DISTRIBUTION OF POLLUTANTS LOADING INTO DIANCHI LAKE

Today, the major pollutants loading into Dianchi Lake are both natural and artificial organic ones, but not heavy metals at the present. The total annual wastewater input (barring rainfall inflow) amounted to 1.854×10^8 m^3, among them 1.34584×10^8 m^3 was domestic, and 0.49804×10^8 m^3 was industrial.

The annual TP loading was 1021 t, TN–8981 t, COD_{Cr}–4167 t into Dianchi Lake in 1995[1]. The distribution of pollutants loading into Dianchi Lake among the industrial, non-point, and domestic sources in 1995 is shown in Figs. 15.16, 15.17 & 15.18. The pollutants loading from industrial wastewater occupied 7–33%, the non-point wastewater occupied 8–46%, and the domestic wastewater occupied the most part among the total loading of 45–70%. The percentage of the non-point pollutants loading changes in relation with rainfall in the year, increasing in the year of abundant rainfall.

The dominant contribution of domestic pollutants loading was of great importance, especially for Caohai Bay, while the contribution of non-point pollutants loading was increasing, particularly for the open waters of Dianchi Lake.

In relation to the rigorous controlling management of the industrial plants and enterprises, the contribution of the industrial pollutants loading in the total loading decreased. A special survey have been made to determine the non-point pollution from the catchment of Dianchi Lake. It was revealed that except for the pollutants from the wet and dry precipitation and industrial plants in the countryside, the non-point pollutants produced as big as 20.576 million tonne, and the release value was 17.917 million tonne.

A great deal of studies have been conducted to reveal the pollution sources and estimate them quantitatively. According to these research results, many measures have been applied during the past decades.

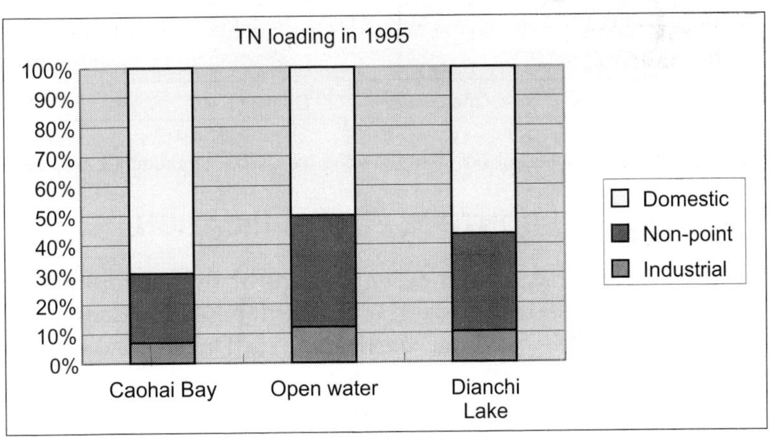

Fig. 15.16 Construction of pollutants loading for TN in Dianchi Lake, 1995

[1]Plan on Protection of Water Pollution in Dianchi Lake Basin during the "9th five-year plan", and the program by 2010.

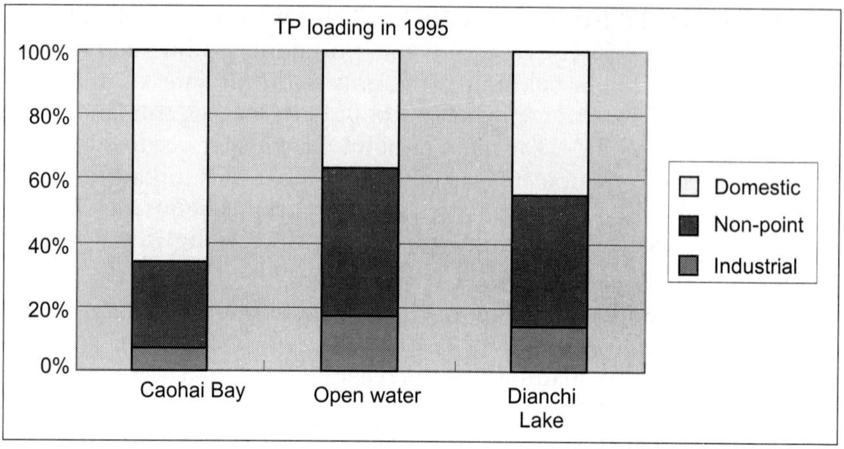

Fig. 15.17 Construction of pollutants loading for TP in Dianchi Lake, 1995

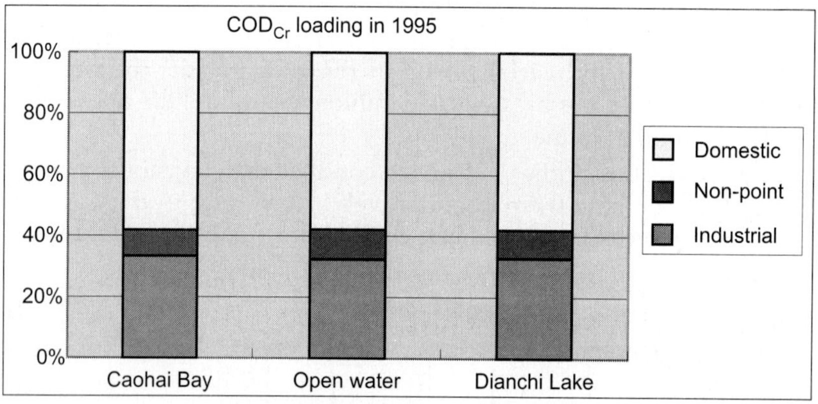

Fig. 15.18 Construction of pollutants loading for COD_{Cr} in Dianchi Lake, 1995

APPROACHES TO CONTROL EUTROPHICATION

A lot of measures have been taken to control the eutrophication in Dianchi Lake from 1980 onwards. There are different approaches applied, tested and developed in China, particularly in Dianchi Lake Basin, as listed below:

- Pollution controlling or transferring;
- Dilution and diffusion of pollutants or flushing with clear water;
- Dredging or covering the lake bottom;
- Adding coagulants or other biochemical substances for killing the algae;

- Biomanipulation;
- Bioremediation;
- Physico-ecological Engineering.

The main measures implemented for controlling eutrophication in Dianchi Lake were:

To control the point pollution sources, the first step was to control the wastewater, both the concentration and total quantity released from the industrial enterprises to meet the state standard. The enterprises unable to maintain the state standard for releasing wastewater should stop their work, make further improvement for wastewater treatment or move to some remote site as a temporary measure. The plants which release the heavy metal pollutants should move away from the city. As a result, the industrial pollutants loading decreased, especially for heavy metal pollution. The state standard industrial wastewater was 300 mg/L a couple years ago, which has now reduced to 80-100 mg/L. The standard of COD_{Cr} for surfacewater of grade III is 8 mg/L. Therefore, the surface river-lake system in the lake basin became the wastewater receiver system. The degradation of the surface waters developed rapidly, and controlling industrial pollution was insufficient to control the eutrophication.

To decrease the domestic wastewater pollution, four wastewater treatment plants were constructed with the aid from the World Bank Foundation. More wastewater treatment plants and the wastewater channel system are under construction and reconstruction. This has partly improved the water quality in the river system in the Kunming City, but the effects were not noticeable, because of the following issues:

a. It takes a lot of money to construct the facilities, but it can treat domestic wastewater only partly;
b. The treated water contained COD_{Cr} 80-100 mg/L, which is much higher than the state standard for surfacewater of grade III (COD_{Cr} 8 mg/L); to improve the efficiency of treatment, more capital is required;
c. This plant could not separate the domestic wastewater and the rainfall. Kunming City is an old city with long history; its wastewater channel net is quite complex and not easy to reconstruct for separating the wastewater and the rainfall.
d. The cost for construction of the wastewater treatment plants is about 1000 Chinese yuan per capita. It means that to treat the wastewater released by the residents of 2.1 million (the residents in the rural also use the flush water system) in the catchment, it needs 2.1 billion Chinese Yuan. Even now, the river system in Dianchi

Lake Basin still receives the water with pollutants concentration exceeding the standard of grade III for surfacewater by a couple of times more.

A big Xiyuan Tunnel with 4760 m length, 4.8 m width, and 2.8 m depth, was constructed during 1994-1995. The designed maximum outflow through the tunnel is 43.9 m/s. The associated dams for regulating the water and for ships to pass through were also constructed between Caohai Bay and the open waters of Dianchi. The major function of the Xiyuan Tunnel was to transport the wastewater to the lower reaches of Dianchi Lake. It could also be used as an added outflow exit for Dianchi Lake in case of flood (there is another natural exit in southwest of Dianchi Lake—Haikou, Fig. 15.1, Fig. 15.3).

The existing main issues are as follows:

a. The pollutants flow with great concentration input into Caohai Bay firstly, then go out through the Xiyuan Tunnel. Therefore, the water in Caohai Bay will be polluted before it goes out.
b. The outflow of polluted water makes a serious pollution issue in the downstream river for a long distance. Some cities and towns have no source of drinking water, which used to be earlier.
c. Dianchi Lake Basin is suffering insufficient water resources. There are no waters to be discharged from Caohai-Dianchi Lake during dry season. The best way is to clear up the wastewater and reuse it, but not to release it from the Caohai-Dianchi Lake system before it be purified.

Hold up the pollutants from the north and west coast of Dianchi Lake: These are two projects, which may stop the wastewater loading into the lake directly. The difficulty is not only the expense, but also how to separate the wastewater and rainfall flow.

Transportation of clear water from upper reaches reservoirs as drinking water sources: The water quality in the upper reaches reservoir is much better than that in Dianchi Lake at present. The downstream water in the river discharged from the reservoirs have become worse by transverse pollution from the catchment. Thereafter, the tubes for transporting the clear water to the city drinking waterworks need to be constructed. This measure can improve the drinking water quality obviously, but will prove to be expensive. In addition, the total water budget cannot increase for the entire catchment.

Green mountain project: This project aims at recovering the forests in mountain regions and to develop the ecological agriculture. This project is quite valuable in decreasing erosion of soil and nutrients lost from the mountain and agricultural fields. This project is currently being implemented.

Algae bloom controlling measures: The algae bloom appeared more and more frequently since the 1980s. A few years later, the algae bloom became a huge cover, firstly in Caohai Bay, and then in northern part of the Dianchi Lake. We have no efficient technology for collecting algae economically to remove the nutrients. Different measures for controlling algae bloom have been tested since the 1990s, especially for preparing the Horticulture Expo '99, Kunming. A lot of research and testing experiments had been done during 1996-1999.

a. Collection ship/machine test: The algae collection ship and machine was constructed (Figs. 15.19 and 15.20). The biomass of blue-green algae were used as fertilizer in liquid state, and feed for animals in dried solid state. These were just experiments. People could not avail of any economic benefit from algae collection. The experiments were supported by national and local governmental funds and now are undergoing development.
b. Some bacteria communities were tested in enclosures in Caohai Bay to transform the nutrients and compete with phytoplankton, such as photosynthetic bacteria and other bacterial communities cultured by different companies in the world. In most cases, the transparency in the enclosure improved, but was not able to improve all the important indices of water quality such as TP, TN, COD_{Cr}. For example, in some cases, the TP increased after dispersing the bacteria carrier. Usually, the effects of bacteria can be prolonged a week to 10 days. The long-term effects were not clear enough. All the experiments were limited in small-enclosed areas.

Fig. 15.19 Algae slime collection ship in Caohai Bay

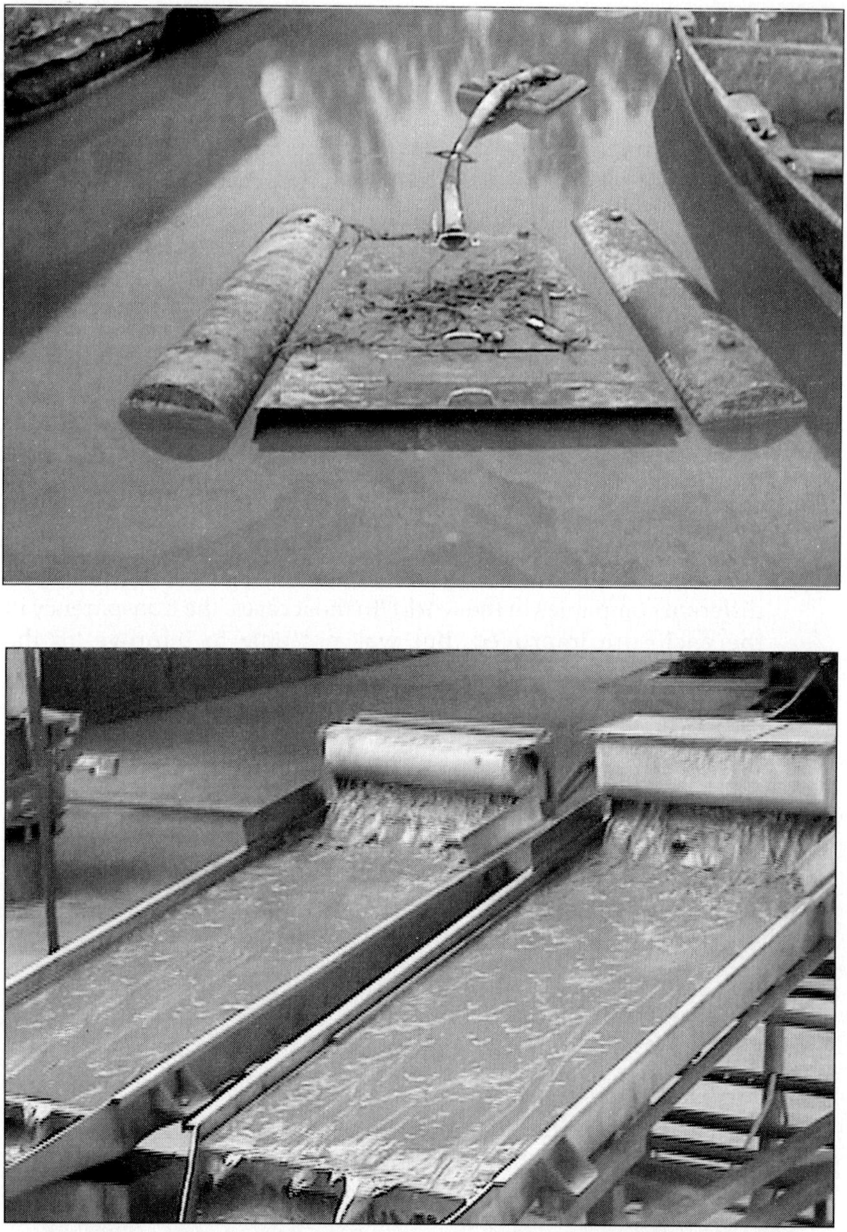

Fig. 15.20 Algae slime collection facility in the northeast bank of Dianchi Lake; Algae input part (above), algae concentration facility (lower)

c. Different artificial or natural coagulators were tested to clarify the possibility of improvement of transparency, and decreasing phytoplankton concentration. The **diatom clay** (the local products) with some natural herb juice, one of the most promising substances, might be used in some cases to improve transparency of the local waters. The more complex situation awaits in case of the moving, open space waters. The natural coagulator substances are much better than the artificial synthetic substances. The latter substances might generate the possible long-term negative effects on the aquaecosystem, clarification of which need long-term investigation. In the drinking water source area, the usage of artificial chemical coagulators should ensure protection of human health. To guarantee the great volume of pure chemical coagulators with low price is another issue of great importance.
d. The chemicals for killing phytoplankton were tested in a laboratory. The issue is the uncertainty regarding their long-term effects on the ecosystem.
e. To use the zooplankton to consume phytoplankton was tested in enclosures. Special species of phytoplanktivore-zooplankton were cultured in laboratories and then released into field enclosure. Transparency in testing enclosure increased dramatically.

Collecting the Water Hyacinths and Closing Down Net-enclosure Fishery

Water hyacinths (WH) were introduced to purify the water and as the feed for fish and livestock. There are suitable climatic conditions for water hyacinth growth in Dianchi Lake Basin. Therefore, WH grow and distribute very quickly in the hypereutrophic Dianchi Lake. The huge floating WH cover influences the waterway transportation. It limits the penetration of air and light into the water body, leading to decrease of dissolved oxygen and death of animals in the water. During the period of WH growth, the concentration of nutrients decreased obviously, and the transparency increased rapidly. But the abundant detritus of WH decomposes and leads to the secondary pollution. To take WH out from the water body is not so easy by manual labor. WH contents about 95% water. A huge amount of fresh WH may be carried to the ship and stored on the bank (Fig. 15.21), and then transported to other places by road. If

Fig. 15.21 Collection of water hyacinths in the northern bank of Dianchi Lake, September 24, 2002

stored on the bank, the decomposed nutrients of WH will return to the water body, again during rainfall. Therefore, it needs a great deal of labor and expense to gather WH.

A special research project had been undertaken to use WH as livestock feed. It was clear that WH is a good feedstuff for fish, pig, cow, sheep and other livestock. The issue is: how to effectively dry the WH and make economic profits from production of WH feedstuff. To develop the technologies for producing the WH products with economic profit, and then develop the Bio-Environmental Enterprise (BEE) is the most important issue to be solved.

The net-enclosure fishery was introduced for increasing the fish product by enhancing feed input into lake. This kind of fishery leads to additional nutrients loading into the lake. The Kunming City Government approved a rule of law for putting down all the net-enclosures of fishery in Dianchi Lake.

Dredging Projects in Rivers and the Northern Part of Caohai Bay

Big projects for dredging the river system in Kunming City were carried out during the last few years. The sediment layer of a couple of decades was taken off from the urban rivers in Kunming. It improved the odor in a short period, but could not improve the water quality indices.

A big project of dredging the bottom sediments in the northern part of Caohai Bay was executed during 1998–1999. The dredged area and volume were 2.828 km^2 and 4.24×10^6 m^3, respectively. The cost of the project was USD $30 million. The tendency of degradation of the ecosystem in Caohai Bay did not cease even now.

Can we control eutrophication by dredging? This issue will be discussed in the next paragraph.

Biomanipulation by silver carp

The approach for controlling eutrophication on the basis of biomanipulation by silver carp was developed and tested in Dianchi Lake in the last few years. The research results in Donghu Lake, Wuhang, China indicated that it has a very slight effect on algae bloom controlling by increasing zooplankton. An intensive culture of silver carp (*Hypophthalmichthys molitrix*) and bighead carp (*Aristichthys nobilis*) after 1971 led to a change of phytoplankton composition. Instead of the dominative *Microsystes, Anabana* and *Oscillatoria*, during 1979-1982, the *Cyclotella* and *Cryplomonas* became the dominative species during 1989-1992 in Donhu Lake. The algae bloom disappeared since in 1985. It was supposed that the increase of the density of Silver carp (*Hypophthalmichthys molitrix*) and bighead carp (*Aristichthys nobilis*) was the main reason for the disappearance of the algae bloom in Donghu Lake. The density of silver carp and bighead carp of ≥ 50 g/m^3 was estimated as the criteria for controlling the algae bloom in the hyper-eutrophic lake (Xie, P. and Liu, J. 2001). This approach is different from the classic biomanipulation approach. The principle scheme, shown in Fig. 15.22 introduce the Piscivore fish to decrease the zooplanktivore fish, and thereafter, to increase the zooplankton, which consume more phytoplankton and can effectively decrease the density of phytoplankton and control the algae bloom (Shapiro et al. 1975). The principle of the new approach is shown in Fig. 15.23. Introducing the planktivore fish with a high density leads to cutting down of the plankton density, including both the phytoplankton and zooplankton, thereafter, control the algae bloom.

A big Complex Project

This biomanipulation approach was introduced in Dianchi Lake in the last few years. A huge complex project for controlling the algae bloom in the two bays with the total water surface area of 6 km^2 was carried out in

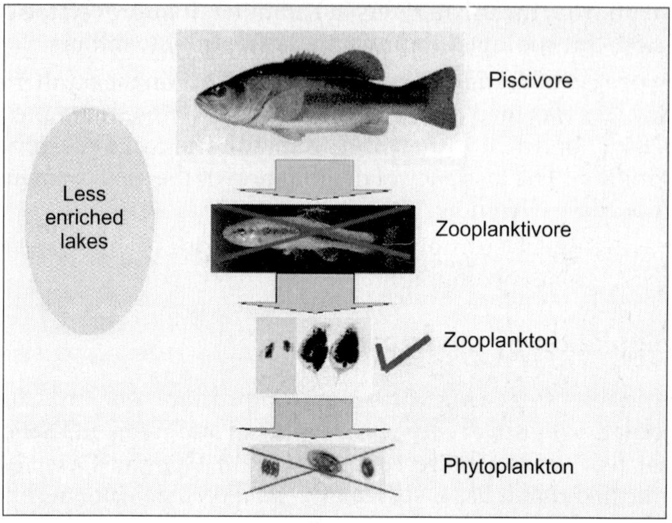

Fig. 15.22 Scheme of classical biomanipulation approach for controlling algae bloom (Xie, P. and Liu, J. 2001)

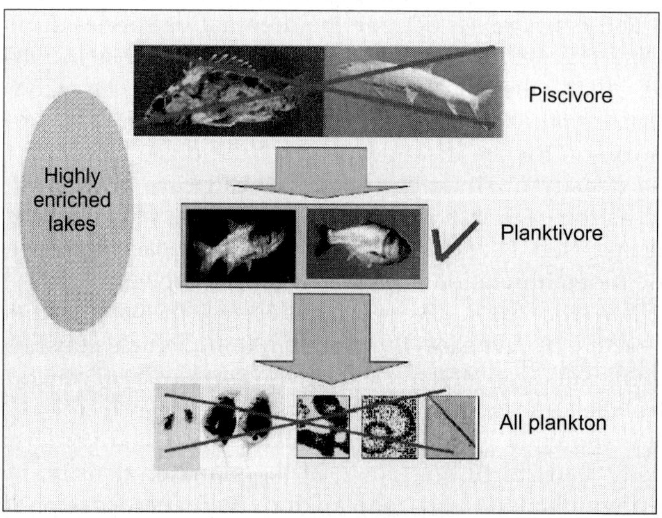

Fig. 15.23 Scheme of new biomanipulation approach for controlling algae bloom by using filter-feeding fish (Xie, P. and Liu, J. 2001)

northeast Dianchi Lake since 1999 (Fig. 15.24). The outer enclosure was constructed for separating the upper layer of waters in the experimental area and the open waters in Dianchi Lake (Fig. 15.25). The lower waters

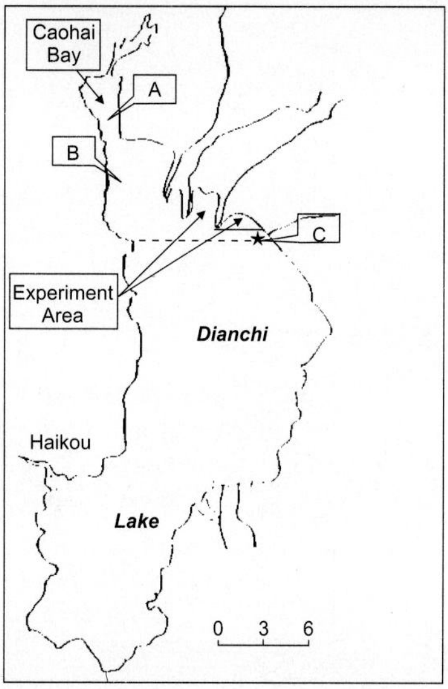

Fig. 15.24 Location of experiment area for controlling eutrophication and algae bloom in Dianchi Lake

were separated by fish net to prevent the fish from escaping. There are some internal enclosures for culture aquatic plants and fishery experiments (Fig. 15.26). According to the experiences (the density of silver carp should be ≥ 50 g/m^3) in Donghu Lake, silver carp and bighead carp were introduced for controlling algae bloom. Different species of emergency plants, submerged plants, and floating in a big way plants were cultured in enclosures. An experiment for collecting and processing algae (Fig. 15.20) was carried out at the costal zone of the experimental bay. The main targets of this project were to decrease the phytoplankton concentration, to increase the transparency (SD $\geq$ 60 cm) and to improve the water quality indices. According to the monitoring data, the annual mean SD were 26.1 cm and 42.1 cm in the experimental engineering and outside the experimental engineering in Dianchi Lake during 2001, respectively.

Two more years have passed since the experiment in Dianchi Lake. Some results had been obtained. But the issues were quite complex and there still remain a lot of difficulties for achieving the specified targets. The project is ongoing.

In summary, the total budget for the directly harnessing Dianchi Lake and controlling its eutrophication during last 20 years was estimated at a couple of billion Chinese Yuan. Even now, the eutrophication-controlling problem is still far from being solved.

Fig. 15.25 View of the outer enclosure (a) and facilities (b) for controlling eutrophication and algae bloom at the experiment site in Dianchi Lake, January 19 2002

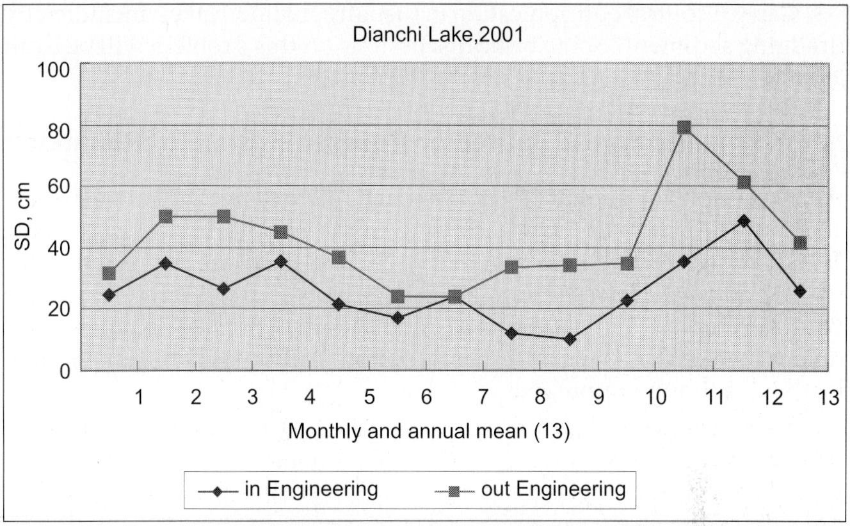

Fig. 15.26 The transparency (SD) in experimental engineering and outside of experimental engineering, in the northeast part of Dianchi during 2001; the SD $\geq$ 60 cm was the target of the experiment

Can we Control Eutrophication by Dredging?

In relation with the rapid increase of population, development of industry and agriculture in the catchment of lakes, the water quality are decreasing obviously not only in Dianchi Lake, but also in many other lakes in China. Many of China's lakes are faced with a serious eutrophication problem. The water quality problem is an important restricting factor for sustainable economic development in many regions. Although some measures for controlling the lake eutrophication have been employed in practice, the deterioration tendency of water quality cannot be efficiently controlled. The lake sediments are rich in nutrients. The concentration of nutrients in the lake sediments is much more than that in the lake waters. Therefore, as one of the possible measures, dredging sediment is applied for controlling eutrophication in some lakes in the world. This measure needs a great deal of funds, but there are no obvious samples for middle-large scale lakes, which indicate the efficiency of dredging as the key measure for controlling the lake eutrophication. For example, for dredging 2.828 km^2 bottom sedimentwith a mean depth of 1.5 m and dredged sediment volume of 4.24 $\times 10^6$ m^3, USD $30 million had been spent. But the tendency of degradation of the ecosystem in Caohai Bay has not stopped even now.

Can we control eutrophication in (usually shallow) lakes, including by dredging sediment? Some basic discussions on this problem will be dealt as follows:

Is the Lake Bottom a Source or Reservoir (Sink) of Pollution?

As an example for Caohai Bay of Dianchi Lake, we have the balance of TP, TN and COD in 1995–1996, according to the monitoring data[1] in Table 15.6. It shows that about 89–96 % TP, TN and COD are deposited at the bottom or partly released into air in a gaseous state. Therefore, Caohai Bay is a reservoir, but not a source of pollutants. Enriched sediments will renew the surface layer of dredged bottom rapidly when the pollutant loading cannot be improved.

There are two kinds of fluxes at the lake bottom: fluid flux and solid flux. The former is upward and the latter is usually downward, and the total fluxes of nutrients are downward. This creates the self-purification capacity of the lake.

Active Layer for Releasing Nutrients from Sediment

The profiles of the nutrient in sediments have a very different pattern. Samples in Caohai Bay are shown in Fig. 15.27. There is an enriched active surface layer covered the bottom. A thickness of 1-cm deposition in a day during a windstorm of 10 m/s was observed in Taihu Lake (PU et al. 1998c). The active layer in a lake is estimated to be a couple of cm. The high concentration of nutrients in the water at the bottom is formed mainly by decomposition of new detritus of organisms, but not by the old one. The nutrient flux may diffuse both upward and downward.

Lake Suwa is a famous lake in Japan with a surface area of 13.3 km^2 and mean depth of 4.1 m. Dredging had been started since 1979. A

Table 15.6 Balance of TP, TN and COD in Caohai Bay of Dianchi Lake during 1995-1996

Item	*TP*		*TN*		*COD*	
	t/a	*%*	*t/a*	*%*	*t/a*	*%*
Into Caohai B.	282	100	3 012	100	16 111	100
Into Dianchi L.	31.1	11.0	241.6	8.0	536.8	3.3
Into water	0.3	0.1	8.5	0.3	34.0	0.2
Deposited/into air*	250.6	**88.9**	2 761.9	**91.7**	15 540.2	**96.5**

* Some parts nitrogen may release into air in gaseous state.

[1]Kunming Monitoring Center of Environments, Kunming Institute of Environmental Sciences, 1997, Basic Data of pollution in Dianchi Lake.

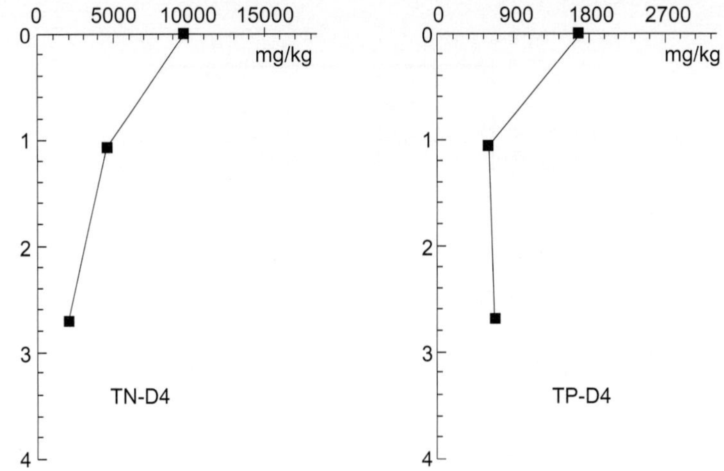

Fig. 15.27 Samples of profile of TN, TP with depth in Caohai Bay of Dianchi Lake, China (ZHKY, 1997)

comparison between the profiles of TP in sediment in 1979 and 1993 was made (Fig. 15.28) (OGIWARA, K, et al., 1995). The TP in the surface layer of dredged bottom decreased at the beginning stage after dredging, but it became the same or a little richer than that in undredged bottom. Therefore, from the point of long-term view, the role of dredging on nutrient distribution in sediment is not essential, when the low outer nutrient loading can be held back.

The Role of Newly Formed Pollutant Cloud of Fine Particles in Releasing Nutrients

It was found that the downward suspended substance flux has the value of 405 $g/m^2/d$ in the hypereutrophic Wuli Bay of Taihu Lake during July 19 to September 2, 1996. Among them, 3/4 were organic substances, which have the decomposition rate (*D*, 1/hr) at moment of *t* hr of (HU, et al. 1998)

$$D = 0.0066241 + 0.005859\text{EXP}\,(-0.0066241\ t). \tag{15.3}$$

About 80% of the organic substances were decomposed during one month. As opposed, the precipitation rate at the same site but inside the Physico-Ecological Engineering (PEEN), where ecosystem was improved and transparency was about 100 cm (three times more than outside PEEN), water quality was much better than that outside, at just 1/14 of that outside PEEN. (HU, et al. 1998). The big upward gradient of organic particles, especially in summer (Table 15.7) (ZHANG, et al., 1999) may lead to a big downward nutrient particle flux by diffusion.

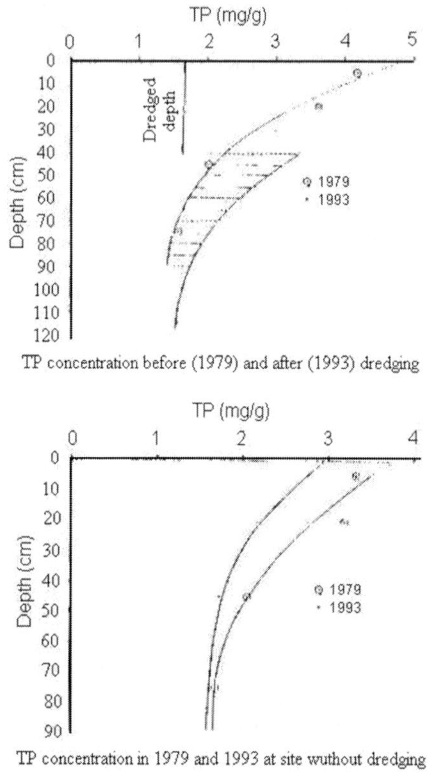

Fig. 15.28 Changes of TP in sediment before dredging (in 1979) and in 1993 at site with (upper) and without (down) in Lake Suwa, Japan (OGIWARA, K, *et al*, 1995)

Table 15.7 TOC content in SS and sediments in Taihu Lake in winter and summer (%) (ZHANG, et al. 1999)

Season	*SS at surface*	*SS at middle*	*SS at bottom*	*Surf. Sediment*	*2 cm below bott. surf.*
Winter	27.3	26.1	27.8	4.5	4.2
Summer	69.1	48.4	28.9	7.2	4.4

Three layers may be divided at the bottom into: the diluted "cloud", movable fluid mud, and consolidated mud. The diluted fractional "pollutant cloud" and movable fluid mud are easy to re-suspend, and are the main internal pollution source in lakes. They are rich in fine and small particles, which is much more rich than the sub-layer of consolidated mud (Van der Does, et al. 1992), (Figs. 15.29, 15.30; Table 15.7) (ZHANG,

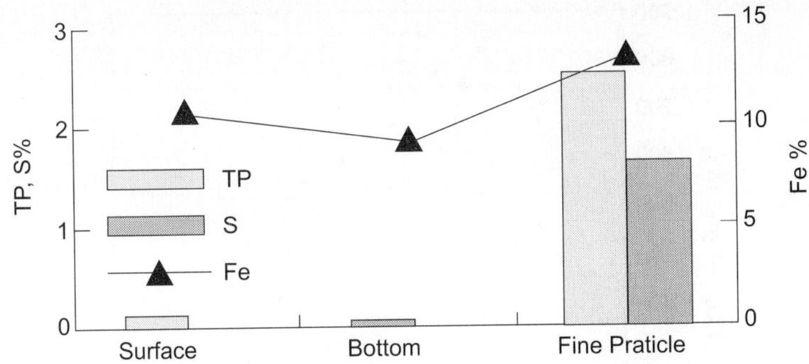

Fig. 15.29 Nutrients in SS in Taihu Lake (ZHANG, et al., 1999)

Fig. 15.30 SEM Photo of fine particle aggregates having fractional pattern (ZHANG, et al., 1999)

et al, 1999). The fine particles of scales of 4, 2, 1 μm have precipitation speed of 1.2, 0.3, and 0.09 m/d, separately (Horne, 1976). They make low transparency of 20-30 cm and giant pollution sources in lakes. They have fractional pattern, easy re-suspend, move place to place and could not be removed by usual dredging; moreover, they are easy renewed after dredging.

Short-term and Long-term Processes of Releasing Nutrients at Water-sediment Interface

The experiments of nutrient release from sediment continue for 3–5 days by using filtered lakewater under dark conditions. These data represent the physico-chemical process of nutrient release in the short term. When the experiments are extended to a longer term, the situation changed, especially when the bacteria also join in the process. The nutrient fluxes become downward, and self-purification processes go forward. Some samples are shown in Figs. 15.31 and 15.32. Some part of dissolved

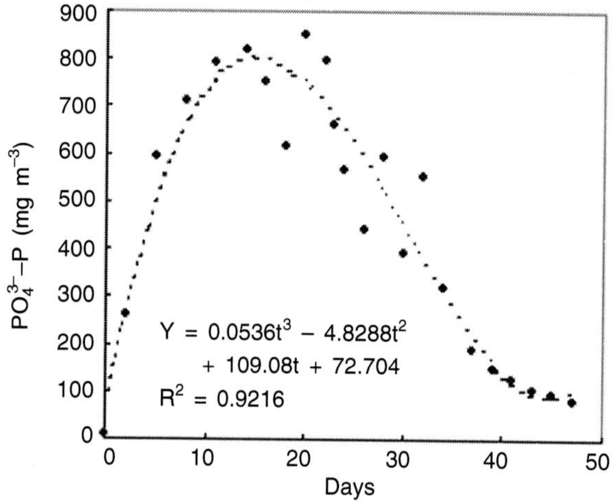

Fig. 15.31 Short-term and long-term changes of phosphorus release from sediment of east Taihu Lake under 25-30°C dark, with aeration condition (LI, et al., 1999)

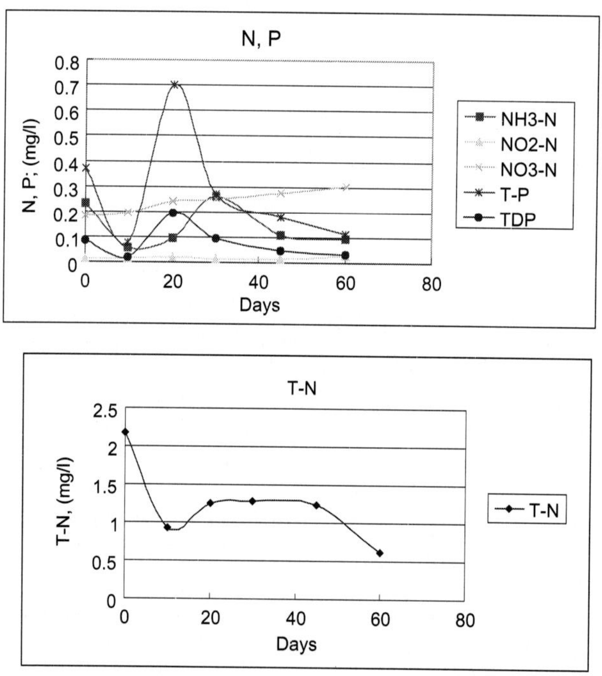

Fig. 15.32 N, P release from sediment of Xihu Lake, Hangzhou City, T = 30, pH = 8, with bacteria[1)]

[1]WU, Genfu, The role of bacteria and environmental elements on transformation of N, P in Xihu Lake, 1996

nutrients may be released into the air (nitrogen) and/or transported to the underlying sediment. The content of N and/or P in water decreased (LI, et al., 1999; WU, 1996)

The Effects of Dredging on Nutrients Release and Its Persistence

Usually, dredging the enriched sediment may decrease the release of nutrients, but not everywhere and everytime. Some samples have been shown in Table 15.8. The release of N and P increased after cutting 30 cm of surface layer of sediment in L. Kasmigaura, Japan (Fan et al. 1995).

Xuanwu Lake in Nanjing, with an area of 3.6 km^2 and mean depth of 1.2 m, was dredged totally in the beginning of 1998. The water in the lake dried up and then the surface layer of sediment was taken off for 30 cm (deposition during last 40-100 a). The renewed water was infused into the lake on March 6, 1998. The phosphorus release decreased one day after infusing the water, but increased to 12 $mg/m^2/d$ (Fig. 15.33), which was a little bigger than that measured in 1993—8 $mg/m^2/d$ (JIN, et al. 1995). It means that the persistence of dredging effects on the nutrient release is limited during first a couple of 10-days or months.

The Efficiency of Dredging on Water Quality Improvement Samples

The influence of dredging the sediment on the water quality improvements can be seen on Fig. 15.34. The SD, COD, TP show no changes and the TN decreased a little after dredging and before loading the settled clear water from the Changjiang River with retention time of

Table 15.8 Nutrient release from the dredged and undredged sludge, L. Kasumigaura, Japan, ($mg/m^2/d$) (Fan, et al., 1995)

Condition			*Aerobic*		*Anaerobic*	
Sampling date			*Feb.1*	*Feb.10*	*Feb.1*	*Feb.10*
PO_4-P		Undredged B	-	0.226	-	1.327
	Actual	Dredged A	0.381	-0.002	0.757	-0.096
		Undredged C	0.486	0.207	3.986	5.534
	Simulated	Dredged C*	-	0.247	-	0.016
NH_4-N		Undredged B	-	41.20	-	31.03
Actual		Dredged A	1.60	34.49	6.09	11.56
		Undredged C	10.02	13.90	41.26	53.37
	Simulated	Dredged C*	-	102.3	-	24.74

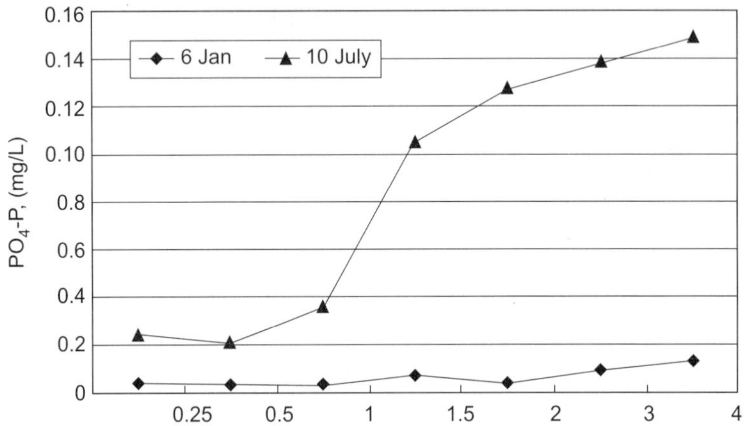

Fig. 15.33 Comparison of PO_4-P release from dredged bottom in Xuanwu Lake, Nanjing, for one day and seven months after dredge. The concentration of PO_4-P in tube with diameter of 30 m on the 4th day multiplied. 75 is the mean release of PO_4-P from sediment during four days in mg/m^2/d

70 days since October 1998. The wastewater loading into the lake was stopped. Every usual measure has been made for controlling eutrophication, but the transparency has still just round 35 cm. Therefore, dredging is not a sufficient condition for eutrophication controlling.

The other samples of two lakes in "Green Heart" of the Netherlands were shown in Fig. 15.35. The concentration of phosphorus in these two lakes decreased after decreasing the phosphorus loading with (L. Geerplas—P-loading decreased from 0.38 to 0.13 g/m^2/a) and without (L. Nieuwkoop—P loading dec-reased from 0.92 to 0.22 g/m^2/a) dredging (Van der Does, et al. 1992). The same results can be reached without dredging. Dredging is not an essential condition for decreasing nutrients in water.

Can we control eutrophication by dredging sediments? From the point of view of algae and phosphorus circulation in algae-type of eutrophic lakes

In comparison with the phosphorus content in the surface sediment, the algae have the similar content of (1–4) $\times 10^{-3}$ in their bodies. But the total value of phosphorus (if it would be decomposed and dissolved) in the water from all the algae bloom with concentration of biomass 5 mg/L (or 10^8 cells/L and Chl a about 400 μg/L, SD<50 cm) (ZHOU et al., 1996) has (2-4) $\times 10^{-9}$. The concentration of TP in water (Caohai Bay) is 6×10^{-7}. The TP at depth of 1 m in sediment is three times less than that in surface layer

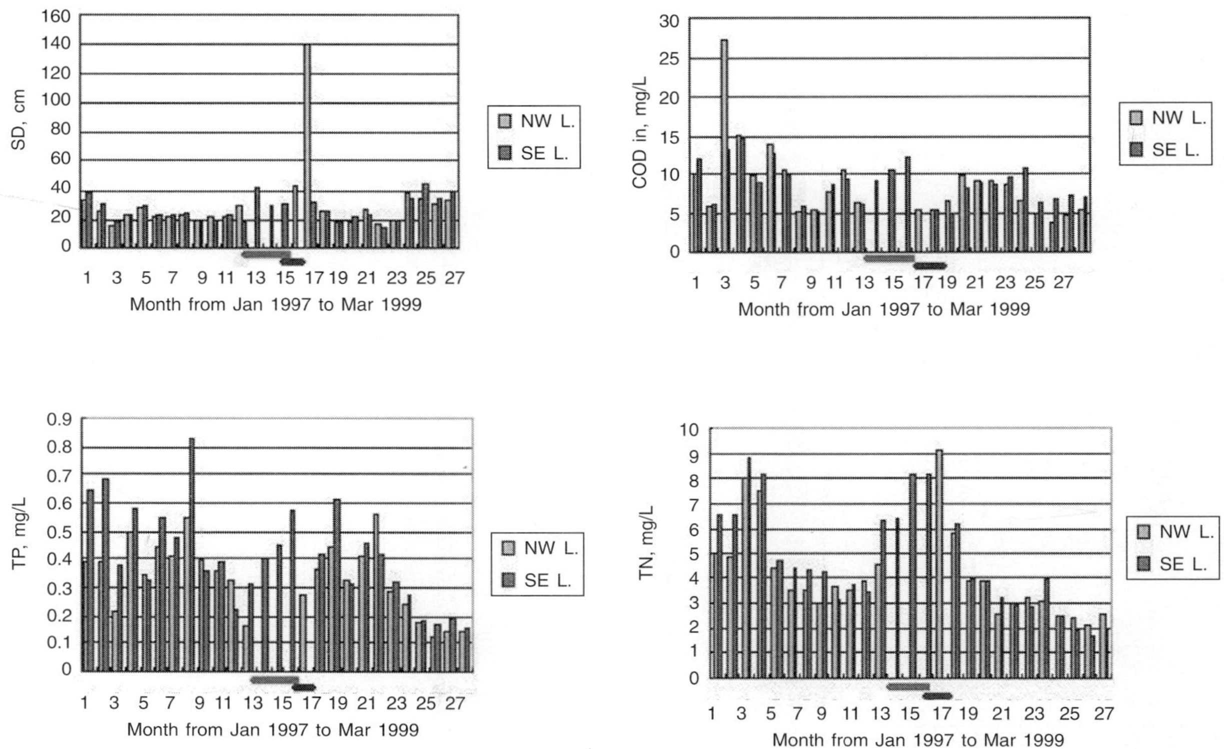

Fig. 15.34 Changes of SD, COD, TN and TP in Xuanwu Lake, Nanjing. Dredging was started in the beginning of 1998 and ended in the end of April 1998. Pumping up all the water in lake take off 30 cm sediments. Loading clear water with retention time of 70 days from the Changjiang River since October 1998

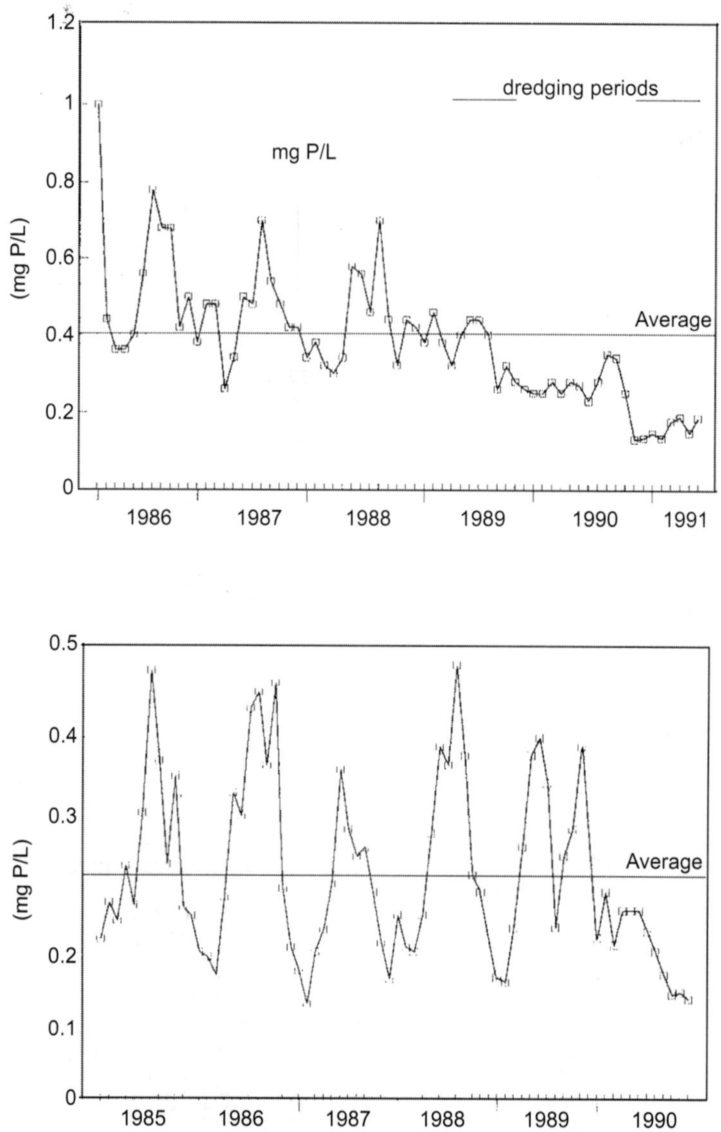

Fig. 15.35 Comparison between the effects with and without dredging on concentration of phosphorus in L. Geerplas (G) and L. Nieuwkoop (N), "Green Heart", Netherlands (Van der Does, et al., 1992)

of bottom sediment 0.6–1 × 10^{-3} (Fig. 15.27). All the factors can be summarized in Fig. 15.36. The horizontal coordinate is the concentration or content of TP in the logarithm scale. The differences between the TP in sediment and that in algae bloom are of 6 orders. It means that these two arguments have a complex non-linear indirect relationship. Dredging sediments can change the TP content of 1 order in surface bottom layer, but cannot be a key moment for controlling algae bloom.

A model of algae growth and phosphorus circulation in an eutrophic lake may be summarized in Fig. 15.37. The mean downward organic particle flux in the middle layer of the water body was estimated as 200 g/m^2/d (200 unit will be used in following) (HU, C. et al. 1998). Among them, the living algae content 170 units (among them 160 units remained resuspended to the upper layer, but 10 units died as detritus in the lower layer), the rest 30 units are organic detritus. The detritus of 40 units was divided into three parts: 32 units move upward to the upper layer, three units precipitated on the bottom (self-purification capacity), five units decomposed and move upward to the upper layer. Two units of detritus among the 32 units decomposed in the upper layer and join the photosynthesis of algae together with the eight units of dissolved nutrients in the upper layer (among them, five units was transported from the lower layer). Therefore, 10 units of new algae grow up in a day,

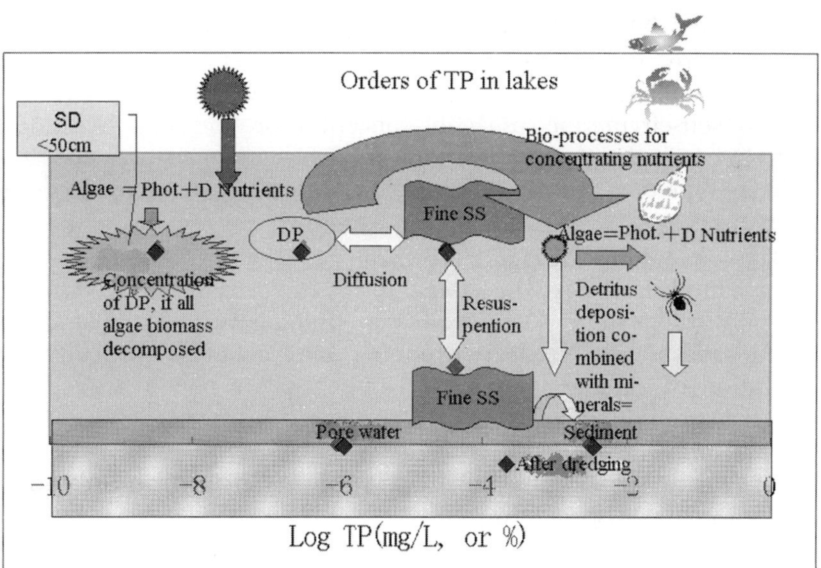

Fig. 15.36 Sketch of the orders of TP in the sediments, pore water, suspended clouds and water body, and their transportation and transformation in lake

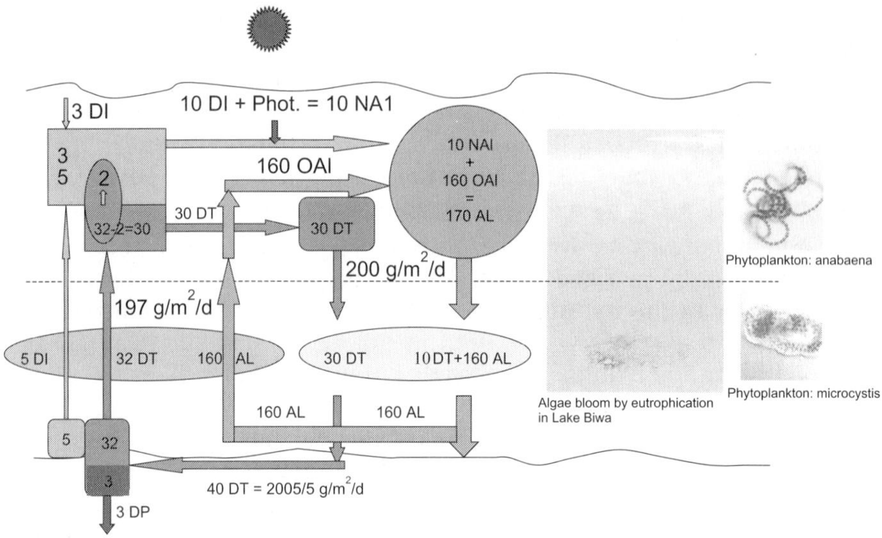

Fig. 15.37 Sketch of the algae growth and phosphorus circulation in an eutrophic lake

and join in the downward flux of 170 units of living algae. The total upward organic flux amounts to 197 units, while the downward one amounts of 200 units. Three units of dissolved nutrients were transformed into living algae and then died, became detritus, and precipitated at the bottom.

Therefore, algae grow up mostly in the upper layer and die and decompose mostly in the lower layer in eutrophic lakes. Intensive vertical mixing, self-circulation of high concentration of algae with low transparency and water quality–these are the basic conception models of the algae-type of an eutrophic lake and cannot be changed by dredging sediments.

The conventional dredging is difficulty to take the diluted cloud and movable mud efficiently. The latter may be recovered the dredged places, especially after a storm. Overall quick dredging may lead to destruction of the ecosystem; too deep dredging may increase difficulty for restoration of submerged plants. These possible negative effects of dredging should to be taken care of.

Water Quality—A Problem of Ecosystem Should be Solved by Ecological Approaches

Water quality indices are a part of the indexes of an aqua-ecosystem. To continually maintain an optimum constructed aqua-ecosystem, it is necessary to maintain the high quality in water. Therefore, to restore an optimum aqua-ecosystem is the aim and main approach for improving the hydro-environments. Many eutrophic lakes with bad water quality had very nice landscape with an abundant plant cover at the bottom and were the resources for drinking water in history. Changtang Lake in Taihu Lake Plain is an example that the water quality can be obviously improved (class II), if there is a vegetation cover at the bottom and a corresponding ecosystem. This lake has surface area about 90 km^2 and mean depth of 1.5 m. The loading water quality (class IV, V) is similar to the other lakes with low water quality in Taihu Lake Plain. The events of aquatic death in the delta exist occasionally, but water becomes clear some distance from the inlet mouth to the lake center. A similar situation is observed in East Taihu Lake. The macrophyte dominated (macrophyte type) lakes have much more self-purification capacity than algae-dominated (algae type) lakes.

The Physico-Ecological Engineering (PEEN) (PU, P. et al., 1993-1998b) limits the pollutant loading; the mass exchange between inside and outside water of PEEN neutrally restores the ecosystem in it with enhanced management by primarily using solar energy. One showed out the bioproducts from PEEN and intend to develop the Biological-Environmental Enterprise (BEE), so that economic benefit may be obtained during improvement of the environment.

Conclusion

In point of view of natural evolution, a lake is a reservoir (sink) of pollutants, but not a source of the same. A lake has giant self-purification capacity pushed by solar energy. Generally, the dissolved nutrient flux is upward, but the solid flux and the total flux are downward at the bottom. The movable active layer of fresh detritus and pollutant cloud having a fractional pattern will form soon after dredging, if the nutrient loading and ecosystem could not be obviously improved. They are the key contributors of nutrient release. Specially designed dredging facilities may take off the active layers and algae species, decrease resuspension of mud and nutrient release, and thereafter, improve water quality in a short period. As a measure of civil engineering, it may expand the channels and reservoirs, and take off (transport) heavy metal pollutants. In the long-term (months or seasons) meaning, dredging is not the sufficient and essential condition for eutrophication controlling.

Water quality—a problem of any ecosystem—should be solved by ecological approaches. In combination with pollution controlling in the catchment, the PEEN and BEE may restore the optimized aqua-ecosystem and historical landscape from the local to all the surface water systems step by step, mainly using solar energy. This is an approach for realizing the eco-environmental and socio-economical benefits with low cost inputs.

Physico-Ecological Engineering (PEEN) for Reconstruction Health Aqua-ecosystem and Improving Water Quality in Local Waters

There was a beautiful landscape with green mountains and clean waters all around in Dianchi Lake Basin, Taihu Lake Basin and many others in China. The surface waters in this basin deteriorated to the "desert waters" during the last decades, because of heavy nutrient loading. The nutrients loading to the surface waters could not be decreased to the historically low level in this area during the past decades. The standard for releasing the industrial wastewater in China on COD_{cr} is 100 mg/L, which is 12.5 times more than the surfacewater standard of grade III. The volume of fertilizer applied in agriculture in China accounts 30 million t/a, which occupies 1/3 of the world usage. The most released domestic wastewater could not be treated as necessary. The nutrient concentration of rainfall added by dry precipitation exceeds the criteria of eutrophication. For a long time, the pollution loading (PL) exceeded the self-purification capacity (SPL) of the waters. All these facts led to a rapid extension of serious eutrophic and hypereutrophic waters and to shrink the clean, and cleaner waters (Fig. 15.14). Therefore, the historical landscape of healthy aquaecosystem could not be restored easily.

How can we save lakes and rivers from progressive environmental deterioration?

To develop a series of most cost-effective and achievable technology for improving the environment in a local water area and improving the water quality for a greater areas is the key point of the new strategy.

The Physico-Ecological Engineering (PEEN, PU, P. et al., 1998) technology had been developed for the reconstruction of a healthy aqua-ecosystem and improvement of water quality in the surfacewaters from local to all the surfacewater systems in the catchment, step by step, mainly using solar energy since 1990. The main contents of PEEN are as follows:

Sufficient and Essential Conditions for Improving Water Quality

Water is the origin of the life. Healthy water is the necessary resource for human society. An ecosystem, which benefits the health of high organisms and human being, is a healthy ecosystem. A healthy ecosystem is the cradle of human civilization. Protection and reconstruction of a stable healthy aqua-ecosystem is the targets facing us.

Water quality is a problem of an aquaecosystem. For sustainable maintainance the good quality in water, it is essential to construct and maintain a correspondent ecosystem in it. Also, certain aqua-ecosystems will associate a correspondent water quality.

Therefore, the sufficient and essential conditions for improving water quality are stable healthy aqua-ecosystems.

The dynamics for maintaining an improving tendency of water quality under pollutants loading is the condition of

$$PL < SPC;$$

where PL–the condition of the pollutants loading; and SPC–Self-purification capacity (Figs. 15.14 and 15.15).

Therefore, the basic concept of the Physico-Ecological Engineering (PEEN) is to protect and reconstruct a healthy aqua-ecosystems in surface waters. The referring systems in use are the existing healthy aqua-ecosystems in history and/ or at the present of this region. The additional subsystems should be constructed according to the concrete conditions under the enhanced environmental pressure.

Disassemble the Complex System into Subsystems

The surfacewater systems in a lake basin, such as Dianchi Lake Basin and Taihu Lake Basin, are quite complex systems. A total solution to the issues of this kind of complex systems is impossible during the limited period with a limited capital. The subecosystems should be divided according to their function in different areas of the waters. For instance, there are drinking water source areas, natural swimming pond areas, mouth of the inflows, wetlands-bays, coastal zones, etc.

In addition, the complex hydrological regime also should to be considered, especially the extreme conditions of flood and drought, and the pollutants loading as well as. For example, we should take into consideration the flood current in the rivers and mouth regions. The reconstructed ecosystem should satisfy the requirement of water conservancy.

To satisfy the reconstruction the aqua-ecosystems in local areas with their specific targets, the soft wall technology was developed to separate the waters, to limit the exchange rate between waters, to filtrate the inflow to the specific areas, and to modify and introduce the current. It is obvious that the restored health aqua-ecosystem with a small area could not remain stable in the surrounding large polluted water badly, if there are no any barriers between the healthy aqua-ecosystem and the surrounding waters. The intensive mass exchange may lead to a strong mixing and destroy the healthy aqua-ecosystem. The larger the scale of the restored health aqua-ecosystem, the more stable this would be. In general, the soft wall technology is vital for reconstruction of the aquaecosystems. The details of the soft wall technology will be introduced in the following headings.

Dilution and/or Concentration

Water is a very good solution and carrier. Most substances may dissolve in it. Many substances, especially small particles, may suspend or re-suspend in it. The dissolved and suspended pollutants distribute from high concentration to lower concentration by dilution and diffusion processes; the eutropy increases in the system. These processes contribute to the purification for the high concentration waters, but contribute to the pollution for the low concentration waters. People use more and more clean water in daily use in relation with the economic development and increased urban living conditions. It means more and more pollutants loading into the water systems. One can improve the water quality in certain local waters with a high concentration of pollutants, but doing so distributes the pollutants to a wider water area. In principle, it is still a process of pollution. Therefore, this approach can be used as a temporary measure in the local area. In the long term and a wide area, dilution is not a good approach. Harnessing water quality is a process to concentrate and enrich the concentration of pollutants. It is opposite to the dilution process. PEEN emphasized on the concentration and enrichment of pollutants by using physico-biological measures. The phytoplankton, macrophytes transform the dissolved nutrients in the water into organic living things by using solar energy. That is the preliminary production in water, enriching the nutrients into living things. The followed food chain-web makes the system a more complex one. The nutrient concentration will decrease stably if the biomass of the aquatic production is removed from the water body, and the total nitrogen in the water released from the water partly in a gaseous state. The organism detritus would be stored in the bottom sediment, and most nutrients would be bound in the bottom sediment.

To reconstruct the macrophytes, especially the submerged plant, is the key point for decrease in the release of nutrients from the bottom sediment. The importance of the macrophytes vegetation is similar the forest on the land. It absorbs organic detritus, mineral particles, and decreases the exchange rate near the lake bottom.

As the preliminary producers, the phytoplankton and macrophytes have competition between themselves. The macrophytes clean the water with high transparency, but the phytoplankton makes the water turbid with a low transparency. We have found that in a close distance from the macrophytes stalks, roots, leaves, and stubs, the Most Possible Number (MPN) of nitrogen cycling bacteria is much higher than that in distance of more 20 cm away (Wang, G. et al. 1998). The nitration and denitration bacteria are rich in the boundary layer of the sediment-water interface originally at the bottom. When the macrophytes do exist in water, they are also rich in the boundary layer of the plant-water interface (Fig. 15.38). The suspended substances also contribute to the nitration and denitration as well. Therefore, the macrophytes promote the purification process in waters not only by their direct absorption and adsorption, but also by the competition with the phytoplankton and the associated bacteria purification process. The aqua-ecosystem with macrophytes domination has more productivity, stronger decomposition power, and more storage capacity for organic detritus and solid substances. The PEEN systems concentrate the substances in water; make water cleaner, more healthy, and add more biodiversity.

Soft Wall and Physico-biological Membrane Technology

To realize the reconstruction of a healthy aqua-ecosystem in a local specific area, it is essential to have a physical substance to separate the

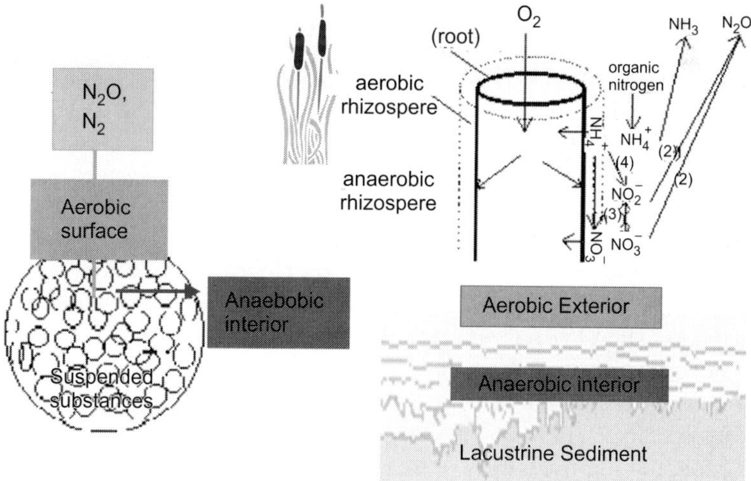

Fig. 15.38 Scheme of sediment-water interface; macrophytes-water interface; and suspended substance-water interface with aerobic and anaerobic conditions; and nitration and denitration processes

waters with different characteristics, and to control the exchange rate between water masses inside and outside the engineering. Because there are strong wind waves, and currents in lakes, any hard facilities with big section areas will suffer from the strong pressure. It may destroy the facilities, will be very expensive. Especially, the hard facilities cannot be removed easily, if it needs to be removed.

If the fetch length is more than 1 km, the wind driven wave will not be limited by fetch length in Taihu Lake, because the depth limitation for this shallow lake. The mean wind driven wave high H is related with wind speed and water depth h (Table 15.9) (WANG, Q. 1987; PU, Peimin, et al. 1993).

Table 15.9 Relationship between mean wind-driven wave high with wind speed and water depth *H*, when the fetch length is more then 1 km-2 km in Taihu Lake (WANG, Q. 1987; PU, Peimin, et al., 1993); wave high *h* in m

Wind speed, m/s	*Water depth H, m*		*Increase rate of wave energy for H = 4 m to H = 2 m, %*
	2	*4*	
10	1.00	1.31	72
15	1.69	2.15	62
18	2.08	2.77	77

The pressure P of wind-driven wave of mean wave high h and wave length of λ onto the hard vertical facilities perpendicular in front of the wave may estimated by the following formula:

$$P = 0.51h + 2.4h^2/\lambda; \qquad (15.4)$$

When $h = 1$ m or 2 m , $\lambda = 12$ m; the wave pressure P will be 0.72 t/m^2 or 1.83 t/m^2 , separately. It is quite big pressure, which may destroy the hard facilities in front of the wave.

The soft wall technology was developed after an experiment in summer 1991 was destroyed, when the depth of Taihu Lake doubled in comparison with the normal value, and the strong wind waves attacked the hard facilities for algae removal experiment. We did a special hydrodynamic experiment for optimizing the soft wall construction in shallow lakes. Then a serial real experiment was performed in Taihu Lake and its bay. The construction of soft wall technology was improved in relation with practical geographical conditions. The principle scheme was shown in Fig. 15.39. A high-tension poly-fiber cloth is used as the basic material for making the soft wall by sewing the combination of different kinds of cloth and belt. The width of the wall is determined by the possible water depth change, including wave height for the enclosures separating the water masses. It has a floating upper part and an anchored

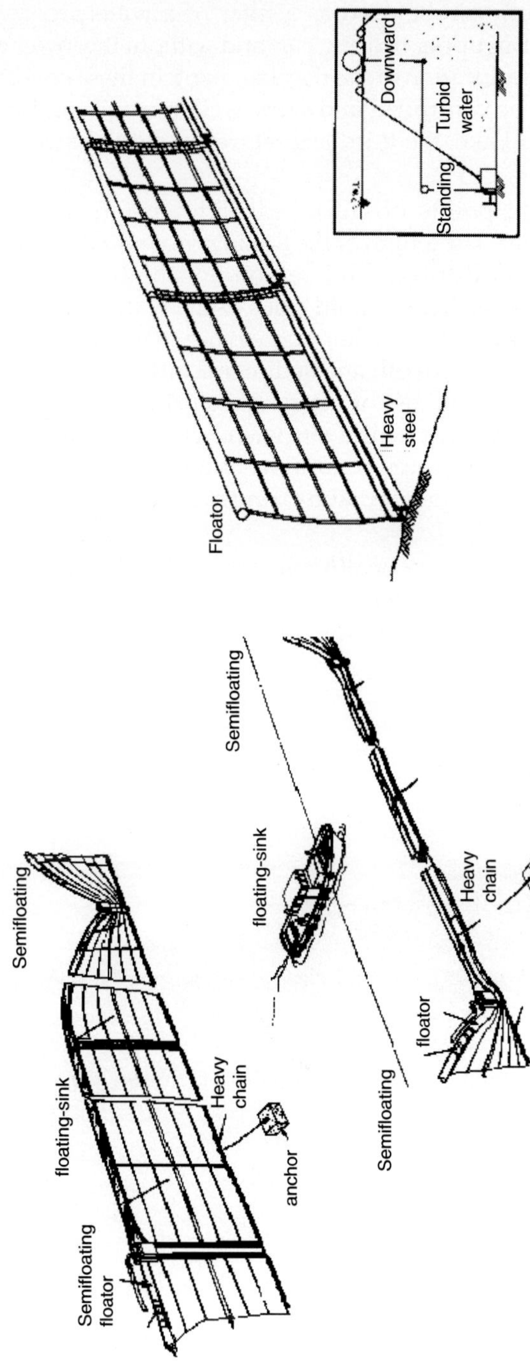

Fig. 15.39 Principle scheme of soft wall technology

lower part. The cloth can be used as a filter or a waterproof wall. The enclosures with just an upper floating part and without the lower anchored part can be used for allocation of the floating plant. In these conditions, the floating plants can be controlled and grow well in wave conditions. This was tested in Taihu Lake, the third largest freshwater lake in China (Fig. 15.40).

There are serial points of enhanced high-tension for linking the anchors at the bottom. The length of the linking rope should guarantee the water level variation during flood season and high wave cases. The anchor points should be kept in both sides of the wall for fixing the wall in the necessary position. The anchor method has many advantages: it can be used in different water depth and sediment conditions; it is easy to be constructed and moved (in case of rearrangement); and it is more feasible and adjustable in different sediment, wave, current and water depth conditions. For example, in the conditions of deep water, soft sediment, and big water level variation, the stub-fixing method is not a suitable one in practice.

For reconstruction of the health aqua-ecosystem in a local area, for example, in the drinking water source area, the soft wall enclosure is essential for protection of the aqua-ecosystem, like a skin for the living

Fig. 15.40 Soft wall technology for culture floating plant in open waters with strong wave conditions, Taihu Lake, China, July 10 1993

substance, or like the membrane for the cells. So, the soft wall enclosure may be named as a **physico-biological membrane** for the aqua-ecosystem.

The pattern of soft wall enclosure should be wave-shaped for decreasing the possible simultaneous attacking force on the wall in front of the wave and current. These soft walls may play a facility for introducing currents.

The soft wall enclosure stores a big volume of water, like a preliminary reservoir, keeping away the algae bloom outside of the engineering, and can avoid the peak value of nutrients and other pollutants. It is of great importance for avoiding the harmful effects on human health (Fig. 15.41). We can protect the drinking water supply area inside the engineering (Fig. 15.42) from the densely populated algae bloom outside the engineering in the upper part (Fig. 15.42).

Concept of "Falsework" and "Cabin Construction"

Obviously, it is impossible to reconstruct a healthy aqua-ecosystem in a big area instantly, especially in the super-degraded condition. A healthy aqua-ecosystem should contain abundant biodiversity and/or the favorite environments for high organisms. The submerged macrophyte is a very important supporting element for a healthy aqua-ecosystem in shallow lakes. There were abundant macrophytes in Dianchi Lake as a typical historical landscape. It guarantees high transparency and high quality of other elements. For reconstructing this kind of an aqua-ecosystem, we should guarantee high transparency, the seed and seedling and other necessary conditions at the beginning stage of reconstruction. Similarly, in the process of construction of a high building, the falsework is necessary for construction of the building. After the accomplishment of the building, the falsework should be backout.

A serial of enclosures with certain smaller scale (according to our experiments in Taihu Lake, the suitable scale is the hexagon with each side length of 100 m, and the enclosed area is 2.6 hm^2 = 0.026 km^2) need to be constructed as the "**falsework**" for reconstruction the health aqua-ecosystem in a large area, dividing the total large area into a serial area of smaller one. The total pattern may be designed according to the concrete natural conditions, such as dominated wind direction, etc. This kind of construction is stronger, and cheaper, which may be named as "honeycomb" (Fig. 15.43) construction-falsework. It provides the necessary conditions for achieving every step during the process for reconstruction in a limited period and area. For example, we should quickly improve the transparency for culturing submerged plant in a couple of days. It is possible only if the treated area is not so big.

Fig. 15.41 Soft wall technology as the physico-biological membranes for protecting the aqua-ecosystem in local areas, Taihu Lake, China, July 6 1994

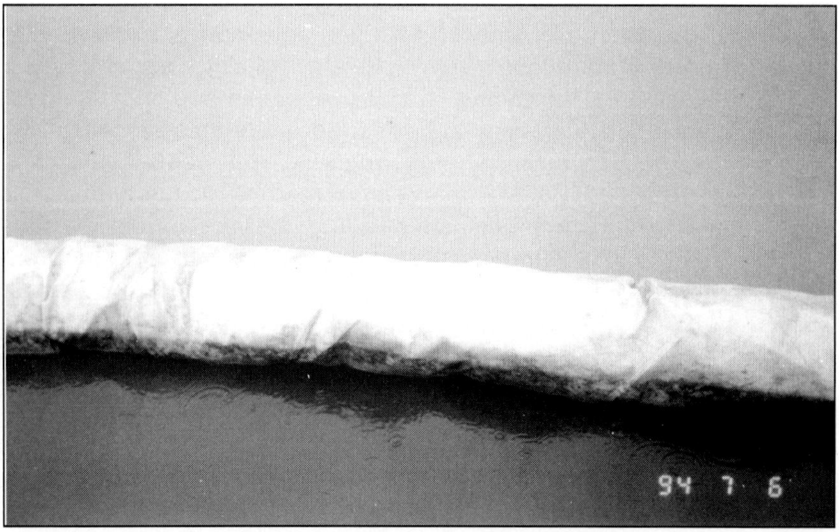

Fig. 15.42 Soft wall technology as the physico-biological membranes for protecting the aqua-ecosystem in local areas, Taihu Lake, China, July 6 1994; inside the engineering was clean water, outside was the densely populated algae bloom

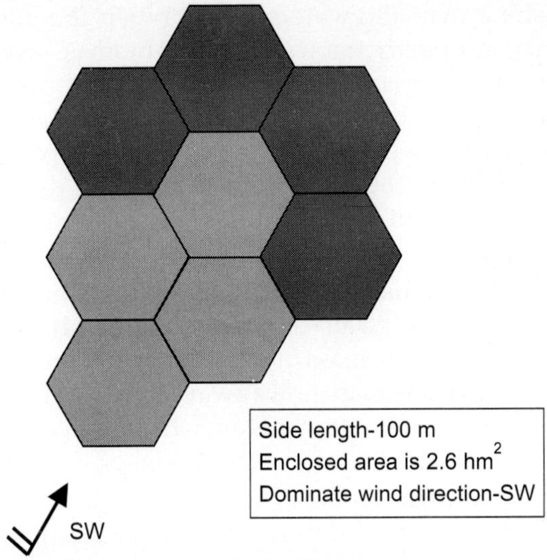

Fig. 15.43 Sample pattern of falsework for reconstruction of a healthy aqua-ecosystem in shallow lakes

Otherwise, the mass mixing process will destroy the water quality in the existing area with improved transparency. The similar measures can be accomplished during the same period, if we have the necessary capital and labor to do so. After the ecosystems in many enclosures are sustainably developed, we may take off the falsework.

For maintaining the reconstructed health aqua-ecosystem in the environments with high pollutants and high possibility of environmental hazards, it is necessary to have not just one enclosure, but also a serial subecosystem. Some of them with independent enclosures are also necessary to prevent akin the environmental hazards, and avoid the possibility of total destruction of the ecosystem. It is akin to a big ship, keeping the different separated cabins for safety. In case of destruction of one or some, but not every cabin, the ship can safely exist and restored. That is the conception of "**cabin construction**" for reconstruction of a sustainable aqua-ecosystem in high pressure of pollutants. Otherwise, there will be high possibility of danger in destruction of the restored ecosystem in case of environmental hazards. The system will suffer degradation gradually, and will not maintain and/or develop sustainably.

Technology for Reducing Algae Plume (TRAP)

In the open waters, phytoplankton might grow, when there are suitable conditions, such as nutrients in water, warm water temperature, and sufficient sunshine over the watersurface. When the algae grow, they absorb nutrients and purify the water. Algae biomass itself is the useful resource. If we can easily take it out from the water, we can purify the water in one hand, and avail of economic benefits in the other. The algae bloom-dominated *microcystis* may form a surface layer of densely populated algae mass, especially in the downward along wind direction bay. How economical is it to take out the algae mass is one of the key difficulties in harnessing the eutrophication.

A Technology **for Reducing Algae Plume (TRAP)** had been developed in combination with the soft wall technology (Fig. 15.44). The currents with algae mass transfer the algae mass into the "TRAP" along the soft wall, the water with much less algae mass flows away from the "TRAP"; the facility concentrate the density of phytoplankton, which then can be removed by pump. Figure 15.45 shows the concentrated algae.

As a fact, that the algae bloom in Dianchi Lake is a very serious issue. It appears more frequently and in more wide areas. In many cases, it is impossible to remove it in a short period during reconstruction of healthy ecosystem. It may appear in the enclosure, because of the inflow of algae mass from outside by some chances, or grow up in the restoration engineering. The algae bloom concentrates in the downward direction of wind direction near the enclosure. This is one kind of collection of algae

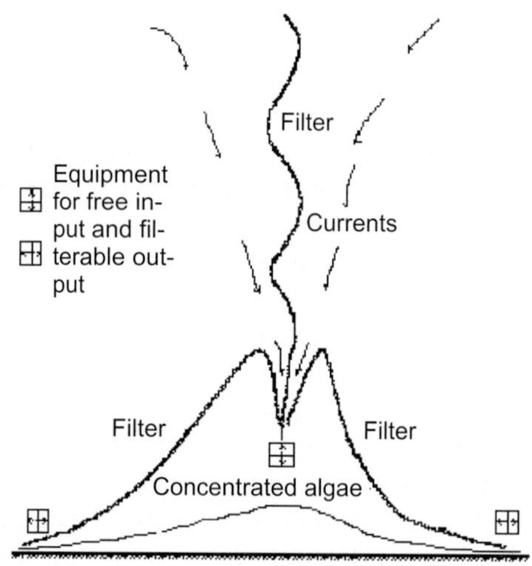

Fig. 15.44 Scheme of the Technology for Reducing Algae Plume (TRAP) in combination with the soft wall technology; the currents with algae mass transfer the algae mass into the "TRAP" along the soft wall, the water with much less algae mass flow away from the "TRAP"; the facility concentrate the density of phytoplankton, which then can be removed by pump

Fig. 15.45 Photo showing the concentrated algae mass by TRAP and soft wall technology in Taihu Lake, July 14, 1993

mass. We should enhance management to put down the floating facility in these area, leading the algae mass out the restoration construction and construct the soft wall to concentrate the algae mass in certain "Trash Can—enclosure for storing and treating algae mass " (treat the algae mass in the "Trash Can", and remove the algae toxicity). This is necessary because we cannot avoid the appearance of algae bloom in the reconstruction engineering completely. This is another important conception for successful reconstruction of a healthy aqua-ecosystem—the conception of the "**Trash Can**".

Technology for Improving Transparency and Removing Vegetarian Fish

Culturing the submerged plant is essential for restoring the aqua-ecosystem in shallow lakes, where there were abundant submerged plants and they disappeared mainly by insufficient light. The macrophytes release some info-chemical substances, which can keep within limits of the algae growth in water. The transparency is good enough for submerged plan growth. The better the growth state, the more efficient the release of info-chemical substances. In association with the increase of nutrients in water, and/or in case when some species of the submerged plants grow badly, the attached algae grow up on the stub of the submerged plant. The latter becomes weaker, its release effect weakens, and the attached algae and phytoplankton grow up more intensively. All the process generate a negative feedback cycle. The submerged macrophytes die and disappear finally. This kind of degradation of aqua-ecosystem has happened in history.

Therefore, for reconstruction of a healthy aqua-ecosystem, it is essential to improve transparency. Some species of floating plants can be used as the frontier plants for improving transparency in a short period. We have tested the floating plants, such as water hyacinths (*Eichhornia crassipes*), *Hydrocharis dubia, Alternanthera philoxeroides*. They can gradually improve the transparency. Even in winter, when the water temperature is lower then 5°C, or close to 0°C, the existing of floating *Alternanthera philoxeroides* can improve transparency. The existing of plant can improve transparency by different effects: absorption, adsorption, and symbiotic bacteria. In winter, the bacteria play a major role in improving transparency.

The physical (electrical-pulse) approaches for improving transparency have been tested in laboratory and field. The effects are obvious (Figs. 15.46 and 15.47).

The grasstivore fish may endanger the newly cultured plants. Therefore, to restore the historical landscape with abundant macrophytes in shallow lakes, to remove the grasstivore fish is also an essential measure. The traditional fish trap net method (Fig. 15.48) and fish net expansion method might be used to take off the grasstivore fish in the enclosures for restoration ecosystem.

Mosaic Technology and Macrobionics

The abundant biodiversity provide a sustainable health aquaecosystem. The seasonal succession of macrophytes is a natural process. Some species of plant grow in winter; some others grow in summer. When the plant is getting decay, its purification capacity is weak; even it can act as a source of secondary pollution. The other growing plant should to support the ecosystem. Therefore, the mosaic macrophyte system is essential for reconstruction of the stable health aquaecosystem.

As an example, fifteen macrophytes were introduced in Taihu Lake experiments. All the population was assembled in terms of their ecological characteristics ZHANG Shenzhao, et al. 1998. The main ecological characteristics of 15 macrophytes are showed in Tab. 15.10. Some species of aquatic plants cultured in PEEN were shown in Fig. 15.49.

Fig. 15.46 Experiment for removing algae and improving transparency in laboratory, July. 15, 2001; the water after electric-pulse treatment becomes clean and the algae cells floating to water surface (cup in center), in comparison with the water without any treatment (left cup), and other treatment of certain bacteria (right cup)

Fig. 15.47 Experiment for removing algae and improving transparency in Mochou Lake, Nanjing, China, Nov. 6, 2001

Fig. 15.48 Scheme of the fish trapper net; the traditional measure for trapping fish (Given by ZHU Xinhua)

Table 15.10 Main ecological characteristics of introduced macrophytes in Taihu Lake PEEN experiments; (ZHANG Shenzhao et al., 1998)

No. and Species name	*Growth period*	*Growth temperature*	*Modes of reproduction*	*Ecotype*
1. Eichhornia crassipes	May to Nov.	20-35	Axillary bud	Floating
2. *Alternanthera philoxeroides*	Mar. to Dec.	20-30	Axillary bud	Floating
3. *Hydrocharis dubia*	June to Nov.	20-30	Creeper	Floating
4. *Trapa maximowiczii*	Apr. to Nov.	20-30	Seed	Leaves-floating
5. *T. Bicornis*	Apr. to Nov.	20-30	Seed	Leaves-floating
6. *Nymphoides peltata*	Apr. to Nov.	15-25	Rhizome	Leaves-floating
7. *N. Indica*	Apr. to Nov.	15-25	Rhizome	Leaves-floating
8. Potamogeton maackinus	Mar. to Nov.	20-30	Axillary bud	Submerged
9. *P. malaianus*	Apr. to Nov.	20-30	Rhizome	Submerged
10. *P. crispus*	Oct. to June	10-20	Winter bud	Submerged
11. *Myriophyllum spicatum*	Mar. to Nov.	15-30	Root tillering	Submerged
12. *Vallisneria spiralis*	Apr. to Nov.	20-30	Corm	Submerged
13. *Hydrilla verticillata*	May to Nov.	20-30	Scaly bud	Submerged
14. *Ctratophyllum demersum*	Apr. to Nov.	20-30	Winter bud	Submerged
15. *Elodea muttallii*	Evergreen	15-30	Axillary bud	Submerged

The vegetation at the bottom sediment not only can keep within limits of the algae growth, but also keep within the release of phosphorus from sediment as well by stabilization of sediment, decrease mixing coefficient near sediment, trap the precipitated detritus, and direct absorption of nutrients. The total decreasing rate of bottom vegetation for phosphorus was estimated as 12 % to 73 % (Reddy and DeBusk, 1987). Therefore, to restore the submerged plant plays a key role for reconstruction of health aquaecosystem.

After the transparency has improved in the shallow lake, the attached algae may grow up at the bottom sediment, if there are no cultured macrophytes. They also may purify the water and improve water transparency. The attached algae also may grow up on the stub of macrophytes, such as on the *Elodea muttallii* in May, when the water temperature increases and *Elodea muttallii* does not grow very well (Fig. 15.50), or on the periphery stub of the macrophyte community (Fig. 15.51). The mixing culture of *Elodea muttallii* and *Trapa maximowiczii* may avoid the growth of attached algae on *Elodea muttallii* (Fig. 15.49).

Except the macrophytes mosaic technology, the combination of mollusks and other organisms-detritustivore are of importance to be

Eichhornia crassipes

Hydrocharis dubia

Elodea muttallii

Ceratpphyllum demersum

Trapa maximowiczii + Elodea muttallil

Nymphoides peltata + Elodea muttallii

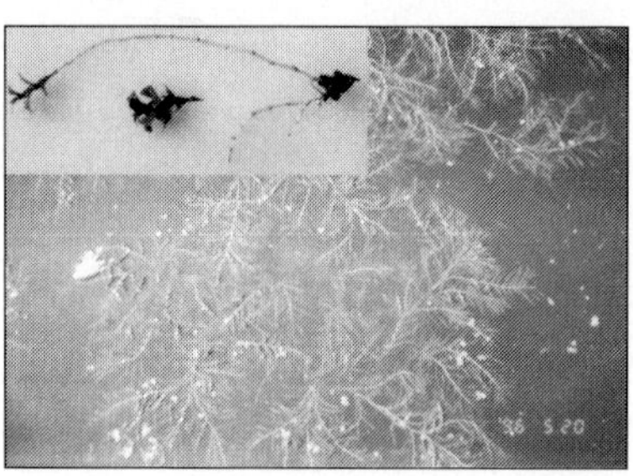

Potrmogeton crispus

Hydrilla varticillata

Fig. 15.49 Some species of aquatic plants cultured in PEEN for purifying the water quality (ZHANG Shenzhao et al., 1998)

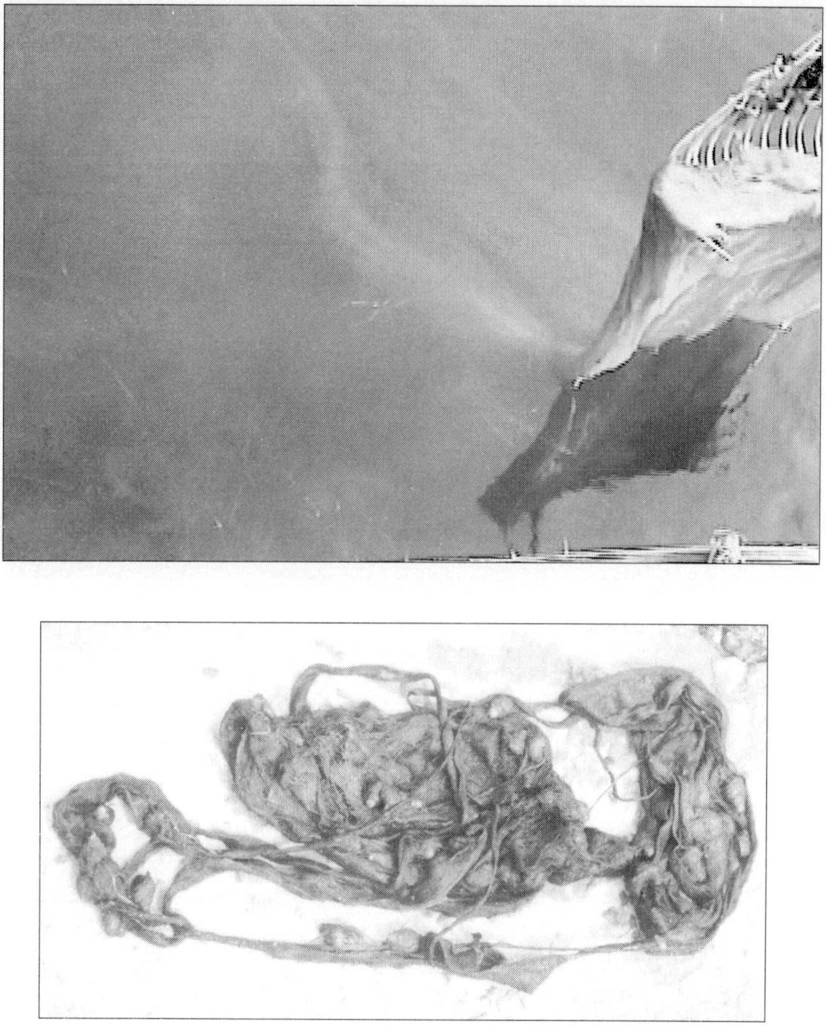

Fig. 15.50 Attached algae may grow up at the bottom sediment, when the transparency has improved but no cultured submerged macrophyte (above); and the snails are living together with the attached algae, Mochou Lake, May 4, 2001

cultured in the aquaecosystem to support the stable maintenance of the ecosystem.

The optimum structure of the aqua-ecosystem should be found in the nature, in the historical landscape and in the existing health ecosystems. This principle might be named as **Macrobionics**.

Fig. 15.51 Attached algae may grow up on the stub of macrophyte, such as *Elodea muttallii*, when the submerged macrophyte was weakly growing or on the periphery of its community, Mochou Lake, May 5, 2001.

Technology of Immobilized Nitrogen Cycle Bacteria (INCB)

The nitrogen cycle bacteria (NCB) are essential for decomposition of organic detritus and purifying the water quality. A technology of Immobilized nitrogen cycle bacteria (INCB) with the specially made carriers was developed (LI, et al. 2000). By using floating macrophytes and INCB technology, a new approach for purifying the water quality and controlling eutrophication was developed and tested in Hongfeng Lake, China. The warm water released from an electric power plant to the lake is used for culture fish with high density. A Physico-Ecological Engineering (PEEN) support for purifying the wastewater of the fishponds was constructed in September 1999.

INCB might be cultured in the laboratory under optimum conditions and set in natural lakes (reservoirs) as the artificial sources with high concentration of NCB. The macrophyte communities and the bottom sediments are the secondary carriers-centers of NCB, which might be transported and diffused to wider water bodies. The suspended particles are the third carriers with enriched NCB. During mineralization, the organic phosphor would be decomposed, transforming inorganic phosphate, which might be absorbed, adsorbed, or/and precipitated to the bottom. The self-purification capacity for nutrients has been increasing, obviously in wider space during longer and longer periods in lakes (reservoirs) since the engineering.

The combined bacteria housing in the carriers are cultured in special conditions. Then half of them are set in the net-boxes in the surfacewater of the purifying pond, and the other half part are cultured still in laboratory, 1/4 of which are replaced every week to the fields in turn.

It was observed that the most probable number (MPN) of the nitrogen cycle bacteria in the water of PEEN site is about 2-3 order-of-magnitudes higher than those in the water of inflow to PEEN site. The Log (MPN) at lower site of 2 km far from PEEN is about 1.5 order more than that of in inflow to PEEN. The MPNs of nitrogen cycle bacteria around the macrophyte are of 1.5 order-of-magnitudes more than that in open waters without macrophyte, especially for ammonifier, nitrosobacteria and denitrifier.

The nitrogen cycle bacteria need both aerobic and anaerobic conditions, alternately. This kind of condition can exist in a small space near the water-sediment interface, the water-macrophytes interface and the water-suspended particles interface. The floating macrophyte, like a cover on the water surface, increases the ammonifier, nitrosofier and denitrifier (Wang, G. et al. 1999). Then, the nitrogen cycle bacteria operate actively in these interfaces. The floating macrophyte and suspended particles extend the mineralization and denitrification processes from the bottom interface to entire water body. The macrophyte and INCB increase the MPNs of nitrogen cycle bacteria in the water body exponentially. Simultaneously, the organic phosphor would be decomposed during mineralization, transforming inorganic phosphate, which might be absorbed, adsorbed, or/and precipitated to the bottom. All these lead to enhance the self-purification capacity remarkably.

The removal rate of INCB for TN, COD and NH_4^- reaches 75 %–90 % with retention time more than eight days (Fig. 15.52). The possibility of water quality over the marks of state standard of most important elements for drinking water source, such as nitrite and non-ion ammonia, Chl *a*, etc., was 39%, but it was zero after the PEEN during the last two years. The lake area around the experiment site becomes the center of improvement of water quality. The most possible number (MPN) of NCB in the immobilized carriers is about 8-9 orders more then in the water of the usual natural condition, and the MPN of NCB round the roots of water hyacinths in natural waters is one order more than that in waters without water hyacinths, and the former is one order less than that round the roots of water hyacinths in waters with artificial INCB. The warmed recycle cooling water from the electric power plant promotes the culture

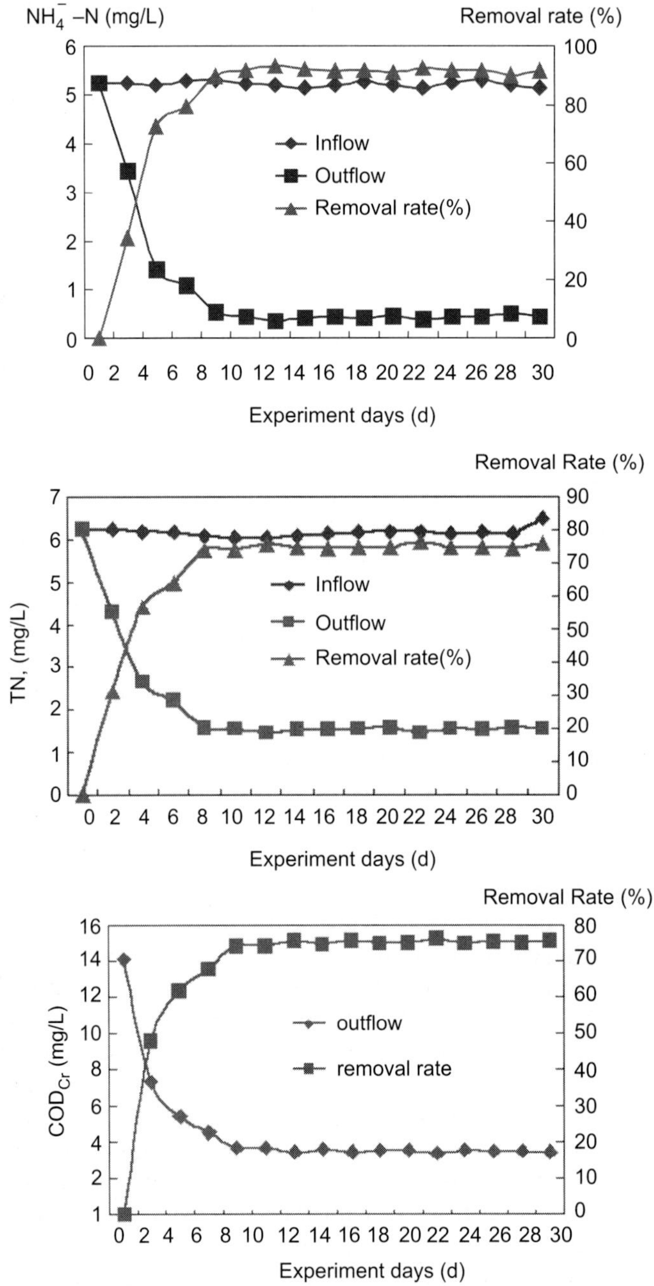

Fig. 15.52 Efficiency of INCB for removing ammonia nitrogen, total nitrogen, and COD_{Cr}

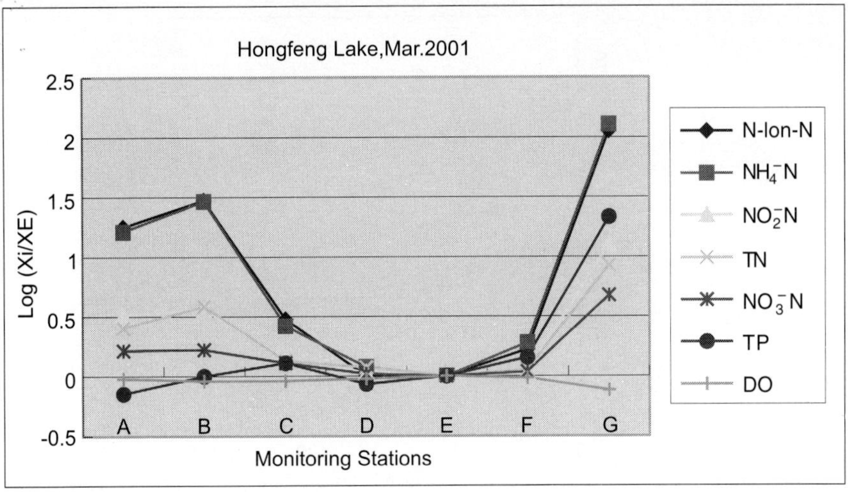

Fig. 15.53 Diffusion effects of PEEN on improvement of water quality in Hongfeng Lake; The point E with the best water quality is the experiment site, others are in distance from E, March, 2001

and distribution of the NCB in Hongfeng Lake. The self-purification of the lake increases, and the water quality of the lake improves. The PEEN experiment site E become the center of the best water quality. An example for March 2001 is shown in Fig. 15.53, which shows the diffusion effects of PEEN for purifying water quality.

Role of molluscs

The snail is an attached algae grazer. It can also release some chemical substances as a biocoagulator, which cleans the water. Some experiments have been done on the purification effect of snail on water quality (WEI Yangchun, et al., 1999). The turbid water with snails at the bottom can effectively clean the water during one hour, while the parallel water could not clean up for more than eight hours (Fig. 15.54).

Zebra mussel (*Dreissena polymorpha*) was rapidly developed during 1992-1993, and reached the density of 1000-4000 idv./m^2, which led to renewal of the water once in 0.8-5 d in Saginaw Bay, Lake Huron. The inner Bay of Saginaw has a surface area of 1 580 km^2, mean depth of 5 m, volume of 79×10^8 m^3, and resident time of 120 d. The water quality changed rapidly. The Chlorophyll a decreased 59-66%; the total phosphorus decreased 43-50%; the transparency (Secchi disk depth) increased 60-100%; other parameters, such as particle nutrients also decreased obviously (Fahnenstiel, G.L. et al., 1995). The molluscs may

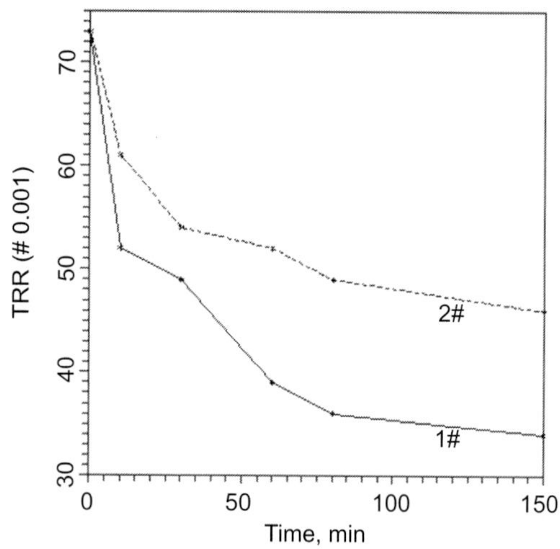

Fig. 15.54 Changing tendency of eliminating rate of transmission light in water with snails (1#) and without snails as the parallel (WEI, Yangchun, et al., 1999)

obviously improve the water transparency, if the total water change rate by molluscs (the time, during which the filtrated water by mollusks in a column equals the water volume in the column) close to the time for doubling the biomass of phytoplankton by its growth.

The similar results were recorded in the eutrophication shallow lakes in west part of Lake Erie Basin. There, the Chlorophyll a decreased 43%; the Secchi disk depth increased 85-100%, after the abundant growth of zebra mussels.

The big lateen mussel has filtration rate as 10 liters per day. Therefore, for filtering the whole water column of 1 square meter and 3 m depth during three days, it needs about 100 lateen mussels/m^2.

According to these data, the molluscs: snails, mussels may improve water quality gradually. We may use the aboriginal species mollusks to purify the water in Dianchi Lake and other lakes in China. The mollusks are the favorite foodstuff for Chinese people. They are good natural resources, which were rich in the history in Dianchi. In addition, for example, the lateen mussel can be used for production of pearl.

Ecological Floating Island

In many cases, the floating plant can be used as a frontier plant for improving the water quality under high environmental pressure. Not only the floating aquatic plant, but also the land-plant/flowers can be used for purifying the water quality in different seasons and for beautification of the watersurface in gardens, parks, or other sightseeing places. The root system of the floating plant is the carrier for mineralization bacteria, which can play a great role in the purification of the water quality. The floating ecological islands technology is very important for controlling eutrophication in reservoirs, where the water level changes rapidly during the year. The widely variation of water depth limits the growth of submerged plant in most parts of the reservoir. Therefore, the floating ecological island technology can find their wide application in reservoirs and in city parks for purifying and beautifying the waters.

Different pattern of Floating Ecological Islands (FEI) were developed during last years. The soft frame- or hard frame-floating islands were then designed. They have the circle, the squareness, the rectangle, the hexagon, the octagon, etc., patterns (Fig. 15.55). The wind-driven rotating FEI—floating ecological islands—were designed for using wind energy to drive the rotation of the island and create water current in ponds, increasing the effects of ecological island on purification of the water quality (Fig. 15.56).

Prevent plant diseases and insect pests by biodiversity

For preventing the plant diseases and insect pests, we cannot use any chemical pesticide in an aqua-ecosystem. Therefore, the physical (management), and biological approaches were tested in practice. The biodiversity is most important in the prevention of the distribution of possible diseases. The diseases of aphides for floating and leaf-floating plant are common. One of the tested measures is shown in Fig. 15.57. The vetiver grass (*Vetiveria zizaniodes*) and *Alternanthera philoxeroides* were mixed and cultured for preventing plant diseases and insect pests in the Physico-Ecological Engineering—PEEN experiment in Wulihu Bay, Taihu Lake during 1995-1999.

5M key

In conclusion, the Physico-Ecological Engineering (PEEN, PU, P. et al., 1998) technology developed during last 10 and more years, includes the 5 M-key taches for reconstruction of a healthy aqua-ecosystem. The 5 M are: abbreviation of Macrophyte, Microorganism, Mollusc, Macrobionics, and Management. The central tache is to enhance the Self Purification Capacity—SPC and the capacity for resisting the possible environmental hazards, even though in local areas, mainly by using the present solar energy. Both the macrophytes and microorganism (for example, the nitrogen cycle bacteria, Fig. 15.52) might increase SPC

Fig. 15.55 Some samples of Floating Ecological Islands (FEI) and their effects for purifying the water quality, in Hongfeng Lake (upper left), Taihu Lake, and Mochou Lake (upper right), China

greatly. Macrobionics is very important for construction of the optimum aquaecosystem. Similarly, the management as the human activity is the most important element for construction of the enhanced optimum aquaecosystem.

Fig. 15.56 Wind-driven rotating FEI in Mochou Lake, Nanjing, China, 2001

Fig. 15.57 The vetiver grass (*Vetiveria zizaniodes*) and *Alternanthera philoxeroides* were mixed and cultured for preventing plant diseases and insect pests, Wulihu Bay, Taihu Lake, May 22, 1998

How Can We Control Eutrophication in Dianchi Lake?

If we can keep the concentration of TN and TP on the level of eutrophication standard, namely 0.2 mg/L and 0.02 mg/L, the contents of TN and TP in Dianchi Lake just have 258 t and 25.8 t, and the annual outputs of TN and TP by outflow just have 80 t and 8 t, respectively. The recent annual loading into Dianchi Lake for TN and TP is about 9 000 t and 500-1000 t, separately, which are 35-19 times for the contents in lake and 112-62 times for the annual outputs by outflow. Therefore, we face an extremely difficult task in controlling the degree of eutrophication of Dianchi Lake.

L. Okeechobee in Florida, USA, has water surface area of 1,730 km^2, mean depth of 2.7 m, catchment area of 12,394 km^2, and population density of 42.2 capita/km^2. The research results showed that there is no correlation between the annual mean TP concentration and annual TP load and water level; no obvious effect for decreasing TP concentration in lake after 40% decreasing of TP outer load, no correlation between annual algae bloom frequency and annual mean TP concentration, and the settling rate of TP was decreasing according to the exponent function with time. Taking these criterions for controlling eutrophication of TP < 40 ppb, and N/P = 30 : 1, the US EPA model for controlling eutrophication gave the results that it needs 200 years (Blasland, 2001). The situation of pollution loading into lakes in China is much more serious than that in L. Okeechobee. The population density in Taihu Lake Basin is 1 079 capita/km^2, and in Dianchi Lake is 720 capita/km^2, which is 25, and 17 times more than that in L. Okeechobee, where it has only 42.2 capita/km^2. Therefore, we can image the difficulty in controlling eutrophication in Dianchi Lake. We should develop the new strategy and approaches for controlling the eutrophication in Dianchi Lake.

The ratio of TN/TP in water of Dianchi Lake was about 12-18 (Table 15.5). The target for controlling eutrophication the ratio TN/TP is about 10 or more. The ratio of TN/TP in water hyacinths is about 6. Therefore, to satisfy the conditions of controlling eutrophication in Dianchi Lake, we can estimate the conditions for controlling the TN, because in that case, the condition for TP would be satisfied without a doubt.

According to the observed data, the annual TN balance and controlling eutrophication in Dianchi Lake by PEEN and BEE may be presented in Fig. 15.58, where PEEN—Physico-Ecological Engineering, including Immobilized Nitrogen Cycle Bacteria (INCB); BEE—Bio-Environmental Enterprise; Number in percentage in basis of the TN annual input of 9000 t/a, the volume of Dianchi Lake of 12.9×10^8 m^3, annual outflow of 4×10^8 m^3, the present TN concentration of 1.8 mg/L, the target TN concentration of 0.2 mg/L. The present relative Self-Purification Capacity (SPC) is 0.92 (among them deposit on bottom occupies 80%, while release to air occupies 20%), the reinforced SPC is 1.23 (among them deposit on bottom occupies 80%, while release to air occupies 20%), the reinforced SPC of 1.23 (the additional 0.23 comes from removing the N contents in the original lake water, decreasing the content of TN in Dianchi Lake) by using PEEN (including INCB) and BEE. It means that PEEN and BEE take off the TN in a year more than the annual TN loading into Dianchi Lake, purifying the original water, decreasing the TN concentration to meet the standard for controlling eutrophication and algae bloom in Dianchi Lake[2].

These results can be summarized in Table 15.11.

Water is a good solvent. Purifying water is a process of concentration of pollutants dissolved and/or suspended in water. The most effective

[2]Here the output of nutrients by fishery is omitted. According to the data in 1988, the TN of 157 t/a, and 12 t/a; the TP of 41 t/a and 49 t/a were taken out from Dianchi Lake by producing of 6000 t/a fish and 2750 t shrimp, respectively, in 1988 (Kunming Institute of Environment Sciences, 1992: 92). In the last years, the fish and shrimp products were much less than that in 1988.

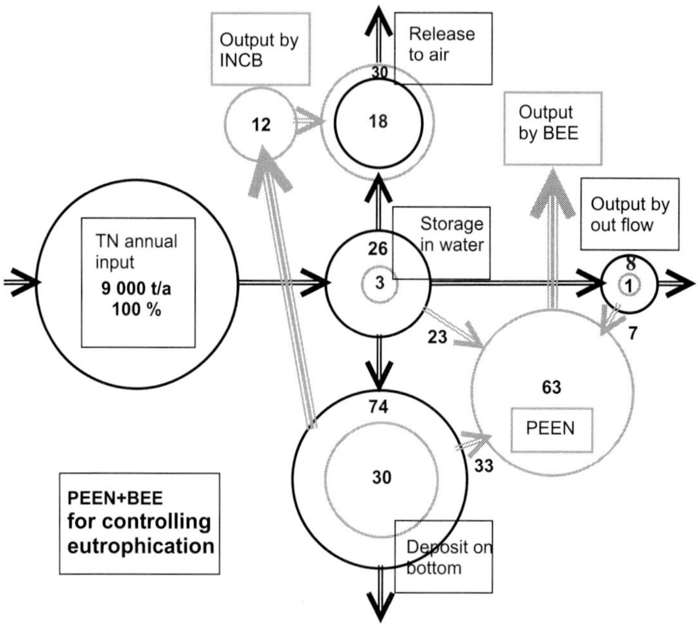

Fig. 15.58 Annual TN balance and controlling eutrophication in Dianchi Lake by PEEN and BEE. PEEN—Physico-Ecological ENgineering, including Immobilized Nitrogen Cycle Bacteria (INCB); BEE—Bio-Environmental Enterprise; Number in percentage in basis of the TN annual input of 9000 t/a, the volume of Dianchi Lake of 12.9×10^8 m^3, outflow of 4×10^8 m^3, the present TN concentration of 1.8 mg/L, the target TN concentration of 0.2 mg/L; The present relative Self-Purification Capacity (SPC) of 0.92 (among them deposit on bottom occupies 80%, while release to air occupies 20%), the reinforced SPC of 1.23 by using PEEN (including INCB) and BEE

Table 15.11 Annual TN balance and controlling eutrophication in Dianchi Lake by PEEN and BEE in relative units. The present TN annual input is 9000 t/a in Dianchi Lake

	Input	*Dep.on bot.*	*Rel. to air*	*Outflow*	*By BEE*	*Stor. in water*
Present	100	74	18	8	0	26
By PEEN	100	30	30	1	63	3
Difference	0	-44	+12	-7	+63	-23

and perspective way is to reinforce the self-purification capacity of the lake, transforming the dissolved abundant nutrients into usable bioresources by using reconstruction of aqua-ecosystems and solar radiation incident onto the wide water surface, from where we might get economic benefits. It means that the new concepts of improving water quality are based on concentration of pollutants, but not dilution of pollutants. There is a SPC in natural waters driven by solar energy.

Dianchi Lake has water surface area of 298 km^2. If we might increase the SPC in the entire water body of Dianchi Lake mainly by using solar energy, transform the nutrients into usable bioresources and take them off from water body, convert them into gases, release them into the atmosphere, and store them at the bottom, we may control eutrophication effectively. When the SPC would be larger than the pollutant loading, the water quality will improve step by step. We may conserve health water resources. In the meanwhile, if we can get economic benefits from bioresources and form Bio-Environmental Enterprise (BEE), it will be much easier for controlling eutrophication–controlling eutrophication with economic benefits. Degradation of surface waters develops from the local area to larger and larger areas. We can remediate aqua-ecosystem and improve water quality also from local to larger and larger areas step by step. The basic concepts for controlling TN were presented in Fig. 15.58 and Table 15.11.

As Table 15.11, proves, for controlling eutrophication in Dianchi Lake, we should take off nitrogen of $9\,000 \times 63\,\% = 5670$ t TN in a year. We should culture water hyacinths (WH) on the water surface of 24 km^2 (8% of the total water surface of Dianchi Lake) during 300 days in a year in combination with INCB technology. There would be grow up 12 000 t wet WH/d. We should make necessary processing for fresh WH and get profits from that. In this case, we may remove also 936 t P/a directly, which is the same order of annual loading of TP. The settling rate will increase by enhanced bacteria process, and may lead decreased concentration of TP to meet the necessary concentration as low as 0.02 mg/L. If necessary, we may culture WH in more water surface to meet the big nutrients loading. Therefore, we may reach the target of controlling eutrophication in Dianchi Lake under the conditions of present environmental pressure by using this approach.

It is obvious that using floating plants, such as water hyacinths, and/or the other species, in combination with bacteria technology for purifying the water quality, and also to get some economic profits from that approaches are the easiest and most feasible approaches. But there are four principle difficulties needed to be overcome:

(1) The floating plant may move from place to place on the watersurface by wind and currents and may influence the waterways transportation and tourisms;

(2) The floating plant will reach the coastal zones, the bays, where the submerged vegetation grow up. The surface cover of floating plant may limit the sunshine into deep waters, and the submerged vegetation will disappear caused by insufficient of light;

(3) It needs to take out the biomass of the floating plants; otherwise, the detritus of the plants may contribute the source of pollution—the secondary pollution, and the enhanced sedimentation rate;

(4) It is not easy to take it out from the water, and to process it into economically usable resource. Special technologies should to be developed. Otherwise, a big budget is needed to perform the collection and transportation the huge amount of WH from the lake.

The soft wall technology of PEEN provides the feasible technology for overcoming the first and second principle difficulties. The concept of "falsework" and "cabin construction", the technology for improving transparency, and the floating ecological islands can guarantee that floating plants grow up in the indicated area. The floating plants can grow up well even in strong wave conditions (Fig. 15.40). The optimum criteria of the density of the floating plant had been found to improve the water transparency and other water quality parameters, including the dissolved oxygen and sunshine in lower layer of the lake. Therefore, the existing of floating plants can not only clean up the water, but also promote the growth of submerged plants. It means that we can more efficiently use the solar energy, and increase the preliminary productivity. The solar energy might be used not only on the water surface by the floating plants, but also be used by the submerged plants under penetrated sunshine, while the transparency increased by the floating plants and bacteria.

The floating plants can be used as the natural resources. The issues are included in the economic feasibleness. The new technologies should be developed to avail of economic profits.

Development of the Bio-Environmental Enterprise (BEE)—the Perspective Way Out for Controlling Eutrophication

Dianchi Lake has surface area of 298 km^2. It has abundant sunshine resources and rich in nutrients at present. These natural conditions lead to growth of serious algae bloom issue. It is suffering the problem of eutrophication and aqua-ecosystem degradation. The urgent target is to reconstruct the health aqua-ecosystem. We could not get the results of decreasing the pollutants loading into the lake to the historical level in the next decades. The vegetation cover at the bottom sediment may effectively limit the nutrients release into water column. The key measures for reducing the nutrient concentration are to transform the nutrients into usable bio-resources and take them off from the water and get profits; to transform the nitrogen into gases and release them into atmosphere; and to precipitate the suspended substances onto the bottom. The floating plants, such as waterhyacinths (*Eichhornia crassipes*), *Alternanthera philoxeroides, Hydrocharis dubia*, etc. can be used for increasing the preliminary productivity and make competition with phytoplankton, and also as the carrier of bacteria, in combination with the purification effect of bacteria on water quality. All these effects may lead to increase the transparency gradually in Dianchi Lake. This was observed in the area covered by water-hyacinths, where the transparency was much more higher than that in the other area with algae bloom. The wide area of water surface can be used to transform the nutrients into plant resources; then take them off from waters. These approaches would be most cost-effective one and may get not only ecological benefits, but also economic profits.

The water hyacinths have the highest productivity in waters. Growth rate of water hyacinth (WH) in Dianchi Lake is greater than that in Taihu Lake, where in summer is 0.5 $kg/m^2/d$. The contents of WH are 1.58 kg N/t, 0.26 kg P/t. The absorbed nutrients by WH are 0.79 t N/ km^2/d; 0.13 t $P/km^2/d$, which are 13 and 21 times of by algae (Fig. 15.59). Therefore, the preliminary production of water hyacinths is about one order higher than that by phytoplankton. The planktivore fish and molluscs are the primary consumers in the lake ecosystem. The molluscs can be controlled more easily than fish, and the molluscs can also release coagulant substances and make the transparency higher. The molluscs are the favorite foodstuff for Chinese people and the lateen mussel can be used for production of pearl. Therefore, to culture molluscs is also a positive measure for taking off nutrients in water and improving water quality.

To culture fish in an enclosed net with a high density and high input of fish food was the main method for increasing production of fish in Dianchi Lake. Fish use the richer dissolved oxygen in the open water, but they also add pollutants into the lake. This kind of fishery has been stopped in Dianchi Lake.

A new technology named as "cloth enclosure fishery—CEF" might be introduced in eutrophicated lakes such as Dianchi Lake in combination with the ecological island technology (Fig. 15.60). A model of cloth enclosure includes a hexagon with each side length of 100 m, and the enclosed area is 2.6 hm^2 = 0.026 km^2 in the center, with 6 similar pattern floating ecological islands with floating plans, such as water hyacinths. The total water surface area of this model is 0.182 km^2. The water hyacinths grow up in the surrounding floating islands till they reach certain density, then take out from these islands on an orderly basis and put the plants into the central cloth enclosure for fish culture as the fish

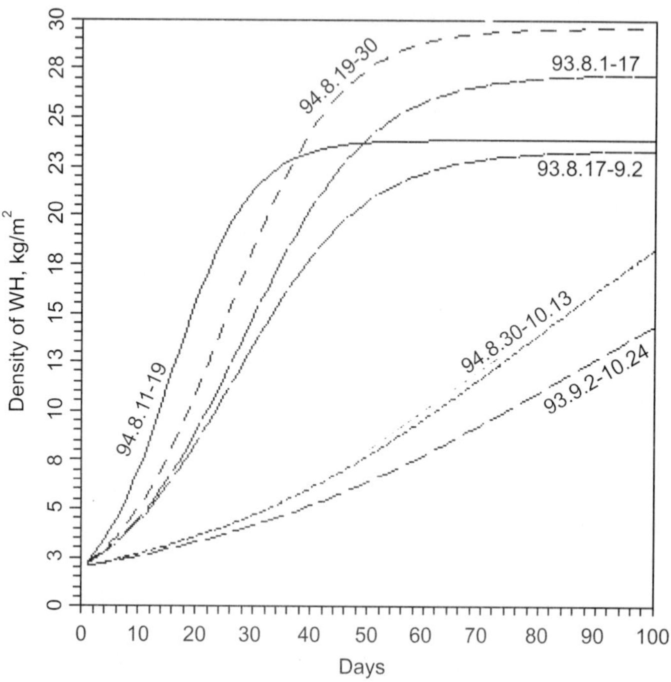

Fig. 15.59 Growth rate of water hyacinths (WH) in Taihu Lake during summer

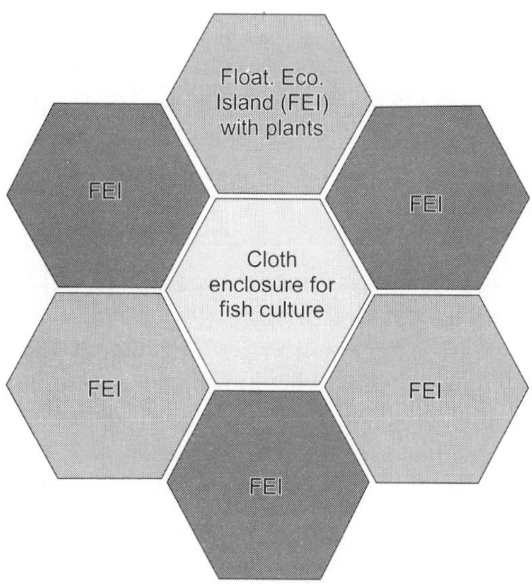

Fig. 15.60 Scheme of cloth enclosure for fish culture model with floating ecological islands for plant culture

feedstuff. It is the easiest way for transforming the dissolved nutrients into plants then to the fish. The remaining detritus and the excretion will remain in the cloth enclosure and would not influence the external waters in lake. The transforming rate for grass fish is about (80-100): 1; namely, by feeding about 80 kg to 100 kg (wet weight) aquatic plant the grass fish can grow 1 kg. A lot of fisherman fished in Dianchi Lake. They stopped fishery for a couple of years to decrease the nutrients loading into lake by feeding in the net enclosures for fish culture. Here, the introduced cloth enclosure may keep all the remains in the enclosure and not effect the surrounding water. An optimized combination of fish species and the quantity ratio of the fish of different ages should be calculated to maximize fish production. Theoretically, the growth rate of water hyacinths in the floating islands can offer feedstuff for growth for 3 kg fish during about 100 days in each square meter water column (the water hyacinths culture area is six times more than the fish culture area and the growth rate is 0.5 $kg/d/m^2$, the grass feed/ production rate for grass fish is 100). The WH grows up in all the 6 hexagons about 78 t/d. This quantity of WH should be taken into the internal cloth enclosure as fish feedstuff on a daily basis. This is an easy process with for one to two pairs of labor (usually they are from one family in China) with a special machine. In the meanwhile, 780 kg adult grass fish should be taken out from the internal cloth enclosure in a day. If the mechanical technology can be introduced, the production rate and economic profits are quite high. A great deal of fish production from harnessing the lake will promote the advanced processing for fish food and lead to the development of Bio-Environmental Enterprise (BEE). In the tropical and low latitude of the subtropical zone, such as Dianchi Lake, where the water hyacinths may grow during most time of the year, this method is very feasible.

The WH, rich in nutrients, is a very rich resource for livestock feed. But it contains about 95% water. After its decomposition, its value will decrease gradually. Therefore, to produce the Low cost-Long term storing WH Livestock Feed, this is the best way out for utilization of WH. To process the WH into livestock feedstuff, which can be stored for a couple of months, we can profit by the high productivity. Also, low price is the other very important approach for solving the problem of eutrophication. Because a huge volume of WH needs to be taken out from the hyper eutrophicated Dianchi Lake, the difficulties for processing with low cost are essential. A small-scale experiment has been done to overcome all the difficulties. The perspective achievements are very recent.

To culture mussels and snails is the other method for controlling eutrophication in Dianchi Lake, but with lower efficiency. According to the data in Saginaw Bay, Lake Huron, we can estimate that it needs culture lateen mussels with density of about 100 ind./m^2 for getting the obvious improvement of transparency and other water quality indices. In practice, the density of mollusks culture might be different from that needed in relation with the concrete situation, but, anyway, it might improve the transparency and prove profitable. This is also a model for biomanipulation and controlling the eutrophication with economic profits.

The Bio-Environmental Enterprise (BEE) should to be developed during harnessing the water quality and reconstruction of health aqua-ecosystem in Dianchi Lake.

Reconstruction of a healthy aqua-ecosystem in Dianchi Lake

As a famous tourism lake in Kunming, Dianchi Lake has been known to possess abundant biodiversity and high quality water. It is a notable recreation site and reputed as "a Sparkling Pearl Imbedded in the Yunnan-Guizhou Plateau". But Dianchi Lake is suffering the serious problem of hypereutrophication. The immediate task is to reconstruct the historical landscape with abundant biodiversity and clean, safe water. How can we reconstruct a healthy aqua-ecosystem in Dianchi Lake? What are the concrete steps to be realized?

As revealed in the previous paragraph, we cannot expect to have a rapid decrease of pollutants loading into Dianchi Lake in the coming years. Even more so, it is almost impossible to reach the pollutants loading as it was in the history of a couple decades ago. The above-recommended approaches for removing nutrients and reconstruction of a healthy aqua-ecosystems by PEEN might be used in Dianchi Lake. The great important steps are: (1) to reconstruct an aqua-ecosystem in Caohai Bay (Fig. 15.24) **(A)** as a nephridium of Dianchi Lake Catchment; (2) to reconstruct a healthy aqua-ecosystem in northeast part (Fig. 15.24) **(B)** of Dianchi Lake; (3) to construct this aqua-ecosystem in the drinking water source area (Fig. 15.24) **(C)** of Dianchi Lake.

Reconstruction of a Healthy Aqua-ecosystem in Caohai Bay as a "Nephridium"

Caohai Bay has a surface area of 7.8 km^2. It was rich in abundant biodiversity, but is now suffering a most serious hypereutrophication problem. There is no strong wind wave in it; therefore, the reconstruction of submerged vegetation is relatively easier. The river systems that can be used for making circulation of the water flow out from the northwest corner of Caohai Bay, along a river northward, then along a manmade (to be constructed) channel of about 15 km eastward, reaching the Panlong River, which flows through the City of Kunming to the east coast of Caohai Bay. Some soft wall would be constructed to introduce the lake current to the open waters in the Bay with abundant macrophytes, mollusks, enhanced nitrogen cycle bacteria, etc., where there is higher self-purification capacity. The Caohai Bay will become the "nephridium" of Dianchi Lake Basin. The clean water flows to the city river system, cleans them, then flows into Caohai Bay for its further purification in the Bay. The water surface in Caohai Bay might be beautified by floating ecological islands with aquatic plants and land plants for sightseeing. In the flood season, when the water from the Xiyuan Tunnel is released, the clean safe water may flow into the lower reaches of Dianchi Lake, instead of the polluted waters, which is dangerous for drinking source.

Reconstruction of Health Aqua-ecosystem in Northeast Part of Dianchi Lake

The most pollutants loading come from the northeast part of Dianchi Lake (Fig. 15.24). Except the Caohai Bay, which receives the pollutants loading from the catchment of Dianchi Lake, the inflow river mouths and bays along the northeast coast of Dianchi Lake also receive the pollutants loading with more volume. There are two projects which may stop the wastewater loading into the lake directly. The difficulty is not only related with the expense, but also the problem of how to separate the wastewater and rainfall flow. In the flood (rainy) season, the inflow water has higher concentration of pollutants than in dry season (Table 15.12).

Table 15.12 Water quality during flood season and dry season in Caohai Bay and Dianchi Lake, 2000 in comparison with the State standard for water of grade III (SSIII), and the treated wastewater water by treatment of class III (WWIII); unit: mg/L

Position	*Caohai Bay*			*Dianchi Lake*		
Parameters	*COD_{Mn}*	*TN*	*TP*	*COD_{Mn}*	*TN*	*TP*
Dry season	7.41	9.06	0.60	5.04	1.73	0.12
Flood season	24.49	21.02	1.65	8.78	2.40	0.25
After flood	8.81	5.81	0.84	9.53	2.23	0.42
Annual mean	13.57	11.96	1.03	7.78	2.12	0.26
SSIII	**8**	**0.3**	**0.025**	**8**	**0.3**	**0.025**
WWIII	20	9	0.5	20	9	0.5
WWIII/SSIII	**2.5**	**30**	**20**	**2.5**	**30**	**20**

The new strategy has two aspects: (1) to use the INCB technology in the upper reaches of the river system and at the entrances of domestic wastewater, releasing the nitrogen cycle bacteria in the beginning of wastewater; and (2) to construct a healthy aqua-ecosystem in the river mouths and bays along the northeast coast of Dianchi Lake. In the other river mouth region, similar reconstruction also can be performed. As estimated in above paragraph, if the total area of WH cover has 24 km^2 (the densely cultured meaning, the real area will about three times more), the most part of coastal zone will covered by submerged and merged aquatic plants, the big loading of TN and TP (as high as the present data of about 9 000 t N/a and about 500~1 000 t P/a) will be removed in combining with the INCB technology. Moreover, the dissolved TN and TP will be partly removed also, decreasing the concentration of TN and TP to the level of 0.2 mg/L and 0.02 mg/L (These are theoretical estimations. At least, the level of 0.5 mg/L and 0.04 mg/L is very possible to be realized), respectively, in Dianchi Lake. The eutrophication problem in Dianchi Lake will be controlled. As a possible approach, a soft wall construction might be established in the northeast part of Dianchi Lake (Fig. 15.61), enclosing an area of about 30 km^2, including Caohai Bay. The most part (79%) of pollutants loading into Dianchi Lake would be included in this area. This area with enhanced purifying capacity might be recognized as the "**nephridium**" of Dianchi Lake. The purified water of grade III might be pumped northward into the city river-channel system, circulating in it and then returning into in the Caohai Bay and Lake Dianchi. In the flood season, the abundant water flowed into the "nephridium" might flow into the main basin of Dianchi Lake with more clean water treated by the nephridium, as showed in Fig. 15.61. In combination with the other measures, the water quality in Dianchi Lake might be improved gradually to meet the need of the remediation of a healthy aqua-ecosystem for the entire Dianchi Lake.

Construction of a Healthy Aqua-ecosystem in Drinking Water Source Area of Dianchi Lake

The Kunming City is suffering from insufficient safe drinking water source, because the water quality in the intake area of drinking waterworks in Dianchi Lake has worsened in relation with the development of eutrophication. To transport the clean water from upper reaches reservoir is expense, as is its maintenance. The urgent issue is to reconstruct a healthy aqua-ecosystem in the intake area of the main waterworks in the east coast zone by using the PEEN technology (Fig. 15.24, C). The principle scheme was shown in Fig. 15.62.

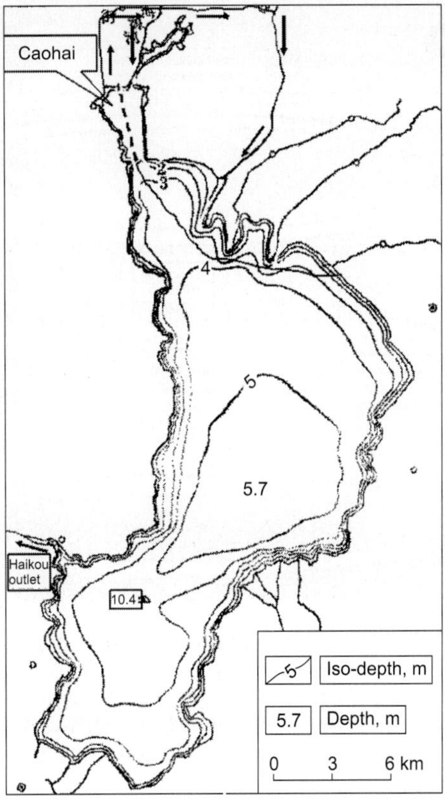

Fig. 15.61 The "**nephridium**" of Dianchi Lake with enhanced purifying capacity in the north-northeast part of the lake, accepting the most pollutants loading

There are two soft wall filters-the physico-biological membranes. The big particles, including the algae, will be partly filtrated by the outer filter. The purified water by the internal health aqua-ecosystem between the two filters will be filtered by the secondary filter near the intake tubes of waterworks. The abundant biodiversity in the PEEN guarantees the purification capacity for supplying sufficient water for drinking water advanced processing. The retention time of eight days is sufficient to guarantee the necessary water quality standard, according to the experiment in Taihu Lake (PU Peimin, et al., 1997). It seems like constructing a large preliminary natural water treatment pond as the retention time is more than eight days in the Dianchi Lake. This approach may be realized in a short period of construction and to meet the urgent need for the citizen, before the obvious improvement of the surrounding environment. In other regions, such as the natural swimming beach at

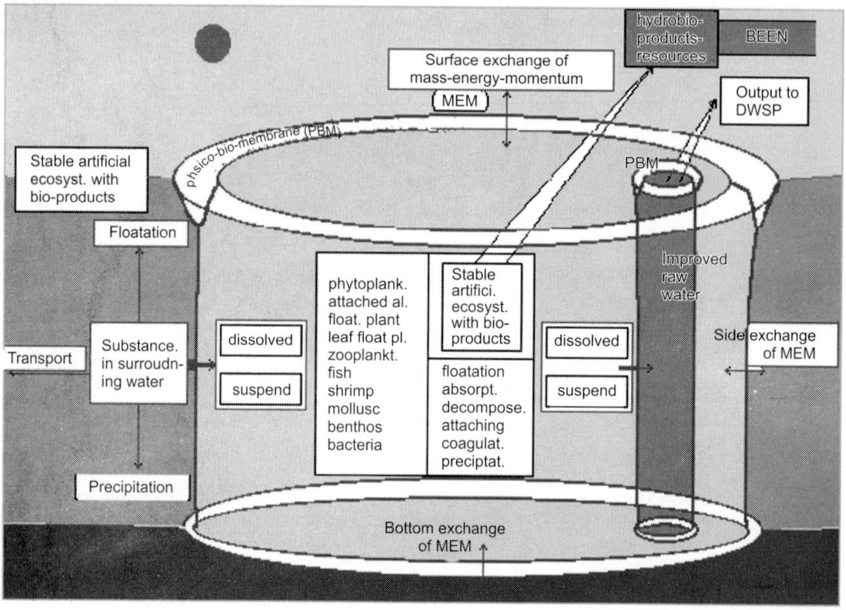

Fig. 15.62 Scheme of the principles of PEEN for a drinking water supply resource

Haigen, the north coastal zone of Dianchi Lake, and/or tourism sites, the PEEN technology can be applied for purifying water and reconstructing a healthy aqua-ecosystem.

Summary

Dianchi Lake has water surface area of 298 km^2, mean depth of 4.4 m, and is the sixth largest freshwater lake in China. It is a notable recreation site, being a highland lake with an elevation of 1886 meters. Getting more intensive sunshine, the highland lake has the higher preliminary production and possesses the characteristics of lakes in the lower latitudes-lakes in the tropical zone.

The northern bay of Dianchi Lake is named as Caohai Bay—a bay rich in aquatic vegetation. In fact, Caohai Bay was covered by profuse macrophytes. The macrophytes are densely distributed in the coastal zone in the history. Before the 1960, 90% of the water surface in Dianchi Lake was covered by macrophytes; the submerged plant distributed almost to the center of the lake, the transparency was a couple of meters. The total macrophytes had more than 100 species, in association with abundant biodiversity. The water quality of Dianchi Lake was quite good before the 1970's. The water was of grade II, suitable for drinking source use. The water became of grade III in 1970's. Beginning from the last years of 1970's, the degradation of water quality became more rapidly. The degradation process was developed in the Caohai Bay firstly and then dispersed to the south, to the open waters and finally, to the entire. The eutrophication process developed rapidly then on to other lakes in China. A lot of measures have been taking for controlling eutrophication, especially since 1980. But the problem for improving water quality, controlling eutrophication and reconstructing a healthy aqua-ecosystem, is still far from over. The possibility of a rapid decrease of pollutants loading into the lake in the next few years is very low.

How can we save Dianchi Lake from progressive environmental deterioration?

The 5 Key technologies of PEEN are:

(1) Soft wall and physico-biological membrane technology;

(2) Technology for removing algae plume (TRAP) and improving the transparency;

(3) Mosaic technology and macrobionics;

(4) Technology of Immobilized Nitrogen Cycle Bacteria (INCB); and

(5) Ecological floating island.

There are 5 M-key taches for reconstruction the healthy aqua-ecosystem?

(1) Macrophyte;

(2) Microorganism;

(3) Mollusc;

(4) Macrobionics; and

(5) Management.

The concept of "falsework" and "cabin construction", the technology for improving transparency, and the floating ecological islands can guarantee that the floating plants grow up in the indicated area. The floating plants can grow up well even in strong wave conditions. The optimum criteria of the density of the floating plant had been found for promoting the reconstruction of submerged vegetation. The existing of floating plants can clean up the water, and promote the growth of submerged plants. It means that we can more efficiently use the solar energy, and increase the preliminary productivity of the whole water body, and also guarantee that the aqua-ecosystem remains stable in face of environmental hazards.

It is essential to remove the nutrients from waters in eutrophic lakes. There are three way to do so: (1) transforming them into biomass resources and take them out from waters as a usable resources; (2)

transforming nitrogen into gaseous state and releasing them into the atmosphere; and (3) store the organic detritus and/or their decomposed nutrients at the bottom and limit their re-entrance into waters. The floating plants such as water hyacinths (WH), *Alternanthera philoxeroides, Hydrocharis dubia*, etc., can be used to increase the preliminary productivity and competetition with the phytoplankton, and also as the carrier of bacteria, in combination with the purification effect of bacteria on the water quality. The WH has the highest productivity in water. According to the data on Taihu Lake, the absorbed nutrients by WH are 0.79 t N/ km^2/d; 0.13 t P/km^2/d, which are 13 and 21 times of by algae. To take off the annual nitrogen loading of 9 000 t N/a into Dianchi and decrease the TN concentration in the water as low as 0.2 mg/L, for controlling eutrophication and avoid the appearance of algae bloom), We should culture water hyacinths (WH) on the water surface of 24 km^2 (8% of the total water surface of Dianchi Lake) during 300 days in a year. There would be grow 12 000 t wet WH/d. In this case, we may remove also 936 t P/a directly, which is the same order of annual loading of TP. The settling rate will increase by enhanced bacteria process, and may lead to a decrease in the concentration of TP to meet the necessary concentration as low as 0.02 mg/L. The PEEN can guarantee the successful controlled culture WH in an open lake with strong wave conditions in association with reconstruc-tion of a healthy aqua-ecosystem dominated by submerged vegetation in Dianchi Lake.

The new technologies of "cloth enclosure fishery-CEF", technology to produce the Low coast-Long term storing WH Livestock Feed, and to culture mussels and snails had been introduced for transforming the nutrients in waters into usable resources, to get not only the eco-environmental profits, but the economic profits also. Therefore, to develop the Bio-Environmental Enterprise (BEE) is the perspective way out for controlling eutrophication.

Water quality problem is a problem of an aqua-ecosystem. To realize the key target of reconstruction of a healthy aqua-ecosystem in Dianchi Lake, and restoring its historical landscape as "a Sparkling Pearl Imbedded in the Yunnan-Guizhou Plateau", three most important steps were proposed by using the approach of PEEN+BEE:

(1) Reconstruction of a healthy aqua-ecosystem in Caohai Bay as a "nephridium";

(2) Reconstruction of a healthy aqua-ecosystem in the northeast part of Dianchi Lake;

(3) Construction of a healthy aqua-ecosystem in the drinking water source area of Dianchi Lake

The eutrophication problem is most serious in the field of limnology in China. The direct economic loss was about USD 15 million during a week for Wuxi City, Jiangsu Province, after an event of algae bloom in Taihu Lake in summer 1990. After this event, a serial research and ex-periment project directly towards for reconstruction of an aqua-ecosystem in the local area, especially in the intake area of drinking water source, was performed under the support of the Chinese State and Provincial Government, the Chinese Academy of Sciences, and the International Organizations as well. The small-scale experiments of a couple of thousands or ten thousands square meters took place in different environmental conditions. The research results have been summarized in this Chapter. The perspective for controlling eutrophication in the surface waters appears with optimism. Of course, there are a lot of issues in this field that still need to be studied and solved in the further.

ACKNOWLEDGEMENTS

The author is grateful to Prof. M. Vikram Reddy, Department of Zoology & Environmental Science Program, Kakatia University, India for his kind invitation to present this chapter. The Chinese Academy of Sciences, the Ministry of Science and Technology, the Bureau of Science and Technology of Jiangsu Province, the Taihu Lake Basin Authority, the European Union Science Commission (under the contract of No. CI1*-CT93-0094), the Hong Kong Polytechnic University (PolyU), the Kahokugata Lake Institute, Japan (JPKLI), supported the projects. The author is indebted to them for their support. The contribution of a great number of persons at Nanjing Institute of Geography & Limnology of Academia Sinica, Nanjing University, Zhejiang University, Nanjing Normal University, Jiangsu Academy of Environmental Science, and the Taihu Laboratory for Lake Ecosystem Research, Academia Sinica, is gratefully acknowledged. Special thanks are due to Prof. Jorgenson S.E., and YAN Jingsong, Profs. ZHANG Limin, PANG Yong, WANG Guoxiang, LI Zhengkui, HU Weiping ZHANG Shengzhao, FAN Yunqi, HU Chunhua, for their direct participating in the research. The author wishes to sincerely acknowledge Prof. FAN Yunqi, Late Mr. JIANG Zhonglu, and Mrs. MAO Benjing for their encouragement.

Corresponding e-mail: pupm1@yahoo.com

REFERENCES

Blasland, Bouck & Lee, 2001. Development of Alternatives for the Lake Okeechobee Sediment Management Feasibility Study, Technical Report, South Florida Water Management District.

Brezonik, P. L. and Fox, J. L., (Eds) 1999. Proceedings of a Symposium on Water Quality Management through Biological Control. University of Florida, Gainesville, pp. 85-89.

Cheng Xiaoying, Guoxiang, Wang and Peimin, Pu. 2001. Purifying Local Eutrophic Lake Water during Autumn and Winter, The 9th International Conference on Conservation and Manage-ment of Lakes, October, 11-17, Otsu, Japan: 440-443.

Cheng Xiaoying, Guoxiang Wang, Peimin Pu. 2002. ZHANG Shengzhao, CHEN Baojun, Restoration and purification of macrophytes in an eutrophic lake during Autumn and Winter, J. of Lake Sciences, 14(2): 139-144.

Du Honshen, Peimin Pu, Shengzhao Zhang, Weiping Hu and Yong Pang. 1995. An experimental study on culture of Eichhornia crassipes (Mart) Solms on the open area of Taihu Lake, J. of Plant Resource and Environment, 4(1): 54-60.

Fahnenstil, G. L., Lang Gregory, A., Nalepa, T. F. and Johengen, T. H. 1995. Effects of zebra mussel (Dreissena polymorphya) colonization on water quality parameters in Saginaw Bay, Lake Huron. J. of Great Lakes Research, 21 (4): 435-449.

Fan Chengxin, et al. 1995. Preliminary study on the evaluation of sludge-dredging work in Lake Kasumigaura using large-size core samples, Proceed. of VI Conference on Cons. & Manag. Of Lakes, Kasumigaura. 1995: 1890-1893.

Horne, R. A. 1969. Marine Chemistry-The structure of water and the chemistry of the hydrosphere, Wiley interscience.

Hu Chunhua and Peimin Pu. 1998. Preliminary study on the precipitation fluxes and decomposition of their organic portion in hypertrophic Wulihu Bay of Taihu Lake, J. of Lake Sciences, Vol. 10, Suppl.: 303-308.

Jin et al. 1995, Environments of China's Lakes, Vol. 2: 621.

Hu Weiping and Peimin, Pu. 1995. Study on the ecological efficiency of soft enclosure in lake water, 6th International Conference on the Conservation and Management of Lakes-Kasmigaura. 1995, 1886-1889.

Hu Weiping, Salomonsen Jorgen, Xu Fu-Liu and Peimin Pu. 1998. A model for the effects of water hyacinths on water quality in an experiment of physico-biological engineering in the lake taihu, china, Ecological Modelling, 107: 171-188.

Hu Weiping and Peimin Pu. 1998. Ecological Indexes for Assessment of Ecosystem Health in Taihu Lake, China, Proceedings of ANSWER'97, Special Issue of J. of Lake Sciences in press.

Hu Weiping and Peimin Pu, 1998. A model on the effects of Trapa Natans Var. Bispinosa on water quality in an experiment of physical-biological engineering, Lake Taihu, China, Proceedings of ANSWER'97, Journal of Lake Sciences, Vol. 10, Suppl. : 527-538.

Huang Yikai, Peimin Pu, Chunhua Hu, Jingyi Zhang. 1998. An experimental study on the decomposition of suspended substances in laks, Water Resources Conservation, : 30-34.

Jin Xiancai, et al. Lake Environment in China (II), Oceanography Press, Beijing, 1995:120-135.

Kunming Institute of Environment Sciences. 1992. Research on the eutrophication of Dianchi Lake, Yunnan Science & Technology Press, pp. 204.

Kunming Institute of Environment Sciences. 1999. Controlling and Prevention of Water Pollution in Dianchi Lake Basin during "the ninth five-year plan" and program for 2010, 1997, Kunming.

Li Wenchao, Yin, C., Chen, K. & Wu, Q. 1999. Macrophytes-dominated phosphorous cycle in Lake Taihu, China, 8th International Symposium on the Interactions between Sediments and Water.

Li Zhengkui & Peimin, Pu. 2000. Dynamic modeling of purifying nitrogen pollutant by immobilized nitrogen cycle bacteria during Autumn and Winter seasons, J. Of Lake Sciences, 12(4): 321-326.

Li Zhengkui, Peimin, Pu., Guoxiang, Wang., Chunhua, Hu., Baojun, Chen., Bo, Li., Xiaoying, Cheng., Shengzhao, Zhang., Yunqi, Fan. 2001. Perification of Taihu Lake Water with Immobilized Nitrogen Cycle Bacteria, The 9th International Conference on Conservation and Management of Lakes, October, 11-17, Otsu, Japan: 344-347.

Ogiwara, Keizo, et al. 1995. The purification of Lake Suwa (Dredging). Proceed. of VI Conference on Cons. & Manag. Of Lakes, Kasumigaura'95: 438-441.

Pang Yong, Peimin, Pu., Yangchun, Wei., Weiping, Hu. 1999. The water quality model of an artificial purification ecosystem near a water plant, Acta Ecologica Sinica,

Pu, P., et al. 1993. An experimental study on the physio-ecological engineering for improving Taihu Lakewater quality in water source area of Mashan Drinking Water Plant, J. of Lake Sciences, 5(3): 171-180.

Pu Peimin et al. 1993. Physic-Ecological Engineering for improving Taihu Lake water quality in intake area of Mashan Drinking Water Plant, Proceedings of 5th International Conference on the Conservation and Management of Lakes, Stresa, Italy: 480-483.

Pu Peimin, et al. 1994. A 'Jellyfish Engineering Experiment' for purifying water quality in large lakes, Proceedings of the 7th International Symposium on River and Lake Environments, Japan, Oct. 1-3, Rep. Suwa Hydrobiol., 9:149-153.

Pu Peimin, et al. 1995. The high-efficient artificial ecosystems in local water bodies and biological enterprises for environment conservation, their roles in eutrophication control for Taihu Lake, China, Proceedings of the 6th International Conservation and Management of Lakes, Kasumigaura' 95, Japan, October 23-27: 1882-1885.

Pu Peimin, 1996. Physical-Biological Engineering for Removing Algae and Purifying Water Quality in Lake Taihu, China, in "Successful Sino-European Science and Technology Cooperation 1981-1995, Department of International Scientific and Technological Cooperation SSTC, China, Tsinghua University Press, Beijing: 35-36.

Pu Peimin, Weiping Hu. 1996. Key Strategies for Controlling Eutrophication of Large Shallow Lakes in China, 28th International Geographical Congress, The Hague, August 4-10.

Pu, P., et al. 1997. A Physico-Ecological Engineering experiment on purifying water quality in a hypereutrophic bay of Taihu Lake, China, (XXVI SIL Congress, Sao Paulo, Brazil, July 23-29, 1995), Verh, Internat. Verein. Limnol. 26: 756-759.

Pu Peimin, Weiping, Hu., Yong, Pang., Yangchun, Wei., Xuefen, Liu., Jingyi, Zhang., Wenchao, Li. 1997. A physico-ecological engineering experiment for purifying raw water quality in a lake, J. of Lake Sciences, 9(2): 150-157.

Pu Peimin, et al. 1998a. Harnessing the hydrosphere environment of Taihu lake, china, in the book of "Research on the environment and ecosystem of Taihu Lake" Selected papers of Taihu Laboratory for Lake Ecosystem Research, Meteorological Press: 188-202.

Pu Peimin, Weiping Hu, Jingsong Yan, Guoxiang Wang, Chunhua Hu. 1998b, A physico-ecological engineering experiment for water treatment in a hypertrophic lake in China, Ecological Engineering, (10): 179-190.

Pu Peimin, Jingsong, Yan. 1998c. Taihu Lake—A Large Shallow Lake in The East China Plain, Journal of Lake Sciences, Vol. 10, Suppl. 1-12.

Pu Peimin, Weiping, Hu., Shengzhao, Zhang., Jingyi, Zhang., Guoxiang, Huang. and Chunhua, Hu. 1998d. The New Strategy for Improving the Aqua-Ecological Environment in Taihu Lake Basin, China—"How can we solve the problem of lack of qualified water and deterioration of environment and natural resources in Taihu Lake Basin?", Proceedings of ANSWER'97, Journal of Lake Sciences, Vol. 10, Suppl.: 47-58.

Pu Peimin, Guoxiang, Wang., Weiping, Hu., Chunhua, Hu., Bo, Li. 1998e. Engineering for purifying water in local areas, Journal of Lake Sciences, Vol. 10, Suppl.: 519-527.

Pu, Peimin, Wang, G. Hu, C. et al. 2000. Can we control eutrophication by dredging? J. of Lake Sciences, 12(3): 269-279.

Pu, Peimin, Wang, G, Li, Z. et al. 2001. Degradation of healthy aqua-ecosystem and its remediation: theory, technology and application, J. of Lake Sciences, 13 (3) 193-203.

Pu Peimin, Zhengkui, Li., Guoxiang, Wang., Chunhua, Hu., Baojun, Chen., Bo, Li., Xiaoying, Cheng., Shengzhao, Zhang., Yunqi, Fan. 2001. Application of floating macrophyte and immobilized bacteria in physico-ecological engineering for increasing self-purification capacity of surface waters, International Conference on Environmental Concerns and Emerging Abatement Technologies, Beijing, October 9-12.

Pu Peimim, Xiaoe, Yang., Guoxiang, Wang. 2001. New strategy for wastewater controlling in Taihu Lake Basin, China, International Conference on Environmental Concerns and Emerging Abatement Technologies, Beijing, October 9-12.

Pu Peimin, Zhengkui, Li., Guoxiang, Wang., Chunhua, Hu., Baojun, Chen., Bo, Li., Xiaoying, Cheng., Shengzhao, Zhang., Yunqi, Fan. 2001. Improvement of Water Quality by Technology of Immobilized Bacteria in Physico-Ecological Engineering—A Case Study in Hongfeng Lake, China, The 9th International Conference on Conservation and Management of Lakes, October, 11-17, Otsu, Japan: 133-136.

Pu Peimin, Guoxiang, Wang., Zhengkui, Li., Chunhua, Hu., Xiaoying, Cheng., Shengzhao, Zhang., Yunqi, Fan. 2001. How can we control eutrophication in Dianchi Lake? Restoration and Management of Eutrophication Lake, November, 18-21, Kunming, China,

Shapiro, J., Lamarra, V., and Lynch, M. 1975. Biomanipulation: An ecosystem approach to lake restoration. In P.L. Brezonik and J.L. Fox (eds.): Water Quality Management through Biological Control. University of Florida, Gainesville, pp. 85-96. 144 Lynch, M. 1977 Fitness and optimal body size in zooplankton populations. Ecology 58: 763-774

Shuleikin, V. V. 1956. Oceanic Physics, Science Press, Moscow.

Xie, P. and Liu, J. 2001. Practical success of biomanipulation using filter-feeding fish to control cyanobacteria blooms - a synthesis of decades of research and application in a subtropical hypereutrophic lake. The Scientific World 1, 337-356.

Van der Does, Joop, et al. 1992. Lake restoration with and without dredging of phosphorus-enriched upper sediment layers, Hydrobiologia 233: 197-210.

Wang Guoxiang, Peimin, Pu., Shengzhao, Zhang., Xiaoying, Cheng., Chunhua, Hu., Baojun, Chen., Bo, Li., Yunqi, Fan. 2001. Mosaic Technique of Aqautic Community for Ecological Remediation of Eutrophic Lake, The 9th International Conference on Conservation and Management of Lakes, October, 11-17, Otsu, Japan: 436-439.

Wang Guoxiang, Peimin, Pu., Shengzhao, Zhang., Wangchun, Li., Weiping, Hu., Chunha, Hu. 1998. The purification of mosaic community of macrophytes for eutrophic lake water, J. of Plant Re-sources and Environment, 1998, 7(4): 35-41.

Wang Guoxiang, Peimin, Pu., Shengzhao, Zhang., Wanchun, Li., Weiping, Hu., Chunhua, Hu. 1998. The purification of artificial complex ecosystem for local water in Taihu Lake, China Environmental Science, 18(5): 410-414.

Wang Guoxiang, Peimin, Pu., Shengzhao, Zhang., Weiping, Hu., Chunhua, Hu 1999, The purification ability of aquatic macrophytes for eutrophic lake water in winter, China Environmental Science, 19(2): 106-109.

Wang Guoxiang, Peimin, Pu., Yikai, Huang., Shengzhao, Zhang. 1999. Distribution and role of denitrifying, nitrifying, nitrosation and ammonifying bacteria in the Taihu Lake, Chinese Journal of Applied & Environmental Biology, 5(2): 190-194.

Wang Guoxiang, Peimin, Pu., Shengzhao, Zhang., Xiaoying, Cheng., Chunhua, Hu., Baojun, Chen., Bo, Li., Yunqi, Fan. 2001. Mosaic Technique of Aqautic Community for Ecological Remediation of Entrophic Lake, The 9th International Conference on Conservation and Management of Lakes, October, 11-17, 2001, Otsu, Japan: 436-439.

Wang, Qidong, ed. By Peimin, Pu. 1987. A preliminary study on momentum transfer at air-water interface and the forming process of wind driven wave and current on Taihu Lake, in Collected Papers of the International Symposium on Thihu Lake and Lake Biwa-A Sino-Japan Collaboration Project, Nanjing, China, 3.1-4.

Wei Yangchun, Peimin, Pu. 1999. Abatement of the nitrogen and phosphorus concentration in water by the snail Bellamya in Lake Taihu, Resources and Environment in the Yangtze Basin, 8(1): 100-105.

Watanabe Yoshito, Yamamoto, Masuo., Hayashi, Hitetake., Okino, Tokio and Tetsujiro Tanaka. 1980. Seasonal changes of the various nitrogen fractions in Lake Suwa, Report of the SUWA Hydrobiological Station, Shinshu University: 47-52.

Yang Longyuan, Haitang, Liang., Weiping, Hu., Jiang, Ji., Weimin, Chen., Peimin, Pu. 2002. The study on natural restoration `of aquatic vegetation in the northern lakeside, Lake Taihu, J. of Lake Sciences, 14(1): 60-66.

Yang Xiaoe and Peimin, Pu. 2001. Theoretical and Practical Approaches to Remediation of Polluted Soil and Water Ecosystems International Conference on Environmental Concerns and Emerging Abatement Technologies, Beijing, October 9-12.

Zhang Chen, Gao, X., Chen, Y. & Gao, G. 1999. Suspended substance and short-time-scale phosphorus circle in shallow lakes, Oceanologia et Limnologia, in Press.

Zhang Shenzhao, Guoxiang, Wang and Peimin, Pu. 1998. The influence of algae-type-eutrophication or aquatic-higher-plants and hypertrophic vegetational reconstruction in Taihu Lake, J. of Plant Resource and Environment, 7(4): 52-57.

Zhang Shengzhao, Guoxiang, Wang., Peimin, Pu., Chigirat T. 1999. Succession of hydrophic vegetation and swampy tendency in the East Taihu Lake, Journal of Plant Resources and Environment, 8(2): 1-6.

Zhky (Chinese Academy of Environmental Sciences). 1997. Atlas of the polluted sediments survey in Caohai Bay, Dianchi Lake.

Zhou, W., Liu, J., Yu, Y. 1996. Status and utilization of lake blue-green algae bloom, Mem. of Nanjing Inst. of Geogr. & Limnol., Academia Sinica, 12: 101-105.

Zhu Haihong, Yingtai, Chen & Peimin, Pu (Ed.) 1989. Environments and Sedimentation of Fault Lakes, Yunnan Province, Science Press, Beijing, pp. 513.

Index